Jachy

Metals, Ceramics and Polymers

H

Metals
Ceramics and Polymers

An introduction to the structure and properties of engineering materials

OLIVER H. WYATT
Senior Principal Scientific Officer
Royal Aircraft Establishment, Farnborough

DAVID DEW-HUGHES
Reader in the Physics of Materials
The University of Lancaster

Cambridge University Press

Published by the Syndics of the Cambridge University Press
Bentley House, 200 Euston Road, London NW1 2DB
American Branch: 32 East 57th Street, New York, N.Y.10022

© Cambridge University Press 1974

Library of Congress Catalogue Card Number: 70-178286

ISBNs:
0 521 08238 2 hard covers
0 521 09834 3 paperback

First published 1974

Printed in Great Britain
by J. W. Arrowsmith Ltd., Bristol BS3 2NT

Contents

Preface

This book is an introduction to the structure and properties of solid materials, which are the starting point of technology. It is intended as a textbook for undergraduates reading engineering in all its branches, metallurgy, materials science or, possibly, solid state physics. It should also be of interest and value to graduate students in these disciplines and to practising engineers.

In one volume we have combined the atomistic, microscopic and macroscopic approaches to materials and covered all the major classes: metals, ceramics, glasses and high polymers. The properties considered are mechanical, electrical and magnetic. For reasons of space, the discussion of some topics has had to be shortened, but this has been clearly indicated, and other subjects ignored completely. Perhaps the most serious omissions are production processes, which are important factors in controlling the structure of the resultant materials but can be regarded as a branch of chemical engineering, and surface phenomena, such as corrosion, friction and wear. However, it was strongly felt that selection was preferable to superficiality, as long as a reasonable overall balance was achieved. Consequently, the chapters on the selected topics should stand comparison with the treatment of them to be found in monographs.

The approach is analytical and endeavours to provide a sound understanding of the principles and philosophy of materials rather than detailed formulations of commercial materials and their properties. The emphasis is on explaining the macroscopic or engineering properties of materials in terms of their structures. The term 'structure' is used in its widest sense to include atomic structure, crystal structure and imperfections, electronic structure and excitation levels, and microstructure, that is, the proportion and arrangement of the phases which are observed with an optical or electron microscope. Nevertheless the practical use of materials in power generation, land sea or air transport, civil construction, communications, computers and many other solid state electronic devices, to name but a few areas of advancing technology, are kept constantly in mind, and typical property values are frequently quoted so that the theory remains tied to practice and the engineering reader will not feel overburdened with pure science.

The opening chapters describe the nature of matter: the atomic structure of the elements and the crystal structure of single phase materials. Thermal energy, equilibrium state and defect structures are included here. Chapter 3 introduces the electron theory of metals and the elements of wave mechanics: Brillouin zone theory is developed to account for the properties of metals, semiconductors and insulators. The microstructure of binary alloys is then discussed in detail with an extensive treatment of equilibrium diagrams; the final section extends this briefly to tertiary alloys. Chapter 5 gives a critical analysis of mechanical properties from a macroscopic standpoint. Emphasis is placed on plasticity in metals as the most important phenomenon in their mechanical behaviour since its onset limits the elastic range and determines the allowable working stress in many structural applications. Creep and fatigue are also discussed, if somewhat briefly. This is followed by the continuum theory of plasticity; slip-line field theory for plane strain is applied to several practical problems. The microscopic approach to plasticity and the theory of dislocations are developed in Chapter 9, and methods of preventing dislocation movement, so as to convert low strength metallic crystals into high strength structural materials, are given in Chapter 11. Probably the most important group of metallic alloys is that based on the iron–carbon system and Chapter 12 is devoted to the microstructure and properties of cast irons, plain carbon steels and heat treated low alloy steels. The electrical and magnetic properties of metals are given in Chapters 13 and 15. These chapters cover the ground usually included in courses on physical metallurgy.

Interwoven with the above are sections devoted to other classes of materials of equal technological importance: ceramics, including cement and concrete, glasses, glass ceramics, and high polymers, including plastics and rubbers (elastomers). Their structures and mechanical properties are discussed in Chapters 7 and 8 and their electrical and magnetic properties, often paramount, in Chapters 14 and 15. These materials are largely brittle and, in contrast with metals, fracture limits the elastic range rather than plastic flow. The physics of fracture, including Griffith theory, is given in Chapter 10, the second part of which is devoted to brittle fracture of mild steel. As is now well known, there are occasions when this standard constructional metal, normally extremely ductile, can fail catastrophically in a brittle manner. A section is devoted to plane strain fracture toughness and the determination of K_{Ic}. Finally mention must be made of composite materials in which high strength fibres of glass, carbon or boron are held in a plastics or metallic matrix. Composites offer the exciting possibility of achieving a step advance in the age old quest for high strength combined with high fracture toughness, without which the high tensile strength achieved in laboratory test pieces has little practical value. The theory of composites is given in the second part of Chapter 11.

We believe that the presentation given here is logical, coherent and intelligible. Any criticisms relating to factual errors or ambiguity will be carefully considered for later editions, if sent care of the publisher.

Our thanks are due to our families for so long enduring the time devoted to this labour of love and to our many scientific colleagues whose original endeavours have contributed to the rapid progress in the field of materials science in the last few decades. It is our hope that their experimental and theoretical discoveries have been correctly and usefully summarised in this book for the benefit of those who have not the time to read their original papers. In particular it has been our aim to bridge the gap between scientists and engineers so that the latter can apply the knowledge which has been so painstakingly created in research laboratories.

O.H.W.
D.D-H.

24 October 1972

1. Atomic structure

1.1 Historical introduction

The study of the nature of matter is concerned with matter's constituents and how the parts combine with one another to form the whole. The subject is best considered by a process of synthesis, starting with three fundamental particles, electrons, protons and neutrons, and seeing how they are combined in atoms, crystals and, finally, polycrystalline matter. Physicists are now discovering many sub-nuclear particles, but these will not be considered here. Historically, there has been an analytical breakdown of the nature of matter, in which the successive theories have interpreted the observed phenomena in terms of progressively fewer and more fundamental units. This development will be traced briefly before considering the synthesis of matter in detail.

In mid-seventeenth century, Robert Boyle postulated that all materials are combinations of a limited number of basic substances, known as elements. All the elements have now been identified: they number just under one hundred, apart from those produced artificially, and are given in the Periodic Table (Table 1.1). At the start of the nineteenth century, John Dalton restated the atomic theory of matter to explain the experimental laws of chemical reactions such as the law of constant proportions. The theory states, *inter alia*, that the atom is the smallest indivisible part of an element and that all the atoms of an element are identical. Later work has caused these statements to be qualified. It has been shown that atoms can be divided into other particles, though by physical rather than by chemical reactions, and that most elements have more than one atomic form or isotope. The isotopes of one element are chemically identical but their masses differ slightly.

At the end of the nineteenth century J. J. Thomson showed by means of the now familiar cathode ray tube that all atoms contain identical negatively charged particles, known as electrons. These were assumed to be embedded in a heavy positive charge, like cherries in a cake, to give an electrically neutral atom. However, in 1913 Rutherford showed, from the scattering of alpha particles by gold foil, that the atom has a planetary structure with a small central nucleus, in which the mass and positive charge reside, and a number of electrons moving in orbits about it. The nucleus was thought to be composed of protons and electrons but, following Chadwick's discovery of the neutron in 1932, it is known to be composed of protons and neutrons with all the electrons outside.

Following Rutherford's concept of the nuclear atom, Bohr invented the quantum theory of spectra to explain why the electrons were found only in certain orbits, and why they did not radiate their energy, as predicted by classical electromagnetic theory for accelerating charges, and fall into the nucleus. Bohr's theory enabled the line spectrum of hydrogen to be explained in terms of the atomic structure but it failed for more complex atoms and contained an arbitrary assumption limiting the electron angular momentum to certain values.

1

Table 1.1. *Periodic Table of the elements (N.B. The names of the elements are given in Table 1.3)*

Group: / Period:	A sub-groups							Transition metals VIII			IB	IIB	B sub-groups					
	IA Alkali metals	IIA Alkaline earths	IIIA	IVA	VA	VIA	VIIA						IIIB	IVB	VB	VIB	VIIB Halogens	0 Inert gases
1 (short)																		2 He 4.00
2 (short)	3 Li 6.94	4 Be 9.03											5 B 10.82	6 C 12.01	7 N 14.008	8 O 16.000	9 F 19.00	10 Ne 20.18
3 (short)	11 Na 22.99	12 Mg 24.32											13 Al 26.98	14 Si 28.06	15 P 31.07	16 S 31.07	17 Cl 35.46	18 A 39.94
4 (1st long)	19 K 39.10	20 Ca 40.08	21 Sc 44.96	22 Ti 47.90	23 V 50.95	24 Cr 52.01	25 Mn 54.94	26 Fe 55.85	27 Co 58.94	28 Ni 58.71	29 Cu 63.54	30 Zn 65.38	31 Ga 69.72	32 Ge 72.60	33 As 74.91	34 Se 78.96	35 Br 79.92	36 Kr 83.80
5 (2nd long)	37 Rb 85.48	38 Sr 87.63	39 Y 88.92	40 Zr 91.22	41 Nb(Cb) 92.91	42 Mo 95.95	43 Ma(Tc) 97.8	44 Ru 101.1	45 Rh 102.91	46 Pd 106.4	47 Ag 107.88	48 Cd 112.41	49 In 114.82	50 Sn 118.70	51 Sb 121.76	52 Te 127.61	53 I 126.91	54 Xe 131.30
6 (3rd long)	55 Cs 132.91	56 Ba 137.36	57–71 Rare† earths	72 Hf 178.6	73 Ta 181	74 W 184	75 Re 186.31	76 Os 191	77 Ir 193.1	78 Pt 195.23	79 Au 197.2	80 Hg 200.61	81 Tl 204.39	82 Pb 207.2	83 Bi 209.00	84 Po (210)	85 At (212)	86 Rn 222
7	87 Fr (223)	88 Ra 226	89 Ac 227	90 Th 232.05	91 Pa 231	92 U 238.07	*	*	*	*								

Legend:
1 – Atomic number Z
H – Element
1.008 – Atomic weight (chemical scale)

Rare earths (lanthanides)							
57 La 138.92	58 Ce 140.13	59 Pr 140.92	60 Nd 144.27	61 Il(Pm) (146)	62 Sm 150.35	63 Eu 152.0	
64 Gd 157.26	65 Tb 158.93	66 Dy 162.51	67 Ho 164.94	68 Er 167.27	69 Tm 168.94	70 Yb 173.04	71 Lu 175.0

* Artificially produced transuranic elements: ^{93}Np ^{94}Pu ^{95}Am ^{96}Cm....

In 1925 Heisenberg and Dirac developed matrix mechanics and later Schrödinger invented wave mechanics, which is mathematically equivalent, to resolve the difficulties of Bohr's theory. A series of equations is used with an entirely new symbolism, which cannot be interpreted in terms of models familiar on a macroscopic scale such as hard spheres with linking arms. The full theory is beyond the scope of this book although some of its ideas will be used qualitatively.

In 1935 Yukewa predicted the existence of pi-mesons (pions) in the nucleus to explain the powerful attractive forces which are necessary to hold the positively charged protons together. The theory predicted that they should be produced in highly energetic nuclear collisions and this was confirmed in 1947 by Powell who observed their tracks in photographic emulsions subjected to high energy protons from cosmic radiation. Since then over a hundred short lived subnuclear particles have been found and a 'unitary symmetry' theory developed which has rationalised much confused experimental data. In 1964 the theory was confirmed when the omega-minus particle whose characteristics it had precisely predicted was observed by Shutt and his co-workers. One recent theory (Gell–Mann's of 1964), called the quark model, postulates that *all* elementary particles can be constructed from the appropriate combination of three 'quarks' with charges zero, one third and two thirds of that of the electron, and their respective antiquarks. These six are then the truly *elementary* particles. This model is able to predict in an elegant fashion all of the observed particles, but to date no universally accepted direct evidence of the existence of quarks has been offered.

1.2 Elementary particles

Three elementary particles of matter are: electrons, protons and neutrons. The electron has by far the smallest mass, 9.11×10^{-31} kg or 5.5×10^{-4} amu,[†] and carries unit negative electric charge (1.6×10^{-19} coulomb). The proton has a mass of 1.7×10^{-27} kg ($1.007\,58$ amu), 1840 times that of the electron, and carries unit positive charge. Finally, the neutron has slightly greater mass than the proton ($1.008\,97$ amu) and is electrically neutral. The values of various physical constants are given in Table 1.2.

Atoms are the next largest unit of matter, consisting of a number of electrons around a central nucleus of protons and neutrons. The nuclear diameter is about 10^{-14} m while that of the atom is about 10^{-10} m; so the atom is mostly empty space, similar to the solar system. The number of protons in the nucleus, called the *atomic number* Z, equals the number of orbital electrons in a complete atom and defines the chemical element to which an atom belongs. For example, all atoms of helium, atomic number 2, have two protons in the nucleus, besides neutrons, and two orbital electrons. Z ranges from one for hydrogen at the start of the Periodic Table (Table 1.1) to 94 for plutonium at the end of the naturally occurring elements. (Uranium-92 is the upper limit for appreciable quantities.) Atoms with higher atomic numbers are produced in nuclear reactors.

[†] amu is the atomic mass unit defined as one sixteenth of the mass of the oxygen-16 isotope.

Table 1.2. *Fundamental physical constants*

	Mass	Charge
Electron	5.5×10^{-4} amu	-1.6×10^{-19} C
	$(9.11 \times 10^{-31}$ kg)	
Proton	1.007 58 amu	$+1.6 \times 10^{-19}$ C
	$(1.67 \times 10^{-27}$ kg)	
Neutron	1.008 97 amu	none
	$(1.67 \times 10^{-27}$ kg)	
Velocity of light (*in vacuo*)	3.00×10^{8} m/s	
Avogadro's number N	6.02×10^{23} molecules/g-mole	
Planck's constant h	6.625×10^{-34} J s	
Boltzmann's constant k	1.380×10^{-23} J/deg (8.617×10^{-5} eV/deg)	
Universal gas constant R	8.301 J g-mole^{-1} deg^{-1} (1.986 cal g-mole^{-1} deg^{-1})	
Gravitation constant G	6.659×10^{-11} m^{3} kg^{-1} s^{-2}	

The mass number A of an atom is the number of protons and neutrons in the nucleus, collectively known as the nucleons. The number of neutrons in atoms of any fixed atomic number is variable and atoms of the same chemical element but different mass number are called *isotopes*. Eighty out of the approximately 100 elements have two or more naturally occurring isotopes and very many more are produced in nuclear reactors. The term *nuclide* has recently been introduced to describe an atomic species as defined by the numbers of protons and neutrons. For example, the species of atom with 92 protons and 146 neutrons constitutes a nuclide – in this case described as $^{238}_{92}$U or $_{92}$U238. Another atomic species is $^{235}_{92}$U with 92 protons and 143 neutrons. These two nuclides have the same atomic number, that is, are of the same element, uranium, and are therefore isotopes of each other. In the past the term isotope has been used loosely to describe any atomic species, when strictly its meaning is limited to two or more nuclides of the same element (i.e. Z constant). (A comparable semantic example would be brothers (for isotopes) and sons (for nuclides) – not every son is a brother although every brother must be a son.)

The atomic weight of an element is the average mass of its atoms relative to other atoms. On the chemical scale, the atomic weight of naturally occurring oxygen is fixed at 16. (In an earlier definition of this scale, hydrogen was fixed at 1.) On the physical scale, the atomic mass unit is taken as one sixteenth of the oxygen isotope with mass number 16. This gives values about 0.03 per cent greater than on the chemical scale. Due to the mixture of isotopes present, most atomic weights are not whole numbers. The ratio of one gramme to unit atomic weight (chemical scale) is *Avogadro's Number $N = 6.02 \times 10^{23}$*. (NB this number is referred to the gramme even in SI units.)

1.3 Radioactivity

Many nuclides are unstable and change spontaneously with the emission of particulate or electromagnetic radiation, a process called *radioactive decay*. There are about 50 occurring naturally, including all the elements with atomic numbers above 83 (bismuth) and a further 700 have been made by nuclear reactions.

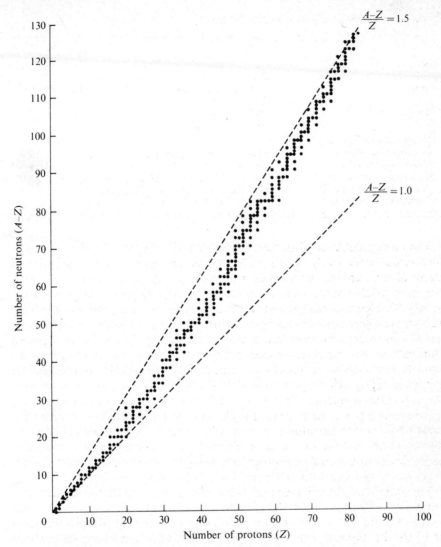

Fig. 1.1. Number of neutrons and protons in stable nuclides. A = mass number, Z = atomic number (number of protons), N = number of neutrons ($= A - Z$).

The ratio of neutrons to protons to achieve a stable nuclide is limited to a narrow range whose mean value increases from 1 to about 1.5 as the atomic number increases, see Fig. 1.1. Nuclides outside this band decay until after one or more steps they fall within it. When there are too many neutrons present, a neutron changes into a proton and an electron, which is emitted at high speed as a β^--ray (or simply β-ray), usually with a photon of electromagnetic energy, called a γ-ray. For energy and momentum reasons, it is necessary to postulate the simultaneous emission of a neutrino, having essentially zero rest mass and no charge, which normally escapes detection. (In relativistic mechanics, mass increases with velocity.) When too few neutrons are present, a proton changes into a neutron and a positron, a positively charged electron or β^+, which is emitted. This can immediately react with a free electron; both particles are annihilated and are converted into two γ-rays. The heavier nuclides, when short of neutrons, emit an α-particle, that is, a helium-4 nucleus, consisting of two protons and two neutrons. Various other radioactive changes involve the emission of α, β and γ-rays, and on rare occasions, neutrons. A nuclide can have several alternative modes of decay.

The number of atoms decaying in a time interval dt is proportional to the number of atoms N present:

$$\frac{dN}{dt} = -\lambda N.$$

Then

$$N = N_0 e^{-\lambda t},$$

where N_0 is the initial number of atoms and λ is the *radioactive decay constant*. The number of original atoms is halved in time $0.693/\lambda$, called the half life. This must be distinguished from the mean life of a nucleus which is $1/\lambda$. The reaction proceeds independently of the temperature and of the state of combination with other atoms. This is in sharp contrast to chemical reactions. Values of the half life range from a microsecond to 10^{15} years.

The best known radioactive nuclide is probably radium-226, isolated by P. and M. Curie in 1898. This is an α and γ-emitter, with a half life of 1620 y, and is one of a long chain of radioactive nuclides commencing with uranium-238 and finishing with lead-206, see Fig. 1.2. The relative quantities of each member of a chain will depend on the various decay constants. Two other radioactive chains run from ^{235}U to ^{207}Pb and from ^{232}Th to ^{208}Pb.

Radioactivity is put to many uses, ranging from killing cancer cells to examining welds in thick steel plate. One important application is in tracing the positions of atoms – radioactive tracers. Using a Geiger counter to detect the radiation it is possible to measure *in situ* down to 10^{-13} kg of a radioactive material, far smaller than can be done after separation with a chemical microbalance, about 10^{-9} kg. The radioactivity is usually brought about by putting the natural element in a nuclear reactor where the stable nuclide captures a neutron. For example, potassium-41, one of the stable natural isotopes, captures a neutron to become ^{42}K, which is a β-emitter with a half life of 12.5 hours, the end product being calcium-42.

Another application of radioactivity is dating of archaeological finds derived from organic matter. The Earth is continually bombarded by ionising radiation from space, called *cosmic rays*. Some of the nitrogen in the atmosphere is transmuted to carbon-14:

$$^{14}_{7}N + n \longrightarrow {}^{14}_{6}C + p.$$

Carbon-14 is a β^--emitter, reverting to nitrogen, with a half life of 5600 years:

$$^{14}_{6}C \longrightarrow {}^{14}_{7}N + \beta^-.$$

An equilibrium is established between production and decay such that the ratio of carbon-14 to other carbon isotopes ($99\% \ {}^{12}_{6}C + 1\% \ {}^{13}_{6}C$) is 1 part in 10^8.

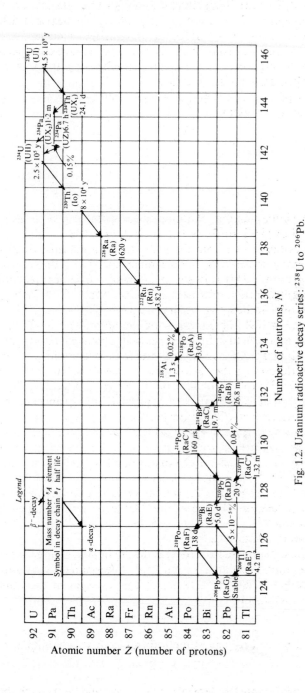

Fig. 1.2. Uranium radioactive decay series: ^{238}U to ^{206}Pb.

In turn this ratio occurs in living plants, which absorb carbon dioxide, and in live animals, which eat the plants. Upon death, or more exactly from the time of absorption, the proportion of $^{14}_{7}C$ and the β^{-}-activity decrease. Thus the age of a specimen can be deduced from measurement of the specific activity, which is initially 16 counts per minute per gramme. Good agreement with other prediction methods confirms the assumptions that the cosmic ray intensity and the ^{14}N content of the atmosphere have been constant for many thousand years. Recent nuclear explosions have so polluted the atmosphere that this method of dating the present will be unavailable to future generations.

1.4 Mass defect and fission

The mass of a nucleus is always less than the sum of the masses of its protons and neutrons, the difference being called the *mass defect*. This is due to the binding energy of the nucleons, which is released on assembly, or has to be supplied to take them apart. Einstein's mass–energy law (1905) states that mass and energy are equivalent according to the relationship:

$$E = mc^2,$$

where c is the velocity of light (3×10^8 m/s), m is the mass in kilogrammes and E is the energy in joules. More useful units are amu for mass and MeV (million electron volts $= 1.6 \times 10^{-13}$ joules) for energy, making

$$E\,(\text{MeV}) = 931m\,(\text{amu}).$$

The mass defect can be obtained from an accurate knowledge of the nuclide mass. For uranium-235:

92 × (proton + electron) mass	92.748	
143 × neutron mass	144.283	
	237.031	
^{235}U mass	235.117	
mass defect	1.914	amu
binding energy	1782	MeV
binding energy per nucleon	7.6	MeV

The binding energy per nucleon varies with the mass number as shown in Fig. 1.3. It rises rapidly from 2.5 MeV for hydrogen to a maximum value of 8.7 MeV, with a slight drop to 7.6 MeV at the higher atomic numbers. These figures are six orders of magnitude (i.e. 10^6) larger than the energies of chemical reactions, which involve only the orbital electrons and are measured in electron volt units. Herein lies the difference between the effects of high explosive and nuclear bombs and between the fuel consumed in oil-fired and nuclear power stations. When a neutron hits the nucleus of a uranium or plutonium atom it may cause fission in some thirty different ways to give products whose atomic numbers lie in the middle of the range where the binding energies are higher by approximately 1 MeV per nucleon. Typically 200 MeV is released per fission. At the same time about 2.5 neutrons are released, which, if carefully conserved, can be used to cause further fission and make the reaction self sustaining.

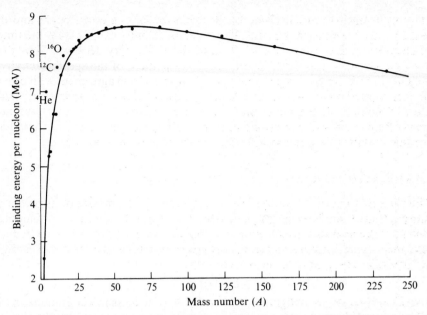

Fig. 1.3. Binding energy per nucleon in stable nuclei. (After *The Elements of Nuclear Reactor Theory* by S. Glasstone and M. C. Edlund. © 1952 by Litton Educational Publishing Inc. Reprinted by permission of Van Nostrand Reinhold Company.)

1.5 Bohr theory of the hydrogen atom

Rutherford is usually credited with having proved (1911) that atoms have a nuclear structure by his experiments on the scattering of α-particles, although earlier (1903) Lenard had come to similar conclusions based on electron scattering. Bohr developed the nuclear model (1913) with electrons in stationary orbits to explain the line spectrum of hydrogen – that is the absorption and emission of electromagnetic radiation. In a hydrogen atom ^{1}H a single electron revolves in an elliptical orbit around a single proton. In the following, the special case of a circular orbit will be considered, with the nucleus at rest.

The electrostatic attraction between opposite charges is given by Coulomb's law : $e^2/4\pi\varepsilon_0 r^2$ newtons, where e is the charge in coulombs, r is the separation in metres, and ε_0 is the permeability of free space ($1/4\pi\varepsilon_0 = 9 \times 10^9$). The centrifugal force is mv^2/r newtons, where m is the mass in kilogrammes and v is the velocity in metres per second. Hence, for equilibrium

$$\frac{e^2}{4\pi\varepsilon_0 r} = mv^2. \tag{1.1}$$

The energy of the system is the sum of the potential energy and the kinetic energy. Taking the energy of the electron as zero at infinite radius, the energy at radius r is given by :

$$E = \frac{1}{2}mv^2 - \frac{1}{4\pi\varepsilon_0}\int_0^\infty \frac{e^2}{r^2}\,\mathrm{d}r.$$

Integrating and substituting from (1.1)

$$E = -\frac{e^2}{8\pi\varepsilon_0 r}. \tag{1.2}$$

Bohr made use of Planck's quantum hypothesis and postulated that only certain radii and energy levels are permissible. Planck had found (1900) that he could explain the distribution of radiant energy from a hot body as a function of frequency if he assumed that the energy was radiated and absorbed in discrete packets, or quanta, whose magnitude was proportional to frequency: $E = h\nu$, where the constant of proportionality h, called Planck's constant, equals 6.63×10^{-34} joule second. Einstein extended this concept, postulating that light always consists of discrete quanta of energy $h\nu$. From these ideas Bohr proposed that the angular momentum, or moment of momentum, of the electron about the nucleus was quantised in integral units of $h/2\pi$ (often written as $\hbar$) that is

$$mvr = \frac{nh}{2\pi}. \tag{1.3}$$

This is known as the quantum condition for the angular momentum.

Eliminating the velocity from (1.1) and (1.3) and substituting in (1.2) gives the allowable energies

$$E = -\frac{me^4}{8\varepsilon_0^2 h^2}\frac{1}{n^2}. \tag{1.4}$$

Putting in the numerical values, the minimum allowable energy, that is for $n = 1$, is -2.15×10^{-18} J (-13.4 eV) and the corresponding radius (from (1.2)) is 5.29×10^{-11} m, which agrees with other estimates. This and the higher energy levels at greater radii are shown in Fig. 1.4. The figure shows the rapid convergence of the energy levels at the higher energies so that the discrete steps of quantum mechanics become indistinguishable from the arbitrarily small steps of classical mechanics. Thus the new mechanics of atoms does not conflict with Newtonian mechanics when applied to macroscopic bodies, a fundamental requirement known as the correspondence principle.

When the electromagnetic radiation emitted by hydrogen atoms after suitable excitation is examined, the radiant energy is found to peak at certain wavelengths to give a typical line spectrum. The wavelengths λ are given by

$$\frac{1}{\lambda} = R_H\left(\frac{1}{n^2} - \frac{1}{m^2}\right) \tag{1.5}$$

where R_H is a constant ($10\,967\,800$ m^{-1}) named after Rydberg, and n and m are integers. For $n = 1$, the lines occur in the ultra-violet region and the series is called after its discoverer Lyman; for $n = 2$ they occur in the visible region and are called after Balmer, and so on.

According to Bohr, radiation is emitted only when the electron moves from one stationary state to another with lower energy, the wavelength being given by

$$\frac{1}{\lambda} = \frac{\Delta E}{hc} \tag{1.6}$$

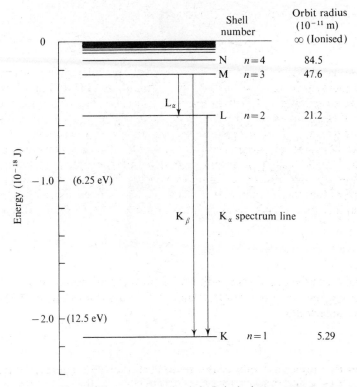

Fig. 1.4. Electron energy levels in Bohr hydrogen atom.

where ΔE is the energy difference of the two states and c is the velocity of light. Hence the line spectrum emitted by an electron dropping into the K-level from higher levels (L, M, N ...) has the wavelengths given by

$$\frac{1}{\lambda} = \frac{me^4}{8\varepsilon_0^2 h^3 c}\left(1 - \frac{1}{m^2}\right)$$

which corresponds to the Lyman series. Similarly for an electron dropping into the L-level:

$$\frac{1}{\lambda} = \frac{me^4}{8\varepsilon_0^2 h^3 c}\left(\frac{1}{4} - \frac{1}{m^2}\right)$$

which corresponds to the Balmer series.

   The Rydberg constant, defined by (1.5), is given by the Bohr theory in terms of the atomic constants as

$$R_{\mathrm{H}} = \frac{me^4}{8\varepsilon_0^2 h^3 c}.$$

Putting in values gives $R_{\mathrm{H}} = 10\,970\,000\ \mathrm{m}^{-1}$, which agrees with the value obtained by spectroscopy. This confirmed Bohr's theory of the structure of the atom and of the origin of line spectra.

1.6 Some elements of wave mechanics

The concepts and equations of motion developed in the classical theory of mechanics were found to be incapable of providing an atomic theory except in the simplest case of a hydrogen atom and one or two similar atoms. A new mechanics, called wave mechanics, was created which is the basis of modern physics and provides an atomic theory capable of explaining the behaviour of electrons in atoms and crystals. A quantitative treatment of wave mechanics applied to electrons in a crystal will be given in Chapter 3.

In wave mechanics the ideas on the nature of light were applied to particles. The dual corpuscular and wave behaviour of light had been resolved by postulating that it was composed of discrete entities, or photons, whose motion was determined by equations of wave motion. These are only convenient ways of visualising the events, no physical waves actually existing. De Broglie proposed in 1925 that particles should also be associated with waves, and then the Principle of Relativity indicated the relationships

$$\lambda = \frac{h}{p} \qquad (a)$$

and
$$v = \frac{h}{E} \qquad (b) \qquad (1.7)$$

where p and E are the particle momentum and energy. (Note that λv gives the wave, or phase, velocity, which can exceed the velocity of light. The particle velocity is associated with the group velocity, that is, the velocity of a finite group of waves. See physics textbooks.) The momentum is

$$p = mv \qquad (1.8)$$

where m and v are the particle mass and velocity. (At velocities approaching that of light, m increases above its rest mass.) These equations were already known to hold for light (cf. (1.6)).

The classical equations of motion were rewritten by Schrödinger during the 1920s in the form of a wave theory. It was shown that for the equations to be mathematically self consistent the preceding relationships are necessary. Thus the equations (1.7) can be deduced by alternative routes. Direct experimental confirmation of (1.7a) and hence of the concepts of wave mechanics was obtained in 1927 by Davisson and Germer, and independently by G. P. Thomson, who showed that electrons are diffracted on passing through a crystal in the same way as light is diffracted by a diffraction grating.

When the associated wavelength is small compared with the dimensions of the particle and the force fields, the wave equations give the same results as classical mechanics, that is the principle of correspondence is observed. However, where the particle size and field of force are comparable with the associated wavelength, many new phenomena occur which are not explicable in terms of the old mechanics. For example, the electron in the lowest energy state of the hydrogen atom, (1.4), has an orbital radius of 5.29×10^{-11} m. The momentum (from (1.3)) is 2×10^{-24} N s and the associated wavelength is 3.32×10^{-10} m. Thus for electrons in atoms the dimensions of the force field and the electron wavelength are of similar order and wave mechanical theory is necessary.

Solution of Schrödinger's equation (see § 3.6), given the potential field in which the electron is moving, is in the form of allowable wave functions $\psi(xyzt)$ where xyz are the coordinates and t the time. Each function corresponds to a definite value of the total energy of the electron. The probability of finding the electron at any position is given by the product of ψ, which is a complex number, and its conjugate ψ^*; $\psi\psi^*$ varies with the coordinates but not with time. It is no longer possible to specify the exact motion (position and velocity) of individual electrons and a statistical approach is the best that can be done. This is in agreement with reality since it is not possible to determine simultaneously the exact position and velocity, called Heisenberg's Uncertainty Principle (see § 3.6).

The electron structure of an atom must now be visualised as an electron cloud of varying density around the nucleus: a high density corresponds to a high probability of finding an electron there. Representation on a two-dimensional page becomes almost impossible since four dimensions are necessary. However, the wave function is always the product of three functions each of which involves only one variable, either the radial distance from the nucleus or one of the angular coordinates (see Fig. 1.5(a)). The electron density per unit volume ρ at any point P with polar coordinates $r\theta\phi$ is then given by

$$\rho = \text{const.} [R(r)R^*(r)][\Theta(\theta)\Theta^*(\theta)][\Phi(\phi)\Phi^*(\phi)], \tag{1.9}$$

where the constant is adjusted so that integration over the whole electron cloud gives unit probability of an electron being present.

The first term RR^*, called the *radial density factor*, can be plotted in rectangular coordinates in the normal manner (Fig. 1.5(b)), and the last two terms, θ and ϕ density factors, in the form of polar diagrams such that the length of the arm in any direction θ or ϕ represents the function. An example is shown in Fig. 1.5(c).

The two-dimensional polar diagrams for $\Theta\Theta^*$ and $\Phi\Phi^*$ can also be combined into a three-dimensional figure so that the length of the arm in any direction $(\theta\phi)$ represents the two functions $(\Theta\Theta^*)(\Phi\Phi^*)$. This will be referred to here as the $\theta\phi$ density factor diagram. An example is given in Fig. 1.6, which combines the θ factor diagram of Fig. 1.5(c) and an assumed constant ϕ factor, that is, axisymmetric about the z-axis. It must not be overlooked that it is necessary to multiply by the radial probability factor to visualise the electron cloud density. However, the outermost shape of an atom is reasonably represented by the outside surface of the three-dimensional polar diagram since the radial factor is normally not varying rapidly.

Besides the electron density *per unit volume,* a second measure of electron density is used; the radial electron density $U(r)$ which is defined such that it gives the number of electrons in a shell of radius r and thickness dr. Curves of this kind show the probability of finding electrons at different distances from the nucleus. They can be misleading unless it is remembered that the surface area of the sphere $(4\pi r^2)$ is involved: a sphere with uniform volumetric electron density ρ has a $U(r)$ function which decreases to zero at $r = 0$. Where there is spherical symmetry

$$U(r) = 4\pi r^2 \rho.$$

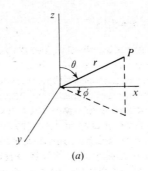

(a)

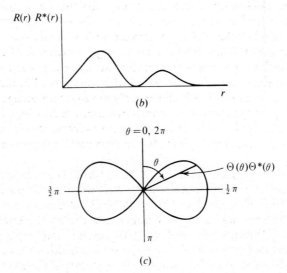

(b)

(c)

Fig. 1.5. Electron state defined by an electron cloud density which is proportional to three factors each involving only one of the polar coordinates r, θ and ϕ: $\rho = R(r)R^*(r) \times \Theta(\theta)\Theta^*(\theta) \times \Phi(\phi)\Phi^*(\phi)$.

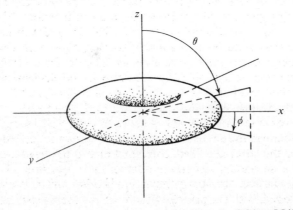

Fig. 1.6. Three-dimensional representation of $\theta\phi$ density factor $(= \Theta\Theta^* \times \Phi\Phi^*)$. The θ and ϕ factors are as given in Fig. 1.5(c).

1.7 Electron states and cloud densities

Solution of the wave equation for an electron moving in the field of the nucleus, modified by the presence of other electrons, gives a series of wave functions Ψ and associated electron cloud densities $\Psi\Psi^*$, which define the allowable states of the electron, as discussed above. Each electron state corresponds to a definite value of the total energy of the electron, called the *eigen values*. However several states may have the same energy, a phenomenon called *degeneracy*.

An electron state is characterised by four parameters or *quantum numbers*, which are given the symbols n, l, m_l and m_s. Only certain values are possible to obtain acceptable solutions of the wave equation. Acceptable means solutions that are finite, single-valued and continuous for the wave function and its derivative. Thus the quantum numbers and discrete energy levels follow directly from the mathematical treatment rather than from an arbitrary restriction, as in the Bohr theory. An analogy in classical mechanics is the equation of motion of a laterally vibrating string, e.g. a harp or violin (see textbooks on mechanics): the solution to the equation of motion includes a parameter (or wave number) which can only have certain integral values such that nodes (i.e. positions with zero vibration amplitude) are formed at the ends of the string. As the wave number is increased by integers the vibration frequency is increased in discrete steps – giving the fundamental note, first harmonic and so on.

The n quantum number, also called the *principal, or shell, number* is an integer $\geqslant 1$, and in practice ranges up to 7. It denotes the number of nodal surfaces on which Ψ and hence the electron cloud density are zero. Nodal surfaces may be spherical surfaces symmetrical around the nucleus or plane surfaces passing through the nucleus. This immediately explains why only positive integral values of n are possible – nodes are either there or not there, and one spherical nodal surface must exist at infinity if the electron is not to vanish. The n quantum number is also a measure of the overall size of the electron cloud, a large n is associated with a large cloud size, and with the l number it largely determines the energy level of a state.

The l number, or *secondary quantum number*, is the number of nodal planes of Ψ or $\Psi\Psi^*$ passing through the nucleus and can therefore take an integral value from 0 to $(n - 1)$ – this allows for the spherical nodal surface at infinity. It is customary to use the letters s p d and f for 0, 1, 2 and 3 respectively and the first two numbers of the state are written as 2s or 3d, for example. Secondary quantum numbers above f (i.e. 3) do not occur in the elements in spite of the principal number going up to 7. It will be seen later (§ 1.8) that f-states are only occupied in the fourth shell and correspond to the rare earths. The d-states, which occur from the third through to the sixth shell, have energy levels above those of the s-state in the next higher shell (i.e. 4s, 5s, 6s and 7s) and only become occupied after the latter are filled, corresponding to the transition metals and actinides. The secondary quantum number is a function of the angular momentum (excluding the spin angular momentum which is covered by m_s) and, as mentioned, with the principal quantum number largely determines the energy level. In the case of the hydrogen atom the angular momentum is equal to $(h/2\pi)\sqrt{l(l + 1)}$. It may seem surprising that an electron whose state is now represented as an electron cloud rather than a point mass moving in an exactly defined manner can still possess a definite angular momentum but each electron state has a function, determined by l, which when electrons collide governs their behaviour in an analogous manner to the angular momentum function in classical mechanics, and the name is therefore retained.

The m_l quantum number is a measure of the orientation of the angular momentum (or nodal planes), and its component in a particular direction, which is defined by a vanishingly small magnetic field. (A completely free atom has no definable directions.) By convention this magnetic field is assumed to be along the z-axis. Values of the m_l quantum number range $-l \ldots 0 \ldots +l$. In the limit the energy level is independent of m_l but with a finite magnetic field m_l is responsible for slight energy differences, which can be neglected here.

The fourth quantum number m_s, which is in a different category from the others, is introduced because an electron behaves as if it were spinning on its own axis, with consequent angular momentum and magnetic moment. The latter gives rise to magnetism (see Chapter 15). The value of m_s is $+\frac{1}{2}$ or $-\frac{1}{2}$. Like m_l, it has only minor influence on the energy level of a state.

From the proceeding it can be seen that the number of electron states (defined by all four quantum numbers) is $2n^2$ in the nth shell and $2(2l + 1)$ in the l-state. The position can be summarised as follows:

Number of electron states

Shell number	l states
1	s^2
2	$s^2\,p^6$
3	$s^2\,p^6\,d^{10}$
4	$s^2\,p^6\,d^{10}\,f^{14}$
5	$s^2\,p^6\,d^{10}\ldots$
6	$s^2\,p^6\,d^{10}\ldots$
7	$s^2\ldots$

The higher l-states in the later shells are not shown as they are not filled by electrons in the elements due to overlapping of the energy levels between shells, as will be discussed in § 1.8.

In considering the electron structure of atoms, that is, the electron cloud density, hydrogen will be considered first since this is the simplest. The electron cloud for the lowest energy state 1s has spherical symmetry (since s = 0) with one nodal surface at infinity. Only the radial density factor varies in the electron density per unit volume (cf. (1.9)) for this and other s-states and the curves of RR^* ($\propto \rho$) versus radius are shown in Fig. 1.7. For the 1s-state ρ is a maximum at the centre and falls rapidly to nearly zero around $r = 0.12$ nm (1.2 Å). On the other hand it can be shown that the most probable radius for the electron to be is 0.053 nm (0.53 Å), which coincides with the first orbit of the Bohr theory. Note again that the similarity ends there since there is no definite orbit or orbit plane in the wave mechanical model but only a spherical ball of electron cloud. In the 2s and 3s-states the number of spherical nodal surfaces increases to two and three respectively, as shown. Considering the 3s-state (Fig. 1.7(c)) there is now a central ball of electron cloud surrounded

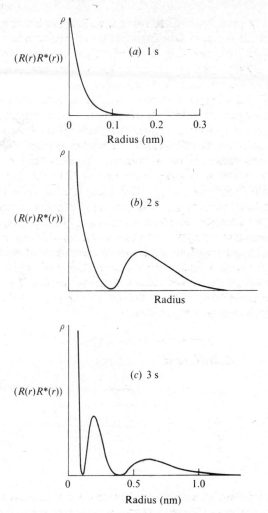

Fig. 1.7. Electron density per unit volume (ρ) for a hydrogen atom in the 1s, 2s and 3s-states. Because s-states are spherically symmetrical only the radial factor RR^* is plotted.

by two shells of decreasing density which extend to a much larger radius, around 1 nm (10 Å). The scale of the ordinate has been greatly magnified as compared with 1s since the overall probability of finding the electron, given by the area under the curve, must remain at unity. The most probable radius is now 0.7 nm (7 Å) in the outermost cloud shell. In each s-state there are two electrons with opposite spin.

Turning now to the p-states, there are six p-states in the second and higher shells, that is, with $m_l = -1, 0$, and $+1$ and two electrons of opposite spin in each. All three factors of the electron cloud density have to be considered. The radial density factor, which is independent of m_l, will have $n - 1$ nodes (i.e. $n - l$) and in addition a node at $r = 0$ (giving rise to a *point* node for ρ, not a nodal *surface* always taken in the preceding discussion). Curves of radial density factor versus radius are shown schematically in Fig. 1.8. On the other

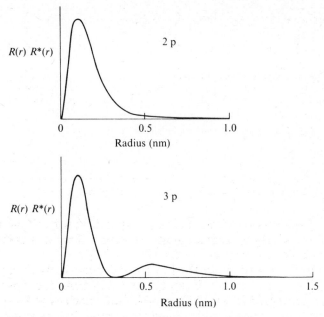

Fig. 1.8. Radial density factors for 2p and 3p-states (not dependent on m_l quantum number).

hand the angular probability factors (θ and ϕ) vary with m_l but are independent of the shell quantum number. The $\theta\phi$ density factor diagrams for $m_l = 0$ and ± 1 for a free atom are shown in Fig. 1.9. A vanishingly small magnetic field is assumed to act along the z-axis but there is no means of defining axes in the xy-plane and the diagrams for $m_l = \pm 1$ are the same, axis symmetric about z. When the atom is in a crystal x and y-axes are identifiable and the $\theta\phi$ diagrams for $m = -1$ and $+1$ are similar but orientated at 90°, see Fig. 1.10. It will be

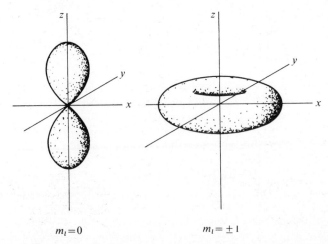

Fig. 1.9. $\theta\phi$ density factors for p-states in a free atom (not dependent on n quantum number). The z-axis is defined by a vanishingly small magnetic field.

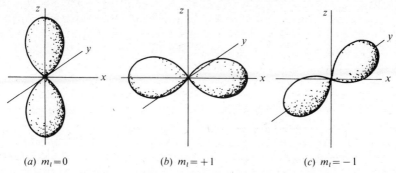

(a) $m_l = 0$ (b) $m_l = +1$ (c) $m_l = -1$

Fig. 1.10. $\theta\phi$ density factors for p-states in an atom in a crystal. (In a crystal x and y-axes can be defined.) Note how Fig. 1.9 is derived by rotating (b) or (c) about the z-axis.

seen that if the electron clouds are superimposed, that is, all the p-states are occupied, the diagram becomes almost spherical. This is a stable structure corresponding to an inert gas (see below).

There are ten d-states in the third and higher shells corresponding to $m_l = -2, -1, 0, 1$ and 2; each can have two electrons of opposite spin. The radial density factor RR^* again has a node at the nucleus and $n-2$ spherical nodes. The $\theta\phi$ density factor diagrams for an atom in a crystal are shown in Fig. 1.11 for the various values of m_l. Again it will be seen that if the five diagrams are superimposed a nearly spherically symmetric cloud is obtained and this is a stable structure. If the atom is free the diagrams for $m = \pm 1$ are the same, obtained by rotation about the z-axis, as in the case of the p-states.

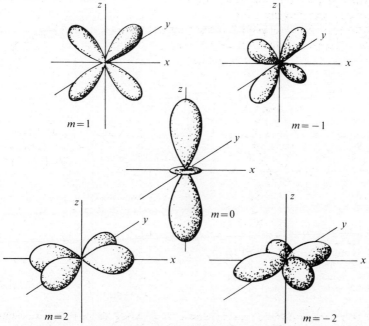

$m = 1$

$m = -1$

$m = 0$

$m = 2$

$m = -2$

Fig. 1.11. $\theta\phi$ density factor for d-states for an atom in a crystal.

Regarding the structure of atoms with atomic number greater than 1, the electron states described for hydrogen are retained in general terms. Determination of the wave functions for each electron becomes complex due to the interaction between the electrons and a process of iteration is necessary. Due to the presence of electrons in lower energy levels, screening the charge in the nucleus, the energy levels are slightly modified. Whereas in the hydrogen atoms the various states were ones into which the single electron could be excited from the lowest energy state (1s), they are now the normal states into which the additional electrons enter as the atomic number increases. The important *Pauli Principle* states that only one electron can occupy any one quantum state as defined by all four quantum numbers. The additional electrons of the higher atomic number elements cannot reside in the lowest energy state 1s but must progressively fill the higher energy states 2s, 2p, 3s, 3p . . . .

The electron distributions of the elements are shown in Table 1.3. Each of the first two shells (with s, and s and p, states respectively) is completed before any

Table 1.3. *Electron structure of atoms*

Period	Atomic number, element, symbol		Principal (n) and secondary (l) quantum numbers									
			n = 1	2		3			4			
			l = s	s	p	s	p	d	s	p	d	f
1st	1 Hydrogen	H	1									
	2 Helium	He	2									
2nd	3 Lithium	Li	2	1								
	4 Beryllium	Be	2	2								
	5 Boron	B	2	2	1							
	6 Carbon	C	2	2	2							
	7 Nitrogen	N	2	2	3							
	8 Oxygen	O	2	2	4							
	9 Fluorine	F	2	2	5							
	10 Neon	Ne	2	2	6							
3rd	11 Sodium	Na	2	8		1						
	12 Magnesium	Mg	2	8		2						
	13 Aluminium	Al	2	8		2	1					
	14 Silicon	Si	2	8		2	2					
	15 Phosphorus	P	2	8		2	3					
	16 Sulphur	S	2	8		2	4					
	17 Chlorine	Cl	2	8		2	5					
	18 Argon	Ar	2	8		2	6					
4th (1st long period)	19 Potassium	K	2	8		8			1			
	20 Calcium	Ca	2	8		8			2			
	21 Scandium	Sc	2	8		8		1	2			
	22 Titanium	Ti	2	8		8		2	2			
	23 Vanadium	V	2	8		8		3	2			
	24 Chromium	Cr	2	8		8		5	1			
	25 Manganese	Mn	2	8		8		5	2			
	26 Iron	Fe	2	8		8		6	2			
	27 Cobalt	Co	2	8		8		7	2			
	28 Nickel	Ni	2	8		8		8	2			
	29 Copper	Cu	2	8		8		10	1			
	30 Zinc	Zn	2	8		8		10	2			

Table 1.3 (cont.)

Period	Atomic number, element, symbol	n = 1	2		3			4				5			6			7
		l = s	s	p	s	p	d	s	p	d	f	s	p	d	s	p	d	s
4th (cont.)	31 Gallium Ga	2	8		8		10	2	1									
	32 Germanium Ge	2	8		8		10	2	2									
	33 Arsenic As	2	8		8		10	2	3									
	34 Selenium Se	2	8		8		10	2	4									
	35 Bromine Br	2	8		8		10	2	5									
	36 Krypton Kr	2	8		8		10	2	6									
5th (2nd long period)	37 Rubidium Rb	2	8		18			2	6			1						
	38 Strontium Sr	2	8		18			2	6			2						
	39 Yttrium Y	2	8		18			2	6	1		2						
	40 Zirconium Zr	2	8		18			2	6	2		2						
	41 Niobium Nb	2	8		18			2	6	4		1						
	42 Molybdenum Mo	2	8		18			2	6	5		1						
	43 Technetium Tc	2	8		18			2	6	6		1						
	44 Ruthenium Ru	2	8		18			2	6	7		1						
	45 Rhodium Rh	2	8		18			2	6	8		1						
	46 Palladium Pd	2	8		18			2	6	10		—						
	47 Silver Ag	2	8		18			18				1						
	48 Cadmium Cd	2	8		18			18				2						
	49 Indium In	2	8		18			18				2	1					
	50 Tin Sn	2	8		18			18				2	2					
	51 Antimony Sb	2	8		18			18				2	3					
	52 Tellurium Te	2	8		18			18				2	4					
	53 Iodine I	2	8		18			18				2	5					
	54 Xenon Xe	2	8		18			18				2	6					
6th (3rd long period)	55 Caesium Cs	2	8		18			18				8			1			
	56 Barium Ba	2	8		18			18				8			2			
	57 Lanthanum La	2	8		18			18				8		1	2			
	58 Cerium Ce (Rare earths)	2	8		18			18			1	8		1	2			
	to										to							
	71 Lutecium Lu	2	8		18			18			14	8		1	2			
	72 Hafnium Hf	2	8		18			32				8		2	2			
	73 Tantalum Ta	2	8		18			32				8		3	2			
	74 Tungsten W	2	8		18			32				8		4	2			
	75 Rhenium Re	2	8		18			32				8		5	2			
	76 Osmium Os	2	8		18			32				8		6	2			
	77 Iridium Ir	2	8		18			32				8		7	2			
	78 Platinum Pt	2	8		18			32				8		8	2			
	79 Gold Au	2	8		18			32				8		10	1			
	80 Mercury Hg	2	8		18			32				18			2			
	81 Thallium Tl	2	8		18			32				18			2	1		
	82 Lead Pb	2	8		18			32				18			2	2		
	83 Bismuth Bi	2	8		18			32				18			2	3		
	84 Polonium Po	2	8		18			32				18			2	4		
	85 Astatine At	2	8		18			32				18			2	5		
	86 Radon Rn	2	8		18			32				18			2	6		
7th	87 Francium Fr	2	8		18			32				18			2	6		1
	88 Radium Ra	2	8		18			32				18			2	6		2
	89 Actinium Ac	2	8		18			32				18			2	6	1	2
	90 Thorium Th	2	8		18			32				18			2	6	2	2
	91 Protactinium Pa	2	8		18			32				18			2	6	3	2
	92 Uranium U	2	8		18			32				18			2	6	4	2

electrons enter the next higher shell. But in the third and subsequent shells, due to the energy of the d-states being higher than the s-states of the next shell, electrons enter the higher shell before filling the d-states. This is the origin of the transition metals (see § 1.8). For similar energy overlap reasons the 4 f-states in the fourth shell are not filled until electrons have entered the 6s-states; actually the filling of the 4f-states intervenes in the filling of the 5d-states, which as previously noted are filled after the 6s-states. This is the origin of the rare earths (see § 1.8).

In the following section the Periodic Table of the elements will be discussed. This arranges the elements in order of their atomic numbers so that elements with similar chemical and physical properties are adjacent to each other. Since these properties stem from the electron structure, the Periodic Table is also a reflection of the electron energy states just discussed, as will now be shown.

1.8 Periodic Table

The Periodic Table is a list of the chemical elements in order of their atomic numbers so arranged that elements with similar properties are adjacent to each other. One form is given in Table 1.1. Several other arrangements are also used in attempts to bring out the rather complex relationships between the elements. The Table was first proposed in 1869 by Mendeléev, who saw that if the elements were arranged in order of their atomic weights there was a periodicity in their chemical properties. There were some irregularities in the early versions which have been eliminated by using the atomic number instead of the atomic weight.

The elements are placed in order of their atomic numbers in seven horizontal lines known as *periods*: the first three are called 'short' periods as distinct from the later 'long' periods. There is a fairly rapid change of properties as one proceeds along a period. The vertical lines of elements are known as *groups* and contain elements with similar properties although a progressive change occurs through a group.

As mentioned in the previous section, the Periodic Table can be related to, and explained in terms of, the electron distribution of the atoms given in Table 1.3. To do so, some appreciation of the nature of the interactions between atoms is necessary, full discussion of which is given in Chapter 3.

The electrons in the outermost shell (i.e. the highest principal quantum number) are called *valency electrons* and they participate in the interactions between atoms in an assembly. A stable electron configuration for an atom is one in which the electron states are occupied up to the level at which any further electron would go into an energy state in a higher, unoccupied, shell. This normally corresponds to eight electrons in the outer shell, occupying the s and p-states (recall that the d-states always have energy levels overlapping the s-states of the next shell); the exception is helium with only two electrons.

Chemical reactions between atoms occur by exchanging or sharing valency electrons so that each atom in the assembly achieves a stable electron configuration. The types of bond which are formed can be classed as follows (see Chapter 3 for a fuller description).

(i) Metallic binding, in which the valency electrons leave their individual atoms and form a free electron cloud (within the solid). The free electrons are responsible for the characteristics of metals – high thermal and electrical conductivity and plasticity. Metallic binding occurs between atoms with one or two electrons in excess of a stable configuration (e.g. Group IA, IIA and transition metals).

(ii) Covalent binding, in which valency electrons are shared between specific atoms so that each partner achieves a stable electron configuration. The bond leads to high thermal and electrical resistivity (since there are no free electrons) and no plasticity (since the atoms are specifically linked). Covalent bindings occur between atoms that are one or two electrons short of a stable configuration (e.g. Group VIIB and VIB).

(iii) Ionic binding, in which valency electrons are exchanged between unlike atoms so that each type of atom gains or loses one or two electrons and achieves a stable configuration. This bond also leads to high thermal and electrical resistivity but some limited plasticity is possible in single crystals (see § 7.2). Ionic bindings occur in compounds between one type of atom with one or two electrons short of a stable configuration and another with one or two electrons in excess of a stable configuration (e.g. NaCl).

(iv) Van der Waals binding. This does not involve exchanging or sharing electrons, but can arise between stable electron configurations. The centres of positive and negative charge in an atom may not coincide so that the atom behaves as a dipole, giving rise to weak interaction forces. The most important example to technology is the binding between the macromolecules of thermoplastics; the binding along the molecule or chain being covalent.

To return to the discussion of the Periodic Table, the elements in the right-hand column, Group 0, called the *inert gases*, all have stable electron configurations, that is, the next unoccupied electron state is in a higher shell, see Table 1.3. Until recently the inert gases were considered to be completely non-reactive. The elements in the adjacent Groups VIIB (*halogens*) and VIB are respectively one and two electrons short of the stable configurations. They form covalent bonds in assemblies of the pure elements (or ionic bonds with metallic elements) and exhibit non-metallic properties. On the other side of the Table, Group IA known as the *alkali metals*, and Group IIA, called the *alkaline earths*, have respectively one and two electrons in excess of stable configurations and form metallic bonds in the elemental state (or ionic bonds with Group VIB and VIIB elements). The elements in the middle of the short periods (IIIB, IVB, VB) have electron structures intermediate between two possible stable configurations and they exhibit metallic or non-metallic properties depending on the circumstances (see Chapter 14 on semiconductors).

In the middle of the three long periods there are twenty-five elements, known as the *transition metals*, which have partially filled d-states and one or two electrons already occupying the next outer shell. This is due to overlapping of the energy levels of the states between shells, as mentioned in the previous section. For example, proceeding along the fourth period, from scandium through to nickel the additional electrons are entering the 3d-states (see Table 1.3); in copper, the 3d-state is completed and this is not included. Similarly, from yttrium to rhodium and from lanthanum to platinum in later periods. With one or two valency electrons the transition elements show metallic properties and there is a close similarity in properties between adjacent groups. The presence of this partially filled inner shell influences both the physical chemical properties, since the d-electrons may on occasions behave as valence electrons; they are also responsible for ferromagnetism (see Chapter 15). A similar series, the *actinides*, beginning with actinium, occurs in the seventh period (4th long), and all the naturally occurring radioactive elements and the artificially produced transuranic elements lie in this series.

In the third long period the filling of the 5d-states is interrupted whilst the 4f-states are filled. This gives rise to the fourteen *rare earth* elements, atomic numbers 58 to 71; these are also known as the lanthanides but it is a slight misnomer since it is now known that filling of 4f-states starts with cerium. The electrons in the incomplete 4f-shell are effectively screened by the 6s and 5d-electrons, and hence the chemical and physical properties of the rare earths are almost identical to each other and to their neighbouring transition elements.

1.9 Classes of materials

In conclusion to this chapter it is convenient to look ahead to the various classes of solid materials to be considered in subsequent chapters. They are:

(1) Metals. The metallic binding, with the valency electrons forming an electron cloud freely moving between the atoms, leads to the characteristic properties: high electrical and thermal conductivity, plasticity and high strength in the presence of defects (see § 5.8 on notch toughness). Metals are the primary structural materials of technology, and include the great range of iron-based alloys (e.g. cast iron, plain carbon steels, alloy steels).

(2) Ceramics. This class may be defined as all inorganic materials (i.e. not based on carbon) with ionic and covalent bindings. It covers a great range of materials including traditional ceramics based on clay, cement and concrete, cermets (ceramic in a matrix of metal), graphite, and pure compounds such as oxides and nitrides. Apart from cement and concrete they have until recently not been regarded as constructional materials, but this is changing as their mechanical properties are improved and they are now meeting critical requirements, especially at high temperature. The electrical and magnetic properties are also of exceptional interest.

(3) Glasses. This class consists of oxides or mixtures of oxides which are non-crystalline. Their characteristic properties are transparency to light, elasticity (stress $\propto$ strain) followed by brittle facture at room temperature and viscous behaviour (stress $\propto$ strain rate) at elevated temperatures. They are usually considered alongside ceramics, which often include a percentage of glassy phase between the crystalline portions.

(4) High polymers. These cover a group of materials with the common feature that the binding is covalent. At one end of the scale are the linear polymers in which the atoms, often carbon, are joined into very long chains (macromolecules) by strong covalent bonds and the binding between the chains is due to weak van der Waals forces. They are never fully crystalline. They are the basis of wood and thermoplastics. At the other end of the scale are the close-network polymers in which three-dimensional covalent molecules are formed by polymerisation of 'monomer' units. Besides plastics, high polymers

are the basis of the paint, rubber and synthetic fibre industries, not to mention life itself.

(5) Semiconductors. This is a special category of non-metallic inorganic material. Definition is difficult to appreciate until the electron theory of crystalline solids has been considered (Chapter 3). Briefly, they are insulators in which the energy gap between the state of the valence electrons and the electron states required for electrical conduction are much smaller than in conventional insulators and can be bridged under certain circumstances; for example, by thermal excitation. These materials are the basis of the current headlong race in the development on transistors, solid state electronics and computers.

QUESTIONS

1. (a) Define: element; atomic number Z; mass number A.
   (b) Approximately how many elements occur in nature?
   (c) 'The atom is the smallest indivisible part of an element and all atoms of one element are identical.' True or bluff?
2. (a) What is the typical ratio of neutrons to protons for a stable nuclide?
   (b) What is the radiation emitted when there are (i) too many neutrons (ii) too few neutrons.
   (c) Sketch the curve of number of atoms of a radioactive element versus time and indicate the difference between decay constant λ and the half life $0.693/\lambda$.
   (d) How would the curves given in (c) be affected by doubling the absolute temperature?
3. The data for a radioactive chain of three nuclides are

	A	B	C
decay constant	λ_A	λ_B	stable
atomic weight	W_A	W_B	W_C

   If there is initially a mass M_A of A present, calculate the maximum mass of B and the time at which it occurs.
4. (a) Define mass defect.
   (b) Sketch the curve of binding energy per nucleon versus atomic number. Hence contrast the fuel consumed in nuclear and fossil power stations.
   (c) Approximately how many electron masses are equivalent to making a cup of tea?
5. (a) In what way does classical mechanics fail to explain the behaviour of a nuclear atom?
   (b) What arbitrary restrictions did Bohr place on electron movement?
6. Derive Bohr's expression for the allowable energy levels in a hydrogen atom and hence obtain Rydberg's constant in terms of fundamental constants.
7. Describe qualitatively the wave mechanical model for the electron structure of an atom.
8. What is the significance of the four quantum numbers n, l, m_l and m_s in defining an electron state?
9. (a) What is meant by radial density factor and angular density factors?
   (b) Sketch the radial density factor and $\theta\phi$ factor diagram for the electron states: 1s, 3s, 2p ($m_s = -1, 0, +1$).
10. What principles govern the filling of the electron states as the atomic number increases?
11. (a) Define: Periodic Table; groups and periods; alkali metals and alkaline earths.
    (b) What are valency electrons and explain their significance to chemical reactions.
12. (a) What is the difference in electronic structure between atoms of metals and non-metals?

(*b*) Explain the connection between d-states and the transition metals.

(*c*) What electron states are associated with the rare earths?

FURTHER READING

M. Born: *Atomic Physics*, 4th edition. Blackie (1946).

W. Hume-Rothery: *Atomic Theory for Students of Metallurgy*. Institute of Metals, London (1947).

A. H. Cottrell: *Theoretical Structural Metallurgy*, 3rd edition. Arnold (1968).

S. Glasstone and M. C. Edlund: *The Elements of Nuclear Reactor Theory*. van Nostrand (1952).

P. J. Durrant: *General and Inorganic Chemistry*, 3rd edition. Longmans (1964).

T. A. Littlefield and N. Thorley: *Atomic and Nuclear Physics: an Introduction*. van Nostrand (1963).

2. Crystal structure

2.1 Crystals

When atoms or molecules come together in the solid state, they arrange them-selves in a regular three-dimensional pattern to form a *crystal*. In layman's language, a crystal must have plane surfaces set at definite angles, as in a cut diamond, but these are only the visible consequence of an underlying orderliness of the atoms. In technical terms, a crystal is any piece of solid matter in which the atoms are in a single continuous pattern, irrespective of the condition of its boundaries. Most solids, including all metals and many ceramics, are crystalline, that is, they are composed of one or more crystals, but there are a few exceptions, notably glasses and some high polymers. The structures of these *amorphous* solids will be considered in Chapters 7 and 8.

Knowledge of the atomic arrangements in crystals has been obtained by X-ray analysis. X-rays, discovered by Röntgen in 1895, are generated when high speed electrons strike a metal target; they are electromagnetic waves with wavelengths below the visible light spectrum, around 0.1 nm (1 Å). Since the spacing of atoms in solids is of the same order, a beam of X-rays is diffracted on passing through a crystal and emerges in several distinct directions. This is similar to a well known effect in optics in which visible light is diffracted when it is passed through a grating of uniformly spaced lines. The diffraction of Röntgen rays was first shown in 1912 by von Laue, who thereby proved that they were electromagnetic waves. The Laue technique was immediately used by W. L. Bragg to determine the crystal structure of sodium chloride (the mineral halite). Since then, X-ray analysis has been used to determine increasingly complex atomic arrangements, reaching a climax with the elucidation of the structure of the giant DNA molecule which carries hereditary information forward into succeeding generations. It is not proposed to discuss the techniques of X-ray analysis here, only to present the results of many experimental investigations.

Solids are normally polycrystalline, that is, they are a collection of crystals or grains which have grown independently, usually during solidification, and have different orientations of their atomic pattern. Where two crystals meet there is a grain boundary and, where this intersects the free surface of a specimen, it can be observed under a microscope after appropriate polishing and etching (see § 4.1). Grain diameters vary widely, from less than 1 μm (e.g. clay particles, see § 7.3) to more than 10 mm (e.g. in brass castings). On many door handles the individual grains are clearly visible to the naked eye, due to the polishing and etchant action from the hand; this is particularly so on squash court door handles due to sweating hands! A typical grain size for wrought metal, for example, mild steel (see § 4.8 and Chapter 12), is in the range 25–50 μm.

A solid may also contain only one large crystal, which may have grown naturally, or have been produced synthetically by man. Crystals of copper sulphate have been grown from supersaturated solution by many schoolboys. The more difficult techniques of growing single crystals of metals and ceramics were originally the province of research laboratories, but now the growing of single crystals for many different applications is an industry. A most important example is the use of single crystals of semiconducting germanium or silicon for transistors and many other solid state devices (see Chapter 14). Crystals of quartz, one form of SiO_2 (see § 7.3), are grown in large quantities, 15 million per annum in the USA alone, for use as precision frequency control devices. This function depends on the piezoelectric properties of the crystal (i.e. the development of electric charges on straining). Quartz crystal is also used for infra-red and ultra-violet windows. Sapphire crystals (Al_2O_3) are grown extensively for gems, low friction, low wear, bearings and record playing needles; large windows up to 125 mm diameter have been made for transmitting infra-red radiation. Finally, metallic single crystals with leading dimensions of several centimetres have been extensively used to elucidate the theory of mechanical properties, including dislocation theory (see Chapter 9).

The arrangement of atoms in a crystal is often demonstrated by one of the three types of model shown in Fig. 2.1. In (a), spheres, usually table tennis balls, are held in their correct places by glueing to each other; in (b), spheres, usually solid plastics balls, are held in place by wire links; in (c), in a rather different category, a two-dimensional close-packed array is obtained with small bubbles floating on the surface of a liquid. Each method has its advantages and disadvantages. Type (a) gives the correct impression of close-packing of the atoms but obscures the crystal planes. Type (b) reveals the crystal planes and directions but gives the incorrect impression that the atoms are widely separated. Type (c) has forces of interaction between the bubbles which obey similar laws to those between atoms and it is possible to study, both qualitatively and quantitatively, the movement of atoms past each other; the disadvantages are that it is only two-dimensional and lacks robustness.

There are various important limitations to all these models. First, real atoms are vibrating about their mean positions due to their thermal energy. Secondly, real crystals contain faults in their structure, both point faults, such as vacant sites, and line faults, such as occur when a plane of atoms ends in the middle of a crystal. The bubble model is an exception in that crystal faults can easily be simulated. Finally, the atoms are not necessarily fixed in position but are capable of moving slowly about the crystal lattice, a phenomenon called *diffusion*.

The first part of this chapter describes the ideal crystal structure, and the final sections are devoted to various departures from the simple model, that is, thermal vibration, lattice faults and diffusion.

2.2 Space lattice

In a crystal, an atom or a group of atoms is repeated regularly in three-dimensions. The cycle of repetition can be shown by an arrangement of points which mark the position of successive groups without indicating anything about the group itself. This regular three-dimensional array of points is called a *space lattice* or *Bravais lattice*. It should be noted that in a space lattice the

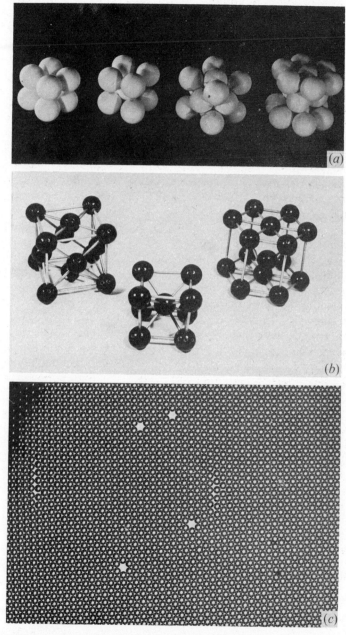

Fig. 2.1. Models of crystal structures: (*a*) close-packed table tennis balls; (*b*) solid balls with connecting rods; (*c*) raft of soap bubbles (two-dimensional) (by courtesy of W. H. Lomer).

surroundings of any point are always indentical with the surroundings of any other point. The group of atoms may consist of only one or two atoms, as in most metals, or of several hundred atoms in many organic substances. The group itself need not be a regular arrangement of atoms but identical groups with the same orientation must occur at each space lattice point.

A common analogy in two dimensions is wallpaper, in which some design, such as a spray of flowers, equivalent to the atomic group, is repeated in a regular pattern (the space lattice). Fig. 2.2(*a*) shows a single atom being repeated on a square lattice; in (*b*) a group of two atoms is repeated on the same square lattice. The two-dimensional space lattice itself is shown in (*c*).

In general crystallography it is important to distinguish between the space lattice and the crystal structure, that is, the arrangement of the atoms associated with each lattice point. The space lattice is fundamental to the idea of a regular array – the regularity is indicated by the existence of a space lattice – and many of the properties of solids are the consequence of this regularity. For example, the diffraction of X-rays by solids is dependent on the existence of a space lattice which determines the direction of the diffracted beams; the atomic groups associated with each is then deduced from the intensities of the beams. However, in two of the three common metallic crystal structures, body-centred cubic and face-centred cubic, there is only one atom per lattice point and the crystal structure, represented by points at the centres of atoms, becomes indistinguishable from the space lattice.

If three sets of equally-spaced parallel planes are drawn to intersect at points of a lattice, it is divided into identical parallelepiped blocks or cells, see Fig. 2.3. Conversely, the lattice can be constructed from such cells by placing them side by side in three directions. By specifying the cell dimensions and the positions of the lattice points in the cell, the lattice is defined. If the planes pass through every point there will be effectively one lattice point per cell; this will be called here the *unit cell*. (N.B. there is no standard terminology for the various types of cell.) There are an infinite number of ways of drawing unit cells in any lattice, but it is usual to choose a cell with nearly equal axes. It is often more convenient to use a cell which contains more than one lattice point in order to indicate the lattice symmetry; this will be called the *lattice structure cell*. (It, also, is sometimes known as the unit cell, although it has more than a single atomic group.) A comparison of the unit and structure cells in face-centred

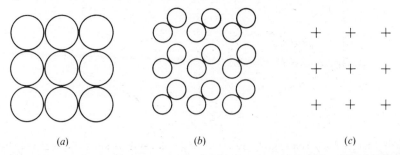

(*a*)   (*b*)   (*c*)

Fig. 2.2. Regular arrays of atoms in two dimensions: (*a*) has one atom per lattice point and (*b*) two; (*c*) shows the space lattice for (*a*) and (*b*).

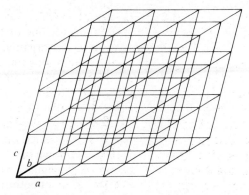

Fig. 2.3. Parallelepiped unit cells in a space lattice.

cubic lattice is shown in Fig. 2.4. The lengths of the sides of a structure cell are called its *unit translations*. When the positions of the atoms associated with each lattice point are shown, the cell will be called the *crystal* structure cell. As mentioned above, in many metals the lattice and crystal structure cells are identical.

Space lattices can be divided into fourteen types according to their degree of symmetry, there being no further ways in which points can be symmetrically arranged. The structure cells are shown in Fig. 2.5. The fourteen space lattices can be further grouped into seven *crystal systems*. The structure cells of any crystal system have the same axes but additional lattice points are added, obviously, only as allowed by the symmetry of the system. For example, in

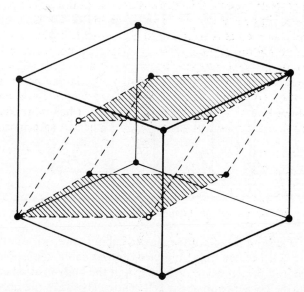

Fig. 2.4. Unit cell and structure cell. The unit cell, shown with a broken line, contains only one lattice point. The structure cell, shown with a full line, contains more than one lattice point but indicates the lattice symmetry. The example here is a face-centred cubic lattice.

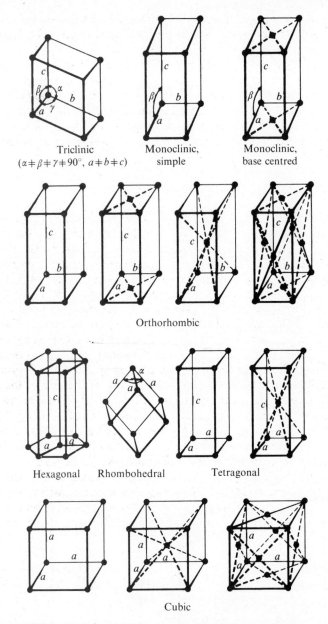

Triclinic
$(\alpha \neq \beta \neq \gamma \neq 90°, \ a \neq b \neq c)$

Monoclinic,
simple

Monoclinic,
base centred

Orthorhombic

Hexagonal   Rhombohedral   Tetragonal

Cubic

Fig. 2.5. The fourteen space lattices and seven crystal systems.

the cubic crystal system there are the simple cubic lattice, with lattice points at the structure cell corners only; the face-centred cubic, with points at the cubic corners and in the centre of each face; and the body-centred cubic, with lattice points at the corners and at the cell centre. These are the only possible lattice positions that have cubic symmetry. The simplest structure cell in each of the seven crystal systems, with one point per cell, is known as the *primitive cell.* The crystal systems are given in Fig. 2.5 ranging from *triclinic*, with no symmetry, through *orthorhombic*, with three axes at right angles but all unequal, to *cubic*, with three axes at right angles and all equal.

In a hexagonal lattice a parallelepiped structure cell fails to indicate the hexagonal symmetry and it is usual and more convenient to regard the hexagonal prism as the structure cell, known as the Miller–Bravais cell, see Fig. 2.5.

It must be recalled that the preceding discussion refers to the space lattice. Although there are only fourteen space lattice types, there is an infinite number of crystal structure cells in which different atom groups are associated with each lattice point.

2.3 Crystal structure of the elements: polymorphism

As regards crystal structure, the elements divide into two main groups and a small intermediate group, as shown in Table 2.1. In Class 1 are the majority of metals, that is, the elements to the left side of the Periodic Table. They have one of three well-packed crystal structures: face-centred cubic (fcc); close-packed hexagonal (cph); and body-centred cubic (bcc). These are drawn in Fig. 2.6. The arrangement is such that each atom has a large number of equally distant neighbours, or nearest neighbours, known as the *coordination number*. This is a consequence of the metallic binding (§ 3.5). As can be seen in Table 2.1, elements in Group IA (the alkali metals: Li, Na, K, Rb and Cs) are entirely bcc, which has a coordination number of 8. The adjacent Group IIA elements (the alkaline earth metals: Be, Mg, Ca, Sr and Ba) are largely cph, with a coordination number of 12, although there is some falling away at the end of the group. The next eight Groups, IIIA through to IB, are the transition metals and these are a mixture of the three well-packed structures, initially mainly bcc and cph and finally fcc. All Group IB (Cu, Ag and Au) are fcc, as are aluminium and many of the adjacent Group VIII, with a coordination number of 12. The lattice constants of the more common metals are given in Table 2.2.

The second main group (Class 3) comprises the non-metallic elements to the right side of the Periodic Table. The crystal structure of each group is such that the coordination number is $8 - N$ where N is the group number. This is a consequence of the formation of covalent bonds between the atoms (§ 3.4).

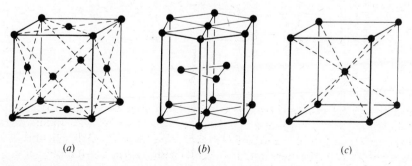

(a) $\qquad\qquad$ (b) $\qquad\qquad$ (c)

Fig. 2.6. Typical metallic crystal structures with high coordination numbers (Class 1): (a) face-centred cubic; (b) close-packed hexagonal; (c) body-centred cubic.

Table 2.1. *The elements grouped into three classes of crystal structure*

IA	IIA	IIIA	IVA	VA	VIA	VIIA	VIII			IB	IIB	IIIB	IVB	VB	VIB	VIIB
Li bcc	Be cph											B hex	C			
Na bcc	Mg cph											Al fcc	Si	P	S	Cl
K bcc	Ca fcc cph	Sc fcc cph	Ti cph	V bcc	Cr bcc cph	Mn cub	Fe bcc fcc	Co cph fcc	Ni fcc cph	Cu fcc	Zn hex	Ga orh	Ge	As	Se	Br
Rb bcc	Sr fcc	Y cph	Zr cph bcc	Nb(Cb) bcc	Mo bcc	Ma cph	Ru cph	Rh fcc cub	Pd fcc	Ag fcc	Cd hex	In fct	Sn	Sb	Te	I
Cs bcc	Ba bcc	La cph fcc	Hf cph	Ta bcc	W bcc cub	Re cph	Os cph	Ir fcc	Pt fcc	Au fcc	Hg rho	Tl cph fcc	Pb fcc	Bi		
			Th fcc		U orh											

Class 1
(High coordination number structures: fcc cph and bcc; metallic binding)

Class 2
(Intermediate)

Class 3
(8 – N coordinating number structures; covalent binding)

Table 2.2. *Crystal structure and lattice constants of common metallic elements at 20 or 25 °C. (From* Structure of Metals, *table 3/e, by C. S. Barrett and T. B. Massalski, McGraw-Hill Book Company (1966))*

bcc		cph				fcc	
	a 10^{-10} m		a 10^{-10} m	c 10^{-10} m	c/a		a 10^{-10} m
Li	3.50	Be	2.27	3.59	1.58	Al	4.04
Na	4.28	Mg	3.20	5.20	1.63	α-Ca	5.56
K	5.33	γ-Ca	3.94	6.46	1.64	γ-Fe	3.56
α-Fe	2.86	Ti	3.00	4.73	1.58	β-Ni	3.52
Mo	3.14	Zn	2.66	4.94	1.85	Cu	3.60
W	3.16	Cd	2.97	5.61	1.89	Ag	4.08
						Au	4.07

Thus, the Group VIIB elements, the halogens, form crystal structures with only one nearest neighbour; the iodine structure in Fig. 2.7(*a*) is an example. Since the binding forces other than between nearest neighbours are weak, these elements are easily broken down to liquids and gases. In Group VIB (S, Se and Te) two nearest neighbours are required to form a stable electronic configuration and the crystal contains chains of atoms, as in the tellurium structure, Fig. 2.7(*b*). The Group VB elements (P, As, Sb and Bi) form crystals with their atoms arranged in layers, each atom having three nearest neighbours within the layer; an example is arsenic, shown in Fig. 2.7(*c*). Lastly, the elements of Group IVB (C (diamond), Si, Ge and Sn (grey form)) have four valency atoms and require to establish covalent bonds with four nearest neighbours to build up stable electron configurations. The atoms arrange themselves at the corners of a tetrahedron with a central atom, forming a strong three dimensional covalent crystal, known as the diamond structure. An alternative crystal structure for carbon contains layers of carbon atoms, known as graphite (see § 7.7).

The few remaining elements, forming Class 2, are in intermediate positions in the Periodic Table and their structures do not fall into either of the main groupings. Zinc and cadmium form hexagonal structures but they are not close-packed – the axial ratio is 1.9 against 1.63 for close-packing (see Table 2.2). Mercury has a simple rhombohedral structure. Gallium has a very complex structure which will not be described here. Lastly, α-thallium (cph), lead (fcc) and indium (fc tetragonal) are similar to metallic structures but evidence on atomic spacing indicates that the atoms are only partially ionized. Finally boron and tin have characteristics of both Class 2 and 3.

About twenty-five elements, including many transition metals, have more than one crystal structure. The phenomena is called *polymorphism* or *allotropy*; several examples are included in Table 2.2. There is usually a preferred structure at any pressure and temperature but whether this is achieved depends on the particular element, and what pressure and temperature cycle it is given – temperature is the normal variable. Thus carbon can exist either as diamond,

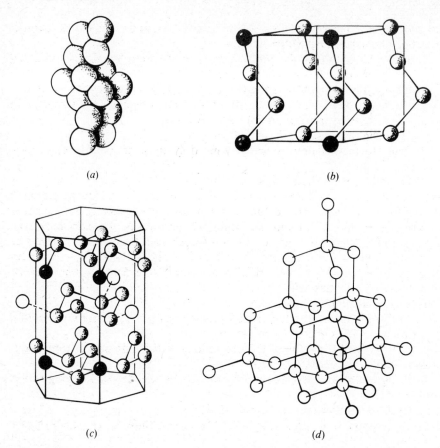

Fig. 2.7. Typical non-metallic crystal structures (Class 3): The coordination number equals $8 - N$ (N = group number in Periodic Table). (*a*) Iodine pairs (Group VIIB); (*b*) tellurium chain (Group VIB); (*c*) arsenic layer structure (Group VB); (*d*) diamond structure of carbon (Group IVB), three-dimensional covalent crystal.

when it has a diamond cubic structure, or as graphite, when it has a hexagonal structure. The change from graphite to diamond is very difficult to bring about, though the reverse is not so difficult above 1000 °C. On the other hand, iron exists as bcc at room temperature (α-iron), as fcc between 906 °C and 1400 °C (γ-iron) and again as bcc between 1400 °C and its melting point at 1535 °C (δ-iron). In this example all the changes occur readily as the material is heated or cooled.

Polymorphism is not confined to crystals of the elements, but also occurs in compounds. For example, silica (SiO_2) has many crystalline forms, the best known of which is quartz (see § 7.3), and zinc sulphide takes up two crystal structures, called zinc blende and wurtzite (see § 2.7).

2.4 Miller indices

The orientation of a plane or a direction in a crystal is specified in relation to the structure cell axes by means of three integers, called Miller indices. These do not indicate the position or line of action, so parallel planes or parallel directions have identical Miller indices.

The rules for obtaining the Miller indices are different for planes and directions. For planes, the steps are as follows:

(1) Determine the lengths of the intercepts of adjacent planes on three axes taken parallel to the sides of the structure cell; measure each intercept in terms of the unit translation on the respective axis.

(2) Take the reciprocal of these numbers.

(3) Bring to the smallest integers by multiplying or dividing through by a common factor.

(4) Enclose in parentheses thus: (h, k, l).

For example, in Fig. 2.8(a), two adjacent planes are shown cutting the three axes of the structure cell, whose unit translations are a, b and c. The first plane is conveniently taken through the origin. The intercepts of the second plane are a, b and $2c$, which in terms of the unit translations become 1, 1 and 2. The remaining steps give: (2) 1, 1 and $\frac{1}{2}$; (3) 2, 2 and 1; (4) (221). The Miller indices of these and all parallel planes are therefore (221). Note that a zero indicates a plane is parallel to that axis.

To specify directions the steps are:

(1) Construct a line from the origin of the structure cell axes, in the required direction and with any convenient length.

(2) Determine the lengths of the projections of this line on the three axes, measured in terms of the unit translations.

(3) Bring to the smallest integers by multiplying or dividing through by a common factor.

(4) Enclose in square brackets thus: $[h, k, l]$.

For example, in Fig. 2.8(b), the line OP has projections $1a$, $1b$ and $\frac{1}{2}c$, which

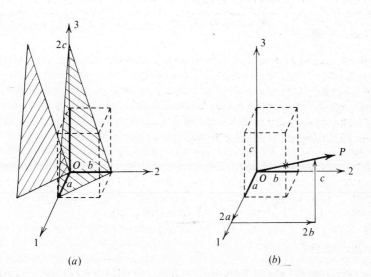

(a)   (b)

Fig. 2.8. Miller indices for (a) plane (221); and (b) direction [221].

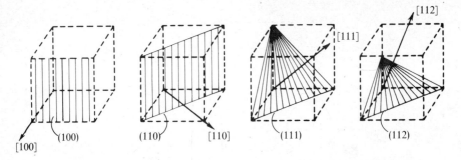

Fig. 2.9. Miller indices of various planes and directions in a cubic lattice.

in terms of the unit translation become 1, 1 and $\frac{1}{2}$. The smallest integers are then 2, 2 and 1, and the Miller indices are written [221]. Note that the reciprocals of the intercepts are taken only for planes.

The positive and negative directions on an axis are arbitrarily chosen but conveniently a right-hand orientation is used, as in the figures here. If more than one set of Miller indices is being considered it is essential that the axes should not be changed once they have been chosen. The Miller indices of various planes and directions in cubic crystals are shown in Fig. 2.9. It will be seen that for cubic axes, planes and directions with the same Miller indices are perpendicular to each other.

It is sometimes desired to refer to a group of crystallographically equivalent planes or directions, irrespective of how the axes happen to be chosen. A shorthand notation is used; for planes, the Miller indices are placed in { } brackets; for directions, ⟨ ⟩ brackets are used. Thus, the planes forming the sides of a cubic structure cell which are crystallographically identical can be indicated by {100}, which stands for the (100), (010) and (001) planes. Similarly the directions of the cube diagonals, which are all equivalent, are specified by ⟨111⟩. This stands for [111], [Ī11], [1Ī1] and [11Ī] (or the reverse directions [ĪĪĪ], [1ĪĪ], [Ī1Ī] and [ĪĪ1]).

In the hexagonal system, planes and directions are often referred to the four axes of the Miller–Bravais cell, see Fig. 2.10. This overcomes the objection that the Miller indices of equivalent planes do not have similar indices. Four numbers are now obtained, known as the Miller–Bravais indices; by convention the first three always refer to the axes in the basal plane and the fourth to the perpendicular (prism) axis. The procedure for planes is similar to above, except that the reciprocals of four intercepts on the axes are obtained and reduced to the smallest integers. If the indices are ($h\,k\,i\,l$), the first three are automatically related by the equation

$$i = -(h + k).$$

In obtaining direction indices, the translations parallel to the four axes are obtained with the provision that they are chosen so that the third index is again the negative of the sum of the first two. Without this condition many alternative indices can be derived for a given direction.

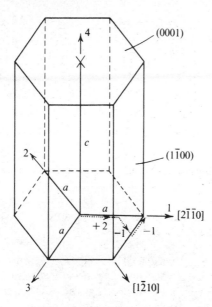

Fig. 2.10. Miller–Bravais indices of various planes and directions in a hexagonal lattice.

2.5 Close-packed crystal structures: atomic diameters

The arrangement of the atoms in face-centred cubic and close-packed hexagonal crystals is close-packed, that is, it is not possible to fit more identical sized hard spheres into a given volume than is done in both these structures. As already mentioned (§ 2.3), most metallic elements adopt either a fcc or a cph structure. They also occur in many ceramics in the sense that large non-metallic ions are in a fcc or cph arrangement, with the smaller metallic ions in interstitial sites. For example, beryllium BeO has the oxygen ions (O^{2-}) in a fcc array with the beryllium ions (Be^{2+}) in certain interstitial positions, the structure being known as zinc blende. These compound structures will be discussed in § 2.7.

The difference between the fcc and cph structures lies in the way close-packed planes of atoms are stacked on top of one another. In trying to close-pack spheres in a plane there is only one solution, the familiar equilateral-triangular array occupied by the atoms in the A positions in Fig. 2.11. In placing a second close-packed layer on top of the first, there are two equivalent sets of hollows into which the spheres can nestle, marked B and C. Suppose the B positions are chosen. For a third layer there are now two distinct sets of hollows: the A positions, lying directly over the first layer atoms, and the C positions, over the unused set of hollows in the first layer.

Two stacking arrangements are thus possible to give crystalline arrays: $ABCABCAB\ldots$, and $ABABABA\ldots$ (or its equivalent $ACACA\ldots$). The first gives the fcc structure with the close-packed planes forming the octahedral planes of the structure cell, that is, the $\{111\}$ planes (see Fig. 2.9). The second gives the cph structure with the close-packed planes forming the basal planes of the structure cell, that is, the (0001) planes (Fig. 2.10). A stacking arrangement

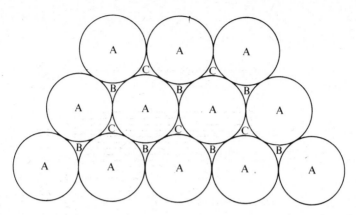

Fig. 2.11. Close-packed structures. For fcc the planes are stacked *ABCABCA* ...; for cph the stacking is *ABABA* ....

such as *ABACAB* ... does not give a space lattice and is not a perfect crystal, although this is a type of fault that is found (see § 2.8).

The close-packed hexagonal arrangement has two atoms to each space lattice point, in contrast with the other common metallic structures (fcc and bcc) which have one. The lattice and crystal structure cells are therefore not identical, compare Figs. 2.6(*c*) and 2.10. The atoms in the mid-cell plane cannot be lattice points since a line drawn from one of the atoms in the lower basal plane through a mid-cell atom does not intersect an atom in the upper basal plane. But lattice points must be equally spaced in any direction.

The significance of the close-packing of atoms in metallic crystals will be apparent in later discussion of the mechanical properties of metals, in particular plastic flow by slip; slip occurs on the close-packed planes in the close-packed directions (see § 9.4). This may be felt to be intuitively correct since the close-packed planes must be the most widely spaced and likely to be weak in shear.

The model of atoms as hard spheres with definite radii has been shown to be a useful concept in the discussion of crystal structures. Fig. 2.12 shows the atomic diameters plotted against the atomic number; these have been obtained from the interatomic distance of nearest neighbours in crystals of the element. There is a relatively small range of sizes when the large range of atomic numbers is considered; between 0.15 nm diameter (1.5 Å) for carbon and 0.5 nm (5 Å) diameter for caesium, with the majority around 0.3 nm diameter. There is marked periodicity in size, in step with the Periodic Table; each period begins with a large atom with a rapid decrease over the next few groups. It will be apparent from the discussion in Chapter 1 that atoms are in many respects not adequately represented by hard spheres and the effective diameter does vary with the environment. Thus, the diameter found in ionic crystals can be very different from that obtained in a metallic crystal; the sizes of various ions are given in Table 2.3. Even with metallic binding, there are small changes in size with coordination number, as found in various polymorphic forms of an element.

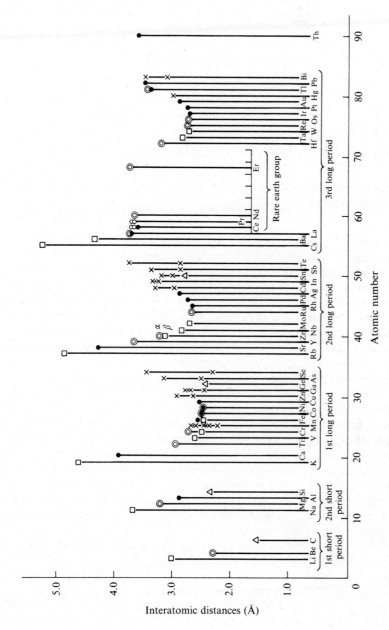

Fig. 2.12. Atomic diameters of the elements (interatomic distances in crystals of the elements).
□ Body-centred cubic structures; ● Face-centred cubic structures; ◎ Close-packed hexagonal structures; △ Diamond type structures; × Complex structures. (After W. Hume-Rothery: *The Structure of Metals and Alloys.* Institute of Metals (1945).)

2.6 Solid solutions

When a crystal is composed of two elements, the structure will either be a solid solution or an intermediate phase. In the former the crystal structure of one of the elements is retained, but in the latter a completely new structure is formed. These will be considered in this and the following section. In Chapter 4 it will be shown how these 'phases' are formed as various types of binary alloys are cooled from the melt.

There are two types of solid solution: substitutional and interstitial. These are shown schematically in Fig. 2.13. As their names imply, in the former the dissolved, or solute, atoms replace the host, or solvent, atoms and in the latter the solute atoms go into sites between the solvent atoms. In both types, the crystal structure of the solvent is unchanged except that there may be some small change in the interatomic distance. The fraction of solute atoms is not fixed by any chemical valency law and may be varied within wide limits, depending on the elements concerned. It may be as high as one hundred per cent for a substitutional solid solution but does not exceed about ten per cent for an interstitial one.

The type of metallic solution and the limit of solubility have been shown by Hume-Rothery to be governed by four factors: atomic size, crystal structure, chemical affinity and relative valency. Similar atomic radii are favourable for a substitutional solid solution and they must differ by less than approximately 15 per cent to allow solubility over the whole range. On the other hand an interstitial solid solution is only possible between the few small atoms (e.g. hydrogen, nitrogen, carbon and boron) which have atomic diameters around 0.15 nm and the large atoms (e.g. the transition metals) which have atomic diameters over 0.25 nm. The same crystal structure for the two elements also favours a wide solubility range and is obviously necessary to achieve solid solution over the entire composition range.

The chemical affinity between two elements is dependent on their relative positions in the electrochemical series or, in effect, in the Periodic Table. The electrochemical series is a list of the elements arranged in order of their potential voltage generated in a standard solution. At one end are the electropositive metals, that is, elements with strong tendencies to form positive ions, *cations*,

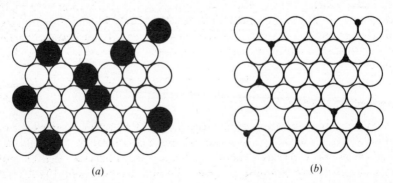

(a)   (b)

Fig. 2.13. Solid solution types: (a) substitutional; (b) interstitial. ○ atoms of *A*; ● atoms of *B*.

by losing their valency electrons, and at the other end the electronegative non-metals, which form negative *anions*. Electronegativity increases as one proceeds up the groups and along the periods, so that the most electronegative elements are in the top right-hand corner of the Periodic Table, e.g. fluorine. The most electropositive elements are in the diagonally opposite corner, for example caesium. The two elements must not be widely separated in the electrochemical series, or Periodic Table, to form solid solutions, and, if they are, a compound will result (see § 2.7).

Finally it is found that a high valency atom can more easily go into solution in a low valency matrix than *vice versa*. In other words, a crystal prefers an excess of electrons rather than a shortage, which tends to upset the binding between atoms.

An example in which all four factors are favourable to a substitutional solid solution is copper and nickel: their atomic diameters are 0.254 nm and 0.250 nm respectively; their crystal structures are fcc; they are both electropositive and have one valency electron each. It is not surprising that they are soluble over the complete range of composition.

An important example of an interstitial solid solution is carbon in the high temperature fcc form of iron, with a maximum solubility of 1.7 weight per cent or 7.4 atomic per cent. The atomic diameters are 0.15 nm and 0.25 nm respectively. It is interesting to note that the carbon is soluble in the fcc form of iron but is almost insoluble in the bcc form at room temperature, in spite of the fact that fcc is a close-packed structure and bcc is not – the voidages are 26 and 32 per cent respectively. This is explained by the shape of the void, which is equiaxial in fcc but a waisted cylinder in bcc. Treating the atoms as hard spheres, it can be calculated that maximum diameter of an interstitial atom is 0.41 times the diameter of the solvent atom in a fcc structure and 0.29 in a bcc structure. The ratio of the carbon to iron atomic diameters is actually 0.6, so it is not surprising that the solubility is limited even in the fcc form.

In a number of solid solutions, atoms of the two elements are not randomly arranged but are ordered into preferred positions, forming a lattice within a lattice. This is known as a *superlattice*. The classic example is the copper–gold alloy with 25 atomic per cent of gold. The crystal structure of both elements is fcc and in the disordered state the atoms are randomly arranged but in the ordered state, which exists at lower temperatures, the gold atoms occupy the cube corners and the copper atoms the cube faces. The effect is detected principally by X-rays but there is also a marked change in many physical properties. It is not proposed to consider further this subject, which has been extensively investigated.

2.7 Intermediate phases

When two elements combine to form a crystal structure which is different from that of either component, this is called an intermediate phase, or compound. There are very many crystal structures of compounds between two or more elements and it is only possible to give a limited account here. Fortunately there are some standard atomic arrangements which occur repeatedly in simple compounds of different elements. These may be formed between two metallic elements and have predominantly metallic bindings, that is, they are alloys, or they may be formed between a metallic and a non-metallic element and have predominantly ionic or covalent bindings. That is, they are ceramics. Unlike crystals of an element, with identical atoms, compound crystals contain two types of atom, which are usually of very different size and, as shown below, this factor is largely responsible for the various structures.

Some common crystal structures are shown in Fig. 2.14: (*a*) to (*d*) represent crystals with composition AX, that is, equal numbers of the two types of atoms in the compound, whilst (*e*) and (*f*) represent crystals with composition AX_2. Note that it is not possible to identify molecules within the crystals and the formulae only indicate the atomic composition. Each structure is named after a particular compound; for example the fluorite (CaF_2) structure is possessed by several other compounds with the Ca^{2+} ions replaced by another metallic ion and the F^- ions replaced by another non-metallic ion.

The choice of structure in compounds with mainly ionic binding, which represent the majority of ceramics (e.g. halides, oxides, silicates), is determined by three rules derived by Pauling (1928). The dominant factor is the size ratio of the positive and negative ions. The anions, that is, the non-metallic ions which have collected negatively charged electrons, are larger than the cations, that is, the positively charged ions which are formed when metal atoms lose valency electrons. Some typical ionic diameters are given in Table 2.3. Pauling's first rule states that each cation surrounds itself with anions so that they are just in contact with each other. Thus, for ionic size ratios near unity (1.0 to 0.7) the cation has eight nearest anions which are at the corners of a cube, see Fig. 2.15(*a*). For smaller cations the number of anions has to decrease if contact is to be maintained with the cation. For ratios around one half (0.7 to 0.4) the cation has six nearest anions which are located at the corners of an octahedron (a regular solid with eight triangular faces). Around one quarter (0.4 to 0.2) only four nearest neighbour anions in contact are possible, located at the corners of a tetrahedron (a regular solid with four triangular faces). Below this ratio, only three or two nearest anions are possible, forming a triangle or line respectively.

Table 2.3. *Diameters of some typical ions*
(in nm)

Anions		Cations	
F^-	0.27	Na^+	0.19
Cl^-	0.36	K^+	0.27
Br^-	0.39	Cs^+	0.34
I^-	0.43	Cu^+	0.19
OH^-	0.31	Be^{2+}	0.06
O^{2-}	0.28	Zn^{2+}	0.15
S^{2-}	0.37	Ca^{2+}	0.20
Se^{2-}	0.40	Mg^{2+}	0.13
		Fe^{2+}	0.16
		Fe^{3+}	0.13
		Al^{3+}	0.10
		Sn^{4+}	0.14
		Si^{4+}	0.08

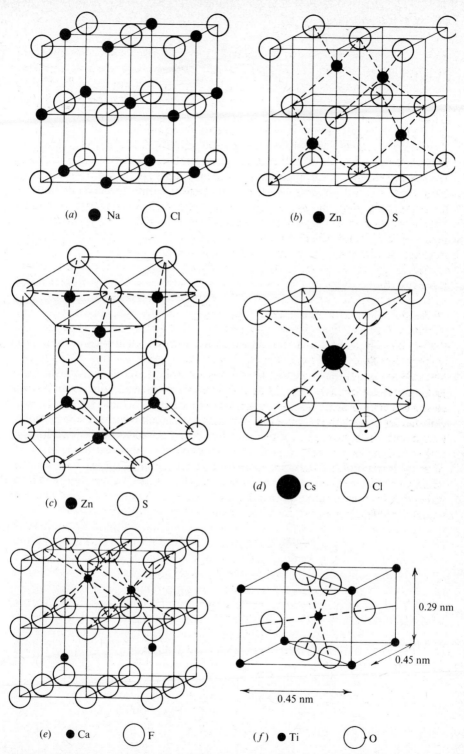

Fig. 2.14. Typical crystal structures of compounds: (a) rock salt (NaCl); (b) zinc blende (ZnS); (c) wurtzite (ZnS); (d) caesium chloride (CsCl); (e) fluorite (CaF₂); (f) rutile (TiO₂). NB In practice the large anions are touching one-another (close-packed) with the small cations in octahedral (a and f), tetrahedral (b and c) or cubic (d and e) interstitial sites.

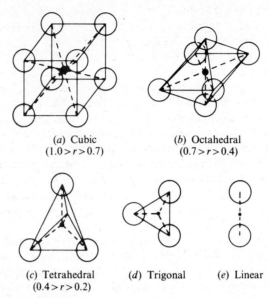

(*a*) Cubic
(1.0 > *r* > 0.7)

(*b*) Octahedral
(0.7 > *r* > 0.4)

(*c*) Tetrahedral
(0.4 > *r* > 0.2)

(*d*) Trigonal

(*e*) Linear

Fig. 2.15. Interstitial sites: small cations between large anions. r = ratio of atomic diameters.

The second rule states that the sharing of polyhedral corners, edges and faces progressively reduces the stability of the structure, since the central cations are being brought closer together. Thus groups around highly charged cations tend to share only corners, for example in silicates. The third rule is that the anions must have sufficient neighbouring cations to satisfy the requirement for neutral net electrical charge. If a bond strength is defined as the cation valency divided by its coordination number, then for each anion the total bond strength to adjacent cations must equal the anion charge. These rules will now be considered in relation to the common structures shown in Fig. 2.14.

Consider first the rock salt (NaCl) structure which is adopted by most other alkali halides, several alkaline earth oxides (e.g. CaO, MgO) and the oxides of the divalent transition metals (e.g. FeO, MnO, NiO). It also occurs in some intermetallic compounds between metallic atoms and weakly electronegative atoms in Group VIB (Se and Te); these have some conductivity indicating the bond is intermediate between ionic and metallic. The crystal structure can be considered in various ways: it is a simple cubic arrangement with alternate A and X atoms along the cube edges; it is two interpenetrating fcc structures with the corners of the A structure cell located at the middle of the edge of the X cells; and it is a close-packed fcc structure of X atoms, with A atoms occupying the octahedral interstitial positions. This last approach is the most useful as it correctly represents the close-packing of the large anions with the small cations located interstitially. Since the size ratio is 0.5 the cation coordination number accords with Pauling's first law. The requirement for electrical neutrality is also met.

The zinc blende and wurtzite structures are closely related and are polymorphic forms of zinc sulphide (ZnS). The former is also found in silicon carbide (carborundum), beryllia (BeO) and many intermetallic compounds between elements equally spaced on either side of Group IVB (e.g. AlSb, AlP, ZnS, ZnSe, ZnTe). The latter is adopted by ZnO, ice, silicon carbide (an alternative form) and some intermetallic compounds (e.g. MgTe and CdSe). The binding is intermediate between ionic and covalent. The zinc blende structure can be regarded either as two interpenetrating fcc structures with the corner atoms of one cell located at the $\frac{1}{4}$, $\frac{1}{4}$, $\frac{1}{4}$ position in the other; or as a close-packed fcc structure of X atoms with half the tetrahedral interstitial sites occupied by A atoms. Since the cation/anion size ratio for ZnS is 0.4, the latter viewpoint is most realistic. The cation coordination number also agrees with Pauling's rule. If the two types of atom are made identical, the structure reduces to diamond and this explains why the structure is chosen by compounds with a tendency to covalent type binding. Wurtzite is the hexagonal analogue of zinc blende with X anions forming a cph structure and the A cations in half the tetrahedral holes.

The caesium chloride structure occurs in numerous intermetallic compounds (e.g. LiAg, LiZn, MgTl, NaIn, NaBi). It illustrates the application of the size rule when the atoms are of nearly equal size. Caesium is an exceptionally large cation and the size ratio to chlorine is 0.94. Pauling's rule calls for a cubic arrangement of anions around the cation, coordination number 8. The bond strength is $\frac{1}{8}$, so the chlorine anion with unit valency requires eight nearest neighbour cations. These are achieved in the simple cubic array of chlorine atoms with caesium in the body-centred position.

Turning to the AX_2 compounds, only fluorite CaF_2 will be discussed here. This structure is adopted by other halides, some oxides (e.g. UO_2, PuO_2, ThO_2) and various intermetallic compounds (e.g. $SiMg_2$, $GeMg_2$, $SnMg_2$, $PbMg_2$). For these, the arrangement must be regarded as a simple close-packed cubic array of large fluorine ions with the smaller calcium ions filling alternate cubic interstitial holes. The size ratio is 0.73, hence the cubic interstitial site; the bond strength is $\frac{1}{4}$, hence the anion (valency one) requires four nearest cations, which it gets. Other oxides (e.g. LiO_2 and Na_2O) take up the 'anti-fluorite' structure. Here the fluoride sites are occupied by the small lithium cations and the calcium sites by the large oxygen atoms. The structure must then be viewed as fcc with all tetrahedral holes filled.

In addition to the preceding valency type structures, metallic intermediate phases form standard structures when the ratio of valency electrons to atoms reaches certain values: 3:2 (1.5) 21:13 (1.6) and 7:4 (1.75). These are known as β, γ and ε-phases respectively. The β-phase has a CsCl type structure, already discussed; the γ-phase is complex with 52 atoms in a structure cell; and the ε-phase is cph. Alloys of Cu, Ag and Au are well known examples that follow this electron concentration rule and this accounts for the similarities in the equilibrium diagrams of Cu–Zn (brass) and Cu–Sn (bronze), which will be considered in Chapter 4 (see Figs. 4.21 and 4.22). The composition of these electron compounds, as they are called, often varies over a wide range, in other words the intermediate phase is able to form a solid solution with the elements concerned.

The structures of various other ceramics, such as alumina, quartz and various aluminium silicates will be treated later in Chapter 7.

2.8 Lattice defects

Many properties of solids depend to a first approximation upon the nature of the atoms comprising the solid and the average, or ideal, way in which they are arranged in the crystals. These *structure insensitive properties* are density, colour, elastic constants, most electrical and thermal properties, magnetic saturation intensity and magnetic susceptibility. In contrast, some properties are *structure sensitive*, that is they are strongly dependent upon small departures from the perfect crystal structure, known as *lattice defects*. These properties all involve large relative motion of atoms, such as plastic flow and diffusion. Any attempt to estimate them on the basis of a perfect lattice gives results which are several orders of magnitude removed from experiment.

Lattice defects are classified by their geometry or shape. *Point defects* are associated with only one lattice point. The simplest of these is the *vacancy*, an unoccupied lattice point (Fig. 2.16(a)). As will be seen in § 2.10, a few vacancies exist in all crystal structures in thermodynamic equilibrium. Additional, non-equilibrium, vacancies may be introduced by plastic deformation, by neutron irradiation, or by rapid cooling from a higher temperature. There is a tendency for non-equilibrium vacancies to agglomerate and form clusters. Divacancies (Fig. 2.16(b)) are also frequently formed.

An atom not on a proper lattice site but lodged between correctly located atoms is known as an *interstitialcy*. Due to their higher energy of formation, the equilibrium density of interstitialcies is very much lower than that of vacancies. They too, however, may result from plastic deformation or neutron bombardment. Vacancies and interstitialcies are often produced in pairs, the atom leaving one lattice site and moving into an interstitial position. This is the major type of damage produced by neutron irradiation.

Point defects of a single type cannot exist in ionic crystals because of the need to preserve charge neutrality. A vacancy is normally regarded as being

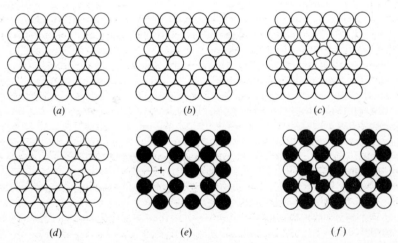

Fig. 2.16. Point defects in a lattice: (a) vacancy, (b) divacancy, (c) interstitialcy, (d) vacancy–interstitialcy pair, (e) Schottky defect, (f) Frenkel defect.

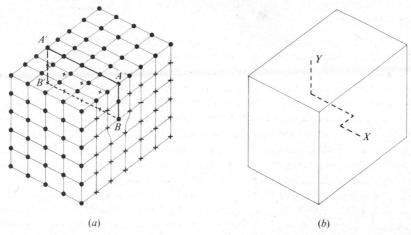

Fig. 2.17. Line defects in a lattice: (*a*) edge dislocation; (*b*) generalised dislocation.

produced by the complete removal of one atom from a lattice site. If this were done in an ionic crystal, for example, in sodium chloride NaCl, by removing a positively charged sodium ion, the crystal would have a net negative charge. Either two oppositely charged ions must be removed, forming a positive–negative divacancy, known as a *Schottky* defect, or the removed ion must be reinserted elsewhere in the crystal as an interstitialcy, producing a vacancy–interstitialcy pair, called a *Frenkel* defect.

Impurity atoms, whether substitutional or interstitial, are also point defects.

Line defects, as their name suggests, lie along a line of lattice points, and are called *dislocations*. Various types occur in crystals: a simple edge type is shown schematically in Fig. 2.17(*a*). Its name follows because it can be regarded as the edge of an additional plane of atoms $AA'B'B$ in one half of the crystal. A more general type of line fault is shown in Fig. 2.17(*b*). Dislocations result from misalignments during crystal growth from the melt or vapour, and from plastic deformation, that is, non-recoverable straining beyond the elastic range (see Chapter 5). The extreme importance of dislocations to the study of the mechanical properties of materials arises from their control of plasticity, which as will be shown later is central to the strength of materials. As shown schematically in Fig. 2.18, plastic flow occurs by the sliding or *slip* of atomic planes over each other in an inhomogeneous manner due to the movement of dislocations. In the figure, the movement of an edge dislocation across the crystal has caused the XX' face to be displaced with respect to the YY' face (i.e. plastic shear strain). A step or *slip line* is created on the surface. The stress necessary to cause inhomogeneous slip in this manner is several orders of magnitude less than that required to cause homogeneous slip in which the shearing is uniform over the slip plane. Plastic deformation by slip and the dislocation theory of plasticity will be dealt with in detail in Chapter 9.

Surface defects are defects which extend over two dimensions in the crystal. The external surface is one such defect, as it represents a cessation of the regular ordering of atoms on lattice points. Atoms lying in the surface are attracted to atoms within the bulk of the crystal, but this attraction is not balanced by

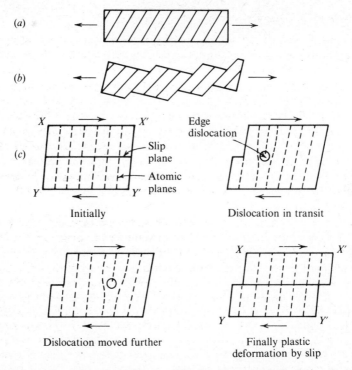

Fig. 2.18. Plastic deformation of metallic crystal by slip: (*a*) potential slip planes; (*b*) slip on a few active slip planes; (*c*) shows the mechanism of slip by the movement of an edge dislocation across the slip plane.

an equal attraction from outside the crystal, resulting in surface energy. Deviations from a planar surface, such as steps, important in growth processes from the vapour, are sometimes called surface defects, though more properly they should be called defects of the surface. Surface defects may also occur within the bulk of the material. Where two dissimilarly oriented crystals meet, a *grain boundary* is formed. Another type of surface defect is a *twin boundary*, separating the twinned part of a crystal from the untwinned matrix (§ 6.12). *Stacking faults* also occur, particularly in close-packed structures. For example, the stacking sequence of close packed layers in an fcc structure, described in § 2.5, may contain a fault:

$$ABCABCBCABCABC$$

This is equivalent to a two atom thick layer of cph inserted into the fcc structure. A similar fault:

$$ABCABCBACBA$$

is equivalent to a twin boundary. Stacking faults may arise after annealing of cold-worked structures, or may be associated with dislocations (see Chapter 9).

2.9 Equilibrium and thermal energy

Due to their thermal energy, the atoms in a crystal are always vibrating at high frequency about their mean positions. Quantum theory shows that some vibration occurs even at zero absolute temperature. The amplitude of vibration increases with temperature, being about one tenth of the interatomic spacing at room temperature; the frequency is almost independent of temperature, in the range 10^{12} to 10^{13} Hz. Vibrational energy is responsible for bringing about many of the changes in condition or structure of materials considered in this book. Examples include diffusion, nucleation and phase changes in the solid state, recrystallization after cold working, creep, and the variation of yield stress with temperature. It also accounts for the specific heat in most cases. Only a very limited study of the thermodynamics of materials will be included here.

The equilibrium state of a mechanical system is a well known concept: a mass on the end of a spring is in its equilibrium state when, for any slight displacement, the change in potential energy of the mass as it moves in the gravitation field gradient equals the change in internal (elastic) energy of the strained spring. The change in the total energy is zero; the sum of the potential energy and the strain energy is a minimum. For a general thermodynamic system, such as a collection of atoms in a solid, equilibrium exists when a thermodynamic function, loosely called the free energy, of the system is a minimum. Two types of free energy are recognised: the Helmholtz function

$$A = U - TS \qquad (2.1a)$$

and the Gibbs function

$$G = H - TS \qquad (2.1b)$$

where U is the internal energy of the system, H its enthalpy, T the absolute temperature and S the entropy.

U and H are thermodynamic functions of state, that is, their value depends only on the current state of the system, not on the route by which the state has been achieved. The enthalpy $H = U +$ generalised work terms, such as PV for gaseous systems, $-FL$ for mechanical systems, $-HB$ for magnetic systems, $-ED$ for dielectrics. In what follows, it will be assumed that $H = U + PV$ only, where P and V are the pressure on, and the volume of, the system. Change of internal energy is given by the first law of thermodynamics

$$dU = dQ - dW \qquad (2.2)$$

where dQ and dW are respectively the heat supplied to, and work done by, the system. In physical terms, U represents the sum of the kinetic energies and potential energies of all the atoms in the system. (N.B. Classical thermodynamics concerns itself only with macroscopic behaviour and parameters.)

The entropy S is another function of state and the change of entropy is given by

$$dS \geqslant dQ/T. \qquad (2.3)$$

The equality applies only for changes made by reversible processes (e.g. no heat flow across finite temperature difference and fully resisted expansions) but since the entropy is a function of state, a series of reversible changes can always be devised to give a change from one state to another, even though this occurred by an irreversible route in practice. For non-equilibrium changes, the inequality applies. Entropy is shown in statistical mechanics (in which the motion of the atoms is considered statistically) to represent the degree of disorder of a system. This disorder is the number of ways $W(E)$ the system can be arranged to achieve a certain energy level E. It can be shown that the entropy is then

$$S = k \ln W(E) \tag{2.4}$$

where k is Boltzmann's constant.

Differentiating (2.1b), with $H = U + PV$, gives:

$$dG = dU + P \, dV + V \, dP - T \, dS - S \, dT$$

and since from (2.2) and (2.3), $dU + P \, dV \leqslant T \, dS$,

$$dG \geqslant V \, dP - S \, dT.$$

At constant pressure and constant temperature, $dP = dT = 0$, and $dG \geqslant 0$. The Gibbs function has a minimum value for equilibrium at constant pressure and temperature, the conditions which correspond to most experimental situations. Treating the Helmholtz function in the same way:

$$dA = dU - T \, dS - S \, dT \geqslant -P \, dV - S \, dT$$

and it can be seen that under conditions of constant volume and temperature, it is the Helmholtz function which must be minimised for equilibrium.

It follows from (2.1) that at low temperatures the equilibrium state is effectively that with the lowest internal energy U, since the other terms are negligible. At higher temperatures the TS term increases in importance and the internal energy and entropy factors act in opposition to each other. The equilibrium state with minimum free energy is then not the one with the lowest internal energy but that with some measure of disorder present, representing increased entropy; this is because the factor TS increases faster than the internal energy for the initial disordering. As an example, consider a solid solution between atoms of two types such that unlike atoms attract each other. In Fig. 2.19 arrangement (a) with unlike nearest neighbours has minimum internal energy, but the entropy is low because it is an ordered state and there is only one possible arrangement with this energy level. Arrangement (b) has some like neighbours so the internal

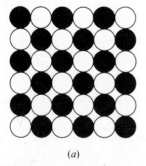

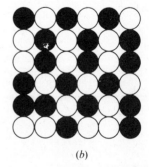

(a) (b)

Fig. 2.19. Equilibrium arrangement of atoms in a solid solution ($\bigcirc$ and $\bullet$ atoms assumed attractive). The minimum free energy $A = U - TS$ is obtained with some disorder since TS increases faster than U initially. (a) Fully ordered: low U and low S. (b) Partially disordered: higher U and S giving minimum free energy.

energy will be higher, but the entropy is also increased because a degree of disorder exists, that is, there are several possible arrangements of the atoms with this energy level. At zero temperature (*a*) will be the equilibrium state, but at higher temperatures (*b*) will take over. (It is assumed that the volume is unaffected by how the atoms are arranged, and the *PV* term remains constant.)

Many systems or materials never reach their equilibrium condition of minimum free energy. For example the equilibrium state of hydrogen and oxygen atoms is combined into water molecules, but if oxygen and hydrogen gases are intermixed no water is formed (in the absence of an ignition source). Again, hardened steel and precipitation hardened aluminium alloys are non-equilibrium structures but, fortunately for technology, they remain unchanged almost indefinitely. Such systems, which are said to be metastable, cannot reach their minimum free energy condition without first passing through a high energy intermediate stage. This is shown in Fig. 2.20. Material initially in the metastable state represented by the point *A* has to pass through the higher energy state *B* to achieve the minimum free energy state at *C*. The energy difference between *A* and *B* is called the *activation energy* and between *A* and *C* the *heat of reaction*.

The change from the metastable to the stable state (i.e. *A* to *C* in Fig. 2.20) can only proceed if the thermal energy is able to provide the required activation energy. It is important to note that the vibrational energy of an atom is continually fluctuating about its mean value ($3kT$ at high temperatures, see below). The occasional high level of energy carries the atom over the energy barrier (*B* in Fig. 2.20) into the equilibrium state of minimum free energy. Often a group of atoms has to pass over an energy barrier, rather than a single atom. Consider, for example, a crystal which is heated or cooled through a phase change, say from one polymorphic crystal structure to another with lower free energy. It is not normally possible for the entire crystal to change phase at one time because the atomic arrangement has to pass through an intermediate stage of higher energy (equivalent to point *B* in Fig. 2.20) and clearly the random fluctuations of the thermal energy about the mean value cannot raise the energy of a whole crystal simultaneously. Instead a small part of the crystal, which has

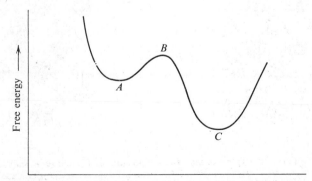

Fig. 2.20. Metastable equilibrium (*A*) and stable equilibrium (*C*). $A_B - A_A$ = activation energy, $A_A - A_C$ = heat of reaction.

momentarily above average thermal energy, is carried over the activation energy barrier into the new phase, forming a *nucleus*. Because the boundary between the nucleus and the original phase represents an increase of internal energy, there will be a minimum size of nucleus which is stable. The decrease of free energy per unit volume of the nucleus is proportional to the cube of the diameter and the increase of free energy per unit area of the boundary is proportional to the square of the diameter. Hence the minimum nucleus size will be smaller the greater the difference in free energy between the new and old phase. A phase change can only occur when the point (e.g. temperature) of equilibrium between the new and old phase has been passed. Subsequently the nucleus will grow by atoms moving over from the old crystal lattice to the new, a process requiring a much lower activation energy.

The thermal rate law was found experimentally by Arrhenius to be:

$$r = A e^{-q/kT} \tag{2.5a}$$

where T is the temperature, q the activation energy per atom, A a constant and k Boltzmann's constant. q and A are parameters of a particular reaction. In SI units, k is 1.38×10^{-23} J/deg and the activation energy is in joules per atom; putting $k = 8.6 \times 10^{-5}$ eV/deg gives q in electron volts per atom. Alternatively the law may be written:

$$r = A e^{-Q/RT} \tag{2.5b}$$

where R is the universal gas constant $(8.30 \text{ J g-mole}^{-1} \text{ deg}^{-1}, 1.98 \text{ cal g-mole}^{-1} \text{ deg}^{-1})$ and Q is obtained in joule (or cal) per mole. For typical activation energies it follows that the reaction rate is very sensitive to temperature. The temperature rise required to double the rate is given by

$$\ln 2 = q\left(\frac{1}{kT_1} - \frac{1}{kT_2}\right) \simeq \frac{q\Delta T}{kT^2}.$$

Putting $q = 0.5$ eV, and at room temperature, the reaction rate doubles every 10 degree Celsius rise.

Where a rate process is considered to be thermally activated this can be confirmed from measurements of the rate taken over a range of temperatures. A plot of the data using axes of log (rate) versus $1/T$ will yield a straight line, whose slope is $-q/k$, and the intercept with the log (rate) axes gives the constant A. An example is given in the discussion of diffusion, which is a thermally activated process (§ 2.11, Fig. 2.24).

The theoretical interpretation of Arrhenius's law is due to Boltzmann. It will not be given here (see textbooks on statistical mechanics). It may be briefly noted that the distribution of energy between atoms of a crystal can be deduced from statistical arguments. For N_T atoms in a crystal, with a mean energy of $3kT$ per atom, the number of atoms with energy $> E$ at any instant is

$$N_{\geqslant E} = N_T e^{-E/kT}. \tag{2.6}$$

If the atoms are exchanging energy levels at a frequency v, the rate at which an energy barrier E can be surmounted is:

$$r = v N_T e^{-E/kT} = A e^{-E/kT}. \tag{2.7}$$

The temperature dependence of the reaction rate is in agreement with Arrhenius's experimental law, (2.5).

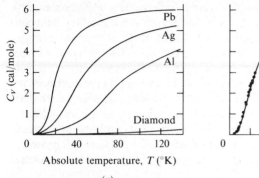 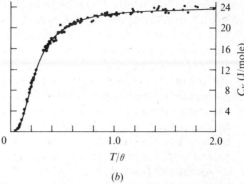

(a)   (b)

Fig. 2.21. Specific heat per mole against (a) absolute temperature T, and (b) T/θ where θ is the Debye temperature of each material. (Curve (b) is based on measurements on eighteen elements and compounds.)

An important parameter of thermal behaviour is the specific heat of a material. This is a subject to which a great deal of experimental and theoretical work has been directed, but it will receive only a passing mention here. The specific heat at constant volume per mole (gramme-atom) of most substances rises from zero at zero absolute temperature to a limiting value of 3R (i.e. 25 J g-mole^{-1} deg^{-1}, 6 cal g-mole^{-1} deg^{-1}) at high temperatures, as first observed in 1819 by Dulong and Petit. If C_V is plotted against T/θ, where θ is a parameter chosen for each material, the curves obtained with different materials can be made to coincide, see Fig. 2.21. θ is called the Debye characteristic temperature and is the temperature at which the specific heat reaches 0.96 of the limiting value.

Debye in 1912 applied quantum mechanics to the oscillations of the atoms in a crystal assembly, allowing for their interaction affecting their frequency (and thus improving on an earlier theory by Einstein), and correctly predicted the shape of the curve, which near zero has the form T^3. The theory shows that the Debye temperature is proportional to the frequency of the atomic vibrations

$$\theta = \frac{hv}{k} \tag{2.8}$$

where h is Planck's constant (6.62×10^{-34} J s) and k is Boltzmann's constant (1.38×10^{-23} J/deg). As might be expected, the highest frequencies and values of θ are found in the light atoms with high elastic moduli, for example, in diamond. Some representative values of θ are given in Table 2.4.

Table 2.4. *Debye temperature* (°K) *of some materials*

Elements		Elements		Ionic crystals	
Pb	88 °K	Cu	315 °K	AgBr	144 °K
K	100	Ni	375	AgCl	183
Na	150	Al	385	KCl	227
Au	170	Fe	420	NaCl	281
Ag	215	Be	1000	CaF$_2$	474
Zn	235	C (diamond)	2000	FeS$_2$	645

The value of the specific heat per mole at high temperatures can be predicted by classical mechanics regarding each atom as an oscillator with six degrees of freedom, that is, kinetic and potential energy along the three axes. From the principle of equipartition of energy (see textbooks on the kinetic theory of gases) each degree of freedom will absorb $kT/2$ of energy. It will be shown in Chapter 3 why the valency electrons in metals which are moving about within the lattice do not participate in the absorption of thermal energy.

2.10 Thermodynamics of defects

As mentioned in § 2.8, simple defects exist in a crystal in thermodynamic equilibrium. At first sight this seems rather surprising, as any departure from the ideal crystal structure is achieved at the expense of an increase in internal energy U. However, as discussed in the preceding section, the quantity which is minimised for equilibrium is the Gibbs function,

$$G = U + PV - TS.$$

The energy required to produce the defects is counteracted by the increase in entropy S their presence produces. Initially S rises more rapidly than U as defects are introduced, and equilibrium is established for a finite number of defects.

In order to calculate this number, it is necessary to know the changes in internal energy, volume and entropy which result from the presence of the defects. Considering the internal energy, it is assumed that there is a change in internal energy, Δu, associated with each defect, and that the defects do not interact with one another. The latter assumption is reasonable as it will be seen that the equilibrium concentration is very small and the defects will be spaced far apart. Each defect also causes a change in volume Δv, and a change in the vibrational entropy Δs_{vib}. The presence of the defects also gives rise to a configurational entropy, or entropy of mixing, ΔS_{mix}. The change in the Gibbs function due to n defects is

$$\Delta G = n\Delta u + nP\Delta v - nT\Delta s_{vib} - T\Delta S_{mix}.$$

The first three terms are equivalent to an effective Gibbs function per vacancy, Δg^*. This is not a true Gibbs function, as it ignores the configurational entropy, and is in fact the change in Gibbs function brought about by the introduction of *one* defect on a *specific* lattice site. Thus the total change in Gibbs function may be written

$$\Delta G = n\Delta g^* - T\Delta S_{mix}.$$

The contribution to the configurational entropy arises from the disorder introduced by the presence of the defects. If there are W ways of arranging the system of crystal plus defects, then by (2.4) the contribution to the entropy $S = k \ln W$. The problem is to determine W and this will now be done for the simplest case of vacancies. A crystal consisting of N lattice sites and N indistinguishable atoms can only be arranged in one way, and $W = 1$. If one atom is removed to form a vacancy, there are now N possible arrangements, as any one of the N atoms could have been removed. If a second atom is removed, $W = N(N - 1)/2$; the factor 2 arising because the removal of atom 1 from site a and atom 2 from site b is indistinguishable from the removal of atom 1 from site b and atom 2 from site a. If a third atom is removed,

$$W = \frac{N(N-1)(N-2)}{2 \times 3}$$

and if n atoms are removed:

$$W = \frac{N!}{n!(N-n)!}. \tag{2.9}$$

The configurational entropy increase due to the presence of n vacancies is thus:

$$\Delta S_{\text{mix}} = k \ln \frac{N!}{n!(N-n)!}. \tag{2.10}$$

And the change in Gibbs function for the n vacancies is

$$\Delta G = n\Delta g^* - kT \ln \frac{N!}{n!(N-n)!}. \tag{2.11}$$

This may be simplified by using Sterling's approximation which states that for large values of N,

$$\ln N! \simeq N \ln (N) - N.$$

This is very accurate for $N > 100$. The Gibbs function then becomes:

$$\Delta G = n\Delta g^* - kT\{N \ln (N) - N - n \ln (n) + n - (N-n) \ln (N-n) + N - n\}$$

$$= n\Delta g^* - kT\{(N-n) \ln (N) + n \ln (N) - n \ln (n) - (N-n) \ln (N-n)\},$$

and

$$\frac{\Delta G}{N} = \frac{n\Delta g^*}{N} + kT\left\{\frac{n}{N} \ln \left(\frac{n}{N}\right) + \left(\frac{N-n}{N}\right) \ln \left(\frac{N-n}{N}\right)\right\}.$$

Now: $n/N = x$, the atom fraction of vacancies and $(N-n)/N = 1 - x$. Hence the change in Gibbs function per atom, $\Delta g_x = \Delta G/N$, due to a concentration x of vacancies, is:

$$\Delta g_x = x\Delta g^* + kT\{x \ln x + (1-x) \ln (1-x)\}. \tag{2.12}$$

The equilibrium concentration of vacancies is determined by minimising this function with respect to x, that is, by putting:

$$\frac{d\Delta g_x}{dx} = \Delta g^* + kT\{\ln x - \ln (1-x)\} = 0,$$

which gives

$$\frac{x'}{1-x'} = e^{-\Delta g^*/kT} \tag{2.13a}$$

and if $x' \ll 1$,

$$x' = e^{-\Delta g^*/kT} = e^{\Delta s_{\text{vib}}/k} e^{-\Delta h/kT} \tag{2.13b}$$

(since $\Delta g^* = \Delta u + P\Delta v - T\Delta s_{\text{vib}} = \Delta h - T\Delta s_{\text{vib}}$).

Thus, the equilibrium concentration of vacancies x' increases exponentially with increasing temperature. Δh is called the energy of formation of the defect, though strictly it is an enthalpy rather than an energy. In addition to the increase in internal energy Δu associated with the presence of a defect, it includes the pressure times the change in volume due to the defect, $P\Delta v$. The sign of Δv

depends upon the nature of the defect. For most simple metals, the energy for formation of a vacancy, Δh, is about 1.6×10^{-19} J (1 eV), and exp $(\Delta s_{vib}/k) \simeq 10$; at 1000 °K, the equilibrium concentration of vacancies is only 10^{-4}!

The calculation for other defects is similar. The energy of formation of an interstitialcy is about five times that of a vacancy, and its equilibrium concentration is therefore very much less. More complicated defects have even larger energies of formation and make smaller contributions to the entropy, so their equilibrium concentrations are so small as to be negligible. The presence of dislocations and surface defects is entirely non-equilibrium.

2.11 Diffusion

If two dissimilar metals are placed in intimate contact at a high temperature, they will be found after a time to have intermingled at their boundary. This is accomplished by a process known as *diffusion*. The effect is shown schematically in Fig. 2.22: atoms of material X gradually move into material Y, and *vice versa*.

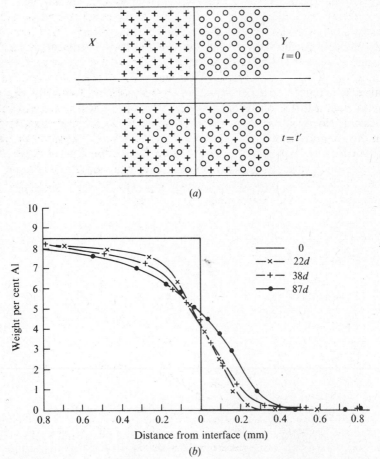

Fig. 2.22. (*a*) Schematic representation of diffusion of two materials X and Y. (*b*) Diffusion penetration curves for Cu and Cu–8.5% Al at 700 °C.

Diffusion is a thermally activated process, in which the atoms migrate by a series of random jumps through the lattice, as described in detail below. Most solid state reactions involve diffusion; examples include the precipitation of a second phase from a solid solution (precipitation hardening, see § 11.7) and the nucleation and growth of new grains in the recrystallisation of cold-worked metal (§ 11.3). Frequently the overall reaction rate is controlled by the diffusion process. The atoms in a crystal of a single element are also moving slowly about in the lattice in a random manner – a process called *self-diffusion*.

The rate at which diffusion occurs is described by a diffusion coefficient D and Fick's law (1855). This states that the mass (m) of material diffusing across plane surface is proportional to the area (A) and to the gradient in concentration of the diffusing atom. Fick's first law is thus:

$$\frac{\mathrm{d}m}{\mathrm{d}t} = -DA\frac{\partial C}{\partial x}, \tag{2.14}$$

where C is the mass concentration per unit volume and D is the *diffusion coefficient*. The minus sign indicates that the direction of mass flow is down the concentration gradient, that is, from the more concentrated to the less concentrated regions. Fick's second law relates to non-steady state conditions.

The diffusion coefficient is dependent on the diffusing atom, the solvent lattice and its defects (possibly introduced by cold work), the concentration of the solute, the grain size if polycrystalline, and, most strongly, the temperature. It is misleading to refer to it as a constant! The variation of D with concentration can be at least a factor of ten, see Fig. 2.23. However, at low concentrations, D varies only slowly and can be treated as a constant with respect to concentration in the integration of Fick's laws. This gives theoretical penetration curves in agreement with experiment; Fig. 2.22(b) indicates the type of results for a one-dimensional case. If the diffusion rates of the opposing materials are unequal, the interface, marked by fine insoluble wires, is observed to move, an effect named (1947) after Kirkendall.

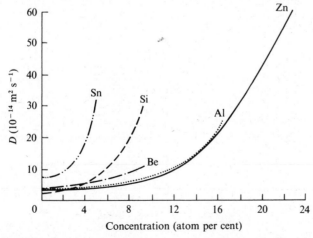

Fig. 2.23. Dependence of diffusion coefficients of various solutes in copper with concentration, at 800 °C. (After Rhines and Mehl.)

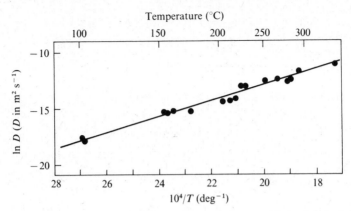

Fig. 2.24. Dependence of diffusion coefficient on temperature: gold diffusing into lead. (After Mehl, *A.I.M.E.* **122**, 11 (1936).)

It is found experimentally that D varies with temperature in an exponential fashion:

$$D = D_0 e^{-E/kT} \qquad (2.15)$$

where D_0 and E are constants for the particular diffusion process. By comparison with (2.5) it is evident that E represents an activation energy which has to be supplied by fluctuations of the thermal energy. By plotting $\ln D$ against $1/T$, diffusion data obtained at different temperatures lie on a straight line whose slope is $-E/k$, and the intercept on the $\ln D$ axis gives $\ln D_0$ and hence D_0. This is shown in Fig. 2.24 for gold diffusing through lead between 100 and 300 °C. The activation energy comes out at 0.61 eV/atom (59 kJ/g-mole), which is on the low side of the typical values given in Table 2.5.

Table 2.5. *Diffusion parameters for some dilute metallic solutions*

	D at 500 °C	D_0	E	
	(m^2/s)	(m^2/s)	eV/atom	kJ/g-mole
Cu in Cu	4.9×10^{-20}	1.1×10^{-3}	2.50	240
Zn in Cu	1.1×10^{-16}	8.0×10^{-5}	1.66	160
Si in Cu	2.2×10^{-17}	5.2×10^{-6}	1.74	168
Be in Cu	4.5×10^{-17}	4.5×10^{-9}	1.22	117
Cu in Ag	5.1×10^{-16}	5.9×10^{-9}	1.08	104
Sb in Ag	3.7×10^{-15}	5.3×10^{-9}	0.94	90
Sn in Ag	6.6×10^{-15}	7.9×10^{-9}	0.93	90
Au in Ag	1.7×10^{-16}	5.3×10^{-8}	1.30	125
Au in Au	3.2×10^{-17}	1.3×10^{-2}	2.23	214
Pd in Au	2.3×10^{-18}	1.1×10^{-7}	1.63	157
Cu in Au	8.2×10^{-16}	5.8×10^{-8}	1.20	115
Pb in Pb		5.1×10^{-4}	1.22	117
Au in Pb		4.9×10^{-5}	0.57	55
Sn in Pb		3.4×10^{-6}	1.06	101
C in α-Fe	4.0×10^{-12}	2.0×10^{-6}	0.87	84

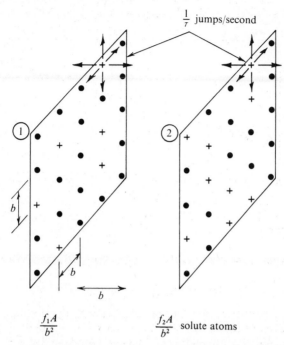

$\frac{1}{\tau}$ jumps/second

$\frac{f_1 A}{b^2}$ $\frac{f_2 A}{b^2}$ solute atoms

Fig. 2.25. Derivation of Fick's first law (separation of planes not to scale).

Fick's law can be derived from purely statistical arguments without resort to details of the mechanism of diffusion. Assume that each solute atom is thermally activated into making a random move on average every τ seconds (i.e. $1/\tau$ jumps per second) and that for an isotropic material the jump may equally well be in any direction. Consider then two adjacent crystal planes, area A and separated by the interatomic distance b, for which the *atomic* fractions of solute atoms are f_1 and f_2 ($f_1 > f_2$), as shown in Fig. 2.25. The number of solute atoms in plane (1) is $f_1 A/b^2$ and the number that jump into plane (2) in time dt is $f_1 A\, dt/6\tau b^2$. (The numerical factor arises from the six possible directions of jumping in a cubic array.) In the reverse direction, the jump rate is $f_2 A\, dt/6\tau b^2$. Hence the net atoms flowing from 1 to 2 are

$$dn = (f_1 - f_2)A\, dt/6\tau b^2$$

or

$$dn = -\frac{\partial f}{\partial x}A\, dt/6\tau b.$$

Multiplying both sides by the atomic mass M gives:

$$\frac{dm}{dt} = -M\frac{A}{6\tau b}\frac{\partial f}{\partial x}. \tag{2.16}$$

Now the *mass* concentration per unit volume C is related to the atomic fraction f by

$$C = \frac{Mf}{b^3},$$

therefore

$$\partial f = \frac{b^3}{M} \partial C.$$

Substituting in (2.16) gives

$$\frac{\mathrm{d}m}{\mathrm{d}t} = -\frac{b^2 A}{6\tau} \frac{\partial C}{\partial x}.$$

Comparison with Fick's exponential law, (2.14), shows that the diffusion coefficient is

$$D = \frac{b^2}{6\tau} \tag{2.17a}$$

or in anisotropic crystals

$$D = \frac{\alpha b^2}{\tau} \tag{2.17b}$$

where α is approximately unity. Determination of τ requires a knowledge of the mechanism of diffusion and the application of Arrhenius's rate law, as follows.

The mechanism of diffusion of interstitial solute atoms is quite simple. All interstitial solutions are dilute, and any interstitial atom will have unoccupied neighbouring interstitial sites into which it can jump. The activation energy is that required for an interstitial atom to force its way from one interstitial site to the next one. It is generally low (a few tenths of an electron volt), and interstitial diffusion is quite rapid.

Diffusion methods in substitutional solid solutions and self-diffusion are more complicated. Various possibilities are shown schematically in Fig. 2.26. One is that the diffusing atom jumps from its lattice site into an interstitial position, leaving behind a vacancy, and then migrates from one interstitial site to another until it finds another vacant site into which it can fall. This process is unlikely owing to the high energy of formation of interstitialcies, but it does occur, for example in the diffusion of copper in germanium (see Chapter 14). Other mechanisms involve the exchange of two atoms directly, or four atoms in a ring motion, suggested in 1950 by Zener. The former involves a very large activation energy and the latter, although requiring less energy, is unable to explain the Kirkendall effect. The diffusion mechanism which is accepted as the most common is by vacancy migration. The diffusing atom remains in its lattice site until a vacancy appears at one of its neighbouring sites. The atom can then jump into the vacant site, leaving behind another vacant site. A flow of diffusing atoms in one direction is thus accompanied by a flow of vacancies in the opposite direction.

The jump frequency $1/\tau$ will now be estimated. As discussed in § 2.9, each atom is vibrating with a characteristic frequency v, which depends on its mass and the interatomic force constants. To make a jump into a neighbouring site an atom must have a vacancy next to it and also sufficient energy (the activation energy for vacancy migration) to force its way between other neighbouring atoms into the vacancy site. The activation energy for migration, Δg_m^*, is the maximum increase in the Gibbs function of the crystal as one atom is in the half-way position between two stable sites. It includes the changes in internal

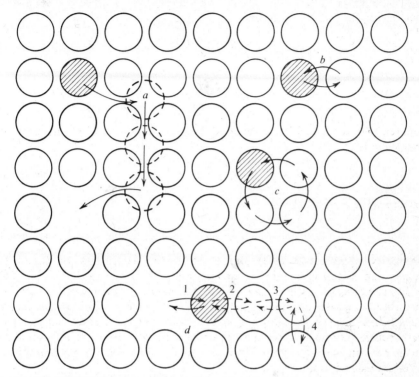

Fig. 2.26. Possible diffusion mechanisms in substitutional solid solutions. (*a*) Interstitial migration; (*b*) two-atom exchange; (*c*) four-atom ring exchange; (*d*) vacancy migration.

energy, volume, and entropy brought about during the movement of the atom. Thus the jump frequency is the product of the factors: v the characteristic frequency; P_v the probability that a neighbouring site is vacant; and P_m the probability that the atom has a vibrating energy equal to or greater than the activation energy Δg_m^*.

The probability of a site being vacant is given by (2.13*b*):

$$P_v = Ze^{-\Delta g_f^*/kT}$$

where the subscript f now indicates that the change in Gibbs function refers to the formation of a vacancy and Z is the number of nearest neighbours. The probability of having the necessary thermal energy to move into the vacancy is given by Boltzmann's equation, (2.6):

$$P_E = \frac{N \geqslant \Delta g_m^*}{N_T} = e^{-\Delta g_m^*/kT}.$$

Hence the frequency of jumps is

$$\frac{1}{\tau} = Zve^{-(\Delta g_f^* + \Delta g_m^*)/kT}$$

$$= Zve^{(\Delta s_f + \Delta s_m)/k}e^{-(\Delta h_f + \Delta h_m)/kT},$$

where the subscripts f and m relate to the formation and the migration of the vacancy.

Substituting into (2.17a), gives the diffusion coefficient

$$D = \frac{b^2}{6} Z v e^{(\Delta s_f + \Delta s_m)/k} e^{-(\Delta h_f + \Delta h_m)/kT}.$$

By comparison with (2.15):

$$D_0 = \frac{Z}{6} v b^2 e^{(\Delta s_f + \Delta s_m)/k},$$

and
$$E = \Delta h_f + \Delta h_m$$

where Δh_f and Δh_m are the activation enthalpies for the formation and motion of the vacancy, respectively; Δs_f and Δs_m are the corresponding vibrational entropies.

Putting in typical values: $b = 0.3$ nm, $v = 10^{13}$ Hz and $\exp(\Delta s_{vib}/k) \approx 10$ gives $D_0 \approx 10^{-5}$ m^2/s. Actual values vary by several orders of magnitude about this figure, indicating that the model is rather too simple or the experimental results inaccurate. The calculated activation energy for vacancy migration in copper is 2.9 eV per atom, which is close to the experimental values, thus confirming vacancy migration as the mechanism of inter-diffusion in substitutional solid solutions and of self-diffusion. The corresponding energies for interstitialcy and pair interchange methods of diffusion are both about 10 eV per atom, which rules them out.

QUESTIONS

1. (a) Define, with two-dimensional representation, crystalline and amorphous material. Give examples.
   (b) Describe various models for demonstrating crystal structures and indicate some of their limitations.
   (c) What is a space lattice? Distinguish it from crystal structure.
2. (a) Draw the body-centred cubic lattice structure cell. Mark in heavy lines a unit cell. Indicate the unit translations.
   (b) How many lattice points are there in the structure cell and in the unit cell drawn in (a)?
   (c) Repeat (a) and (b) for the close-packed hexagonal lattice, using a Miller–Bravais structure cell.
3. (a) Distinguish between the crystal systems and space lattices.
   (b) How many crystal systems and space lattices are there?
   (c) Distinguish between a primitive cell, unit cell and structure cell.
4. (a) Draw the three crystal structure cells common in the metallic elements (Class 1).
   (b) Are these identical with the lattice structure cells?
   (c) What rule governs the crystal structure of the non-metallic elements (Class 3)? Give examples.
5. (a) Define polymorphism.
   (b) Sketch the polymorphic forms of carbon and ZnS.
6. (a) Draw a bcc structure cell and mark the following: (100) (010) (111) [110] [111]. Identify the close-packed directions.
   (b) Draw a fcc structure cell and mark the {100} planes. How many directions (positive and negative) are represented by ⟨111⟩? Identify the close-packed planes.
   (c) Draw a Miller–Bravais cell of a cph space lattice and mark the indices of the basal plane, prism faces and cell axis direction.

7. (*a*) Calculate the atomic diameter in terms of the unit translations for fcc and bcc crystal structures.

(*b*) What dilation ($\Delta l/l$) occurs when iron changes from fcc to bcc?

(*c*) From the lattice parameters (Table 2.2) and atomic weights (Table 1.1), calculate the density of tungsten, beryllium and titanium.

8. (*a*) Distinguish between interstitial solid solution, substitutional solid solution and an intermediate phase.

(*b*) Discuss the factors required to form solid solutions over a wide composition range.

9. (*a*) Sketch the following interstitial positions: cubic, octahedral, tetrahedral and trigonal.

(*b*) Calculate the ratio of atomic diameters for each.

(*c*) Sketch a fcc structure cell and mark the octahedral interstitial sites.

10. (*a*) Calculate the void fraction in bcc and fcc structures.

(*b*) Calculate the maximum interstitial atom ratio in each.

(*c*) What is the relevance of (*a*) and (*b*) to steel (iron-carbon)?

11. (*a*) State Pauling's rules for the crystal structures of ionic compounds.

(*b*) Sketch and describe three crystal structures of AX compounds which are formed when the size ratio is near unity, around $\frac{1}{2}$, and about $\frac{1}{4}$.

(*c*) What are the conditions for a compound to adopt a fluorite structure?

12. Distinguish between valency-type and electron-type intermetallic compounds.

13. (*a*) Distinguish between the various types of lattice defect (e.g. point, line and surface). Give two examples of each.

(*b*) What processes can increase the defects above the number determined by thermodynamic equilibrium?

14. (*a*) For thermodynamic equilibrium of a system (isothermal, constant volume) is the internal energy of the system at a minimum?

(*b*) If not, what is? Starting from the first and second laws of thermodynamics, prove your answer.

(*c*) Illustrate the application of Helmholtz free energy function by considering a solid solution of two types of atoms which are attractive to each other.

15. (*a*) Explain why systems do not always achieve their equilibrium state in a finite time.

(*b*) Quote Arrhenius's empirical law for the reaction rate of thermally activated processes. How is the activation energy determined?

16. (*a*) Sketch the curves of specific heat per kg mol versus temperature of silver and iron (using the data on Debye temperatures, Table 2.4).

(*b*) Calculate the limiting values of the specific heat per kg of silver and copper.

(*c*) Calculate the frequency of atomic vibration of copper, silver and iron from their Debye temperatures.

17. (*a*) Given that the entropy $S = k \ln W$ where W is the number of ways of arranging the atoms in a system with a given internal energy, derive a formula for the concentration of vacancy defects in thermodynamic equilibrium (E_f = energy of defect formation).

(*b*) Plot a graph of equilibrium defect concentration versus temperature for $E_f = 1.0$ and 10 eV. Which value is more realistic for vacancies?

18. (*a*) What is 'diffusion' in a solid? Define the diffusion coefficient and Fick's first law.

(*b*) Is the diffusion coefficient a constant, given the solvent and solute atoms?

(*c*) How does the diffusion coefficient vary with temperature? Obtain D_0 (in m^2/s) and the activation energy (in eV per atom) for copper diffusing in silver from the data:

Temp (°C)	400	500	600
$D(m^2/s)$	5.0×10^{-17}	5.2×10^{-16}	3.4×10^{-15}

19. (*a*) From statistical arguments, derive Fick's first law and hence obtain an expression

for the diffusion coefficient in terms of the interatomic distance and jump frequency.
(b) Sketch the various mechanisms of diffusion which have been proposed. Which is accepted as the most common?
(c) Assuming diffusion by vacancy migration, derive an expression for the jump frequency and hence (from (a)) the diffusion coefficient in terms of the atomic parameters.

FURTHER READING

F. Seitz: *Physics of Metals*. McGraw-Hill (1943).

W. Hume-Rothery: *The Structure of Metals and Alloys*. Institute of Metals, London (1947).

C. S. Barrett and T. B. Massalski: *Structure of Metals*. McGraw-Hill (1966).

A. H. Cottrell: *Theoretical Structural Metallurgy*, 3rd edition. Arnold (1968).

W. D. Kingery: *Introduction to Ceramics*. Wiley (1960).

R. C. Evans: *An Introduction to Crystal Chemistry*. Cambridge University Press (1964).

W. L. Bragg: *The Crystalline State*: vol. 1 *A General Survey*, vol. 4 *Crystal Structures of Minerals*. G. Bell (1965).

D. W. Budworth: *An Introduction to Ceramic Science*. Pergamon Press (1970).

3. Electron theory of crystalline solids

3.1 Introduction

In Chapter 1, the electronic structure of individual free atoms was described in terms of atomic theory. It is now proposed to consider electron theory, which treats the interactions between the electronic states when atoms are brought together into an aggregate. This is not so involved as it might seem, because the behaviour of only the outermost, or valence, electrons needs to be considered, the inner electrons being screened by the outer ones and therefore not interacting with neighbouring atoms. Also wave mechanical theory will be used which is concerned with the average behaviour of all electrons rather than their individual behaviour.

The nature of the interaction between the valence electrons determines most of the physical properties of an aggregate. In metals, for example, the valence electrons are not bound to individual atoms but are free to move within the metal. The free electrons are directly responsible for the high electrical and thermal conductivities and the plasticity which are characteristic of all metals. On the other hand insulators have their valence electrons closely associated with the individual atoms, accounting for the low electrical and thermal conductivities and for the brittle behaviour of these materials. In order to understand the physical properties of solids it is therefore necessary to have some knowledge of the forces which hold atoms together.

The binding forces between atoms in a solid must be of two signs: an attractive force to hold the atoms together and a repulsive force to prevent them coalescing. Both forces increase as the distance between atoms decreases, the repulsive force rising more rapidly than the attractive force as the interatomic distance gets very small. The forces and their resultant will vary with interatomic distance a, as shown schematically in Fig. 3.1(a). At the equilibrium position, a_0, the resultant force is zero; this is equivalent to the minimum in the potential energy curve shown in Fig. 3.1(b). Values of a_0 for most solids are in the region of 0.3 nm (3 Å) (cf. Fig. 2.12).

The elastic properties of a solid are clearly related to the shape of the force versus interatomic distance curve. The bulk modulus K may be derived as follows. The elastic energy per unit volume for a fractional change in volume $\Delta V/V$ is equal to $\frac{1}{2}K(\Delta V/V)^2$. If the change in volume is small

$$\frac{\Delta V}{V} = \frac{3\Delta a}{a}.$$

Let the energy per atom be $E(a)$, where a is the interatomic distance. The change in energy per atom as the interatomic distance changes from a to $a + \Delta a$ is then

$$\Delta E = E(a + \Delta a) - E(a).$$

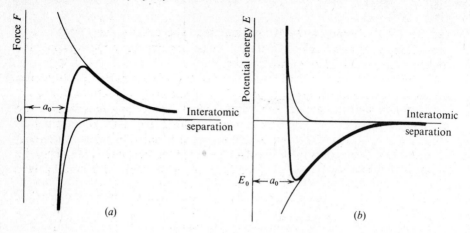

Fig. 3.1. (*a*) Force, (*b*) potential energy versus interatomic distance for a crystalline solid.

Expanding by a Taylor series:

$$\Delta E = \frac{\mathrm{d}E}{\mathrm{d}a}\Delta a + \frac{1}{2}\frac{\mathrm{d}^2 E}{\mathrm{d}a^2}(\Delta a)^2 + \cdots .$$

Now the change in energy per unit volume will be $\Delta E/a^3$, since the number of atoms per unit volume is $1/a^3$, i.e.

$$\frac{9}{2}K\left(\frac{\Delta a}{a}\right)^2 = \frac{1}{a^3}\frac{\mathrm{d}E}{\mathrm{d}a}\Delta a + \frac{1}{2a^3}\frac{\mathrm{d}^2 E}{\mathrm{d}a^2}(\Delta a)^2.$$

Evaluating this at $a = a_0$, the equilibrium spacing,

$$\left.\frac{\mathrm{d}E}{\mathrm{d}a}\right|_{a=a_0} = 0,$$

and

$$K = \frac{1}{9a_0}\left.\frac{\mathrm{d}^2 E}{\mathrm{d}a^2}\right|_{a=a_0}.$$

Since K is known to have a constant value for most solids, at least over a range of small strains, the potential energy versus distance curve must be parabolic in the region of $a = a_0$. The shear modulus G may also be related, in a slightly more complicated way, to the curvature (second derivative) of the potential energy curve.

The curvature of the potential energy function is determined mainly by the repulsive component of the interatomic force, which therefore dictates the elastic behaviour of the solid. It originates when atoms approach sufficiently close that their outer electron clouds overlap. The positive charges on the nuclei are no longer being completely screened and a Coulombic repulsion is experienced between them.

The cohesive properties of the solid, its melting and vaporisation behaviour, are determined by the magnitude of the maximum binding energy E_0, which is governed by the attractive component of the interatomic force. Unlike the repulsive component which has a common origin for all solids, this may be thought of as arising in several different ways, producing four types of binding, van der Waals, ionic, covalent and metallic. The interatomic forces in any one solid are not always clearly definable as belonging to any one type but may approximate to a mixture of two or more.

Except for the van der Waals binding, the attractive forces can be considered as due to the need for atoms to achieve stable electron configurations by acquiring, losing or sharing valency electrons (cf. § 1.8). The stable configurations are those in which the next available electron state, or energy level, lies in a higher shell. This corresponds to the magic number of 8 electrons in the outermost occupied shell (except for the first two elements of the Periodic Table where the number is 2). The stable configurations are, of course, those of the inert gases:

Shell number	1	2	3	4	5	
	2					helium
	2	8				neon
	2	8	8			argon
	2	8	18	8		krypton
	2	8	18	18	8	xenon

This 'chemist's' approach to binding is simple and in the main descriptive rather than quantitative, but it does give a useful picture of the nature of interatomic forces and will briefly be described in the following four sections. A quantitative 'physicist's' approach based on wave mechanics will be given in the second part of this chapter.

3.2 Van der Waals binding

The van der Waals forces which lead to this type of binding are present between all atoms. They are relatively weak and are only important when other binding forces are absent, for example between atoms with stable electronic configurations or between molecules which have been formed by electron sharing to give a stable electronic configuration for the molecule (see § 3.4). Van der Waals forces also become appreciable between large organic molecules and are responsible for the overall cohesion of many long chain polymers (see Chapter 8). Because the forces are weak, substances which depend on them are usually gases down to quite low temperatures.

Van der Waals forces arise from the fact that atoms act as small dipoles. Though on a time average the spatial distribution of electrons in an atom is symmetrical about the nucleus, at any instant the centre of negative charge of the electrons may not coincide with the positive charge of the nucleus. The interaction between the resulting dipoles gives rise to a force which is the attractive component of van der Waals binding. The interaction is further strengthened by the fact that the presence of a dipole in one atom will tend to induce a dipole in a neighbouring atom. As already explained, the repulsive component arises when the atoms are so close that their electron clouds begin to overlap.

The strength of van der Waals binding increases as the polarisability of the atoms increases, that is, the ease with which the electron distribution can become asymmetrical about the nucleus. Polarisability increases with atomic number, and radon, the heaviest of the rare gases, has a binding energy per atom of 1.6×10^{-19} J (1 eV). This is ten times that of neon in the crystalline state, though it is only about a fifth that of a solid formed by other types of binding.

3.3 Ionic binding

Ionic binding occurs only in compounds formed between dissimilar atoms. Electrons are transferred from the atoms of one type to the atoms of the other type, such that each atom achieves a stable electronic configuration, and as a result becomes electrically charged. This can be illustrated by reference to sodium chloride, NaCl. Sodium, in the first group of the Periodic Table, has one electron in its outermost shell; chlorine in the seventh group has seven electrons. The removal of the one electron from sodium leaves a positively charged sodium ion with the stable configuration of neon; the addition of this electron to chlorine gives a negatively charged chloride ion with the stable configuration of argon. Similarly for magnesium oxide, MgO: magnesium in the second group donates two electrons to oxygen in the sixth group.

The interatomic forces in an ionic solid are electrostatic or Coulombic forces between charged ions. They vary as the inverse square of the distance between the ions and are attractive for unlike ions, repulsive for like ions. The repulsive force at small distances due to the overlapping of the electron clouds varies as a large negative power (~ 10) of the distance.

Cohesion in a crystal results only if the structure is such that a positive ion has only negative ions as nearest neighbours and a negative ion has only positive ions for nearest neighbours. This is the principal condition determining crystal structure which can be quite complicated for materials in which the atom species have different valencies. A secondary factor is the relative size of the ions. In sodium chloride the sodium ion is not large enough to have more than six chlorine ions around it, but in caesium chloride the larger caesium ion can have eight chlorine ions. The two structures are shown in Fig. 3.2.

The attractive force between two oppositely charged ions, with z_1 and z_2 unit electron charges respectively, at a distance r apart, is given by $z_1 z_2 e^2 / r^2$. Thus, binding should be strongest between small ions of high valency. The effect of ion size on strength of binding is shown by the variation in melting point for the sodium halides:

	NaI	NaBr	NaCl	NaF
Interatomic distance, nm	0.318	0.294	0.279	0.231
Melting point, °C	660	740	801	988

The effect of valency is shown by comparing the above with:

	MgO	BN	NbC
Valency	Divalent	Trivalent	Quadravalent
Melting point, °C	2640	3000	3500

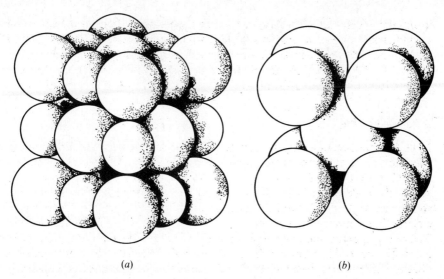

(a)   (b)

Fig. 3.2. (*a*) Structure of NaCl; fcc lattice. (*b*) Structure of CsCl; simple cubic lattice.

As valency increases there is, however, a tendency to move towards a more covalent type of binding. Ionic solids are usually strong and brittle, with fairly high melting points. The electrons are localised to individual atoms, resulting in electrical and thermal insulators.

3.4 Covalent binding

Covalent, or as it is sometimes called, homopolar, binding occurs as the result of the production of stable electronic configurations by the sharing of electrons between specific atoms. The attractive force produced can only be properly understood in terms of quantum mechanical equations.

The simplest example of covalent binding is a diatomic gas molecule formed between two atoms with seven electrons each in their outermost shells. The stable configuration is obtained by each atom sharing one of its electrons with the other atom, as shown in Fig. 3.3. In oxygen, two electrons from each atom have to be shared, as oxygen has only six electrons in its outermost shell.

Covalent binding is found in many crystalline solids, particularly for elements at the right hand side of the Periodic Table (cf. Table 2.1, Class 3). The crystal structure is determined by the $(8 - N)$ rule where N is the number of electrons

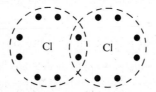

Fig. 3.3. The chlorine molecule; two chlorine atoms, each with seven valence electrons, share one electron each to produce a stable configuration.

in the outermost shell. Each atom in the crystal shares one electron with its nearest neighbour, and therefore has a coordination number equal to the number of electrons that are needed to produce a stable configuration. When $N = 7$ the atom has only one neighbour, and a diatomic molecule is formed. This is why the elements in Group VIIB are usually gases. With $N = 6$ the atom requires two neighbours and a continuous chain molecule can be formed. Three-dimensional crystals with purely covalent binding can only be built up if the value of N is less than 6. An example is diamond, a covalent crystalline form of carbon; carbon has four valence electrons and each carbon atom shares one electron with each of four neighbouring atoms. (Note that N is never less than 4).

It is only in the covalent structure that there is a definite bond or link between specific atoms, the shared electrons spending most of their time in the region between the two sharing atoms. Thus the bond consists of a concentration of negative charge and neighbouring bonds repel one another. For this reason, when an atom has several bonds they occur at equal angles to one another, for example at 109° in diamond. The covalent bond is formed in definite directions; it is a directed bond.

Covalent binding may be very strong, as in diamond, or quite weak as in bismuth, and the melting points of covalent solids correspondingly cover a wide range of temperature. Mechanical strength is generally high, due to the directed nature of the bond. The local distortions which are necessary to allow plastic deformation are not easily accommodated by the rigid bonds. Electrical and thermal properties also vary over a wide range, from those of insulators to conductivities almost as high as those of metals. The latter is because during the sharing process, electrons may move from one atom pair to the next, and are not rigidly associated with given atoms, as in ionic solids.

Though ionic solids and covalent solids have been treated as distinct types, the binding in a real solid may often be a mixture of the two.

3.5 Metallic binding

The first attempt to describe the metallic state resulted in the free electron theory, proposed by Drude in 1900 and developed by Lorentz. In this theory a metal crystal is regarded as a lattice of positive ions through which the valence electrons may wander at will. The lattice is permeated by an electron 'gas' and each valence electron is shared by all the atoms in the crystal. The binding forces arise in the following way. The free electrons screen the positive ions from one another, thus preventing their mutual repulsion. In their turn, the ions form a region of low potential energy (a potential box) for the electrons. It is this reduction in the potential energy of the electrons which gives rise to the binding forces.

The free electron model qualitatively explains many of the properties characteristic of the metallic state. The high electrical and thermal conductivities follow directly from the free movement of the valence electrons. The Wiedemann–Franz Law, which states that at a given temperature the ratio of thermal to electrical conductivity is a constant for all metals, had already been established empirically and the application of classical Newtonian mechanics to the motion of the electrons gave it theoretical justification. The ability of the free electrons to absorb energy of all wavelengths accounts for the opacity of all metals; the emission of this absorbed energy similarly explains the high reflectivity of metallic surfaces.

The Drude–Lorentz theory failed in three important respects:

(1) It predicted that the specific heat of a metal should be higher than it actually is.

(2) The way in which electrical resistivity varies with temperature could not be explained.

(3) The difference in electrical conductivity between metals of different valency could not be explained.

The specific heat of a metal should consist of two parts, the specific heat of the lattice and the specific heat of the electron gas, which in this theory is postulated to obey the kinetic gas laws. According to kinetic theory of matter,[†] the specific heat of the lattice, with six degrees of freedom per atom, should have a value of $3R$ per gramme atom at high temperatures, where R is the gas constant. If the free electron gas obeys the gas laws, its gramme-atomic specific heat, with three degrees of freedom, should be $3R/2$ per valence electron. The specific heat of a monovalent metal should then be $9R/2$ per gramme atom, and for a divalent metal $6R$ per gramme atom. Experimentally the specific heat of a metal is never larger than $3R$, and it must be concluded that this is due to the lattice only, the electrons making no contribution, in contradiction to the theory.

The electrical resistivity of a metal is given (see Chapter 13) by the formula

$$\rho = \frac{mv}{ne^2 l} \tag{3.1}$$

where m is the electron mass, v their mean velocity, n their number per unit volume and e the charge per electron; l is their mean free path, that is, the average distance travelled between collisions. On the free electron theory, the resistance arises from the electrons undergoing collisions with, and being scattered by, the positive ions of the metal lattice. To a first approximation the only quantity in (3.1) to vary with temperature is the mean velocity. Treating the electrons as a perfect gas, from kinetic theory:

$$\tfrac{1}{2}mv^2 = \tfrac{3}{2}kT.$$

Substituting in (3.1) gives the resistivity proportional to $T^{\frac{1}{2}}$. Actually the resistance of metals varies linearly with temperature.

The theory also predicts that a divalent metal, with twice as many free electrons as a monovalent metal, will have an electrical conductivity approximately twice as high as a monovalent metal, and similarly for metals of higher valency. In practice the monovalent metals have the highest conductivities, then the trivalent metals, followed by the divalent metals; the transition metals, to which it is rather difficult to ascribe a valency, have the lowest conductivities of all. A comparison between different metals is only meaningful if the conductivities are measured at a characteristic temperature for each metal. The appropriate characteristic temperature is the Debye temperature, θ_D, at which the thermal vibration spectra are similar for all crystals, since thermal vibrations make the

---

[†] See textbooks on thermodynamics.

largest contribution to the resistivity of pure metals (cf. § 2.9). Conductivities are usually quoted for 0 °C or 20 °C, and these values must be divided by θ_D to give a quantity which is comparable for different metals. This is done in Table 3.1, and it can be seen on average that for divalent metals it is roughly $\frac{1}{3}$, and for trivalent metals roughly $\frac{1}{2}$, of the value for monovalent metals. (Li and Ga have anomalous values, and should be ignored.)

These contradictions indicate that the classical free electron theory is incorrect. The theory breaks down because it tries to treat the motion of the free electron in terms of classical equations of motion. In 1928, Sommerfeld applied the new concepts of wave mechanics to the motion of electrons and was able to explain the first two of the points discussed above; the third was not explained until the development of the theory of Brillouin zones.

The approach to solids described in this first part of the chapter, as has already been mentioned, is essentially descriptive. It can be extended, with reasonable success, to account for van der Waals and ionic solids, but is very limited in its description of covalent solids and metals. The alternative approach, that of wave mechanics, was originally developed to account for the behaviour of metals, but is, in fact, applicable to all solids. The remainder of this chapter is concerned with wave mechanics and its application to crystalline solids.

Table 3.1. *Electrical conductivities of various pure metals*

Period	Metal	Conductivity at 0 °C, σ $(10^6\,\Omega^{-1}\,m^{-1})$	Debye temp. $\theta_D(°K)$	σ/θ_D
		Monovalent metals		
2nd	Li	11	430	0.026
3rd	Na	21	150	0.13
4th	K	13	100	0.13
	Cu	59	315	0.19
5th	Ag	62	215	0.29
6th	Au	46	170	0.27
		Divalent metals		
2nd	Be	28	1000	0.03
3rd	Mg	23	325	0.07
4th	Ca	28	225	0.12
	Zn	17	235	0.07
5th	Cd	14	165	0.085
		Trivalent metals		
3rd	Al	37	385	0.096
4th	Ga	7	240	0.03
5th	In	16	120	0.13
6th	Tl	6	100	0.06

Data from T. G. Meaden: *Electrical Resistance of Metals*, Heywood (1966).

WAVE MECHANICAL THEORY OF ELECTRONS IN SOLIDS

3.6 Schrödinger's equation

The dual nature of electrons, particle and wave, has already been referred to in § 1.6. The duality can be reconciled by regarding the electron as a particle whose behaviour is only describable in terms of wave mechanics, rather than classical mechanics.

Classical mechanics cannot be used for the following reason. In that theory it is assumed that the position and momentum of a body are known exactly at a given time. This assumption is actually an approximation, but it is a very adequate one for many macroscopic systems, for example, in studying the motion of billard balls. In such a problem the position and velocity of a ball may be determined in several ways, but all will involve an interaction with some intermediary agency, usually a beam of light. The light is reflected from the ball, the photons which comprise the beam undergoing a change of momentum. The billard ball must undergo a similar change in momentum, but because its mass is many orders of magnitude larger than that of the photons, its velocity is changed only slightly, probably by an amount much less than the accuracy of the experiment. If the same experiment is carried out on an electron, the mass of the electron is now comparable to that of the photon, and it must undergo a very large change in its velocity. It is therefore impossible to observe the behaviour of an electron (or for that matter any small particle) without disturbing the electron. The energy, momentum or position of a single electron cannot be known at any given time; all that can be done is to talk in terms of the most probable behaviour of the electron, which is best described by means of wave mechanics.

Uncertainty in the behaviour of individual particles was first postulated by Heisenberg in 1925. His Uncertainty Principle may be stated mathematically:

$$\Delta p \Delta x \geqslant h \quad \text{or} \quad \Delta E \Delta t \geqslant h,$$

where Δp, Δx, ΔE and Δt are uncertainties in momentum, position, energy and time respectively, and h is Planck's constant (6.625×10^{-34} J s).

The suggestion that electrons are describable in terms of waves stems from the work of Planck and Einstein (1901–5) who showed that electromagnetic radiation must not be regarded as a continuous wave but as being divided up into discrete packets of energy, called quanta or photons. Planck proposed that the energy E of a single quantum be related to the frequency of the radiation

$$E = h\nu$$

where h is Planck's constant. The radiation can exchange energy with a system only by an integral number of these quanta. The evidence for this hypothesis was very strong, particularly with regard to the photoelectric effect.

In 1925, de Broglie suggested that if electromagnetic radiation exhibited both wave and particle nature, then could not the converse be true, and particulate matter also display a similar duality? Shortly after de Broglie's proposal, experiments by Davisson and Germer and independently by G. P. Thomson on diffraction of electrons by crystals indicated its correctness. The mathematical basis was developed in 1926 by Schrödinger, and is given below.

De Broglie proposed that a particle moving with a momentum p has an associated wavelength λ given by

$$p = h/\lambda.$$

This is of immediate application in diffraction experiments, where the wavelike properties of the particles can be demonstrated directly. However it is necessary to develop the theory further, so that it can be applied to more general problems.

The equation which describes a generalised wave is of the form (in complex number notation)

$$\Psi = Ae^{i(kx - \omega t)} \tag{3.2}$$

where k is the *wave number* $(= 2\pi/\lambda)$, ω is the angular frequency $(= 2\pi v)$, A is a constant, x and t are position[†] and time, and Ψ is some generalised property of the wave.

Born postulated in 1926 that the probability density of electrons is given by $|\Psi||\Psi^*|$, that is, that the probability of finding an electron in a volume dx, dy, dz situated at coordinates x, y and z at a time t is given by $|\Psi||\Psi^*|\,dx\,dy\,dz$ evaluated at x, y, z and t.

In order to proceed further, it is necessary to relate the wave number k to the frequency ω of the wave. This is equivalent to determining the velocity of the wave. If it is assumed that the wave equation is describing the behaviour of only one quantum, or particle, then the total energy of the particle is given by Planck's equation:

$$E = hv = \frac{h\omega}{2\pi}.$$

The total energy of an electron is the sum of its kinetic energy $p^2/2m$ and its potential energy, which is equal to the potential field $\phi(x, y, z, t)$ through which it moves

i.e.
$$E = \frac{p^2}{2m} + \phi,$$

but
$$p^2 = \frac{h^2}{\lambda^2} = \frac{h^2 k^2}{4\pi^2}$$

giving
$$\frac{h\omega}{2\pi} = \frac{h^2 k^2}{8\pi^2 m} + \phi. \tag{3.3}$$

It is to be expected that the potential field will regulate the behaviour of the electron, whether in a simple atom or in a crystal. What is now required is an equation relating Ψ to ϕ. This can be obtained by substituting in (3.3) expressions for ω and k in terms of Ψ. Differentiating (3.2) with respect to time

$$\frac{\partial \Psi}{\partial t} = -i\omega Ae^{i(kx - \omega t)} = -i\omega\Psi.$$

Therefore
$$\omega = \frac{i}{\Psi}\frac{\partial \Psi}{\partial t}.$$

[†] Position of course, should be represented by three variables, e.g. x, y and z. Only one, x, is used here, but it should be regarded as standing for all three.

Differentiating (3.2) twice with respect to x:

$$\frac{\partial \Psi}{\partial x} = ikAe^{i(kx - \omega t)}$$

and

$$\frac{\partial^2 \Psi}{\partial x^2} = -k^2 Ae^{i(kx - \omega t)} = -k^2 \Psi,$$

which gives

$$k^2 = -\frac{1}{\Psi}\frac{\partial^2 \Psi}{\partial x^2}.$$

Substituting these expressions in (3.3) gives

$$\frac{ih}{2\pi\Psi}\frac{\partial \Psi}{\partial t} = -\frac{h^2}{8\pi^2 m\Psi}\frac{\partial^2 \Psi}{\partial x^2} + \phi,$$

which can be rewritten:

$$\frac{ih}{2\pi}\frac{\partial \Psi}{\partial t} = -\frac{h^2}{8\pi^2 m}\frac{\partial^2 \Psi}{\partial x^2} + \phi\Psi. \tag{3.4}$$

This is the time-dependent Schrödinger equation, with Ψ and ϕ both functions of space and time.

Providing ϕ is known, the wave mechanical description of the behaviour of any electron is obtained by inserting the appropriate function for ϕ into (3.4) and obtaining a solution for Ψ in terms of space and time. The solution is given only in terms of the probability function $|\Psi||\Psi^*|$, as suggested at the beginning of this section.

In many problems ϕ is independent of time, and the space and time variables may be separated. Suppose $\Psi(x, t)$ may be written $\psi(x)\xi(t)$. Substituting in (3.4):

$$\frac{ih}{2\pi}\psi\frac{d\xi}{dt} = -\frac{h^2}{8\pi^2 m}\xi\frac{d^2\psi}{dx^2} + \phi(x)\psi\xi.$$

Dividing by $\psi\xi$ gives

$$\frac{ih}{2\pi}\frac{1}{\xi}\frac{d\xi}{dt} = -\frac{h^2}{2\pi^2 m}\frac{1}{\psi}\frac{d^2\psi}{dx^2} + \phi(x).$$

The left-hand side is a function of t only, and the right-hand side a function of x only, and as x and t are independent variables, both sides of the equation must be equal to some constant, B, say.

Then, the LHS,

$$\frac{ih}{2\pi}\frac{1}{\xi}\frac{d\xi}{dt} = B$$

has a solution

$$\xi = e^{-(2\pi i/h)Bt}$$

which is a simple oscillation in time with a frequency $v = B/h$. But, from Planck, $v = E/h$ and therefore B is equal to the total energy E. For the RHS,

$$-\frac{h^2}{8\pi^2 m}\frac{1}{\psi}\frac{d^2\psi}{dx^2} + \phi(x) = E,$$

which may be rewritten to give

$$\frac{d^2\psi}{dx^2} + \frac{8\pi^2 m}{h^2}(E - \phi(x))\psi = 0. \tag{3.5}$$

(Note that $E - \phi(x)$ is equal to the kinetic energy of the particle.)

This is the time-independent Schrödinger equation. For an electron in an atom or in a crystal, the potential ϕ arises from the positive charge on the ions, and does not vary with time. This is therefore the appropriate equation to use for such problems.

In three dimensions, Schrödinger's equation is formulated

$$\nabla^2\psi + \frac{8\pi^2 m}{h^2}(E - \phi)\psi = 0. \tag{3.6}$$

The difficulty of solving this equation depends entirely upon the complexity of the function ϕ chosen. The solution is assisted by the appropriate choice of boundary conditions. It is assumed that everywhere the function ψ and its first derivative are continuous and, for a problem with a finite number of electrons, as $x \to \infty$, $\psi \to 0$. The constant of integration is determined by a normalisation procedure

$$\iiint_{\text{all space}} |\psi||\psi^*|\,\mathrm{d}x\,\mathrm{d}y\,\mathrm{d}z = \text{total number of electrons in problem.}$$

The application of this procedure, to the simple problem of the electron in a potential box, is given in the next section.

3.7 Electron in a box: the Sommerfeld theory of the metallic state

After the failure of the free electron theory Sommerfeld, in 1928, applied Schrödinger's equation to the valence electrons moving in a solid crystal. He took the simplest possible function for the potential ϕ by assuming it to be uniform and zero everywhere within a piece of metal and equal to infinity everywhere outside. The latter condition ensures that the probability of finding an electron outside the metal is zero. Such a potential variation is called a 'potential box'. The first condition can only be an approximation as the potential must have a value at the centre of an ion different from that in regions between them. How good an approximation this is will be seen as the theory is developed.

Consider, for simplicity, the one-dimensional case of electrons in a piece of metal of length L. Schrödinger's equation is

$$\frac{\mathrm{d}^2\psi}{\mathrm{d}x^2} + \frac{8\pi^2 m}{h^2}[E - \phi(x)]\psi = 0. \tag{3.7}$$

For the assumed function $\phi(x)$, this becomes, inside the metal:

$$\frac{\mathrm{d}^2\psi}{\mathrm{d}x^2} + \frac{8\pi^2 m}{h^2}E\psi = 0$$

and outside the metal: $\qquad \psi = 0.$

A possible solution is of the form:

$$\psi = A\cos kx + B\sin kx \tag{3.8}$$

where A and B are constants and k, from (3.3), is given by

$$k^2 = \frac{8\pi^2 m}{h^2}E \tag{3.9}$$

Because ψ must be a continuous function, the boundary conditions are

$$\psi(0) = \psi(L) = 0;$$

hence $$A = 0$$

and $$\psi(L) = B \sin kL = 0.$$

$B = 0$ is a trivial solution, hence $\sin kL$ must be zero. This can only be so if $kL = n\pi$ where n is any positive integer; $n = 0$ also gives a trivial solution and is thus not allowed.

The immediate consequence of the use of wave mechanics is that the wave number k and the energy E of the electron (from (3.9)) are not continuous functions but have only certain values, known as eigenvalues. These are given by:

$$k_n = \frac{n\pi}{L}$$

and $$E_n = \frac{h^2 n^2}{8mL^2}.$$

The energy is said to be 'quantised', and n is the quantum number. The principle of quantisation of the energy of electrons in individual atoms has been introduced in Chapter 1 and the Sommerfeld theory shows, as might be expected, that it also applies to the valence electrons in a solid crystal. The valence electrons belong to the whole crystal and are free to move throughout the crystal, as in the free electron theory. The energy levels are a property of the whole crystal, not of individual atoms; this is the source of the binding energy.

The lowest energy level, known as the ground state is

$$E_1 = \frac{h^2}{8mL^2}$$

and the higher energy levels may be written

$$E_n = E_1 n^2.$$

Substitution of the eigenvalues into the solution of Schrödinger's equation (3.8) gives the ψ eigenfunctions:

$$\psi_n = B \sin \frac{n\pi x}{L}.$$

The value of the constant B is determined by normalisation: the probability of finding the electron within the metal must be unity. Integrating $\psi\psi^*$ over the whole length of the piece of metal, summing over all values of n, and putting the result equal to unity, allows B to be evaluated.

The eigenfunctions are solutions which are similar to those for a vibrating string, or a column of air in an organ pipe, and may be regarded as a fundamental ($n = 1$) with higher order harmonics ($n > 1$). The first three eigenfunctions are illustrated in Fig. 3.4.

To be of any practical value, the simple one-dimensional picture must be extended to three dimensions. This can be done without involving any new

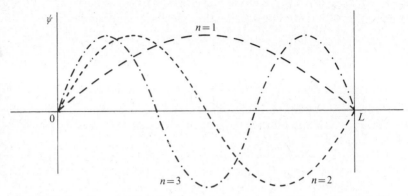

Fig. 3.4. The first three eigenfunctions for an electron in a one-dimensional potential box.

principles, and for a piece of metal of sides L_1, L_2, and L_3, the eigenfunctions are

$$\psi_{xyz} = \text{const} \cdot \sin\frac{n_x\pi x}{L_1} \sin\frac{n_y\pi y}{L_2} \sin\frac{n_z\pi z}{L_3}.$$

The eigenvalues of k are

$$k_x = \frac{n_x\pi}{L_1}, \qquad k_y = \frac{n_y\pi}{L_2}, \qquad k_z = \frac{n_z\pi}{L_3},$$

and the energy eigenvalues are

$$E_{xyz} = \frac{h^2}{8m}\left[\left(\frac{n_x}{L_1}\right)^2 + \left(\frac{n_y}{L_2}\right)^2 + \left(\frac{n_z}{L_3}\right)^2\right].$$

If the piece of metal is taken as a cube such that $L_1 = L_2 = L_3 = L$, then

$$E_{xyz} = \frac{h^2}{8mL^2}(n_x^2 + n_y^2 + n_z^2);$$

or using the same definition for E_1 as was used in the one-dimensional case,

$$E_{xyz} = E_1(n_x^2 + n_y^2 + n_z^2).$$

The energy now depends upon three quantum numbers, n_x, n_y and n_z. In the one-dimensional case the quantum number n could only take on positive integral values, but in the three-dimensional case, n_x, n_y and n_z may be positive integers or zero, provided that all three quantum numbers are not zero at the same time. It is possible to arrive at the same value of E by using different combinations of n_x, n_y and n_z. For example, both the quantum numbers (221) and (300) will give $E = 9E_1$. An energy level which may be represented by more than one set of quantum numbers is said to be degenerate, and the number of different ways of arriving at the same energy is the 'degeneracy' of the level. The separation between energy levels is normally equal to E_1, since almost any whole number can be made up of the sum of three squares. (There are some exceptions, the first three of which are 7, 15 and 23, and at these points the separation between energy levels is $2E_1$). An order of magnitude calculation for a one centimetre cube of metal gives $E_1 = 10^{-33}$ J (6×10^{-15} eV). This is extremely small, equivalent to a temperature rise of less than 10^{-10} °K. Thus the energy may be considered

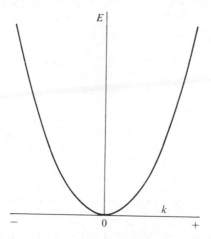

Fig. 3.5. Relation between energy and wave number for a free electron.

to be distributed over a continuous range; similarly the parabolic curve of E versus k, (3.9), which is composed of discrete points may be regarded as continuous, see Fig. 3.5.

There are about 10^{29} atoms in a cubic metre of crystal, and the number of electrons that must be accommodated in the quantum states is equal to 10^{29} times the valency of the metal atom. At the absolute zero of temperature the states are filled from the lowest energy level according to the Pauli Exclusion Principle (see § 1.7), which allows only one electron per state. A quantum state is defined by four quantum numbers, which are the three quantum numbers n_x, n_y, n_z and a spin quantum number m_s, which takes the value $\pm\frac{1}{2}$. A non-degenerate energy level may hold two electrons, providing they are of opposite spin, and a degenerate level may hold $2g$ electrons where g is the degeneracy of the level, again providing that electrons are paired with opposite spin. The highest filled level at absolute zero is called the Fermi level, and its energy E_f found in the following way. Energy states with the same energy may be represented by

$$E = E_1 R^2,$$

where
$$R^2 = n_x^2 + n_y^2 + n_z^2.$$

If n_x, n_y and n_z are imagined to be measured along the x, y and z axes, then R may be regarded as the radius of a sphere centred on the origin, and all points on the surface of this sphere are states of equal energy. As positive values only of n_x, n_y and n_z are allowed, the surface is in fact one octant of the sphere. All states with energies $\leqslant E$ are contained within this octant, and the volume of the octant, $= \frac{1}{6}\pi R^3$, is equal to the number of states with energies $\leqslant E$. Since electrons are placed into states in pairs with opposite spin, N electrons will occupy $N/2$ states, and as these are filled at absolute zero from the lowest state up, the energy of the highest filled state, the Fermi energy E_f, is given by

$$E_f = E_1 R^2,$$

where R is given by
$$\tfrac{1}{6}\pi R^3 = N/2.$$

Substituting for E_1 yields:

$$E_f = \left(\frac{3N}{\pi V}\right)^{\frac{2}{3}} \frac{h^2}{8m},$$

where N is the number of valence electrons in the metal and $V(= L^3)$ is the volume of the metal. This may be rewritten as

$$E_f = 5.78 \times 10^{-20}(n_0/v_0)^{\frac{2}{3}} \text{ J}$$

where n_0 is the number of valence electrons per atom, and v_0 is the atomic volume in nm^3. For copper, the Fermi energy is 1.14×10^{-18} J (7.1 eV).

At finite temperatures electrons are excited by thermal energy into vacant quantum states, still according to the Pauli Principle. The thermal energy of kT per electron ($kT \simeq 4 \times 10^{-21}$ J at room temperature) can affect only those few electrons with energies which are initially near the Fermi level and thus within reach of empty states. Less than one per cent of the total valence electrons are excited. This can be illustrated by plotting the probability of occupancy of a state, $P(E)$, versus the energy of the state, E, at $0\,^\circ$K and at some finite temperature $T\,^\circ$K, see Fig. 3.6. The curve at any temperature is determined by Fermi–Dirac statistics, a discussion of which is not necessary here. Thus the failure of the free electron theory to predict that the electron gas has zero specific heat, is immediately resolved in the Sommerfeld theory.

The Sommerfeld theory can also account for the observed temperature variation of electrical resistivity. The free electron theory assumed that the resistance to the motion of electrons was due to their being scattered by the ions of the lattice and that the temperature dependence of the resistivity was due to a variation of the velocity with temperature. In the Sommerfeld theory the velocity v is independent of temperature and the uniform potential assumed would not scatter electrons. Even were this replaced by the correct periodic potential, as is done in zone theory, the electrons will not be scattered, providing the periodic potential is perfectly regular. Scattering can only arise from disruptions in this periodicity, due to impurities, crystal imperfections or thermal vibrations. In an otherwise perfect lattice the scattering is proportional to the mean square displacement of the thermal vibrations, which in turn is directly proportional to the

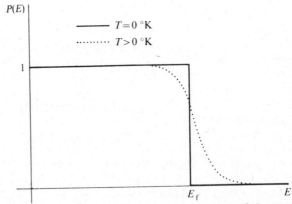

Fig. 3.6. Probability of states being occupied at $0\,^\circ$K and at a finite temperature.

temperature. The electron mean free path decreases as the scattering increases, and is inversely proportional to the temperature, in agreement with experiment. From values of conductivity the mean free path is found to be about 100 inter-atomic distances at room temperature. This is in reasonable agreement with mean free paths calculated by wave mechanical theory from thermal displace-ments, and is in strong contrast with the value of one or two interatomic dis-tances predicted by free electron theory. Vacancies, dislocations, grain bound-aries and impurities will also disturb the lattice sufficiently to scatter electrons, but their temperature effect is negligible.

The Sommerfeld theory is not able to explain why the electrical conductivity of divalent metals is generally lower than that of monovalent and trivalent metals. It is only applicable to the metallic state, and a theory which can describe the electron behaviour and binding forces of all solids is far more desirable. The final development of electron theory, the theory of Brillouin zones, is able to meet these objections.

3.8 Brillouin zone theory

A correct solution to Schrödinger's equation can only be determined if the proper potential function ϕ is known. Sommerfeld's assumption that the potential is uniform in a crystal gives reasonable results for some metals, is rather poor for others, and does not hold at all for non-metals. Electrons moving through a crystal lattice will experience a varying potential which is periodic with the interatomic spacing, due to the positively charged ions of the crystal lattice. The form of this potential is shown schematically in Fig. 3.7 for a metal, an ionic solid and a covalent solid.

There are two important features of these potential functions, the periodicity of the variation in ϕ, determined by the interatomic spacing, and the amplitude of the variation, determined by the strength of the interaction between an electron and a positive ion. This is weak for metals, and hence the amplitude of the variation in ϕ is small, but is strong in covalent solids, leading to a large amplitude. The shape of the function in an ionic solid is complicated by the fact that the electrons are strongly attracted to one type of ion and repelled by the other. (This theory is not normally applied to van der Waals solids, which can be quite adequately treated in terms of the attraction between dipoles.) The exact shape of the ϕ function is not important, and provided the two quanti-ties above are known, reasonable answers can be produced by using rectangular functions (Kronig–Penney model – see Fig. 3.7(d)).

The introduction of a periodic function for ϕ into Schrödinger's equation gives a result which differs from the uniform potential case only near certain values of the wave number k at which there is a discontinuity in the distribution of energy levels. The critical k values are determined solely by the crystal struc-ture, as will be seen shortly. On a one-dimensional plot of E versus k, the parabola obtained on the Sommerfeld theory is distorted as shown in Fig. 3.8. There are now gaps in the range of energy levels, which represent forbidden energy. As the wave number of an electron is increased (by small discrete steps) its energy increases until at critical values of the wave number ($k_1, k_2 \dots$) the energy suddenly jumps to much higher values. Electrons may have energies within the

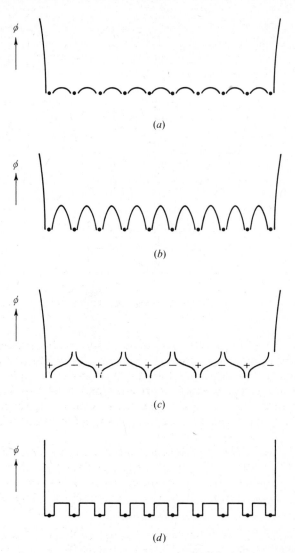

Fig. 3.7. Periodic variation of potential energy ϕ of electrons in different types of crystalline solid (schematic). (*a*) Metal. (*b*) Tightly bound covalent crystal. (*c*) Ionic solid. (*d*) Square well (Kronig–Penney) model.

ranges $0 - E_1$, $E_2 - E_3$, $E_4 \ldots$, but not within the ranges $E_1 - E_2$, $E_3 - E_4, \ldots$.

The allowed energy ranges are called energy bands. The widths of the energy gaps are determined by the amplitude of the function ϕ, the greater the variation in ϕ the larger the energy gaps: a uniform potential, as has already been seen in the Sommerfeld theory, gives no gaps. In practice the ϕ function is rarely known initially, but is estimated from the magnitude of the energy gaps, which can be determined experimentally.

The presence of an energy gap indicates a localisation of electrons to individual atoms and the theory thus brings the various types of binding (metallic, ionic,

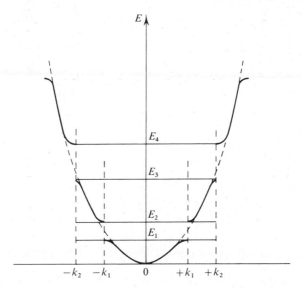

Fig. 3.8. Energy versus wave number for a periodic potential, showing forbidden gaps in the energy band structure.

covalent) into a continuous spectrum. Qualitatively it can be said that in metals the ϕ function has a small amplitude, there is no energy gap, and as a result, the valence electrons are not bound to individual atoms. In ionic crystals the amplitude of ϕ is large, the energy gaps are large, and the electrons are strongly bound to the individual atoms. Covalent crystals cover a range from small gap characteristic of metals to gaps larger than in any ionic solid, for example, in diamond.

The critical values of k at which energy gaps occur are determined by the reflection of electrons from the crystal lattice. Diffraction theory shows that electrons which would move in a direction at an angle θ to a set of crystal planes of spacing d are totally reflected if their wavelength satisfies Bragg's Law:

$$n\lambda = 2d \sin \theta.$$

An electron which fulfils this condition is reflected back and forth between the same set of crystal planes; averaged over a relatively long period of time the electron appears to be stationary. An electron which moves through the crystal cannot therefore have a wave number

$$k_n = \frac{n\pi}{d \sin \theta} \tag{3.10}$$

and this equation defines the values of k at which gaps occur in the energy spectrum. For a one-dimensional lattice of interatomic spacing a, $d \sin \theta$ can only equal a, and $k_n = n\pi/a$.

A three-dimensional medium, k-space, can be defined such that any point P represents a wave number whose magnitude and direction are given by the vector OP, where O is the origin of the coordinate system. For any crystal structure, the points in k-space corresponding to the critical values of k at which energy gaps are found form a series of plane surfaces completely enclosing

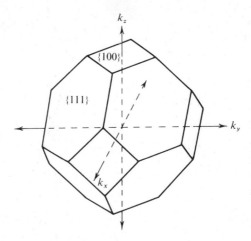

Fig. 3.9. The first Brillouin zone for the fcc lattice.

volumes of *k*-space, centred on the origin. These volumes between successive surfaces are called Brillouin zones; the first Brillouin zone being the volume enclosed by the surface or zone boundary nearest to the origin. The size and shape of the zones are determined solely by the crystal structure and lattice parameter, and the zone can be readily constructed by the use of (3.10). The planes which make up the zone boundaries are parallel to those crystal planes which reflect any electron whose wave number *k* is represented by a vector from the origin to any point upon the zone boundary. The first Brillouin zone for the fcc crystal structure is shown in Fig. 3.9.

Each point in *k*-space has an associated energy value. Equienergy surfaces are spherical well away from a zone boundary where the one-dimensional E–k curve is still parabolic, but become distorted as the zone boundary is approached. The value of the energy on the zone boundary varies from point to point. The highest energy level in the first zone may be higher than the lowest energy level in the second zone, though these extreme values must occur at different points on the boundary. This is illustrated by drawing the E–k curves for different crystallographic directions, as in Fig. 3.10. In (*a*) the highest energy level in the first zone is below the lowest in the second zone and there is a gap in the overall distribution of energy levels irrespective of direction. In (*b*) the highest level in the first zone is above the lowest level in the second zone, and there is no overall discontinuity in the energy spectrum. This second case is called zone overlap, and the valence electrons enter the second zone before the first zone is completely filled.

For an electron to assist in carrying a thermal or electrical current it must be excited into a higher energy state. The electrons able to contribute are therefore those with energies close to empty states. A crystal with a completely filled zone and no overlap offers no empty states for conduction electrons, and is a non-metal. A metal has either an incompletely filled zone, or, if filled, the zones overlap; the situation is not much different from that described by the Sommerfeld theory.

Brillouin zone theory is able to explain the difference in electrical conductivity

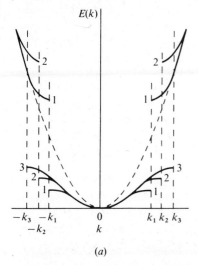

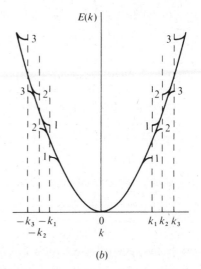

(a)   (b)

Fig. 3.10. Energy versus wave number in three different crystallographic directions. (a) Lowest energy level in second zone is above highest energy level in first zone; an energy gap appears and the material is an insulator. (b) Lowest energy level in second zone is below highest energy band in first zone; there is no energy gap and the material is a metal.

between metals of differing valencies. The conductivity σ is (from (3.1))

$$\sigma = \frac{ne^2l}{mv_f},$$

where n is now the number of electrons available to carry current, that is, the number of electrons in states near the Fermi surface, which in turn is dependent on the density of allowable states at this level. v_f is the velocity of the electrons at the Fermi level, and l is the mean free path determined by disruptions to the periodic potential caused by imperfections as enumerated for the Sommerfeld theory.

The density of states as a function of the energy can be calculated on the theory and is plotted schematically in Fig. 3.11. It is high in the middle of the

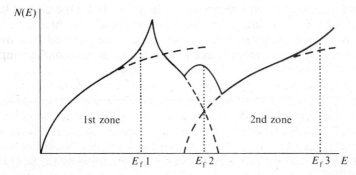

Fig. 3.11. Density of states for a metal, showing the Fermi levels for mono-, di- and trivalent atoms ($E_f 1$, $E_f 2$, $E_f 3$).

Brillouin zones and falls to zero at the edges. The number of quantum states in a zone is equal to twice the number of unit cells in the crystal, and for a simple structure such as fcc or bcc it equals twice the number of atoms in the crystal. For a monovalent metal, the first zone will be half filled, and the Fermi level will have a high density of states, producing high conductivity. The situation is similar in a trivalent metal but for a divalent metal the Fermi level occurs in the region of overlap between the first and second zones, where the density of states is much lower. This accounts for the lower conductivity of divalent metals as compared to mono- and trivalent metals.

3.9 Energy band treatment

The concept of allowed and forbidden energy levels can be arrived at by a different approach. This can be appreciated by considering first of all what happens when two hydrogen atoms are brought together to form a hydrogen molecule. The wave function ψ for an isolated hydrogen atom will look something like Fig. 3.12(*a*), but since the important quantity is $\psi\psi^*$ it could equally be the negative of this, as shown in Fig. 3.12(*b*). As the atoms are brought closer together the wave functions will not be affected until the electron clouds begin to overlap. After overlap occurs, provided the overlap is small, it can be assumed that the two wave functions do not affect one another, and the wave function for the molecule is thus a straightforward addition of the two original wave functions. If the two atoms originally had positive functions (as in Fig. 3.12(*a*)) their sum will be symmetric as in Fig. 3.12(*c*), but if one was positive and one was negative, then their sum will be antisymmetric as in Fig. 3.12(*d*).

The symmetric solution leads to a finite probability of finding the electrons between the two nuclei, and when one of the electrons is in this position the attraction of each of the positive nuclei to this electron produces the binding which gives rise to a hydrogen molecule. The energy of the two atoms is lowered by the symmetric wave function. Conversely the antisymmetric function causes

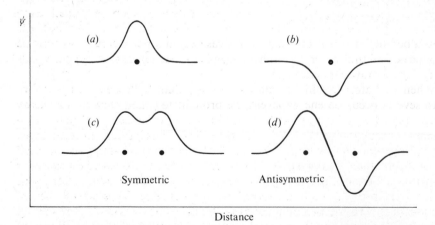

Fig. 3.12. Symmetric and antisymmetric wave functions for the hydrogen molecule.

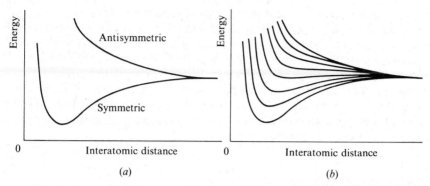

Fig. 3.13. (a) Potential energy versus interatomic distance for the symmetric and antisymmetric wave functions of a hydrogen molecule. (b) Multiple splitting of energy levels in a hypothetical hydrogen crystal.

an increase in energy as the probability of finding an electron between the nuclei falls to zero.

The energy versus interatomic distance for the two functions is shown in Fig. 3.13(a). In a real hydrogen molecule, if the electrons are of opposite spin, then they can both be in the quantum state described by the symmetric function, and binding occurs. If they both have the same spin, then by the Pauli Exclusion Principle, one electron must enter the antisymmetric state. As the energy in this state rises more rapidly than it falls in the symmetric state, there is a net rise in energy and therefore such a molecule is unstable and binding does not occur. It is now possible to understand why helium does not form molecules. The wave functions will be similar to those for hydrogen, but now four electrons must be accommodated, two in each state. Again this produces a net increase in energy and therefore the molecule does not form.

If a large number of hydrogen atoms were brought together to form a crystal, then there would be as many wave functions as atoms, and between the original symmetric and antisymmetric wave function there would now be a continuum of energy states (Fig. 3.13(b)), that is an energy 'band' has been formed. Each state would accept two electrons of opposite spin, and half of the energy levels would be filled. It turns out that in this case again a net increase in energy is produced, and thus no crystal of hydrogen is formed (except by van der Waals forces at very low temperatures).

When two atoms of higher atomic number than hydrogen, and therefore with several occupied energy levels, are brought together, each of the energy levels splits into symmetric and antisymmetric wave functions. If many such atoms are brought together, then each energy level splits into a band as described for the hypothetical hydrogen crystal. If the electrons are strongly localised on the atom, then the bands are narrow, and do not overlap, and an electrical insulator is formed (Fig. 3.14(a)). Loosely bound electrons produce wide bands which overlap (Fig. 3.14(b)), and metallic properties result. These energy 'bands' are identical with the Brillouin zones of the previous section. A combination of both zone theory and the energy band model is useful in giving a full picture of the behaviour of electrons in crystalline solids.

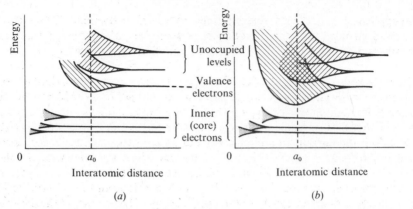

Fig. 3.14. Potential energy versus interatomic distance, showing the broadening of the levels as the interatomic distance is reduced. (a) Insulator; (b) metal.

3.10 Metals, semiconductors and insulators

The Brillouin zone or energy band theories are able to account for the electronic behaviour of metals. They have, however, a much wider validity. A material can be placed, in terms of its electrical properties, into one of three categories: metal, semiconductor or insulator (see Chapter 13). The existence of these three classes of material is readily explainable in terms of their Brillouin zone structure.

Metals, as has already been mentioned, have overlapping and incompletely filled zones. Semiconductors have a completely filled zone separated from the next empty zone by an energy gap, but the width of the gap is small enough for the electrons to be excited thermally across it into the empty zone where they can carry a current. The number of excited electrons is much less than the number of electrons with energies near the Fermi level in metals, and hence the conductivity is lower. Increasing the temperature excites further electrons, and hence increases the conductivity, in contrast with metals. Insulators have a completely filled zone separated from the next empty zone by a large energy gap. Electrons are localised in the region of the positive ions and conduction is difficult. When subjected to very high field strengths an insulator may suddenly develop a low resistance, due to various processes known collectively as breakdown. The field strength at which this occurs is called the breakdown or *dielectric strength* of the material.

The *dielectric constant* is a measure of a different phenomenon. Though the electrons in an insulator normally undergo no large scale translation, under the action of an applied field they may be displaced slightly from their equilibrium distribution about the positive ions. The amount of displacement (or polarisation), for a given applied field, depends upon the polarisability of the material. The effect of this displacement is to produce a large number of electric dipoles, whose presence enhances the electric field within the material. The dielectric constant is the ratio of this enhanced field within the material to the applied field. In structures in which electrical charge is not homogeneously distributed among the atoms, such as in ionic solids, and in polymers with polarisable side groups, the application of a field induces dipoles by atomic movement. These dipoles are much stronger than those produced by electronic polarisation, the dielectric constants may be large (~ 2–10) and such materials are known as *dielectrics*.

When an alternating field is applied to a dielectric, the charge polarisation will try to keep time with the field. However, resistance to the motion of atoms within the material introduces a delay between changes in the field and changes in the polarisation. This delay, which will be a function of the frequency of the applied alternating field, may be expressed as a phase difference, or *loss angle*, δ. δ is zero if the current in a circuit, for example a capacitor, leads the electric field by $90°$, and is equal to $\pi/2$ when the current is in phase with the field. Tan δ is called the *loss tangent* (sometimes *dissipation factor*) and is equal to the energy absorbed (the loss) per cycle by the dielectric from the field divided by $2\pi \times$ the maximum energy stored in the dielectric. The product of the dielectric constant and the loss tangent, the *loss factor*, is proportional to the energy absorbed per cycle divided by $2\pi \times$ the maximum stored energy in the electric field in the absence of any dielectric. Sin δ is called the *power factor*. Semiconductors, insulators and dielectrics are discussed in greater detail in Chapter 14.

QUESTIONS

1. The potential energy of a pair of ions in an ionic crystal is given by:
$$E(a) = -\frac{e^2 A}{a} + \frac{B}{a^8}$$
where a is the interionic spacing, e is the electronic charge (4.8×10^{10} statcoulomb) and A is the Madelung constant. The Madelung constant has a value 1.75 for the NaCl structure. The equilibrium interionic distance in NaCl is 0.279 nm; what is the value of its bulk elastic modulus?

2. The interaction energy between two atoms, at a distance a apart, is given by
$$E(a) = -\frac{\alpha}{a^m} + \frac{\beta}{a^n}$$
where α, β, m and n are constants.
(a) At what value of $a = a_0$ is a stable compound formed?
(b) What is the total potential energy of the stable compound?
(c) What force is required to break up the stable compound?

3. Distinguish between the four types of interatomic binding, and describe the different properties to which they give rise.

4. Planck's constant has the value of 6.63×10^{-34} J s. Discuss the consequences for the physical universe if its value were 6.63×10^{-20} J s.

5. In an electron microscope, electrons are accelerated through a potential of 100 kV.
(a) What is their final velocity?
(b) What is their effective wavelength?
(c) How does the wavelength vary as the accelerating voltage is changed?

6. Define the term 'degeneracy'. Calculate the degeneracy of the first twenty energy levels.

7. Aluminium has three valence electrons per atom, and is face-centred cubic, with a lattice parameter of 0.36 nm. Calculate the Fermi energy for aluminium.

8. At any finite temperature T, the occupancy of energy levels does not drop off suddenly at the Fermi level; instead the decrease in occupancy is spread out over an energy range of $\sim kT$. How many energy levels will show partial occupancy, at room temperature (300 °K) in a 1-cm cube of metal?

9. How is the electrical conductivity of metals accounted for in terms of quantum theory?
10. Why do electrons make little contribution to the specific heat of solids, except at very low temperatures?
11. Differentiate between Fermi level and Brillouin zone boundary. Derive and sketch the first Brillouin zone for (*a*) a bcc crystal; (*b*) a cph crystal.
12. Explain how quantum theory accounts for the classification of solids in terms of their electrical properties.

FURTHER READING

W. Hume-Rothery: *Atomic Theory for Students of Metallurgy*. Institute of Metals, London (1955).
L. Pauling: *The Nature of the Chemical Bond*. Cornell University Press (1960).
T. S. Hutchinson and D. C. Baird: *The Physics of Engineering Solids*. Wiley (1963).
R. L. Sproull: *Modern Physics*. Wiley (1963).
C. Kittel: *Introduction to Solid State Physics*, 3rd edition. Wiley (1966).
A. H. Cottrell: *Theoretical Structural Metallurgy*, 3rd edition. Arnold (1968).

4. Equilibrium diagrams and microstructure

4.1 Introduction

This chapter describes the constitution, or structure, of alloys; that is, the nature, proportions and arrangement of the phases formed when two or more elements are combined. The crystal structures of the phases have already been described in Chapter 2. In this chapter the structure will be examined on a coarser scale since the discussion is about crystals in a polycrystalline material rather than atoms in a single crystal.

For a particular alloy system the structure will depend on the composition, or relative quantities of the elements (or components), and the temperature. Information about the structure can be recorded simply and concisely in an *equilibrium diagram*, also called a constitutional, or phase, diagram. For binary alloys, this has the composition plotted along the abscissa and the temperature along the ordinate: a number of lines divide the diagram into areas in which certain structures and phases exist: see, for example, Fig. 4.1. Although at first sight equilibrium diagrams may appear complex and forbidding, they are built up from only a few basic components, which can be fairly easily understood. Common to all the equilibrium diagrams to be considered here is the condition that the elements are completely miscible in the liquid state; there are a few metallic systems in practice where this is not true, such as lead and iron. The ability to read equilibrium diagrams is fundamental to understanding the science of materials. They provide data on the equilibrium microstructure, from which the macroscopic properties are derived, and also give guidance on the possibility of forming non-equilibrium structures by heat treatment.

The chapter is largely devoted to binary alloys with some discussion of tertiary alloys in the final section. The majority of references to actual systems are to metallic alloys since these have received widespread attention for their technological application. However, ceramic phase diagrams have also been extensively studied. They are constructed in a similar manner to metallic systems, obeying the same rules, and some examples of ceramic systems are included; further examples occur in Chapter 7.

Phase diagrams show the phases and their proportions existing in equilibrium at any temperature, that is, in the lowest free energy state. It must be emphasised that the time taken to reach equilibrium is not indicated. It will depend on various factors such as atomic diffusion rates, free energy differences between phases and interphase surface energies. These have been discussed briefly in Chapter 2. In metallic systems equilibrium structures are fairly easily achieved, nevertheless non-equilibrium structures can be obtained by special heat treatments, which remain stable indefinitely. The hardening of steel by quenching is an important example (see Chapters 11 and 12). In ceramic systems the development of the equilibrium structure is generally slower on account of lower atomic mobility, and frequently non-equilibrium phases can exist indefinitely. For example, all glasses are in a non-equilibrium state (Chapter 7).

It is not the intention here to discuss the experimental methods of obtaining equilibrium diagrams and observing microstructures. However, in view of the fundamental importance of the end results a few paragraphs on these topics are desirable. Fuller details are available in other textbooks.

To observe a metallic microstructure with an optical microscope, following Sorby's method of 1863, a flat area of undisturbed specimen has first to be prepared by grinding and polishing. A small specimen is mounted at the surface of a plastics matrix, say a 30 mm diameter right cylinder. Even if a larger specimen is available, it often reduces the time spent on grinding to mount a small specimen. The specimen is now ground on successively finer emery papers; these are of special quality with no oversize grit, which would cause a last minute scratch. After the finest grade paper (4/0), polishing is continued with suspensions of very fine grit such as alumina, chromic oxide and diamond dust. A soft cloth ('Selvyt') is mounted on a rotating disc and wetted with the suspension. Care must be taken not to press the specimen too hard on to the cloth as this will cause over-heating and affect the microstructure. The essential objective of all this procedure is to progressively eliminate distortion due to the previous grinding or polishing stage until finally a brightly polished and flat surface is formed with only a thin surface layer of amorphous atoms, known as the *Beilby* layer, overlaying virgin material. A chemical, or etchant, is used to remove this layer and preferentially attack the components of the microstructure. The surface is now ready for examination under a metallurgical microscope.

Metallic specimens are completely opaque to light even in thin sections (cf. geological and biological specimens) and the microstructure can only be observed with reflected light. A metallurgical microscope is provided with an internal light source and the beam of light is directed on to the specimen in one of two methods. Either it passes through the objective lens to form a parallel beam orthogonal to the surface, so that a polished unetched surface reflects the light back onto the objective, called *bright field* illumination. Or, less common, the light is fed around the objective to strike the specimen surface obliquely, so that a polished surface appears black, hence the term *dark field* illumination.

The choice of etchant is a vital step, which is still largely empirical. The etchant is required to reveal various details of the microstructure such as the grain boundaries or the arrangement of certain components. Experience has shown which etchants will reveal particular details for each material. The information is recorded in the books on metallography (e.g. see Smithells, vol. 1). Given a new material, research is required to determine suitable etchants and to interpret the resultant pictures.

An alternative to mechanical polishing is electrolytic polishing. In this, the surface is removed in an electrolytic cell, the reverse of electroplating. By adjusting the current density and correct choice of electrolyte, the protrusions on the surface are attacked preferentially to give a bright polished surface with reasonable flatness. It is particularly useful with soft metals. Changing the current density will now cause the surface to be etched. Alternatively a chemical etch can be used. The method was developed by Jacquet during the 1940 decade. It is an essential component of specimen preparation for thin film electron microscopy.

Optical microscopy is possible up to magnifications of about × 1000. Above this the wavelength of light (∼ 20 μm) limits the resolving power. Electron microscopy must then be used, which will give magnifications up to × 50 000 and above. The highest resolution is achieved with a field ion microscope (∼ 1960) which is capable of revealing individual atoms. Electron microscopy can be broadly divided into three types: transmission through replicas of the surface to be examined – the original technique; transmission through thin films of the actual material (∼ 1957); and scanning by an oblique beam which is reflected from the surface (1965). The theory will not be described here except to note that electron motion is determined by wave theory just as photons of light are (see § 1.6).

The principal techniques of determining equilibrium diagrams are as follows. In the cooling curve method, a specimen is allowed to cool slowly from a high temperature, possibly in the liquid state, whilst the temperature is measured against the time. Any discontinuity in the curve of time–temperature or, better, temperature versus time per unit temperature interval will reveal the existence of a phase change, due to the release of internal energy. In the dilation method, the length of a specimen is measured and a graph of length versus temperature obtained. Again a discontinuity develops at any phase change. Another technique is to observe the microstructure, either using a hot-stage microscope or by rapid quenching of the specimen so that the high temperature structure is maintained to room temperature. Finally X-ray measurements to detect the crystal structures present can be used. As already mentioned, the temperatures at which changes occur on heating and cooling are displaced from the true equilibrium change point and allowance must be made for this either by using very slow rates of change of temperature or by observing the changes on raising and lowering the temperature.

Before considering the basic units of binary equilibrium diagrams, the important term *phase* will be defined, together with the phase rule which is an important thermodynamic law governing the number of equilibrium phases, first stated by Gibbs.

4.2 Phase rule

Willard Gibbs in 1876 defined a phase as 'a portion of matter which is homogeneous in the sense that its smallest mechanically isolatable parts are indistinguishable from one another'. For example, ice is a single phase: it is homogeneous and no part can be cut from it that is distinguishable from the remainder, differences of lattice orientation being ignored. Similarly water is a single phase. But at the freezing point when both water and ice are present there are two phases; the matter is not homogeneous in the sense that dissimilar parts can be mechanically isolated.

In the following sections it will be seen that some very refined means of mechanical separation are assumed to exist which will permit the separation of parts that can only be distinguished under a microscope. However, the separation of individual atoms is not permitted and a solid solution of two elements is definitely classed as a single phase.

A fundamental law of thermodynamics governs the equilibrium of phases and the form of equilibrium diagrams.[†] The Gibbs phase rule says that the maximum number of phases P which can coexist in a chemical system plus the number of degrees of freedom F equals the number of components C plus 2:

$$P + F = C + 2.$$

The degrees of freedom F are the external conditions of temperature, pressure and composition which can be independently controlled. If other variables are capable of influencing the equilibrium, the constant in the equation must be increased. C is the number of independent constituents, which may be taken here as the number of elements.

In binary alloys C is 2 and, since the pressure is always fixed at one atmosphere in metallurgical systems, the phase rule becomes:

$$P + F = 3,$$

where F covers temperature and composition only. It follows that:

(*a*) It is not possible to have four phases coexisting, since a negative degree of freedom has no meaning.

(*b*) If three phases are coexisting there are no degrees of freedom, that is, equilibrium occurs at a single temperature between phases of definite composition. It always appears in the equilibrium diagram as a horizontal line which touches the three phases, two at the ends and one at an intermediate point along the line, thus

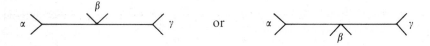

The first represents a eutectic type of reaction and the second a peritectic type. Such isotherms are keystones of equilibrium diagrams and are discussed in detail below.

(*c*) If two phases are present there is one degree of freedom and either the composition of one of the phases or the temperature has to be selected to determine the equilibrium state. The compositions of the two phases which can co-exist at some temperature, called *conjugate* phases, are obtained by drawing an isothermal tie-line across the two-phase region of the equilibrium diagram (see § 4.3). Thus the sides of two-phase regions must contact single-phase regions and the upper and lower temperature bounds must be either isothermal lines or points.

(*d*) If only one phase is present there are two degrees of freedom, meaning that both the temperature and composition can be varied independently, or must be specified to define the state. Single-phase regions can only touch at points, not along a boundary line.

4.3 Complete solid insolubility (eutectic system)

The first basic type of equilibrium diagram to be considered is for an alloy system between two elements that are completely insoluble in each other in the solid

[†] The remainder of this section may be more easily understood after reading the following sections on the construction of binary equilibrium diagrams.

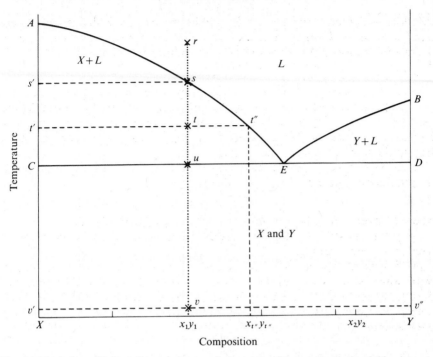

Fig. 4.1. Typical equilibrium diagram for an alloy system of two mutually insoluble elements (eutectic system).

state. As mentioned already, complete liquid solubility is assumed. This is known as a *eutectic system*, and the characteristic equilibrium diagram between elements X and Y is shown by the continuous lines in Fig. 4.1. The related microstructures are shown schematically in Figs. 4.2 and 4.3, based on a surface preparation which makes X appear white, and Y and the grain boundaries dark.

To interpret any equilibrium diagram and determine the various alloy structures it is necessary to consider the changes which occur in specimens of

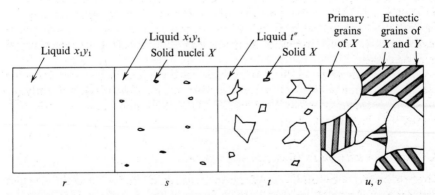

Fig. 4.2. Schematic microstructure of alloy composition x_1y_1 (see Fig. 4.1) during cooling from the liquid state. The etchant is attacking and darkening the grain boundaries and the Y phase. (Letters refer to similar points in Fig. 4.1.)

different composition as they are slowly cooled from the homogeneous single phase liquid state.

Take first a specimen with composition $x_1 y_1$ and consider the changes on cooling from state r to state v, as marked in Fig. 4.1. At r the specimen is a single liquid phase and would appear completely homogeneous under a microscope. When the temperature is reduced to s, that is, to the first intercept with a line of the diagram, element X begins to be precipitated out of the liquid in the form of minute crystals, or nuclei, which are seen in the microstructure, Fig. 4.2 (s). The line AEB marking the temperatures above which alloys are completely liquid is called the *liquidus*. On further cooling, more solid phase, element X, will be formed by the growth of the original nuclei and the formation of new ones. At the same time the composition of the remaining liquid changes towards a higher proportion of Y, since the overall composition is remaining constant. The locus of the liquid phase is the liquidus from s to E.

In any two-phase region of an equilibrium diagram, to define the structure at a particular temperature and overall composition, it is necessary to know the compositions of the conjugate phases, their relative quantities, and, when both are solid, what is their arrangement. The compositions and proportions of the phases are found by drawing an isothermal tie-line through the point representing the state of the alloy, for example, the point t in Fig. 4.1. The first intercepts to the right and left with lines of the equilibrium diagram indicate the phases present and their compositions can be read off the abscissa. In the example, the two intercepts are t' and t'', representing a solid phase with composition pure X and a liquid phase with composition $x_{t''} y_{t''}$ respectively.

The relative quantities are given by a lever rule which states that the mass ratio of the two phases equals the reciprocal of the ratio of the amounts by which their compositions differ from the overall composition. This rule follows from the conservation of mass of each element. For the alloy state represented by the point t, let the weights of the t' and t'' phases be $m_{t'}$ and $m_{t''}$ respectively. Then the lever rule states that

$$\frac{m_{t'}}{m_{t''}} = \frac{tt''}{tt'} \quad \text{or} \quad \frac{m_{t'}}{M} = \frac{tt''}{t't''} \quad \text{and} \quad \frac{m_{t''}}{M} = \frac{tt'}{t't''},$$

where M is the total weight of alloy present. At this stage the microstructure will be as shown in Fig. 4.2 (t).

On cooling composition $x_1 y_1$ until the overall state is at u, the liquid is then at the cusp in the liquidus, marked E, known as the *eutectic point*. From this point the temperature of the liquidus rises with increasing enrichment with Y and a new stage of solidification is forced to occur: all the remaining liquid now solidifies at constant temperature by the formation of X and Y as two separate solid phases. These may be arranged in several ways dependent on their relative proportions in the eutectic composition. A common form is a lamellar one with grains composed of alternate plates of X and Y, as shown schematically in Fig. 4.2 (u). (These plates are drawn too thick for reality.) They are called *eutectic grains* as distinct from the *primary grains* which are formed during the first stage of solidification, from s to u. The use of the term 'grain' to describe the eutectic is misleading since there is not a single phase with a continuous crystal lattice but two distinct phases, each with its own crystal lattice, mixed together in alternate layers. These are microscopically thin but not so thin that the eutectic grain is 'homogeneous in the sense that its smallest mechanically isolatable parts are indistinguishable from one another'.

The alloy is now completely solid. The line on the equilibrium diagram marking the temperatures below which the alloys are completely solid is called the *solidus*, here *CED*.

At v, on the room temperature ordinate, the alloy is still in a two-phase region and the rules for determining their compositions and relative quantities apply. A horizontal tie-line through v intercepts the boundaries at v' and v'', so that the phases present are pure X and pure Y. This must be so since it was postulated that X and Y are mutually insoluble in the solid state. The relative quantities are given by the lever rule:

$$\frac{m_X}{m_Y} = \frac{vv''}{vv'} \quad \text{or} \quad \frac{m_X}{M} = \frac{vv''}{v'v''} \quad \text{and} \quad \frac{m_Y}{M} = \frac{vv'}{v'v''}.$$

The arrangement of the phases is known from the preceding consideration of the changes during cooling. It is still as shown schematically in Fig. 4.2 (u); that is, all the Y phase is in the eutectic grains, while the X phase is divided between the primary and eutectic grains. The division can be determined from the equilibrium diagram by applying the lever rule just above and below the solidus:

$$\text{Just above } CED, \text{ proportion of primary } X = \frac{uE}{CE}.$$

$$\text{Just below } CED, \text{ proportion of total } X = \frac{uD}{CD},$$

$$\text{and proportion of } Y = \frac{uC}{CD}.$$

$$\text{Hence, proportion of eutectic } X = \frac{uD}{CD} - \frac{uE}{CE}.$$

No changes occur on cooling from the solidus to room temperature in this type of alloy system.

The alloy structure over the whole range of compositions from pure X to pure Y can be deduced from similar considerations of the changes which occur during slow cooling from the single-phase liquid state. The microstructures at room temperature are shown schematically in Fig. 4.3, drawn on the assumption that X appears white, and Y and the grain boundaries dark: (a), which is pure X, shows only the grain boundaries; (b) has composition x_1y_1, already considered, and has white primary grains of X and eutectic grains containing lamellae of white X and black Y; (c) has the eutectic composition, and the microstructure consists entirely of lamellar eutectic grains; (d) is for a composition on the Y side of the eutectic (x_2y_2 in Fig. 4.1) and the primary grains are now black Y alongside eutectic grains. A micrograph of pure Y would, of course, appear completely dark.

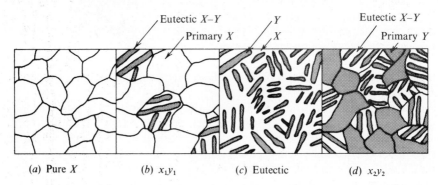

Fig. 4.3. Schematic microstructure of alloys of different compositions in Fig. 4.1, at room temperature. (Same etchant action as for Fig. 4.2.)

It is instructive at this point to consider how the micrographs appear if, as is more usual, the etching technique causes only the grain boundaries to show up dark, whilst both X and Y grains are white. The microstructures (*a*) to (*d*) of Fig. 4.3 are repeated for the new conditions in Fig. 4.4. Since the X and Y phases are now indistinguishable, alloys on either side of the eutectic composition look similar, and the eutectic grains show only the boundaries between the lamellae. Frequently the lamellae in the eutectic grains are so fine that the boundaries are not resolved at normal magnifications (say $\times 250$) and they appear as dark areas although neither component is darkened by the etchant.

Although the pure eutectic system is a useful component in the analysis of equilibrium diagrams, in practice there is usually some mutual solubility between the solid phases, see § 4.5. One example of a eutectic alloy with only slight terminal solid solutions is aluminium–silicon, for which the equilibrium diagram is given in Fig. 4.5. The eutectic composition (11.7 per cent Si) is a common composition for light alloy castings, the structure usually being modified by the addition of a small amount of sodium. More eutectic systems with negligible solid solubility will be found in alloys involving intermediate phases (§ 4.6).

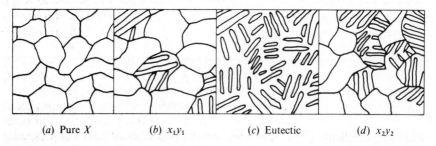

Fig. 4.4. Schematic microstructures of alloy compositions as for Fig. 4.3 but etchant is now only attacking the grain boundaries.

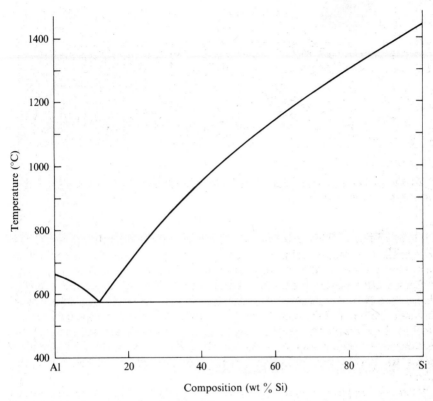

Fig. 4.5. Aluminium–silicon equilibrium diagram – example of a eutectic system. (Slight solid solubility neglected.)

A limiting case of a eutectic system occurs when the eutectic composition lies very close to one of the components. An example is shown in Fig. 4.6 for beryllium–aluminium. These alloys, containing up to 30 per cent aluminium, are currently being developed on account of their high strength–weight ratio combined with much greater ductility than pure beryllium.

Very similar to the eutectic reaction is the eutectoid reaction. In both, a single phase changes isothermally on removing heat into two other phases, represented thus on the diagram

$$\beta \rangle \!\!\!\!\!-\!\!\!-\!\!\!-\!\!\!\underset{\alpha}{\vee}\!\!\!-\!\!\!-\!\!\!-\!\!\!\langle \gamma$$

As has been indicated, in a eutectic reaction α is a liquid phase and β and γ are solid phases. In a eutectoid reaction all three phases are solid; this is discussed further in § 4.8.

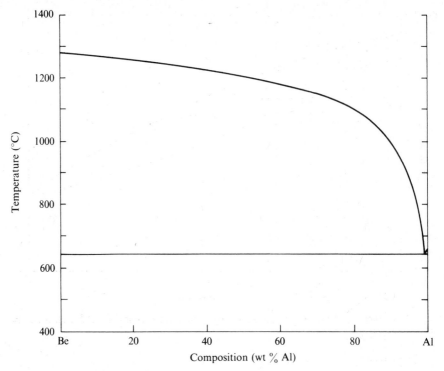

Fig. 4.6. Beryllium–aluminium equilibrium diagram. A limiting example of a eutectic system with the eutectic composition approaching one component.

4.4 Complete solid solubility (isomorphous system)

The second basic type of equilibrium diagram is for two elements that are completely soluble in each other in the solid state. This is called an *isomorphous system*, and the characteristic equilibrium diagram is shown by the continuous lines in Fig. 4.7.

As before, consider the changes in an alloy of composition xy during slow cooling from the liquid state. The corresponding microstructures are drawn schematically in Fig. 4.8. Initially, at r, there is a single-phase liquid. At s, on the liquidus, solidification begins and the alloy enters a two-phase region. Applying the rules for a two-phase region and drawing an isotherm, the solid phase is represented by s', that is, a solid solution with composition $x_{s'}y_{s'}$. Further cooling increases the quantity of solid phase, decreases the quantity of liquid phase, and changes both their compositions. Thus, when the overall state is given by t, the solid phase is represented by t', composition $x_{t'}y_{t'}$ and the liquid phase by t'', composition $x_{t''}y_{t''}$. The original nuclei of composition $x_{s'}y_{s'}$ will have absorbed additional amounts of Y out of the liquid phase to bring their composition into line with the new equilibrium value for the solid phase. This process takes place by diffusion in the solid crystals and is a slow one; if cooling proceeds too fast, *coring* will occur in which the composition varies from centre to edge of each grain. The relative quantities of the two phases are obtained by

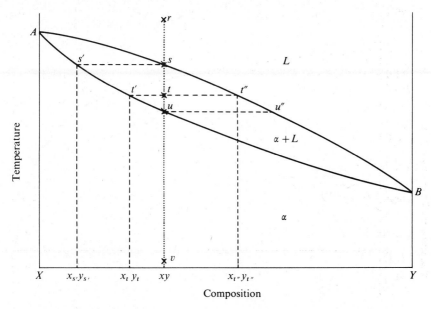

Fig. 4.7. Typical equilibrium diagram for an alloy system of two mutually soluble elements (isomorphous system).

the lever rule:

$$\frac{m_{t'}}{m_{t''}} = \frac{tt''}{tt'} \quad \text{or} \quad \frac{m_{t'}}{M} = \frac{tt''}{t't''} \quad \text{and} \quad \frac{m_{t''}}{M} = \frac{tt'}{t't''}.$$

Solidification is completed at u on the solidus, the last liquid to exist being represented by u'. The alloy now consists of a single solid phase of composition xy, and no further changes occur on cooling to room temperature. The microstructure shows only the boundaries between grains which have identical composition.

Similar changes occur in all alloy compositions from this system and identical microstructures are obtained. There are thus many points of contrast with the previous type of alloy system.

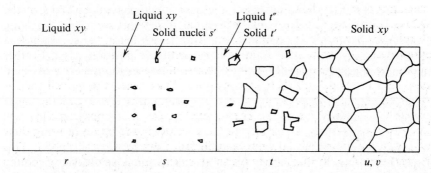

Fig. 4.8. Schematic microstructures of alloy composition xy (see Fig. 4.7) during cooling from the liquid state.

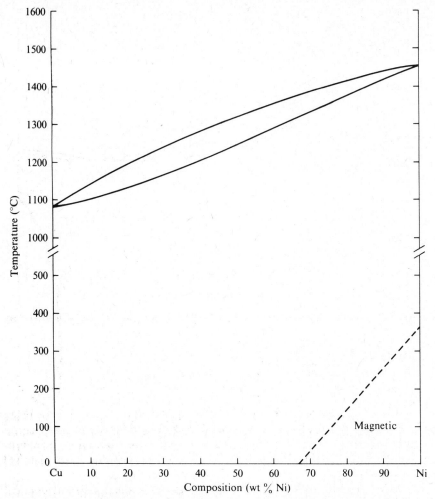

Fig. 4.9. Copper–nickel equilibrium diagram – example of an isomorphous system. (Broken line indicates magnetic transformation, see Chapter 15.)

An example of this type of diagram is the copper–nickel system, shown in Fig. 4.9. It was mentioned in § 2.6 that copper and nickel satisfied the Hume-Rothery conditions for forming a substitutional solid solution over the whole composition range. Alloys from this system are widely used on account of their good mechanical properties combined with high resistance to corrosion. For example, the 30 per cent nickel alloy is used for marine steam condensers and many chemical plants. 75 per cent copper–25 per cent nickel is used for coinage in various countries since its silvery appearance is deceptive! An ore found in Canada yields an alloy ('Monel') containing 67 per cent nickel and 30 per cent copper with some manganese and iron; it is widely used in steam and process plants, and for marine applications such as propellers and shafts.

The liquidus and solidus in some equilibrium diagrams of this type do not run smoothly from the melting point of one component to that of the other, but

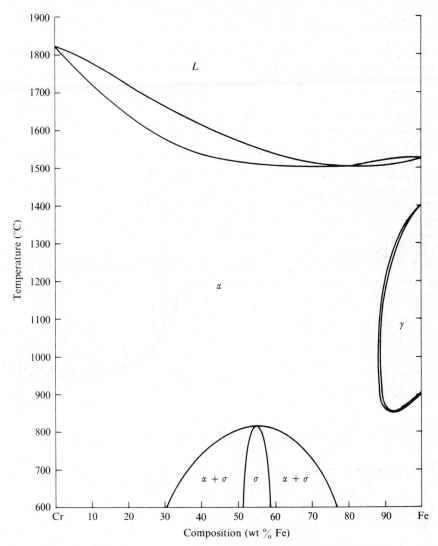

Fig. 4.10. Chromium–iron equilibrium diagram – example of an isomorphous system on which the liquidus and solidus lines converge at an intermediate composition.

pass through a maximum or minimum at which they meet tangentially. An example is found in the chromium–iron alloy system, see Fig. 4.10. At such a point the composition melts *congruently*, that is, one phase changes into another of the same composition without any intermediate stage involving phases of different compositions. The Fe–Cr system includes three congruent phase changes, excluding those of the pure elements, which must clearly be congruent. The lower part of the diagram involves polymorphic changes in the iron (α and γ-phases) and an intermediate phase (σ); these components of equilibrium diagrams are considered below.

4.5 Partial solid solubility

Combination of the first two types of equilibrium diagram gives one which is characteristic of alloy systems between two elements that are soluble in each other in the solid state only over limited ranges of composition. A typical diagram is shown in Fig. 4.11, with the region of solid solution of Y in X extending from the X-ordinate to the line ACv' and that of X in Y from the Y-ordinate to the line BDv''. These terminal solid solutions will be referred to as the α and β-phases respectively.

Take first a composition x_1y_1 just outside the α-phase range at room temperature; the corresponding microstructures during cooling are shown schematically in Fig. 4.12. Initially the changes are the same as for alloys of the second basic type: solidification starts at s with nuclei of the α-phase represented by s'; further cooling increases the amount of the solid phase, and alters its composition; at t, the liquid phase is eliminated and the alloy is entirely α-phase, composition x_1y_1. A new phenomenon now occurs. At u, nuclei of β-phase solid solution, represented by u'', are formed either at the grain boundaries or on certain crystal planes of the existing phase. Further cooling increases the precipitation of β-phase and the composition of the original nuclei and the new

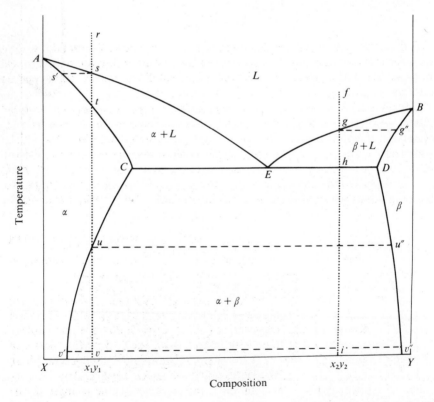

Fig. 4.11. Typical equilibrium diagram for an alloy system of two elements with limited solid solubility.

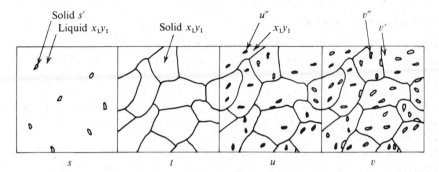

Fig. 4.12. Schematic microstructures of alloy composition x_1y_1 (see Fig. 4.11) during cooling from the liquid state.

precipitate changes along the line $u''v''$; at the same time the α-phase is changing its composition along the line uv'. Finally, at room temperature, when the overall state is represented by v, by the rules for a two-phase region the phases present are: α-phase, a solid solution of Y in X with composition v'; and β-phase, a solid solution of X in Y with composition v''. Their relative quantities are given by the lever rule

$$\frac{m_{v'}}{M} = \frac{vv''}{v'v''} \quad \text{and} \quad \frac{m_{v''}}{M} = \frac{vv'}{v'v''}.$$

If the composition had been chosen nearer to the pure X-ordinate, so that the room temperature state is represented by a point within the region of terminal solid solution, there would have been no solid phase precipitation and no changes on cooling once solidification was completed.

Consider next a specimen with composition x_2y_2, clear of the terminal phases, for which the microstructures at various temperatures are shown schematically in Fig. 4.13. At f there is a single-phase liquid solution of X and Y, composition x_2y_2. At g, the alloy enters a two-phase region with the precipitation of β-phase, a solid solution of X in Y with a composition given by g''. At h, the solid phase forming the primary grains has composition D, while the remaining liquid phase has reached the eutectic point E. The relative quantities

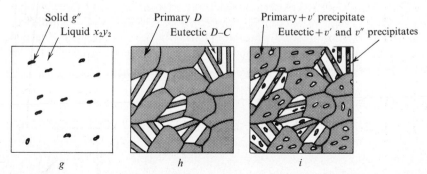

Fig. 4.13. Schematic microstructures of alloy composition x_2y_2 (see Fig. 4.11) during cooling from the liquid state.

are, by the lever rule:

$$\frac{m_D}{m_E} = \frac{hE}{hD} \quad \text{or} \quad \frac{m_D}{M} = \frac{hE}{ED} \quad \text{and} \quad \frac{m_E}{M} = \frac{hD}{ED}.$$

The liquid phase now solidifies at constant temperature into eutectic grains of alternate layers of α-phase, composition C, and β-phase, composition D. The overall quantities of the two phases then become

$$\frac{m_D}{m_C} = \frac{hC}{hD} \quad \text{or} \quad \frac{m_D}{M} = \frac{hC}{CD} \quad \text{and} \quad \frac{m_C}{M} = \frac{hD}{CD}.$$

Since the fraction of β-phase in the primary grain is already known the fraction in the eutectic grain can be deduced:

$$\frac{(m_D)_E}{M} = \frac{hC}{CD} - \frac{hE}{ED}.$$

All the α-phase is, of course, in the eutectic grains.

Further cooling will cause α-phase to be precipitated out of the β-phase, which alters its composition along the line $Du''v''$, and at the same time β-phase will be precipitated out of the α-phase, which alters its composition along the line Cuv'. When the overall state has reached i the phases present are α-phase with composition v' and β-phase with composition v''. The overall quantities are:

$$\frac{m_{v'}}{m_{v''}} = \frac{iv''}{iv'} \quad \text{or} \quad \frac{m_{v'}}{M} = \frac{iv''}{v'v''} \quad \text{and} \quad \frac{m_{v''}}{M} = \frac{iv'}{v'v''}.$$

The distribution of these phases between the primary and eutectic grains and the precipitates within them can be determined quantitatively from the calculations of the structure at various stages along the cooling line, as outlined above; this will be left to the reader to pursue.

A simple example of this type is the equilibrium diagram between tin and lead, given in Fig. 4.14. These alloys are used as soft solders. The eutectic composition 63 % Sn–37 % Pb has the lowest melting point (183 °C) and has the best wetting and running properties. Higher proportions of lead are used for wiped joints on lead pipe and cable where solidification over a wide temperature range is required for making the joint. Lower melting point solders, down to 70 °C are obtained by forming a quaternary eutectic with the addition of bismuth and cadmium (Woods metal).

Many other examples of eutectic reactions between two solid solution phases will be found in the more complex diagrams which follow. The precipitation of a second phase from a solid solution is particularly valuable since it provides a powerful method of strengthening, known as precipitation hardening. This will be discussed in Chapter 11.

4.6 Intermediate phase

The third important component of equilibrium diagrams is the intermediate compound, or intermediate phase, which occurs when the two elements combine to form a completely new crystal structure in accordance with the laws described in § 2.7. The lead–magnesium equilibrium diagram shown in Fig. 4.15 contains a typical example.

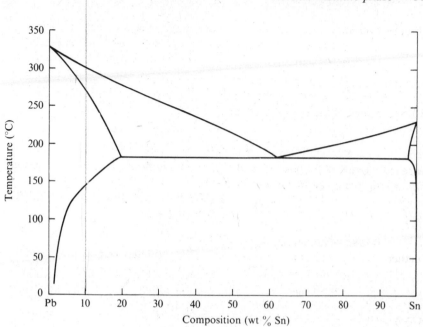

Fig. 4.14. Lead–tin equilibrium diagram (soft solder) – example of an alloy system with limited solid solubility.

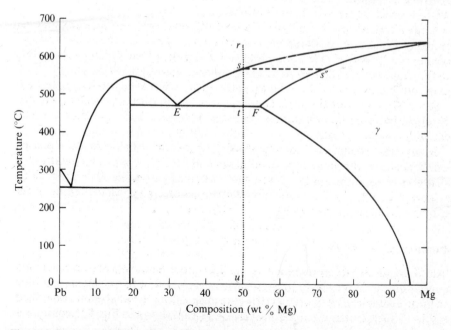

Fig. 4.15. Typical equilibrium diagram involving an intermediate phase: lead–magnesium alloy

The intermediate compound appears in the diagram as a vertical line running at the appropriate composition from room temperature to the melting point of the compound, which is sharp. It has the effect of dividing the equilibrium diagram into two parts, which are similar to the diagrams already described except that one of the elements is replaced by the intermediate compound. In Fig. 4.15, one part is a simple eutectic system with no solid solubility and the other is a eutectic system with some solid solubility of the intermediate phase in the magnesium.

The empirical formula of the intermediate compound can be deduced from the equilibrium diagram. Suppose it is X_rY_s, where r and s will be small integers (except with some electron compounds), and let the atomic weights of X and Y be respectively a_x and a_y. If the relative quantities of X and Y, as determined from the equilibrium diagram, are p and q, then

$$\frac{ra_x}{sa_y} = \frac{p}{q}.$$

For example, in the lead–magnesium alloy system the intermediate compound occurs at about 82 % Pb–18 % Mg and the atomic weights of lead and magnesium are 207 and 24.3 respectively. Hence, if the empirical formula is Pb_rMg_s

$$\frac{r \times 207}{s \times 24.3} = \frac{82}{18}$$

and therefore $r = 0.53s.$

Making allowance for errors due to the small scale of the equilibrium diagram, the smallest integral values of r and s which satisfy the equation are 1 and 2. The empirical formula is therefore $PbMg_2$.

In determining alloy structures, the parts of the diagram to the right and left of the intermediate compound can be treated as separate. Thus an alloy with a composition between the compound and the element Pb will change on cooling in the same way as a specimen from the eutectic alloy system discussed in § 4.3, except that for X and Y now read Pb and $PbMg_2$. The part of the diagram to the right of the intermediate compound will not enter into the consideration. On the other hand, if the composition is between $PbMg_2$ and Mg, the structures will be dependent only on the diagram to the right of the intermediate compound. For example, consider the cooling of 50 % Pb–50 % Mg alloy, r to u in Fig. 4.15. As usual there is initially a single-phase liquid (r). At s, nuclei are formed of magnesium solid solution (γ-phase) with a composition given by s''. This can be regarded for the moment as the intermediate compound in solid solution in the element magnesium. At t, the remaining liquid has reached the eutectic point E, composition 68 % Pb–32 % Mg, and the solid phase is now represented by the point F, 45 % Pb–55 % Mg. The liquid now solidifies at constant temperature into two phases, the intermediate compound $PbMg_2$ and the γ-phase at F. By drawing isothermal tie-lines slightly above and below the eutectic temperature and applying the lever rule, the relative proportions of the phases and their disposition can be obtained. Cooling to u causes the intermediate compound to be precipitated from the γ-phase in the primary and eutectic grains. The final composition of the γ-phase is 4 % Pb–96 % Mg and the second phase remains as PbMg$_2$. The overall proportions are then

$$\frac{\gamma}{\text{PbMg}_2} = \frac{50 - 18}{96 - 50} = 0.7.$$

The γ-phase solid solution was described above as a solid solution of the intermediate compound PbMg$_2$ in magnesium. This is not strictly accurate as

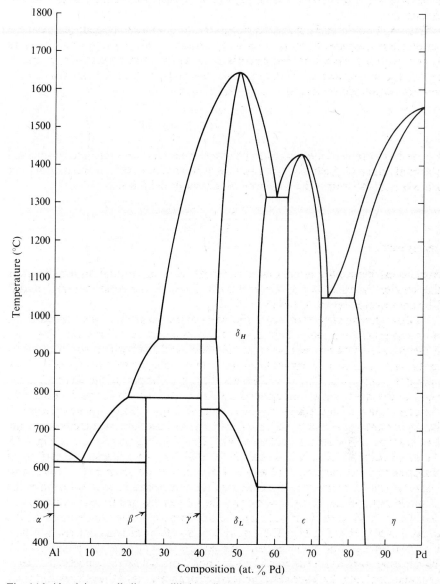

Fig. 4.16. Aluminium–palladium equilibrium diagram. The system has four intermediate phases: PdAl$_3$(β); Pd$_2$Al$_3$(γ); PdAl(δ) and Pd$_2$Al(ϵ). The β and γ phases break down on heating (peritectic reaction); the δ phase shows high and low temperature polymorphic forms. (*NB* Composition scale is *atomic* per cent palladium.)

X-ray analysis has shown that the structure of an intermediate phase disappears when it is dissolved in one of the elements, and the correct description here is a solid solution of lead in magnesium. In this respect the intermediate compound does not divide the diagram into two entirely separate parts.

A development of this type of diagram, though not really involving any new principles, is the alloy system with several intermediate compounds, each of which is represented by a vertical line. The diagram is then divided into three or more parts which may be treated as separate entities.

Another step is when the intermediate compound itself forms a solid solution and the line is broadened into an area, which comes to a point at the melting point. In the aluminium–palladium system, see Fig. 4.16, there are four intermediate phases, of which the δ and ε-phases form solid solutions.

Intermediate phases are frequently not stable over the whole temperature range covered by the diagram, that is, from about 0 °C up to the liquid state. They may react at some temperature, either on heating or cooling, into two other phases or into another single phase solid. For example, both the β and γ-phases of the aluminium–palladium system (Fig. 4.16) break down on heating into another solid phase and a liquid phase. This is a *peritectic* reaction. If they had been two solid phases it would be called a *peritectoid* reaction. These reactions will be described in detail in the next section. (The diagram also shows that the δ-phase has high and low temperature polymorphic structures – see § 4.8.)

From the concept of intermediate phases involving just two elements it is a

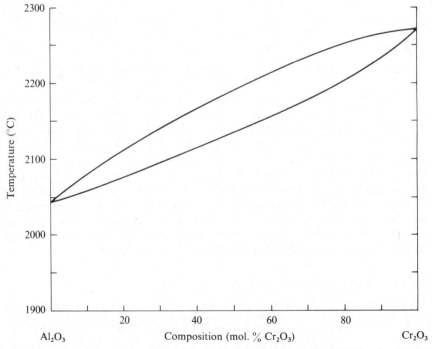

Fig. 4.17. Aluminium oxide–chromium oxide equilibrium diagram. Example of an isomorphous ceramic system (cf. Fig. 4.9 Cu–Ni system).

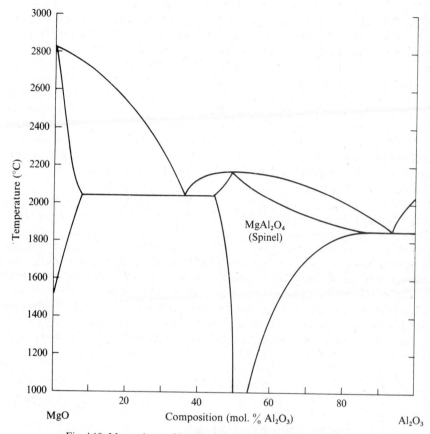

Fig. 4.18. Magnesium oxide–aluminium oxide equilibrium diagram.

simple step to phase diagrams between two compounds, as occurs frequently in ceramic systems. A simple example is the isomorphous system between alumina Al_2O_3 and chromium oxide Cr_2O_3, given in Fig. 4.17. This is very similar to the copper–nickel system (Fig. 4.9) except that instead of two metals with fcc structure the components are now two metallic oxides with identical crystal structures, since a solid solution is formed over the whole composition range. A further example of a ceramic phase diagram, involving no new ideas, is that for the magnesium oxide–alumina system shown in Fig. 4.18. It will be seen that in both these ceramic systems, although there are three elements involved, the two compounds are behaving as single components. This may not always be true and a ternary diagram would then be required (see § 4.9).

4.7 Peritectic and peritectoid reactions

Two further components of equilibrium diagrams are the peritectic reaction and the very similar peritectoid reaction, which appear on a diagram as the isotherm:

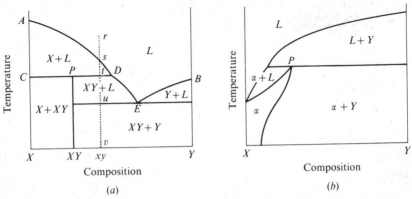

Fig. 4.19. Typical equilibrium diagrams involving a peritectic reaction (at point *pr*): (*a*) breakdown of an intermediate phase, (*b*) breakdown of a terminal solid solution.

In a *peritectic* reaction a solid phase β breaks down on heating into a liquid phase and another solid phase γ. The original phase is usually an intermediate compound but it can be a terminal solid solution. In a *peritectoid* reaction two solid phases are formed on heating an intermediate compound.

The characteristic equilibrium diagrams for a peritectic reaction are shown in Fig. 4.19. In (*a*), the line representing the intermediate compound XY no longer extends through to the liquid phase but finishes at P on an isothermal line which indicates an equilibrium exists between the intermediate compound, solid X, represented by C, and a liquid phase, represented by D. The changes on cooling an alloy with a composition xy as marked in the figure are as follows. At r, there is a single-phase liquid solution. At s, nuclei of X form and grow until, at t, the liquid phase is represented by D. A peritectic reaction now occurs at constant temperature between the solid phase and some of the liquid phase D to yield intermediate compound XY and excess liquid phase D. The proportions are obtained by drawing isothermal tie-lines just above and below the peritectic temperature:

before reaction, $\dfrac{m_D}{m_C} = \dfrac{tC}{tD}$ or $\dfrac{m_X}{M} = \dfrac{tD}{CD}$ and $\dfrac{m_D}{M} = \dfrac{tC}{CD}$

after reaction, $\dfrac{m_D}{m_P} = \dfrac{tP}{tD}$ or $\dfrac{m_{XY}}{M} = \dfrac{tD}{PD}$ and $\dfrac{m_D}{M} = \dfrac{tP}{PD}$.

Further cooling deposits more XY and changes the liquid composition along the line DE, until, at E, the remaining liquid undergoes a eutectic reaction with the simultaneous formation of XY and solid Y. There are no further changes on cooling to room temperature.

Examples of intermediate phases with peritectic reactions have already been pointed out in the aluminium–palladium system (Fig. 4.16). In the antimony–tin system, see Fig. 4.20, both the terminal solid solution (β-phase) and the intermediate phase (δ) undergo peritectic reactions.

Examples of the peritectoid type of reaction, in which an intermediate compound changes on heating into two solid phases, occur in the copper–tin system (bronze), see Fig. 4.21: the δ and ζ-phases break down at 590 °C and 645 °C

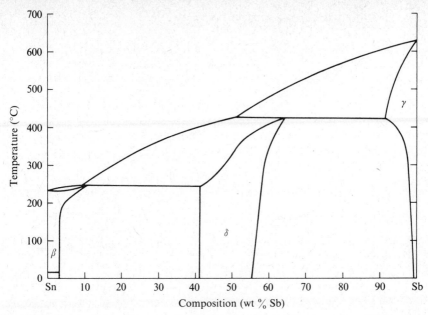

Fig. 4.20. Antimony–tin equilibrium diagram – example of an alloy system involving peritectic reactions.

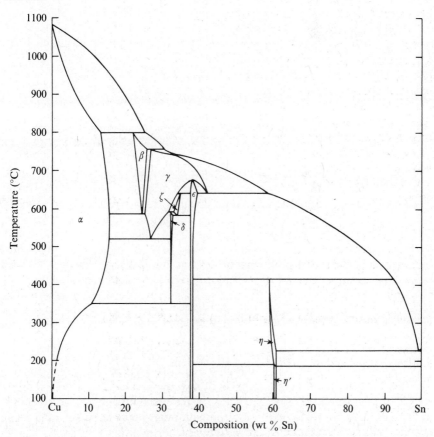

Fig. 4.21. Copper–tin equilibrium diagram (bronze).

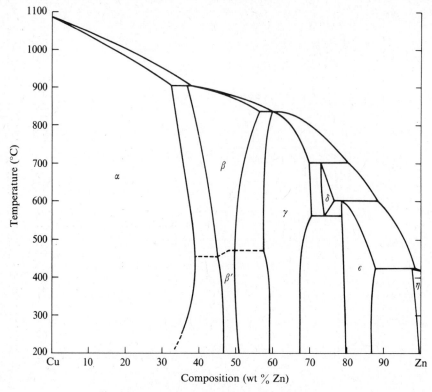

Fig. 4.22. Copper–zinc equilibrium diagram (brass).

respectively. The diagram also includes an example of an intermediate phase (ε) which changes congruently on heating into another solid phase (γ), see § 4.8, and several peritectic reactions.

The equilibrium diagrams of two further copper alloy systems are given in Fig. 4.22, copper–zinc, known as brass, and Fig. 4.23, copper–aluminium. These and Fig. 4.21 for bronze are as complicated as any that exist, but the reader should now feel that all parts of them are intelligible. Each of the diagrams is for commercially useful alloys. The brasses are fairly cheap, easily cast or fabricated and corrosion resistant. Bronzes have rather similar properties, plus a resistance to wear which makes them useful in sliding bearings; they are more costly. Copper rich alloys with aluminium are called aluminium bronzes, although they contain no tin. They have similar properties to bronze besides being heat treatable. The aluminium end of the system provides the basis of the standard heat treatable aluminium alloy (§ 11.7).

4.8 Polymorphic phase change: Fe–Fe₃C system

The final component of equilibrium diagrams arises from polymorphism (cf. § 2.3). The change of crystal structure of an element when heated or cooled in the solid state is shown on the diagram in the same way as liquid–solid phase changes and similar characteristic patterns are formed.

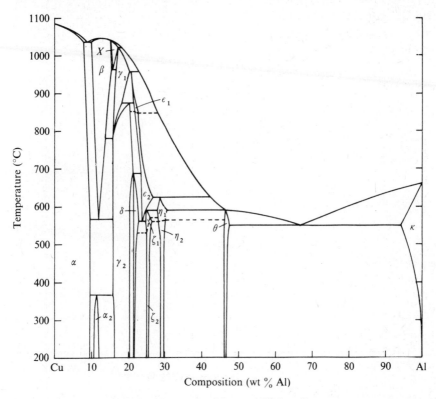

Fig. 4.23. Copper–aluminium equilibrium diagram.

Several possibilities are shown in Fig. 4.24. In (*a*) and (*b*) both elements are polymorphic, with high and low temperature crystal structures. The high temperature forms are mutually soluble over the whole composition range so the pattern on freezing is identical with that already considered in § 4.4 (isomorphous system). In (*a*) the low temperature crystal structures are also soluble so that the pattern of lines due to the polymorphic phase changes of the elements at temperatures T_{px} and T_{py} is similar to that during freezing. A solid phase is now precipitated from a solid solution instead of from a liquid solution. The mobility of the atoms is very much reduced but the principle is the same. Nuclei of the new phase form and grow until the phase change is complete. In (*b*) the low temperature crystal structures are mutually insoluble and the pattern is that already considered in § 4.3 for the freezing of two insoluble elements (eutectic system). At the *eutectoid* point *E* the high temperature solid solution breaks down isothermally into two separate phases which may be arranged as alternate lamellae, as occurs in the eutectic reaction. A eutectoid reaction always appears in an equilibrium diagram in the form

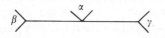

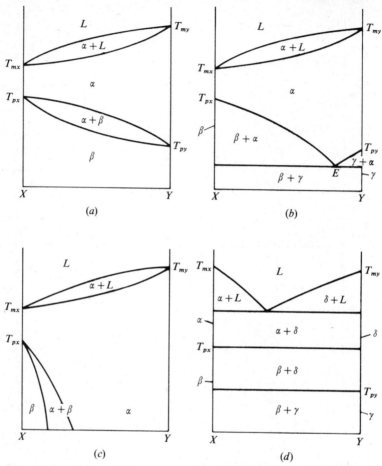

Fig. 4.24. Typical equilibrium diagrams involving polymorphic phase changes of one or both elements.

where all three phases are solids (cf. a eutectic reaction where α is a liquid phase).

Figure 4.24(c) is a typical diagram for an alloy system in which only one of the elements is polymorphic. If the second element is assumed to undergo a polymorphic change at a temperature below the range of the diagram, it will be seen that the diagram can be regarded as part of (a). Finally (d) shows a new pattern which is produced when both the high and low temperature crystal forms are insoluble. The changes on freezing are the familiar eutectic system (cf. Fig. 4.1) but the polymorphic changes now give rise to horizontal lines across the diagram as each phase undergoes its polymorphic change independently.

The most important example involving polymorphic changes is undoubtedly the phase diagram for the iron–carbon alloy system, the useful part of which is given in Fig. 4.25. It covers the constitution of plain carbon steels and white cast irons. Iron is bcc at room temperature (α-phase or *ferrite*) and changes at 910 °C into fcc crystal structure (γ-phase or *austenite*). There is a further polymorphic

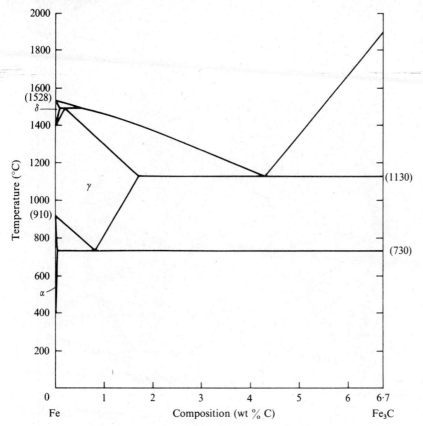

Fig. 4.25. Iron–iron carbide equilibrium diagram.

change at 1400 °C, when the iron reverts to bcc structure (δ-phase), and melting occurs at 1528 °C. These temperatures are, of course, marked on the diagram, but the crystal structures and the names of the phases need to be remembered. Some diagrams also mark the Curie point 768 °C at which ferrite finally loses its magnetic properties (see Chapter 15), but since this does not involve a change of crystal structure it is omitted here. Austenite is non-magnetic.

Iron and carbon form an intermediate compound Fe_3C, called *cementite*. This divides the alloy system in two parts at 6.7 per cent carbon and only the iron-rich side is of practical interest. This part is sometimes called the iron–iron carbide diagram, for obvious reasons. Iron carbide is strictly only a metastable phase and will break down into iron and graphite if held for several years at 650–700 °C or more quickly when other elements, notably silicon, are also present. This occurs in most cast irons (e.g. grey cast iron, see § 12.3). The crystal structure of cementite is orthorhombic (three unequal axes at right-angles) with 12 iron atoms in a fairly close-packed arrangement and 4 carbon atoms in interstitial positions, see Fig. 4.26.

The various forms of iron are able to take different amounts of carbon into solution. It is correct to speak of carbon in solution rather than cementite since

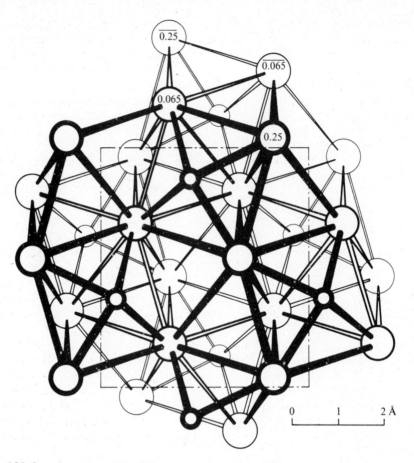

Fig. 4.26. Crystal structure of Fe_3C (cementite). The larger circles represent iron atoms and the smaller ones carbon. The atoms lie on four planes ($z = -0.25$, -0.065, $+0.065$, $+0.25$ Å) as indicated by the bolder outlines of the higher atoms. (After H. Lipton and N. J. Petch, *J.I.S.I.* **142**, 95 (1940).)

the latter breaks down and loses its identity when dissolved. Bcc iron forms only a very limited solution, with a maximum of 0.025 per cent carbon. Although this small amount can be ignored in calculating proportions of the phases present at any temperature and composition, it has a large effect on the mechanical properties of ferrite and steel (see § 5.5 and § 11.5 on the sharp yield point). Fcc iron is able to take up to 1.7 per cent carbon into interstitial solid solution. This is taken as the dividing line between steels and cast irons. It was explained in § 2.6 how the shape of the voids is responsible for the paradox that the close-packed form of iron is able to dissolve more carbon than the non-close-packed bcc iron.

The temperature ranges over which the various forms of iron are stable are dependent on the carbon content, as indicated by the diagram. For example, fcc iron is stable down to 910 °C but combined with 0.8 per cent carbon, as austenite, it is stable down to 720 °C, when it undergoes a eutectoid reaction. The line joining these two points through the *upper critical temperatures* is called the A_3 line and above it there is the pure austenite region. The isothermal line through the eutectic point is the A_1 line, or *lower critical temperature*, below which all austenite has been transformed. The range between these lines is the *critical temperature range*, over which austenite is conjugate with ferrite or cementite. The microstructures of annealed or normalised steels, that is, when cooled in the furnace or in still air respectively from just above the critical temperature range, are discussed below. Those of heat-treated steels are given in Chapter 12, where the properties of steels are also discussed.

Three invariant points should be noted in the diagram. The eutectic point at 4.3 per cent carbon, 1130 °C has the liquid phase in equilibrium with austenite (at 1.7 per cent carbon) and cementite. This composition is called *ledeburite*, and with the lowest melting point it is the most easily cast. The eutectoid point occurs at 0.83 per cent carbon, 720 °C, at which austenite is conjugate with ferrite and cementite. The latter are formed on cooling in typical alternate lamellae, called *pearlite*, on account of its reputed pearl-like appearance at low magnifications. Pearlite is present in all steels together with either ferrite or cementite, see Fig. 4.25, and it is easy to be misled into thinking it is a single phase. It is not. Finally there is a peritectic reaction at 0.18 per cent carbon, 1492 °C: on heating austenite of this composition, it changes into δ-phase, with 0.1 per cent carbon in solution, and liquid iron with 0.5 per cent carbon.

It should be emphasised that the lines shown on the diagram represent change points under theoretically perfect equilibrium conditions. At finite rates of heating or cooling, the 'arrest' points are displaced upwards or downwards, typically by 10–20 °C. This must be allowed for in practice. The nomenclature is that points obtained on heating are given the subscript c and on cooling r (from *chauffage* and *refroidissement* respectively). The diagram is frequently shown with both sets of lines but this adds to the complication of an already complex picture.

Typical microstructures of plain carbon steels and white cast irons are shown in Fig. 4.27, using the standard etchant of nital (nitric acid and methyl alcohol). These are for the metal in the normalised condition, that is, after slowly cooling in still air from a high temperature; for steels this means from the single-phase austenitic region, typically just above the critical temperature range. Commercially pure iron (ingot, or Armco, iron) consists entirely of ferrite grains and only the grain boundaries are visible in the micrograph (*a*). Mild steel with 0.1 to 0.3 per cent carbon consists of ferrite and pearlite, see (*b*). As has been explained, pearlite is a lamellar mixture of ferrite and cementite; and at low carbon contents and normal magnifications the lamellae are not resolved, so the pearlite appears uniformly dark between the grains of ferrite. The structure develops on cooling from the single-phase austenite region as follows. At the upper critical temperature, ferrite is nucleated at the grain boundaries of the austenite. If the grains are large, ferrite is also formed on preferential planes within the austenite grains. This is called *Widmanstätten* structure and is undesirable; it is found in castings and welds. Over the critical temperature range ferrite grows at the expense of austenite until at the lower critical temperature the remaining austenite transforms isothermally to ferrite and cementite lamellae, pearlite.

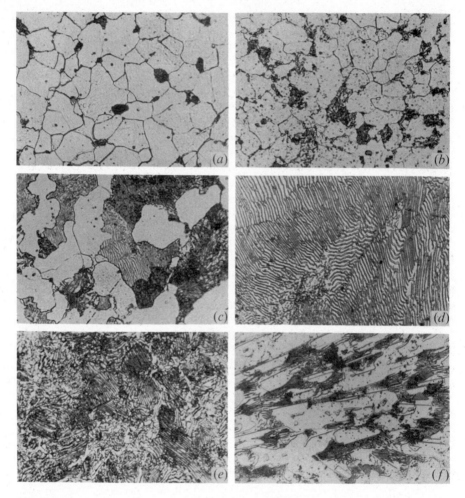

Fig. 4.27. Microstructures of iron–iron carbide alloys. Normalised condition. (Etched in 2 per cent nital. × 375.)

(a) 0.05 per cent carbon. Ingot iron. Ferrite (α-iron) grains with some inclusions.

(b) 0.2 per cent carbon. Mild steel. Ferrite (white) and pearlite (dark). The ferrite–cementite (Fe₃C) lamellae in pearlite are not resolved.

(c) 0.4 per cent carbon. Medium carbon steel. Ferrite (white) and pearlite (dark). Lamellae of ferrite and cementite in pearlite grains are visible.

(d) 0.8 per cent carbon. (Eutectoid carbon) Tool steel. Thick layers of ferrite and thin layers of cementite, interfaces dark.

(e) 1.2 per cent carbon. Tool steel. A few small grains of proeutectoid cementite (white) in a pearlite matrix.

(f) 3 per cent carbon. White cast iron. Massive cementite grains (white) and pearlite (dark). Pearlite is formed from proeutectic (primary) austenite and eutectic austenite (4 per cent picral etchant).

Higher carbon steels will contain a greater proportion of pearlite (see Fig. 4.27(c)) and at the eutectoid composition, around 0.85 per cent carbon, only pearlite is present (d). The individual lamellae of ferrite and cementite can now be easily resolved. Hypereutectoid steels (carbon greater than the eutectoid composition) consists of cementite surrounding pearlite, see (*e*). The cementite is nucleated at the grain boundaries of the austenite and the austenite shortly changes to cementite and ferrite lamellae. The microstructure just above and below the eutectoid composition can appear the same, although in one case it is ferrite and in the other cementite around the pearlite. There is a natural tendency to regard cementite as a dark constituent (after etching with nitric acid in alcohol) since in low carbon steels, ferrite is white and ferrite and cementite in pearlite grains is dark. This is incorrect. The pearlite appears dark because of the multitude of unresolved lamellae boundaries. Steels with carbon content above about 0.4 per cent carbon are usually heat treated by rapid cooling from the austenite state to develop their optimum properties and different microstructures are developed from those discussed, as will be discussed in Chapter 12.

At very much higher carbon contents, in the range 2–4 per cent carbon, white cast iron is formed. On cooling from the liquid state austenite is nucleated (see the phase diagram, Fig. 4.25) and grows until the composition of the remaining liquid reaches the eutectic composition 4.3 per cent carbon (ledeburite) when the solidification occurs to austenite and cementite. Cooling of the solid causes the austenite, both in the primary and eutectic grains, to precipitate cementite until it reaches the eutectoid composition, 0.85 per cent carbon, when it undergoes a eutectoid reaction to pearlite. In the photomicrograph (Fig. 4.27(*f*)), the pearlite appears grey and all the cementite is white. White cast iron is very hard and brittle and has limited use. In the common forms of cast iron, for example grey iron, the cementite is caused to dissociate to graphite and ferrite, with improved properties. These will be discussed in § 12.3.

A summary of the terms used in connection with the iron–carbon (or iron–iron carbide) system is given in Table 4.1.

4.9 Ternary equilibrium diagrams

When three elements are combined, the equilibrium diagram requires three dimensions, two for composition and one for temperature, and its presentation on a two-dimensional page becomes rather difficult. The pressure is still usually fixed at one atmosphere. Before proceeding to details, it may be noted that small additions of further elements to a binary metallic alloy are frequently made, not to introduce new phases, but to control the kinetics of reactions and hence the distribution of the phases.

The most convenient axes for composition are in the form of an equilateral triangle, proposed by Gibbs. In Fig. 4.28 the elements (or compounds) A, B and C are represented by the apices and the composition of an alloy represented by point P has the composition

$$\frac{aB}{AB} \left(\text{or } \frac{a'C}{AC}\right) \text{ of } A; \qquad \frac{bC}{BC} \text{ of } B; \qquad \frac{cA}{CA} \text{ of } C.$$

That is, along any side, for example AB:

$$\%A = \frac{aB}{AB} \times 100; \qquad \%B = \frac{b'A}{AB} \times 100; \qquad \%C = \frac{ab'}{AB} \times 100.$$

The construction lines in the figure are drawn parallel to the sides of the triangle.

Table 4.1. *Summary of terms used in iron–carbon (iron–iron carbide) system*

α-iron ⎱ Ferrite ⎰	bcc form of iron existing at lower temperatures (e.g. room temperature). Strictly, α-iron is pure bcc iron, and ferrite is a solid solution, usually of carbon, in α-iron (C ≯ 0.05 per cent).
γ-iron ⎱ Austenite ⎰	fcc form of iron. Strictly, γ-iron is pure fcc iron and austenite is a solid solution, usually of carbon, in γ-iron (C ≯ 1.7 per cent).
δ-iron	bcc form of iron existing at high temperatures (1400 °C to m.p. at 1528 °C).
Iron carbide Fe_3C ⎱ Cementite ⎰	The intermediate compound formed between iron and carbon.
Ledeburite	Eutectic composition between Fe and FeC_3, i.e. alloy with lowest melting point (1130 °C). It solidifies at constant temperature to alternate layers of austenite (with 1.7 per cent C) and cementite. Important in cast irons.
Pearlite	The microstructure consisting of alternate lamellae of ferrite and cementite which is formed on cooling austenite through the eutectoid point (720 °C, 0.83% C).
Armco iron ⎱ Ingot iron ⎰	Alternative names for commercially pure iron (0.02–0.04 per cent C).
Mild steel	$Fe–Fe_3C$ alloys with ∼0.05 to 0.3 per cent C.
Medium carbon steel	$Fe–Fe_3C$ alloys with 0.3 to 0.5 per cent C.
Tool steel	$Fe–Fe_3C$ alloys with ≯ 0.5 per cent C, usually heat treated to increase hardness (see Table 12.2).
Hypoeutectoid steel	Carbon content less than eutectoid composition (∼0.85 per cent C).
Hypereutectoid steel	Carbon content above the eutectoid composition (∼0.85 per cent C).
Cast iron	Fe–C alloys with carbon content greater than 1.7 per cent C. Carbon may be in form of cementite or graphite flakes (see Chapter 12).
Upper critical temperature:	temperature above which only austenite phase is present for a particular composition.
Lower critical temperature:	temperature below which no austenite is present, only ferrite and cementite.
Critical temperature range:	temperature range over which austenite is formed on heating from ferrite and cementite.

It is a simple exercise in geometry to show that the same procedure can be used for a scalene triangle (i.e. non-equilateral). This is useful where compounds are formed between the primary components, dividing the phase diagram into two or more composition triangles (see below). Note that along a composition line drawn parallel to one of the sides of the triangle, say line XX in Fig. 4.28(*b*), the proportion of one element is constant, here A. Also, along a line through an apex, say AY in Fig. 4.28(*c*), the ratio of the other two elements is constant, here $B:C$.

The temperature axis is taken perpendicular to the composition plane and for each composition the temperature of various change points can be marked. The surfaces so formed divide the temperature–composition continuum into three-dimensional regions, or spaces, in which certain phases exist.

Only three examples will be considered here out of the many possible complex types. Following the procedure used for building up binary diagrams, Fig. 4.29 shows two basic space diagrams, for an isomorphous system and a eutectic system. In the former, liquid- and solid-solution phases are formed at all compositions; in the latter, the liquid phases are assumed to be soluble but the solid phases are insoluble.

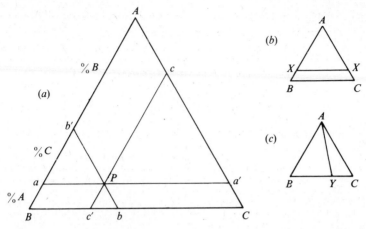

Fig. 4.28. Gibbs' composition triangle for ternary alloys. In (*b*), constant proportion of *A*; in (*c*), constant ratio *B*:*C*.

Consider the cooling from the liquid state of composition *P* in the isomorphous system (points 1 ... 5 in Fig. 4.29(*a*)). Initially (1) a single-phase liquid exists. On reaching the liquidus surface (2), nuclei are formed composed of a solid solution represented by S_2. At a slightly lower temperature (point 3, between the liquidus and solidus surfaces) the liquid phase is represented by L_3 and the solid phase by S_3. Clearly these points lie on the liquidus and solidus surfaces at the same temperature as point 3. To determine their composition an

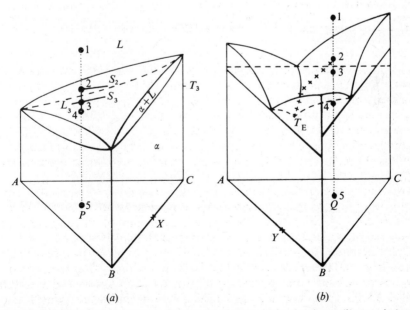

Fig. 4.29. Equilibrium space diagram for ternary systems: (*a*) isomorphous; (*b*) eutectic (× × × marks the locus of the liquid phase during crystallisation).

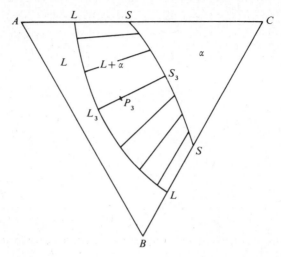

Fig. 4.30. Isotherm (horizontal section) at T_3 through the isomorphous phase diagram of Fig. 4.29(a).

isothermal section is required, called an *isotherm*, see Fig. 4.30. This isotherm, which is simple by comparison with some, consists of only three areas: the liquid phase, the solid phase and the conjugate liquid and solid phases. In the two-phase region, several tie-lines indicate the compositions of the various liquid phases (on *LL*) and the solid phases (on *SS*) which are conjugate at this temperature. For the composition represented by point P_3, the liquid phase is L_3 and the solid phase S_3. The lever rule can be used to determine the quantities of each phase. Note that tie-lines are not parallel and do not intersect the apex C, except by chance. Owing to lack of information, many isotherms do not indicate the tie-lines in two-phase regions and the conjugate phases can then only be guessed. (This is easiest near the boundaries to which the adjacent tie-line will be parallel.) Further cooling completes the solidification at the solidus surface (point 4) and the single-phase solid is then stable down to ambient temperature.

Turning to the ternary eutectic system (Fig. 4.29(b)), a series of isotherms at reducing temperatures are given in Fig. 4.31. For composition Q as marked, solidification begins at the intersection with the liquidus surface. The isotherm at this temperature (a) indicates that nuclei of C (or γ-phase) are formed. At a slightly lower temperature the liquid phase has composition L_b as a result of γ-crystals growing. It will be realised that where a pure component is conjugate with another phase of variable composition, no tie-lines are necessary – the liquid has the same proportions of A to B as the overall composition and must therefore lie on the line CQ_b extended. The temperature of the isotherm (b) is below the highest of the binary eutectic temperatures (here between B and C) so the diagram has a three-phase region. These are always triangular. If the composition fell in this area the phases present would be represented by the apices of the triangle, that is, $B(\beta)$, $C(\gamma)$ and liquid of composition X. The proportions of each phase are determined by an extension of the lever rule principle to two dimensions: the centre of mass of the three phases positioned at their respective apices must be at the overall composition point. Returning

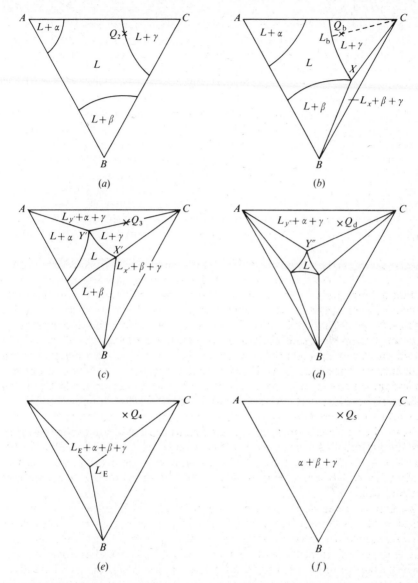

Fig. 4.31. Isotherms at progressively lower temperatures for the eutectic system shown in Fig. 4.29(b). (a) Intersection with liquidus for composition Q. (b) Below B–C eutectic temperature. (c) Below A–C eutectic temperature. (d) Below A–B eutectic temperature. (e) At ternary eutectic temperature. (f) At room temperature.

to composition Q, at a lower temperature the isotherm shows the point Q_d falling within a three-phase region of liquid Y'', α and γ. This means that a binary eutectic of α and γ is now being precipitated. The locus of the composition of the remaining liquid is along the valley formed between the two liquidus surfaces surrounding A and C (see Fig. 4.29(b)). At the ternary eutectic temperature the liquid composition has reached L_E (Fig. 4.31(e)) which is at the junction of the

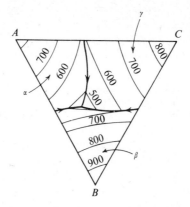

Fig. 4.32. Plan phase diagram for ternary system with no solid solubility (Fig. 4.29(*b*).)

three valleys starting from the three binary eutectic points. According to the phase rule (§ 4.2), with three components and the pressure fixed at one atmosphere

$$P + F = 4.$$

Thus with four phases present there are no degrees of freedom; that is, the compositions of the phases and the temperature are all determined. They are α, β, γ and the liquid composition L_E which now forms a ternary eutectic of $(\alpha + \beta + \gamma)$ at constant temperature. Solidification is now complete and no further changes occur down to room temperature, when the isotherm is as in Fig. 4.31(f). The microstructure contains primary γ, binary eutectic $\alpha + \gamma$, and finally tertiary eutectic $(\alpha + \beta + \gamma)$.

The information in the six isotherms of Fig. 4.31 can be summarised into (or, conversely, deduced from) a single diagram, as in Fig. 4.32. This will be called here a 'plan' equilibrium diagram and it shows in plan view the valleys or boundaries which separate the primary phase fields. It must not be confused with an isotherm since the boundary lines are not at a constant temperature. A primary phase is the phase that crystallises first on cooling a melt. Compositions on the boundary lines crystallise into two phases simultaneously and where they meet, at the ternary eutectic, three phases are formed. The isothermal lines indicate the temperature contours of the liquidus surfaces and by interpolation the temperatures of the eutectics and points along the boundary lines. The arrows marked on the latter show at a glance the directions of decreasing temperature and thus the crystallisation path.

Vertical sections through equilibrium space diagrams are called *isopleths*. They are useful for showing concisely the temperature intervals over which different phases exist for a range of compositions – usually with a constant proportion of one element (i.e. parallel to one side of the composition triangle) or with a constant ratio of two elements (i.e. a line through an apex). Two examples are given Fig. 4.33, taken through the ismorphous and eutectic systems of Fig. 4.29. It should be apparent from the preceding discussion that in multiphase regions it is not usually possible to determine their compositions from the isopleth because they will lie out of the plane of the diagram.

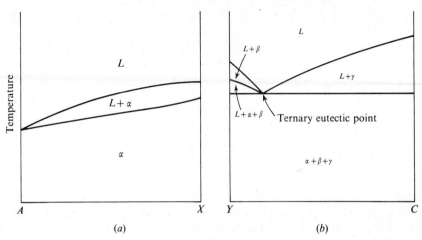

Fig. 4.33. Isopleths: vertical sections through (*a*) isomorphous system (Fig. 4.29(*a*)), and (*b*) eutectic system (Fig. 4.29(*b*)).

Thus the similarity of some isopleths to binary phase diagrams can be misleading.

As with binary diagrams, the equilibrium space diagram for a ternary system in which each component shows limited solid solubility is obtained by combining the characteristics of the isomorphous and the eutectic systems. A typical diagram is given in Fig. 4.34. It is left as an exercise to draw isotherms at the four temperatures marked.

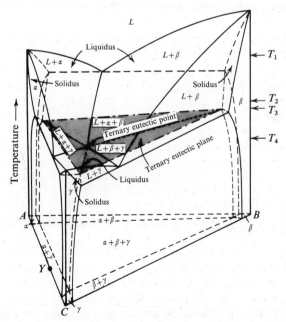

Fig. 4.34. Equilibrium space diagram for a ternary eutectic system with solid solubility. (A combination of Fig. 4.29(*a*) and (*b*)).

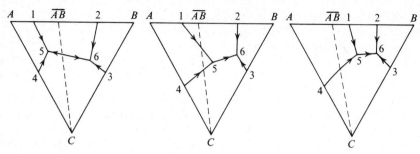

Fig. 4.35. Typical plan equilibrium diagrams for a ternary system with a binary compound $\overline{AB}$. No solid solutions are formed.

When a binary or tertiary compound is formed in a ternary system, its composition is marked on the phase diagram. Figure 4.35 shows several possible plan diagrams given a binary compound $\overline{AB}$ in a system with no solid solutions formed. The straight line joining $\overline{AB}$ to the component C is called a *composition*, or *Alkemade*, *line*. In (*a*), the compound melts congruently and the composition line $\overline{AB}$ to C does not cross the primary fields of the other components. Then both the points 5 and 6 are ternary eutectics, the intersect of the line C–$\overline{AB}$ with 5–6 is the highest point on 5–6, and the diagram is divided into two independent ternary systems. The line C–$\overline{AB}$ also forms a true binary system. If, however, the composition line C–$\overline{AB}$ crosses another primary field, as in (*b*), only the point 6 is a ternary eutectic. In (*c*) the compound melts incongruently by peritectic breakdown and the composition $\overline{AB}$ lies outside its own primary field. (An example is found in porcelain, $K_2O.Al_2O_3.SiO_2$ system, see § 7.3 and Fig. 7.4.)

QUESTIONS

(Components are to be assumed miscible in the liquid state.)
1. (*a*) Define *phase* and *phase diagram*.
   (*b*) Distinguish between *composition* and *constitution* of an alloy.
   (*c*) What is the 'lever rule'? Prove it.
   (*d*) What is the use of equilibrium diagrams?
2. Sketch the general form of the equilibrium diagram for a binary system in which the two elements
   (*a*) are mutually soluble in the solid state (isomorphous system),
   (*b*) are mutually insoluble over the whole composition range (eutectic system),
   (*c*) form an intermediate compound which is mutually insoluble with either element.
   Describe the changes during cooling from the liquid state of one composition in each system.
3. (*a*) Which compositions in the aluminium–silicon system (Fig. 4.5) have sharp melting points?
   (*b*) What is the constitution of 60% Al–40% Si at 800 °C and 500 °C. (Determine the proportions quantitatively.)
   (*c*) Over what temperature range are there two phases present for 95% Al–5% Si.
4. Outline briefly the technique of observing the microstructure of a metallic specimen.
5. What are the compositions of the phases and their proportions by weight:
   (*a*) 70% Cu–30% Ni at 1200 °C and 800 °C (Fig. 4.9),

(b) 45% Mg–55% Pb at 600 °C, 500 °C and 400 °C (Fig. 4.15),

(c) 90% Pb–10% Sn at 300 °C, 200 °C, 100 °C (Fig. 4.14)?

Sketch the microstructure at each stage.

6. Construct the equilibrium diagram for the silver–copper system from the following data: pure Ag and pure Cu freeze at 960 °C and 1080 °C respectively; 95% Ag–5% Cu has change points at 925 °C, 875 °C and 685 °C; 81.5% Ag–28.5% Cu freezes at 780 °C; 40% Ag–60% Cu has change points at 900 °C and 780 °C; 5% Ag–95% Cu has change points at 1065 °C, 1000 °C and 700 °C. Describe the changes during the eutectic reaction.

7. (a) What are the compositions of the intermediate phases in the aluminium–palladium system (Fig. 4.16)?

(b) Describe the changes in microstructure as each of them is heated from 0 °C to the finally liquid state.

8. (a) By means of an isotherm connecting three phases, demonstrate the connection between the following reactions: eutectic, eutectoid, peritectic and peritectoid.

(b) Identify as many of them as possible in the Fe–Fe$_3$C system (Fig. 4.25).

(c) Repeat for the Cu–Sn (bronze) system (Fig. 4.21).

9. Describe the changes on cooling 75% Sb–25% Sn from the liquid state (Fig. 4.20).

10. (a) What are the polymorphic crystal forms of iron and over what temperature range do they exist (Fig. 4.25)?

(b) What is the maximum amount of carbon which each can take into solution? What type of solid solution is formed?

(c) If it is not in solid solution, where is the carbon located?

11. (a) Describe the changes that occur as a mild steel with 0.25 per cent C is cooled from the austenitic state. Determine quantitatively the constitution at room temperature.

(b) Describe the changes in room temperature microstructure with increased carbon content of an Fe–Fe$_3$C alloy (0 to 3 per cent C).

(c) Which of these are phases: austenite, pearlite, cementite, ledeburite? What are the others?

12. Construct the titanium–uranium phase diagram from the following data. (Use atomic per cent scale.) α-U is stable up to 670 °C, β-U between 670 °C and 775 °C, and γ-U between 775 °C and its m.p. at 1140 °C. δ-Ti is stable up to 880 °C, γ-Ti between 880 °C and its m.p. at 1640 °C. The γ-U and γ-Ti phases are mutually soluble over the whole composition range. An intermediate compound U$_2$Ti changes congruently to γ-phase at 900 °C. Eutectoid reactions occur at 725 °C for 93at.% U–7at.% Ti and at 655 °C for 7at.% U–93at.% Ti. β-U forms a solid solution up to a maximum of 4at.% Ti.

13. Draw a hypothetical equilibrium space diagram for an isomorphous ternary system, together with three isotherms before, during and after solidification. Mark the phases present in each area. Describe the solidification of any composition.

14. Draw a hypothetical equilibrium space diagram for a eutectic ternary system (no solid solutions present). Draw the isotherm at some temperature between the ternary eutectic point and the lowest binary eutectic temperature and mark the phases present in each area. Describe the solidification of a representative composition.

15. Draw the plan phase diagram for a ternary system with no solid solutions from the following data. The components melt congruently at: A, 1050 °C; B, 980 °C; C, 900 °C. The eutectic compositions and corresponding temperatures are: 40% A 60% B, 900 °C; 50% B 50% C, 800 °C; 80% C 20% A, 850 °C; 15% A 25% B 60% C, 750 °C. Describe the solidification of 55% A 25% B 20% C, giving approximate temperatures of the change points.

16. Using Fig. 4.34, draw the isotherms at T_1, T_2, T_3 and T_4, as marked.

17. When a binary compound BC is formed in a ternary eutectic system (A–B–C), in what circumstances will the isopleth through A and BC represent a binary system?

FURTHER READING

G. L. Kehl: *The Principles of Metallographic Laboratory Practice.* McGraw-Hill (1949).

W. Hume-Rothery *et al.*: *Metallurgical Equilibrium Diagrams.* Institute of Physics, London (1952).

M. C. Smith: *Alloy Series in Physical Metallurgy.* Harper (1956).

F. N. Rhines: *Phase Diagrams in Metallurgy.* McGraw-Hill (1956).

M. Hansen and K. Anderko: *Constitution of Binary Alloys.* McGraw-Hill (1958).

R. P. Elliott: *Constitution of Binary Alloys*, First supplement. McGraw-Hill (1965).

W. D. Kingery: *Introduction to Ceramics.* Wiley (1960).

J. Nutting and R. G. Baker: *The Microstructure of Metals.* Institute of Metals, London (1965).

E. M. Levin, C. R. Robbins and H. F. McMurdie: *Phase Diagrams for Ceramicists.* The American Ceramic Society (1964).

R. H. Greaves and H. Wrighton: *Practical Microscopical Metallography*, 4th edition. Chapman and Hall (1967).

R. M. Brick, R. B. Gordon and A. Phillips: *Structure and Properties of Alloys*, 3rd edition. McGraw-Hill (1965).

C. J. Smithells: *Metals Reference Book*, vols. 1 and 2. [Vol. 1 covers metallography, Vol. 2 phase diagrams.] Butterworth (1967).

5. Mechanical properties

5.1 Introduction

The mechanical properties of a material relate to its response when it is loaded or deformed, that is, when it is subjected to stress (load/cross-sectional area) or strain (extension/initial length). The load may be constant in magnitude or fluctuating continuously, and be applied for a fraction of a second or many years. The environment may be air at high, ambient, or low temperature; or it may be one of a great number of possible fluids, many of which are corrosive. The response of a material to such loading conditions will be elastic strain, or plastic strain (not recovered on unloading) or fracture. These may develop immediately on applying the stress, or over a period of time.

Structural materials, that is, those materials able to support loads without appreciable deformation or fracture, have been utilized by man throughout the ages: examples include stone axes, wooden dug-out canoes, reinforced concrete bridges and aluminium alloy airframes. At the same time higher loads, or similar loads applied at higher temperature, can often be used to shape materials, particularly metals and thermoplastics, to desired shapes and sizes, for example by forging, rolling or extrusion.

The three successive stages in the response of materials to increasing stress are: elastic strain; plastic strain; and fracture. Elastic strain is, by definition, recovered on unloading. It is due to the stretching of the atomic bonds and the change of internal energy (see § 3.1). Maximum elastic strains in metals are around 0.1 per cent before plastic straining occurs, and in most other materials around 0.1 per cent before fracture in tension. Certain high polymers, called elastomers or rubbers, show a further type of elastic straining due to rotation of the covalent bonds linking the atoms into chains. This is called high, or entropy, elasticity (see Chapter 8), and the maximum elastic strains may be as high as 500 per cent.

Plastic strain is defined as strain not recovered on unloading. It is mainly due to the sliding of planes of atoms over each other; the process occurs inhomogeneously by means of lattice faults called dislocations, already noted in Chapter 2. Dislocation theory, which is central to the study of mechanical properties of materials, will be discussed in detail in Chapter 9. The maximum plastic strain varies greatly from one class of material to another due to their different atomic bindings and the ease or difficulty with which dislocations are able to move in the crystals. Metals are noted for their ability to deform plastically, up to 100 per cent local strain in tension and higher under some combined stresses (i.e. stresses along three axes). Ceramics have little or no plastic range in the polycrystalline state. With glasses and polymers, the plastic strain is dependent on the temperature and loading rate; under normal conditions they show negligible plasticity (see Chapters 7 and 8).

Finally, at fracture the atomic planes are separated and new surfaces are formed. The fracture stage will be discussed in detail in Chapter 10.

Table 5.1. *Average mechanical properties of typical metals, ceramics, glasses,*

Material	Melting point (°C)	Specific gravity	Young's modulus E (GN/m²)*	Yield stress (0.1% set) (MN/m²)*
Metals				
Iron whisker				(10 000)
Zone refined iron (20 ppm C)	1527	7.87	215	35
Electrolytic iron (0.006% C)				55
Armco iron (0.04% C)				140
Mild steel (0.15% C)	1510			200
Structural alloy steel { medium strength				600
{ high strength				1200
Austformed alloy steel				1500–2000
Carbon tool steel (quenched)				
Cast iron (2–4% C) { low strength	1150		117	115
Cast iron, pearlitic grey { high strength			145	250
Cast iron modular { normalised	to		178	450
{ heat treated	1350		168	600
Aluminium { annealed	660	2.7	72	35
(99.99%) { hard rolled				
Aluminium alloy { solution treated				275
(4.5% Cu–MnMgSi) { precipitation hardened				400
Titanium alloy (Ti–4% Al–4% Mn)	~1700	4.5	110	800
Beryllium	1285	1.8	310	400
Ceramics				
Concrete		2.4	15–35	
Building brick (clay)			10–17	
Fire brick 25% porosity				
Alumina (99.5% Al₂O₃) <0.5% porosity	2000	3.98	400	
1% porosity		3.93		
5% porosity				
Tungsten carbide −6% cobalt cermet	1400	14	600	
Glass ceramic		2.6	80–140	
Glasses	*Softening point*			
Silica (fused quartz: >99.5% SiO₂)	1600	2.2	72.5	
Soda lime { 70% SiO₂ 15% Na₂O 10% CaO	735	2.46	69	
{ thermally toughened				
{ chemically toughened				
Fibre E-glass { freshly drawn		2.5	72.3	
{ as reinforcement				
High Polymers			*Initial modulus*	
Wood { white pine (air dried 12% H₂O)		0.36	8.8	40
{ white oak (air dried 12% H₂O)		0.67	12.2	55
Nylon 66 { dry as moulded		1.1	2.8	80
{ at equilibrium in 50% RH (2.5% H₂O)			1.8	60
Polycarbonate	230	0.9	2.3	60
Elastomers (rubbers)		1–1.4	$1–7 \times 10^{-3}$	
Composites				
70% glass fibre } epoxy plastics matrix		2.2	40	
70% carbon fibre }		2	500	

The ability of materials to deform plastically before fracturing is invaluable in structural applications, although stresses are nominally in the elastic range. The use of brittle materials is not completely ruled out: for example, many large cast iron bridges were built in the last century and still stand. However, in ductile materials the safety factor (e.g. ratio of yield stress/allowable working stress) can be very much lower because plastic flow prevents local higher stresses

high polymers and composites

Tensile strength TS (MN/m²)*	Elongation to fracture ($L_0 = 5d$) (per cent)	Indentation hardness DPN (Kg/mm²)	Remarks
10 000	<0.1		Contains few 'dislocations' and approaches theoretical strength $E/10$
100			
140			
280	40		
400	25	120	
800	20	265	3 % Ni, 0.3 % C. Quench 830, temper 600 °C (En21)
1300	15	360	$2\frac{1}{2}$ % Ni–Cr–Mo, 0.4 % C. Q830 °C, T600 °C(En26)
2–3000	10		N.B. Very high strength with ductility
	<0.1	900	Q750 °C T150 °C
175	0.5	175	Grey cast irons have compressive strengths 3 or 4 times the TS
350	0.5	250	
900	6	295	
800	6.5	260	
50	50	20	
100	5	40	
430	15	135	
460	8	180	
900	15		
550	1.5		High strength/weight ratio but little ductility

Modulus of rupture	*Compressive strength* MN/m^2	*Knoop*	
2.5	30–40		
2.10	15–40		
0.3–5	20–40		
700		~2000	Translucent oxide
600	3100	~1600	Cutting tool material (3 μm grain size)
			Typical high grade alumina
1600	4300	~1500	Standard cutting tool
100–600		~620	≤1 μm grain size. Zero porosity

Tensile strength		*DPN*	
100		725	Median strength values for glasses – wide scatter.
70		560	Standard glass for windows, bottles etc.
210			{ Residual compressive surface stresses
390			{ suppress weakening effect of cracks
3800			55 % SiO_2, 22 % CaO, 8.5 % B_2O_3, 14.5 % Al_2O_3
2000			

100	30		Values parallel to grain low compressive strength due to buckling of fibres
190	45		
80			
77			
65			Sharp yield point
7–10			

1200			
7000			Very high stiffness (E) and strength

* 1 MN/m² = 145 psi.

developing at internal faults, holes, changes of cross-section, and surface irregularities. These stress concentrators, if unrelieved, are capable of initiating a crack although the average stress is within the elastic range. If a crack should form, the high energy absorption by plastic deformation around the front of the advancing crack prevents it growing rapidly into a catastrophic fracture (see Chapter 10).

Thus, as regards mechanical properties, materials divide sharply into two groups, *ductile* and *brittle*: ductile materials, notably metals, deform plastically before fracturing; and brittle materials, notably ceramics, glasses and high polymers, fracture with little prior plastic flow. As will be shown subsequently (Chapter 11) the problems in improving the strength of these groups are quite separate. Ductile materials need the movement of dislocations to be restrained so that the stress for the onset of plastic flow is raised closer to the fracture stress. Brittle materials need to have some degree of plastic flow introduced to prevent stress concentrators causing sudden fracture at low nominal stresses; alternatively they may be used in composite materials.

This chapter discusses the phenomenological behaviour of materials, mainly ductile metals, as measured in a number of standard laboratory tests. These have been designed to reproduce the possible loading conditions in structures. They include tensile, compression, hardness, notched bar impact, fatigue and creep tests. Ideally, only one test would be required to determine all the mechanical properties, but this has not so far proved possible. The first three tests to be discussed, tensile, compression and hardness, are closely related, since they measure the same basic physical property, the resistance of materials to gross plastic deformation. However, the remaining tests are quite independent and their results are not predictable from one another.

Engineers use the test results in the design of structures. For specified loads, the stresses everywhere are determined by experimental or theoretical stress analysis and compared with the allowable stress, which will be based on the test results for the particular material to be used. On the other hand, metallurgists are producing materials to meet the requirements of the standard tests, and they are concerned with the relationships between microstructure and the test results. Thus mechanical testing stands on the dividing line between materials and engineering. It may be argued that the engineer needs no knowledge of the microstructure of materials and that the metallurgist or materials scientist can be ignorant of structural applications, but in practice an understanding of the other's problems and language is necessary.

Typical values of the parameters obtained in the standard tests of some representative materials are given in Table 5.1.

5.2 Tension

The parameters obtained in a tensile test are the commonest measures of mechanical properties and they are used in determining the allowable stress in many engineering designs. The principle of the test is simple: a test piece, such as a length of round bar, is held vertically by the top end and weights are attached to the bottom end, measurements of the length being taken as the load is increased up to fracture. The test is made at room temperature over a period of about five minutes. In practice, a fairly costly testing machine may be required which can apply loads of many tons. The test piece is prepared with a central portion of reduced cross-section so that deformation and fracture occurs there rather than in the grips, which will be designed to apply the load accurately along the axis of the test piece. Finally, some sophisticated methods of measuring the length may be employed, such as rotating mirror (Martens) or electrical

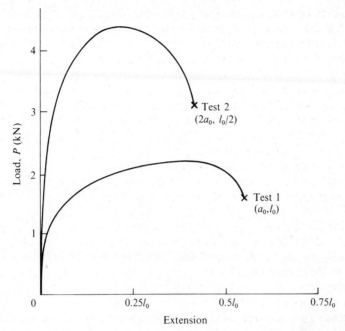

Fig. 5.1. Load versus extension in a tensile test. Two specimens of different geometry, one ductile material.

resistance strain gauges. The experimental details will not be considered further here, only the test results and their significance.

Typical tensile test results for polycrystalline metal are shown in Figs. 5.1 and 5.2. In Fig. 5.1 the load is plotted against the extension and the curve is dependent on the specimen size and shape, but by plotting nominal[†] stress (load/initial cross-sectional area) versus linear[†] strain (extension/initial length) a single curve is obtained for a given material up to the maximum stress, see Fig. 5.2. Thereafter the curve is again dependent on specimen geometry because of non-uniform extension.

There are several stages in the test. Initially the stress σ is proportional to the strain ε, called Hooke's Law (1678):

$$\sigma = E\varepsilon, \tag{5.1}$$

where E is Young's modulus.

The value of E for metals is between 35 and 350 GN/m² (5–50 × 10⁶ psi). The corresponding values are higher for ceramics, up to 1000 GN/m² (140 × 10⁶ psi) in diamond, and lower for high polymers, around 3.5 GN/m² (0.5 × 10⁶ psi). These differences can be traced back to the different types of atomic binding (§ 3.1). A high value of Young's modulus can be important: in gas-lubricated bearings even small elastic strains can upset the design clearances; in very high strength materials, such as glass fibre resins, significant distortions of a structure can take place within the working load (see § 11.12); and, of general importance,

---

[†] The terms 'nominal' and 'linear' are usually omitted, especially in an engineering context.

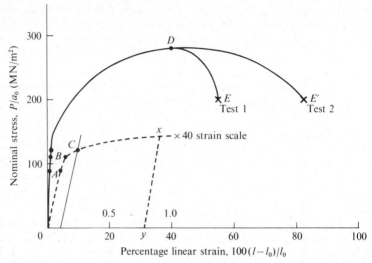

Fig. 5.2. Nominal stress versus linear strain. Replot of data in Fig. 5.1. A = limit of proportionality, B = elastic limit, C = 0.1 per cent proof stress (or yield stress), D = tensile strength, E = fracture.

the buckling capacity of compression members (struts) is proportional to the Young's modulus.

The axial extension is accompanied by lateral contraction: the lateral strain is $v\sigma/E$, where v is called Poisson's ratio. Values of v are around 0.3. In the plastic range, v is 0.5, which means there is no volumetric change.

Hooke's Law is obeyed up to some point A called the limit of proportionality. Another point which is sometimes noted is the elastic limit, B, beyond which plastic strain begins. These points can usually be treated as coincident. In fact, both of them, and the whole concept of proportional and elastic ranges, are mathematical idealisations of the behaviour of real materials, adequate for many purposes. Small deviations from elastic behaviour, such as anelasticity and microplasticity, will be discussed in later sections of this chapter.

For the purpose of identifying a stress at which gross plastic flow commences, a method is required which is independent of the sensitivity of the strain gauge employed and the judgement of the experimenter. It is usual to define a *proof stress*, loosely called the yield stress, at which a finite amount of plastic strain has occurred, normally 0.1 per cent. The 0.1 per cent proof stress is determined by drawing a line parallel to the tangent to the stress–strain curve at the origin but off-set by 0.001 strain; the intercept between this line and the yield curve, point C in Fig. 5.2, gives the proof stress. At the proof stress the plastic strain is of the same order of magnitude as the elastic strain (10^{-3}) and above this stress level departures from the ideal elastic behaviour rapidly become very significant.

The yield stress is a structure-sensitive property, that is, it depends on the details of the microstructure such as the grain size, and the arrangement of any second phase. Typical values range from about 35 MN/m^2 (5000 psi) for pure polycrystalline aluminium to 1.8 GN/m^2 (2.5×10^5 psi) for very high strength steel alloys. In brittle materials the yield stress is, by definition of brittle, very close to the fracture stress, up to 10 GN/m^2 ($\sim 1.5 \times 10^6$ psi).

In steels and a few other metals there is a sharp yield point at the onset of major plastic flow and some one or two per cent strain occurs at a constant stress, below that required to initiate the yield (Fig. 5.8). Yielding here is easily determined. It will be discussed further in § 5.5.

In the plastic range, the yield stress increases with plastic strain, an important effect, called *work*, or *strain*, *hardening*. If the load is removed at some point, say *x* in Fig. 5.2, the strain approximately follows the straight line *xy*, parallel to *OA*, and the *set* is *Oy*. On reloading, the material is elastic up to *x* and then yields along the original strain hardening curve *CDE*. The yield stress can now be very much greater than the primitive yield stress of the annealed material. The work done on the material, equal to the area under the yield curve, is mainly converted to heat. A fraction is stored internally, some 15 per cent at low strains dropping to 5 per cent at high strains. Strain hardening may be useful as a method of strengthening (see Chapter 11), or be troublesome, as in metal-forming processes such as wire drawing or rolling, where it causes the loads and power to increase. It can be removed by heating above the material's recrystallisation temperature, when the crystal grains are reformed in the annealed state (see § 11.3).

The nominal stress rises to a maximum value at *D*, called the *tensile strength* (TS). This was formerly called the ultimate tensile strength (UTS) but this term is now obsolete since it could mistakenly be taken to indicate the fracture stress. At *D* a tensile specimen becomes unstable in that any section which extends, and hence contracts in cross-sectional area, becomes the weak link in which all further deformation is concentrated, a process called *necking* (Fig. 5.3). The condition for necking in terms of the balance between strain hardening and the geometric factor is derived in the next section. Although lacking any physical importance the TS is the usual measure quoted for a material's strength. Typical

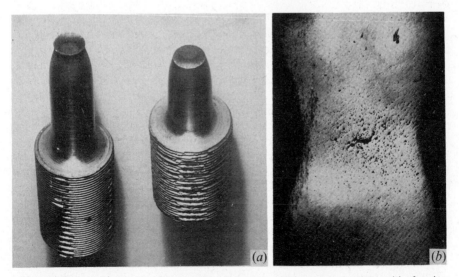

(a)   *(b)*

Fig. 5.3. Cup and cone fracture of ductile metal: (a) mild steel; (b) copper, showing cavities forming at inclusions (× 6). (From K. E. Puttick, *Phil. Mag. Ser. 8*, **4**, 946 (1959).)

values for ductile materials range from 50 MN/m^2 (7000 psi) for soft aluminium to 3 GN/m^2 (4.5 × 10^5 psi) for high strength steel alloys (Table 5.1).

Once necking has started, the extension of a specimen is independent of the gauge length and the stress–strain curve dependent on the specimen geometry (curves (i) and (ii) in Fig. 5.2). Barba's similarity law states that the shape of the neck is independent of the specimen diameter. The extension of the neck region is therefore proportional to the diameter and the stress–strain curves will only coincide if the ratio of the gauge length (l_0) to the diameter (d) is fixed. It has recently been agreed internationally to use the relationship $l_0 = 5.65\sqrt{a_0}$, which for round sections gives $l_0 = 5d$. (Formerly, in Britain $l_0 = 4\sqrt{a_0}$, and in USA $l_0 = 4.5\sqrt{a_0}$.)

At E, fracture occurs. A crack forms in the centre of the neck and extends slowly along a zig-zag path in a plane perpendicular to the axis, finishing by shearing at 45° to the axis to form a typical cup and cone fracture, see Fig. 5.3. Small inclusions have recently been shown to be important in this stage; cavities form around them and eventually coalesce. The final reduction of area (at the minimum section) and percentage elongation (of the gauge length) are measured by putting the two parts together after fracture.

The actual stress and strain in a neck progressively differ from the nominal stress and overall linear strain. Although one of the oldest problems in plasticity, there is still no agreed solution. Initially a reasonable approach is to use the 'true' stress, defined as the load/actual area at neck. The local strain is obtained from the area at the neck; since $la = $ constant ($v = 0.5$ in plastic straining)

$$\varepsilon = \frac{l - l_0}{l_0} = \frac{a_0 - a}{a}.$$

The resultant curve of linear strain versus true stress can often be represented by a simple power function:

$$\sigma_t = b\varepsilon^x \tag{5.2}$$

where b and x are constants for the material. The value of b is very dependent on the structure but x, called the *strain-hardening index*, is fairly constant within one type of material. Values of the strain-hardening index vary between around 0.2 for steels to 0.5 for copper and brass (some values appear in Fig. 5.16). The strain hardening is thus continuously positive until fracture intervenes and does not, as the conventional curve suggests, go into reverse at high strains. The curve also under-emphasises the extent of the plastic strain taking place prior to fracture; local linear strains of 100–300 per cent, corresponding to reduction of areas of 50–75 per cent, appear reduced by a factor of ten because the local extension in the neck is averaged over the full gauge length. More exact estimates of the stress distribution in a neck will be given in § 5.4.

The fracture load is not usually measured in the tensile test of ductile material, only the maximum load, and the fracture stress is thus unknown. Values range from 140 MN/m^2 (20 000 psi) for pure aluminium (cf. 50 MN/m^2 TS) to 3.5 GN/m^2 (5 × 10^5 psi) for drawn steel wire. Higher values, near the theoretical value $E/10$ (see Chapter 10), occur in fibres and whiskers.

The stress–strain curve considered so far is typical of *ductile* materials, that is, those that deform plastically before fracturing. As discussed in § 5.1, some useful structural materials such as cast iron, bricks, granite and concrete are *brittle*, that is, they are almost elastic up to fracture. A convenient dividing line is to call materials ductile if they neck before fracturing, and hence their tensile strength is a measure of the plastic flow strength; and brittle if they fracture without any neck forming. Whenever possible brittle materials are used in compression as their fracture strength is an order of magnitude higher than in tension. This is in contrast with the yield or plastic strength of ductile materials, which is independent of the sign of the applied stress (see § 5.6).

The term *ductility* refers in a general way to the ability of a material to deform plastically, usually in tension, without fracturing. Two measures of it often quoted are the elongation and reduction of area to fracture. However, the suitability of a material to undergo the various metal-forming operations, such as cupping, drawing, and pressing, is not simply related to these parameters. As already noted (§ 5.1) ductility is also considered a desirable property of a material to be used in an engineering structure, for which the stress level is predominantly in the elastic range, since it permits regions of unavoidable stress concentration to flow plastically rather than fracture. A quantitative figure for the necessary ductility has not yet been established, possibly because this may not be the real parameter which determines whether a structure will fracture at a point of stress concentration. The main use of the figures for reduction of area and elongation is in the quality control of production processes.

The time factor has not yet been mentioned. In the later stages of a tensile test the strain becomes noticeably dependent on the time. The strain rate is high after applying a load, and then slows down to a constant value. The phenomenon is called *creep*. With sensitive strain gauges creep can be detected at much earlier stages and, if the test temperature is high relative to the material's melting point, it may be significant at stresses below the yield stress. Creep is discussed in § 5.11. It is not important in the usual tensile test, which is carried out at room temperature over a period of about five minutes, with a fairly constant rate of loading or straining.

In recent years interest has revived in the phenomenon of *superplasticity* which is found in some metallic alloys specially prepared with a very fine grain size (~ 1 μm, cf. 50–500 μm). The effect was observed in 1920 by Rosenhain *et al.* and studied by Pearson in 1934. In these materials the plastic flow in a tensile test is quasi-viscous rather than strain hardening and the yield stress is a power function of the strain rate

$$\sigma_t = \eta \dot{\varepsilon}^m \cdot$$

The index m lies between 0.4 and 0.9. Large uniform plastic straining in tension, up to 2000 per cent, can now be produced because the tendency to form a neck is counteracted by the higher local strain rate and, hence, higher yield stress. This is shown mathematically in § 7.9 in connection with the drawing of glass, which at elevated temperatures exhibits Newtonian viscosity, that is $m = 1$. Confusion may arise between creep and superplasticity since both are strain rate effects. Creep is mainly a low strain rate phenomenon (10^{-6} to 10^{-12} s^{-1}) and occurs in conventional metals which are primarily strain hardening in a tension test; superplasticity is a high strain rate phenomenon (10^{-1} to 10^{-4} s^{-1}), typical of tensile testing rates, and occurs in only a few alloys. There appear to be no microstructural changes during superplastic flow in contrast with normal plasticity (see Chapter 9). The possibility of commercial application of superplasticity is now being explored, but the subject will not be considered further here.

Many points have been discussed in this section which would not be considered in a routine tensile test (see BS18 for details). In a few minutes a specimen is put into a test machine and loaded up to fracture, with readings taken of the yield point or proof stress, TS, reduction of area, and elongation. If these exceed the minimum values specified the material is passed as satisfactory.

5.3 Calculation of the TS of a ductile material

It has been explained in the preceding section how the tensile strength (TS) of a ductile material is dependent on the plastic, rather than the fracture, properties. If the strain-hardening curve is given graphically or analytically, the TS can be deduced, as shown by Considère (1885).

In a tensile specimen, the load P at any instant is given by

$$P = \sigma_t a,$$

where σ_t is the true stress and a is the instantaneous cross-sectional area. As the area is reduced by straining, the yield stress increases due to strain hardening. The maximum load is obtained by differentiation:

$$\frac{dP}{da} = a\frac{d\sigma_t}{da} + \sigma_t = 0$$

i.e.
$$\frac{d\sigma_t}{\sigma_t} = -\frac{da}{a}. \tag{5.3}$$

This equation is saying that at the maximum load the fractional increase of the yield stress due to strain hardening is just balancing the fractional decrease of cross-sectional area. Beyond this point, necking commences and the load decreases.

It has been found experimentally that plastic flow is at constant volume:

$$al = \text{const.}$$

Therefore
$$\frac{da}{a} = -\frac{dl}{l} = -\frac{dl/l_0}{1 + (l - l_0)/l_0} = -\frac{d\varepsilon}{1 + \varepsilon},$$

where l_0 is the initial gauge length, l the instantaneous length and ε the linear strain. (This equation only applies as long as the extension is uniform.)

Substituting in (5.3) gives

$$\frac{d\sigma_t}{d\varepsilon} = \frac{\sigma_t'}{1 + \varepsilon'}, \tag{5.4}$$

where the prime indicate values at the maximum load point.

If the strain-hardening law is given as a curve of true stress versus linear strain, as in Fig. 5.4, the point on it which satisfies (5.4) is found by drawing a tangent to the curve from $\varepsilon = -1$. The tensile strength is then

$$\text{TS} = \sigma_t'\frac{a'}{a_0} = \frac{\sigma_t'}{1 + \varepsilon'}.$$

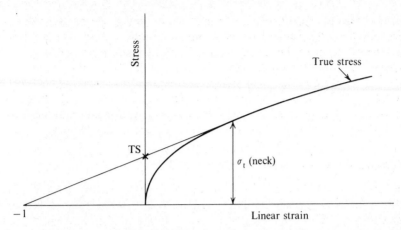

Fig. 5.4. Considère's construction for determining the tensile strength of a ductile material.

By similar triangles, the tensile strength is given by the intercept of the tangent on the stress axis.

When the strain-hardening curve is represented analytically by (5.2):

$$\sigma_t = b\varepsilon^x$$

where b and x are constants of the material, a simple analysis shows that the tensile strength is:

$$\text{TS} = bx^x(1 - x)^{1-x}. \tag{5.5}$$

This equation is of use in the correlation of tensile strength and hardness, which will be described in § 5.7. For a group of materials, such as steels with varying carbon content, or brasses with varying zinc content, the value of the strain-hardening index x is fairly constant and variation of the tensile strength is due to changes in the parameter b. The value of b can be quickly established from a hardness test, which is simple and non-destructive, and a good estimate of the tensile strength obtained by means of (5.5).

5.4 Stress system in a necked tensile specimen

Although the 'true' stress (load/actual area) is a good approximation for the axial stress in the neck of a tensile test piece, more accurate estimates of the stress system have been made by Nadai and others. According to Bridgman it can be analysed into two components, a uniform axial stress $\sigma_z(a)$ and a hydrostatic tensile stress (i.e. equal stress along three axes) $\sigma_h(r)$, which increases from zero at the surface to a maximum on the axis, see Fig. 5.5.

At radius r, the hydrostatic stress component is

$$\sigma_h(r) = \sigma_z(a) \ln \left[1 + \frac{1}{2}\frac{a}{R} - \frac{1}{2}\frac{a}{R}\left(\frac{r}{a}\right)^2 \right], \tag{5.6}$$

where a is the radius of the neck cross-section and R is the radius of curvature at the neck in a plane containing the specimen axis. The function a/R must be measured, or use can be made of an empirical relationship between it and the

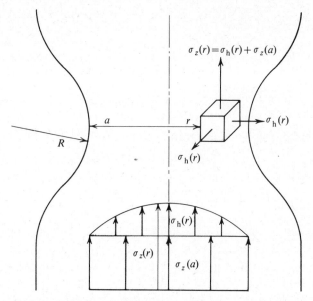

Fig. 5.5. Stress system in neck of tensile test piece, according to Bridgman.

reduction of area at the neck, which is more easily measured. This is given in Fig. 5.6. It was found by Bridgman to apply to many steels and non-ferrous metals such as brass and bronze, although the considerable scatter of the points will be noted.

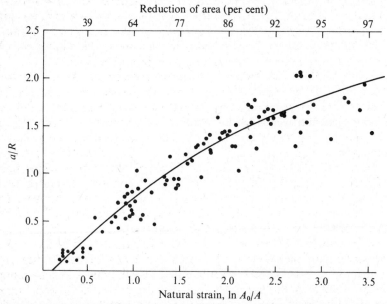

Fig. 5.6. Empirical relationship between neck curvature R and the reduction of area. (Based on steels, bronze, and brass.) (From P. W. Bridgman, *Large Plastic Flow and Fracture*, McGraw-Hill, 1952.)

The average stress over the neck cross-section, that is, the true stress as defined above, is

$$\sigma_t = \overline{\sigma_z(r)} = \sigma_z(a) + \overline{\sigma_h(r)},$$

where $\overline{\sigma_h(r)}$ is the average hydrostatic stress component, given by

$$\overline{\sigma_h(r)} = \frac{1}{\pi a^2} \int_0^a \sigma_h(r) 2\pi r \, dr.$$

Taking $\sigma_h(r)$ from (5.6) and integrating by means of the substitution

$$y = 1 + \frac{a}{2R} - \frac{a}{2R}\left(\frac{r}{a}\right)^2$$

and

$$dy = \frac{r}{Ra} dr$$

gives

$$\overline{\sigma_h(r)} = \sigma_z(a)\left[\left(1 + \frac{2R}{a}\right) \ln\left(1 + \frac{a}{2R}\right) - 1\right].$$

Hence

$$\sigma_t = \sigma_z(a)\left(1 + \frac{2R}{a}\right) \ln\left(1 + \frac{a}{2R}\right). \tag{5.7}$$

In Fig. 5.7 are plotted, relative to the true stress, the uniform component of axial stress, the maximum hydrostatic component (at $r = 0$) and their sum, the maximum axial stress on the axis. It will be seen that with increasing necking the hydrostatic component and the maximum axial stress on the axis increase. However at 100 per cent natural strain, equivalent to 64 per cent reduction of area, typical of the more ductile engineering materials, the maximum hydrostatic stress is under one third the true stress and the maximum axial stress is only 15 per cent above the true stress. The figures for greater strains were obtained by subjecting the test piece to a large hydrostatic pressure which suppresses the tendency to fracture without affecting the plastic flow behaviour.

A comparison of the experimental and theoretical curves of the stress system in the neck of a steel specimen (0.25 per cent carbon) is given in Fig. 5.8. The experimental results were obtained by Parker, Davis and Flanigan, in 1946, by straining the specimen (57.2 mm dia.) until a neck had developed, and un-loading. Strain gauges were stuck on the narrowest section and the relaxation strains determined as the centre of the bar was bored out. From these the stress system under load was deduced. The curves given, which relate to the point of rupture at a true stress of 900 MN/m² (130 × 10³ psi) and a natural strain of 107 per cent, were obtained by extrapolation of results at lower stresses. The theoretical curves have been obtained using the preceding Bridgman theory. The parameter a/R was taken from Fig. 5.6, given the natural strain 1.07, and knowing σ_t, the uniform component of axial stress $\sigma_z(a)$ was obtained from (5.7). The hydrostatic component followed from (5.6). It will be seen that the measured radial and hoop stresses were not quite equal and the stress system cannot therefore be treated as the sum of an axial stress and a hydrostatic stress. However, the calculated $\sigma_h(r)$ is in rough agreement with the hoop and radial stresses. Similarly the measured total axial stress σ is only roughly approximated by the Bridgman theoretical curve $\sigma_z(r)$. Since the experimental curves

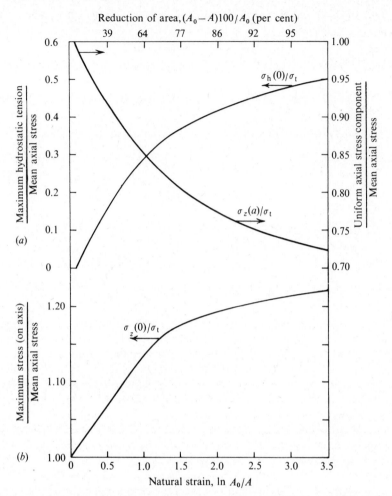

Fig. 5.7. Stress system in neck relative to the mean or 'true' stress. (*a*) Maximum hydrostatic stress component (at $r = 0$) and uniform axial stress component against natural strain in neck ($\ln A_0/A$). (*b*) Maximum stress (on axis) against natural strain.

were not obtained directly but involve assumptions and calculations it is not easy to say exactly what is the true stress system.

5.5 Yield point phenomenon and strain ageing

In low carbon steels, and some other materials, there is a discontinuity at the onset of plastic flow which is called a yield point phenomenon.

The effect on the tensile stress–strain curve of mild steel is shown in Fig. 5.9. The stress rises initially according to Hooke's Law until, at around 0.1 to 0.2 per cent strain, it drops sharply and remains constant for a strain of about 2 per cent. The curve then rises again and continues in the normal manner for a ductile material. The limit of the initial elastic range, or start of major yielding,

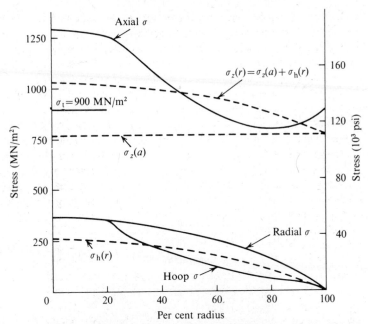

Fig. 5.8. Comparison of experimental and theoretical stress systems in a neck : —— experimental results by Parker, Davis and Flanigan on 0.25 per cent carbon steel ; – – – theoretical curves obtained using Bridgman theory.

is called the upper yield point and occurs at the upper yield stress. The horizontal portion of the curve occurs at the lower yield stress, and is called the Lüders strain.

The upper and lower yield stresses are independent of the sensitivity of the strain gauge (cf. the yield stress of most materials) but the true value of the upper

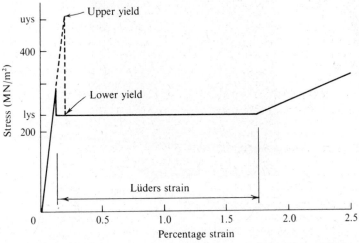

Fig. 5.9. Yield point phenomenon on mild steel. Dashed curve is only obtained if great care is taken to avoid stress concentrators.

yield stress, which may be twice the lower one, is easily missed if the specimen is not uniformly stressed. Surface scratches or rough gripping will cause premature local yielding, which can spread across the specimen at only slightly above the lower yield stress, and this is what happens in a conventional tensile test. As a result, it is common to refer to the lower yield stress as the yield stress, and, owing to the universal use of mild steel, it is often not realised that a sharp yield point phenomenon is confined to a few materials.

If a bright specimen is used, one or more surface marks are seen to develop at the moment the stress passes the upper yield point. These are called Lüders bands (Hartmann lines or stretcher-strain marks) and are parts of the specimen which have undergone plastic deformation to the full extent of the Lüders strain, that is, the horizontal portion of the yield curve. At this stage the specimen is

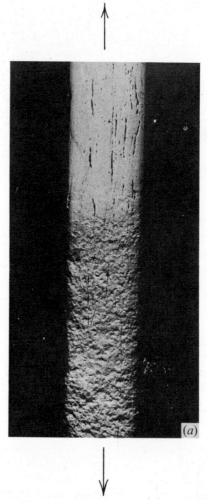

Fig. 5.10(*a*). Lüders bands: Band front in a coarse grained mild steel tensile specimen; the rough material has undergone the Lüders strain (from E. O. Hall, *J.I.S.I.* **170**, 331 (1952)).

Fig. 5.10(*b*). Several Lüders bands in the base of a pressed steel dish (from E. O. Hall, *Yield Point Phenomena in Metals and Alloys*, Macmillan 1970).

inhomogeneous, see Fig. 5.10(*a*). Further straining causes the boundaries of the Lüders bands to advance through the remaining material, a process requiring only the lower yield stress. To observe this effect it is necessary to use a 'hard' testing machine in which the applied load decreases rapidly as the specimen extends; a deadweight load causes the bands to spread immediately throughout the specimen on reaching the upper yield point. When the whole specimen is plastic the stress again rises due to normal strain hardening. Lüders bands are a source of trouble in pressing mild steel sheet since they ruin the surface finish, see Fig. 5.10(*b*). They are useful, however, in experimental stress analysis as they will indicate the points of stress concentration.

The upper and lower yield stresses are dependent on the rate of loading, more so than in normal yielding processes. The time required to initiate yielding in mild steel is plotted against the stress in Fig. 5.11. This is not important in conventional testing but under shock loads and in the propagation of fractures at high velocities the upper yield stress may be significantly raised.

A phenomenon closely related to the yield point is strain ageing. If a material, say mild steel, which initially has a sharp yield point is given some plastic deformation (*OABC* in Fig. 5.12), the yield point is destroyed and on reloading there is a smooth transition from the elastic range to further yielding (*CB'D*). But if the material is now left for a few days, or about thirty minutes at a slightly elevated temperature (60–100 °C), a new yield point phenomenon is developed.

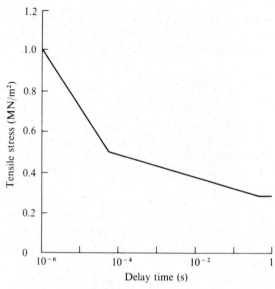

Fig. 5.11. Delay time for the yielding of mild steel as a function of the applied stress.

This may consist of a new upper yield point dropping back to the former lower yield stress, or a new lower yield stress and Lüders strain, or a complete raised strain-hardening curve (*CEF* in Fig. 5.12). The effects have been extensively investigated in recent years and explained theoretically in terms of 'dislocation locking' by solute atoms, notably nitrogen in steel. This work is discussed in § 11.5.

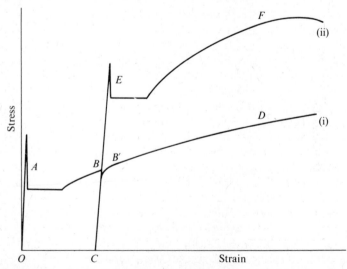

Fig. 5.12. Strain ageing of mild steel. Curve (ii) is obtained after prestraining the annealed metal (curve (i)) to *C* and ageing for some days at room temperature (or 30 minutes at 100 °C).

There are two important practical considerations. First, strain ageing will reintroduce the possibility of stretcher-strain marks in any subsequent press work. Second, the transition temperature for brittle fracture is raised significantly, see Chapter 10.

5.6 Compression

A compression test subjects a specimen of material to an increasing compressive load along a single axis and the length is measured as a function of the load. A typical metallic specimen might be a cylinder, 15 mm in diameter by 20 mm long, and it is compressed between parallel plates in a modified tensile test machine. The specimen must be short in relation to the diameter if it is to compress uniformly rather than buckle sideways (see textbooks on the theory of struts). It is then difficult to ensure uniform uniaxial stress since the ends tend to stick to the compression plates, restraining the radial expansion which accompanies the axial compression. A lubricant will help but does not always entirely overcome the trouble.

The results of tension and compression tests on aluminium, a typical ductile metal, are compared in Fig. 5.13. The axes chosen are true stress and *natural* (or *logarithmic*) *strain*, which is defined as

$$e = \int_l^{l_0} \frac{\mathrm{d}l}{l} = \ln \frac{l}{l_0}.$$

By using these axes, which have a sounder physical basis than nominal stress and linear strain, the similarity of the elastic and plastic behaviour in tension and compression is revealed. The figure shows that points obtained from tension and compression tests lie on a single curve, which can be represented by a simple power function of the natural strain. (N.B. The index will be slightly different

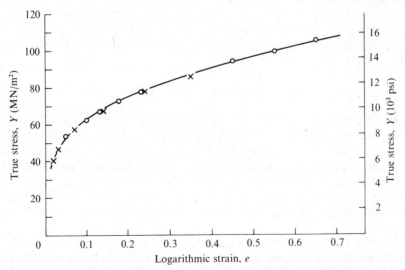

Fig. 5.13. Correlation of yield curve in tension and compression. Results on aluminium. ○ tension, × compression, —— $Y = 12e^{1.28}$.

from that obtained using the linear strain, as was done in (5.2).) There is the same approximately proportional and elastic range followed by yielding, with the yield stress rising with strain. Unlike tension, no instability leading to necking develops in compression, and the strain is always uniform. Very large compressive strains are possible. The limiting factor is often the surface tensile stresses which arise from friction on the end plates.

Since it reveals little about ductile materials that is not found from a tensile test, the compression test is not a standard requirement in their specifications. On the other hand for brittle materials, for example concrete, it is often used. Brittle materials have up to ten times greater strengths in compression than in tension, and may show some plastic flow before fracturing on the maximum shear planes at 45° to the compression axis. This is due to the important part played by minute cracks in the fracture of brittle materials, a mechanism which is dependent on the sign of the applied stress (see Chapter 10).

Brittle materials are also frequently tested in bending which combines to some extent both the tensile and compressive behaviour of a material. The maximum longitudinal (fibre) stress occurs at the surface and its value at failure, calculated from elastic theory, is called the *modulus of rupture*, or *flexural strength*. Even in brittle materials there is some slight plastic flow before fracture and, since this is not allowed for in the analysis, the modulus of rupture is higher than the tensile strength. For example, in concrete and other ceramics it is about double.

If a ductile specimen is plastically strained first in tension and then in compression, the strain hardening is found to have operated in both directions. However there is some initial discrepancy between two yield curves in the forward and reverse direction, see Fig. 5.14. This is known as the Bauschinger effect (1881). Although the yield stress is initially rather lower in the reverse direction of straining it soon rises to the former value as if no change in the direction of straining had occurred. The effect was originally explained by Heyn in terms of the residual stresses which may be expected to develop between grains of different orientation. However a similar effect is found in single crystals so this cannot be the whole explanation. A low temperature anneal will remove the Bauschinger effect without reducing the yield stress.

5.7 Hardness

Hardness is property which is not uniquely defined. Several tests which claim to measure 'hardness' are actually measuring different properties. The tests can be grouped according to whether they principally measure a material's resistance to plastic deformation under static loads (static indentation hardness), to deformation under impact load (dynamic hardness), to scratching, or to wear by abrasion. Only the first group will be considered here and the term hardness will be confined to this sense. For a fuller treatment, reference should be made to Tabor.

The principal tests of static indentation hardness are named after Brinell (1900), Vickers (developed from Smith and Sandland, 1922) and Rockwell (1922), the last being used mainly in the USA. In each, an indenter is forced by a known load into the surface of the material and the hardness is calculated from

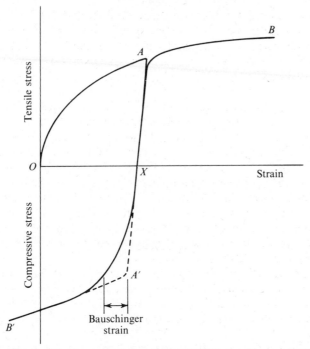

Fig. 5.14. Bauschinger effect. If a tensile test is interrupted and the specimen strained in compression the yield stress is initially less, but it soon rises to $A'B'$, the image of the original strain-hardening curve AB.

the size of the indentation, typically 0.5–5 mm across. The Knoop test (1939) which is used on very hard materials, is a miniature form of the Vickers test, with an indentation about 100 μm across. The various tests are outlined in Fig. 5.15.

The Brinell test uses a hardened steel ball, normally 10 mm diameter, and the BHN (Brinell Hardness Number) or H_B is the load (kg) divided by the actual curved area of the impression (mm^2). This value varies with the relative size of the impression and, to be independent, it is necessary to choose a load, in the range 30–3000 kg, such that the ratio of diameter of impression to diameter of ball lies between 0.25 and 0.5. The load is applied for about 20 seconds. Meyer suggested in 1908 that the yield pressure, defined as the ratio of the load to projected area of the impression, which is also the mean pressure acting over the ball in the absence of friction, would be a better criterion of hardness, called the Meyer hardness, but this is not much used. Meyer's Law relates the load W and the impression diameter d:

$$W = kd^n, \qquad (5.8)$$

where k and n (Meyer index) are constants for a material. k depends on the ball size such that

$$kD^{n-2} = \text{const.} \qquad (5.9)$$

The Meyer index n is independent of the ball size and can be correlated with the strain-hardening index x (5.2): $n = 2 + x$, as shown in Table 5.2 for various metals.

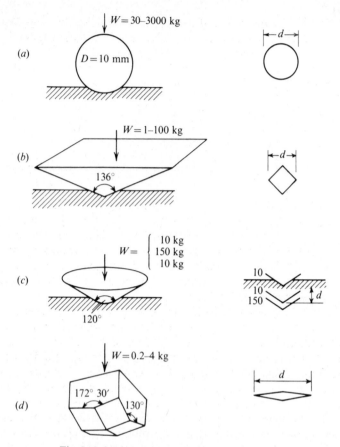

Fig. 5.15. Static indentation hardness tests.
(a) Brinell and Meyer. W is adjusted to make $d = 2.5$–5 mm. $H_B = W$/curved area = $2W/\pi D[D - \sqrt{(D^2 - d^2)}]$ kg/mm². $H_M = W$/projected area = $4W/\pi d^2$ kg/mm².
(b) Vickers. W is adjusted to make $d = 0.5$–1 mm. $H_V = W$/pyramid area = $1.854W/d^2$ kg/mm².
(c) Rockwell C. $H_{RC} = $ const. $- d/2$.
(d) Knoop. $H_K = W$/projected area = $W/0.0703d^2$.

The Vickers test uses a diamond pyramid with an included angle of 136° (representing the mean Brinell impression angle) and the DPN (Diamond Pyramid Number) or H_V is again the load (kg) divided by the actual area of the impression (mm²). The value is independent of the depth of penetration, as may be expected since all indentations are geometrically similar, and the load, which is automatically applied for ten seconds, is chosen between 1 and 100 kg to give an indentation which can be easily measured, about 1 mm across. In microhardness tests, loads in the range 1–100 g are used and the impression is only a few microns (10^{-3} mm) across, requiring a more powerful microscope for measurement. Good surface finish is important with such small indentations, but the hardness of smaller regions, even of individual grains (50 μm across), can be plotted.

A development of the Vickers is the Knoop test (1939) in which the diamond pyramid is no longer square based, but asymmetric, see Fig. 5.15(d). It forms

Table 5.2. *Correlation of Meyer index n and strain-hardening
index x. (From D. Tabor*, The Hardness of Metals,
Clarendon Press (1951))

Material	n	$n - 2$	x	$(n - 2)/x$
Mild steel	2.25	0.25	0.269	0.93
Brass (70 Cu–30 Zn)	2.44	0.44	0.404	1.09
Copper	2.40	0.40	0.38	1.05
Nickel	2.50	0.50	0.43	1.16
Aluminium	2.20	0.20	0.15	1.33

an indentation with the shape of a parallelogram in which the larger diagonal is
seven times the shorter one. Loads are in the range 0.2 to 4 kg and only the longer
diagonal is measured, typically around 100 μm. The Knoop hardness is defined
as load over the projected area. It is claimed that the Knoop indenter is easier to
make than the Vickers and makes a satisfactory indentation in even the hardest
materials, such as glass and diamond. Hardness values approximately correspond on the two scales.

In the Rockwell test a steel ball is used for soft metals (B scale) and a diamond
cone with a hemispherical tip for hard metals (C scale). A setting load (10 kg) is
first applied, and then a test load of 100 kg (B) or 150 kg (C), before reducing
to the setting load again. The hardness is derived from the difference in depth of
the indentation between initial and final application of the setting load.

Typical hardness values of representative construction materials are included
in Table 5.1; values for some other classes of material are given in Table 5.3.

In spite of differences in detail, the principal hardness tests measure largely
the same physical property, a material's resistance to plastic deformation under
combined stresses, which is related to its yield stress in simple tension (see § 6.5).
It is therefore possible to correlate the various hardness numbers with each other,
and also with the tensile yield curve. For ductile materials the hardness can then
be related to the tensile strength. The analysis will be shown below for the diamond pyramid test (Vickers), for which it is simplest because of the geometric
similarity of the impressions at all loads.

The conversion between hardness scales is given in Table 5.4. It will be seen
that the values of the DPN and BHN are fairly close, as might be expected
from the similarity of the tests; Rockwell C values are on a very compressed
scale by comparison, very roughly by a factor of 10. It should be noted that there
is not a linear relationship between the DPN and R_C values: between DPN
700 and 800 (typical of hardened steels) the R_C value changes only 4 points;
between DPN 300 and 400, R_C changes 10 points. The DPN and BHN values
diverge at the higher end of the scale due to the deformation of the Brinell ball,
which should be of tungsten carbide for values above 500. The correlation has
been obtained experimentally with numerous carbon and alloy steels and the
same relationship will not necessarily hold for other materials. In particular the
Rockwell reading is affected by the elastic recovery, and two materials with the
same yield stress but different values of Young's modulus will give different
Rockwell values.

Table 5.3. *Hardness of some materials (BHN $\simeq$ DPN $\simeq$ Knoop)*

Metals	Condition	Hardness
99.5% Aluminium	Annealed	20
	Cold rolled	40
Aluminium alloy	Annealed	60
(Al–Zn–Mg–Cu)	Precipitation hardened	170
Mild steel	Normalised	120
(0.2% C)	Cold rolled	200
Ball bearing steel	Normalised	200
(1% C–Cr, En31)	Quenched (830 °C) (fully martensitic)	900
	Tempered (150 °C)	750
Ceramics		
Tungsten carbide WC	Sintered	1500–2400
Cermet WC–6% cobalt	20 °C	1500
	750 °C	1000
Alumina, Al_2O_3		~1500
(hard anodising)		400–500
Boron carbide		2500–3700
Boron nitride (boron azide, cubic)		7500
Diamond		6000–10 000
Glasses		
Silica		700–750
Soda-lime		540–580
Borosilicate (low expansion)		550–600
High polymers		
Polystyrene		17
Polymethyl-methacrylate ('Perspex')		16

The correlation of the hardness and the tensile yield curve is logical since both are measures of the plastic behaviour of a material, although the stress patterns are clearly very different. For heavily work-hardened materials, the tensile stress–strain curve has an elastic range followed by yielding at nearly constant stress. This type of material is known as elastic–ideally plastic and is amenable to the mathematical theory of plasticity. It has been calculated (see § 6.9) that the hardness (DPN) is related to the yield stress (Y kg/mm^2) by:

$$DPN = 2.8Y.$$

It is found experimentally that the best fit is given by:

$$DPN = 3.0Y. \tag{5.10a}$$

With strain-hardening material, the tensile stress–strain curve has an elastic range followed by yielding at increasing stress. Since the amount of deformation varies around indentation, the yield stress will now vary with position. However, it is found experimentally that it is possible to use an average yield stress, corresponding to a strain of 8 per cent, to fit the previous equation, that is:

$$DPN = 3.0Y_e, \tag{5.10b}$$

where Y_e is now the effective yield stress (in kg/mm^2) corresponding to a strain of 8 per cent. This equation also holds for partially work-hardened material. Experimental results confirming these relationships are given in Table 5.5.

Table 5.4. *Correlation of indentation hardness tests: Brinell, Vickers and Rockwell (C)*

DPN	BHN	R_C	TS (MN/m²)*	DPN	BHN	R_C	TS (MN/m²)*
100	95			420	397	42.7	1380
110	105			440	415	44.5	1450
120	114			460	433	46.1	1520
130	124		426	480	448	47.7	1585
140	133		455	500	471	49.1	1650
150	143		490	520	488	50.5	1720
160	152	(0)	516	540	507	51.7	1790
170	162	(3.0)	545	560	525	53.0	1850
180	171	(6.0)	579	580	545	54.1	1920
190	181	(8.5)	600	600	564	55.2	1990
200	190	(11.0)	634	620	582	56.3	2060
210	200	(13.4)	668	640	601	57.3	2130
220	209	(15.7)	695	660	620	58.3	2200
230	219	(18.0)	730	680	638	59.2	2270
240	228	20.3	765	700	656	60.1	
250	238	22.2	799	720	670	61.0	
260	247	24.0	834	740	684	61.8	
270	256	25.6	868	760	698	62.5	
280	265	27.1	902	780	710	63.3	
290	275	28.5	936	800	722	64.4	
300	284	29.8	970	820		64.7	
320	303	32.2	1040	840		65.3	
340	322	34.4	1110	860		65.9	
360	341	36.6	1170	880		66.4	
380	360	38.8	1240	900		67.0	
400	379	40.8	1310	920		67.5	
				940		68.0	

* 1 MN/m² = 145 psi.

Table 5.5. *Relationship between hardness (DPN) and yield stress for various materials in the annealed to fully strain-hardened condition. (From D. Tabor,* The Hardness of Metals, *Clarendon Press (1951))*

	Yield stress Y (kg/mm²)	Hardness DPN (kg/mm²)	Ratio: $\dfrac{DPN}{Y}$
Full strain hardened			
Tellurium–lead alloy	2.1	6.20	2.95
Aluminium	12.3	36.6	2.98
Copper	27	81.5	3.02
Steel	70	210	3.00
Annealed	$Y(\varepsilon = 8\%)$		
Mild steel	55	156	2.84
Copper	15	45	3.0
Partially strain hardened	$Y(\varepsilon = \varepsilon_0 + 8\%)$		
Mild steel (10% initial strain)	66	187	2.84
Mild steel (25% initial strain)	73	209	2.86
Copper (12.5% initial strain)	23.3	69	2.96
Copper (25% initial strain)	26.6	81	3.04

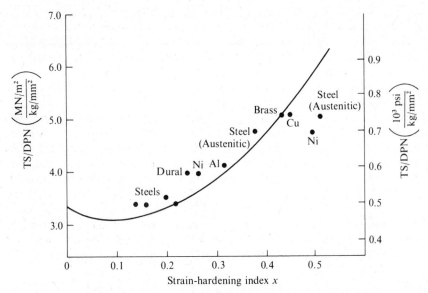

Fig. 5.16. Ratio of tensile strength (MN/m^2) to indentation hardness (DPN kg/mm^2) versus strain-hardening index. Continuous line is the analytical function, (5.11).

The relationship between hardness and tensile strength is derived as follows. Let the yield curve in tension be represented by

$$\sigma_t = b\varepsilon^x$$

where σ_t is the true stress, ε the linear strain, and b and x (strain-hardening index) are constants of the material. Then, from (5.5):

$$TS = bx^x(1 - x)^{1-x}.$$

Now, by (5.10b)

$$DPN = 3b(0.08)^x$$

Hence

$$\frac{TS}{DPN} = \frac{1 - x}{3}\left(\frac{12.5x}{1 - x}\right)^x. \tag{5.11}$$

For values of the strain-hardening index between 0.1 and 0.3 this function does not change rapidly and for similar types of material which will have about the same work-hardening index:

$$\frac{TS}{DPN} = \text{const.}$$

The value of the constant may be determined directly by experiment, or by inserting the value of the strain-hardening index in the analytical function. This may be determined in a tensile test or from a Meyer analysis of the Brinell test. The analytical function is plotted against x in Fig. 5.16, which also shows various experimental points.

5.8 Notch toughness

Notch toughness is a measure of a material's strength in the presence of a stress concentrator such as a notch or a crack. The term toughness is often used alone, but this can confuse since it may also refer to the energy absorbed in a tensile test, which is a separate matter. In recent years the term 'fracture toughness' has been used. It is common experience that brittle materials such as glass or ceramic will fracture readily at a notch or surface scratch but have quite high strengths in the absence of any stress raiser. A similar low notch toughness may occur in metals, and the chief problem in developing high strength alloys is to do so without destroying their notch toughness.

While it is true that all brittle materials, that is, those that fracture with negligible plastic deformation in a straight tension test, will have very low notch toughness, the converse is unfortunately not true. It has been found that even though a material is ductile in a tensile test, with high elongation and reduction of area, it may fracture in the presence of a notch or crack without appreciable plastic flow and at a low average stress. This phenomenon was found first (around 1900) in heat-treated alloy steels, and various notched bar impact tests, such as the Izod and Charpy tests, were developed to measure a material's ability to withstand stress concentrations. Typical figures for an alloy steel (En 13 BS970) subjected to slightly different heat treatments are as follows:

	Temper °C	Yield point MN/m² (10^3 psi)	TS MN/m² (10^3 psi)	El %	RA %	Izod J (ft lb)
(a)	400	700 (100)	880 (1280)	22	63	16 (12)
(b)	450	700 (100)	850 (1230)	23	65	88 (65)

The Izod value is the energy absorbed in the bending of a notched bar, described in detail below.

It was only during the Second World War that the notch toughness of mild steel, the highly ductile, standard, constructional material, was questioned following the catastrophic fracture of many Liberty ships (in the absence of enemy action). It was found then that under some circumstances a crack can propagate through mild steel, and other body-centred cubic metals, with little plastic deformation and at a low applied stress (75 MN/m^2, 11 000 psi). In the course of intensive research into the many complex aspects of this effect, the Izod and Charpy tests have been shown to be inadequate for the full assessment of a material's behaviour in the presence of a running crack and other tests have been devised which will be described in Chapter 10. However the Izod and Charpy tests are still the standard tests of notch toughness, though their value is probably only comparative, and they will be briefly described here.

In the Izod test, which is chiefly used in Britain, the specimen is a square bar (10×10 mm) in which a 45° notch with root radius 0.25 mm is cut across one face to a depth of 2 mm (see Fig. 5.17(a)). It is held vertically in a vice, the top of which is level with the notch, and the free end struck on the notched side by a 27.2 kg (60 lb) pendulum moving at 3.5 m/s (11.5 ft/s). The energy (J or ft lb) absorbed in fracturing the specimen or bending it to allow free passage of the pendulum, is obtained from the decrease in amplitude of the pendulum. The Charpy test, which is favoured in the USA, uses a similar sized specimen with a notch cut at mid-section. It is simply supported at both ends and struck with a 27.2 kg (60 lb) pendulum at 5.3 m/s (17.4 ft/s) in the middle on the opposite side to the notch (Fig. 5.17(b)). Again, the energy absorbed is measured. Originally a

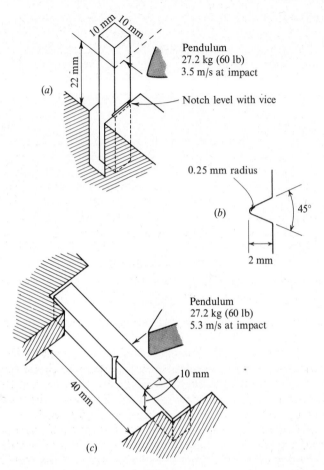

Fig. 5.17. Notch toughness (impact) tests: (a) Izod, and (c) Charpy. Pendulum swing after impact is measured to obtain the energy absorbed. (b) shows notch detail (used on Izod and Charpy).

keyhole notch, with a 1 mm root radius, was used but this was not severe enough and it is now usual to use an Izod V-notch (see BS131).

For a long time, with regard to heat-treated steels, the energy absorbed in a test at room temperature was regarded as a sufficient criterion of a material's resistance to fracture under notch-impact conditions. In recent years, since the tests have been applied to mild steels, the nature of the fracture surface as well as the energy absorbed have been measured as functions of the temperature. Typical curves of these parameters against temperature are shown in Fig. 5.18. Over a transition temperature range they change fairly rapidly from a low value to a high one. It is considered that a material is satisfactory in service if used above the transition temperature range, and unsafe below it. Unfortunately the transition temperature range is dependent on details of the test, such as notch acuity, and on the particular parameter chosen. A consequence is that it is strongly debated which is the best test for obtaining an accurate assessment of a material's behaviour in service. It is now realised that much of the earlier

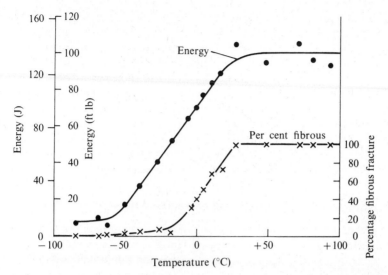

Fig. 5.18. Temperature transitions in the energy absorbed and fracture appearance in a notched bend test.

work on heat-treated steels is incomplete because the importance of the tempera-ture was not appreciated.

What new factors have been introduced that are responsible for behaviour which is not revealed by the tensile test? It is popularly held that the speed of loading is the important variable and the tests are often referred to as impact tests. This is incorrect, for in most materials the energy absorbed is nearly independent of the speed. The variation of load with angle is informative, see Fig. 5.19(a). Up to the maximum load, there is little difference between the curves for material in the notch tough state and in the notch brittle state, but after the crack has started the latter curve drops to zero. This explains why the Izod value is not entirely satisfactory as a measure of notch toughness, that is, of resistance to sudden crack propagation : the energy absorbed prior to the forma-

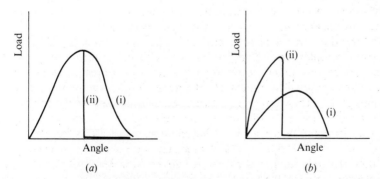

Fig. 5.19. Load versus angle in Izod-type bend test for notch toughness. (a) compares a single material above (curve (i)) and below (curve (ii)) its transition temperature range. (b) compares two different materials with similar values of energy absorbed: curve (i) is notch tough and safe while curve (ii) is notch brittle and unsafe.

tion of the crack may mask the low energy absorbed during crack propagation. Of the two curves in Fig. 5.19(*b*), curve (ii) represents a thoroughly unsound material whilst curve (i) is quite satisfactory, yet the Izod values would be identical. Thus high Izod values in high strength heat-treated steel may be delusive. Another variable is obviously that the specimen is now in bending instead of in tension, but this also is of little significance and notch tensile tests can also be used with the same results.

The most important factor seems to be the concentration of triaxial stress just below the root of the notch. Under these circumstances a crack is able to form which spreads across the specimen without further absorption of external energy. This subject will be discussed further in Chapter 10.

5.9 Anelasticity, static hysteresis and microplasticity

In the preceding sections the behaviour of ductile materials has been idealised into two distinct ranges: elastic, obeying Hooke's Law, and plastic. This model is adequate for many purposes but, as was pointed out in defining the proof, or yield, stress (§ 5.2), plastic flow does not develop rapidly at a well defined stress but increases slowly as the stress is increased from very low values, within the nominally elastic range. The final sections of this chapter consider various small departures from Hookean elasticity at low stresses; they are tabulated in Table 5.6.

Table 5.6. *Diagram showing sub-divisions of idealised elastic–plastic–fracture model of materials*

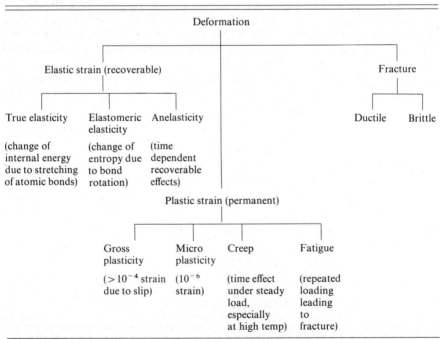

One type of deviation from ideal elasticity is called *anelasticity*. Following the sudden application of a stress, there is an immediate strain ε_0 (according to Hooke's Law) and an anelastic strain ε_a which only achieves its final value after some time interval. The total strain develops according to the time law

$$\varepsilon = \varepsilon_0 + \varepsilon_a(1 - e^{-t/\tau}), \tag{5.12}$$

where τ is a constant called the *relaxation time*. Several atomic mechanisms, discussed later, can produce time laws of this type. On removing the stress the immediate strain is recovered at once and the anelastic component returns to zero slowly, with the same relaxation time as on loading. If the value of τ is greater than a few seconds, the effect can be observed in quasi-static experiments and is known as *recoverable creep* on loading and *elastic after effect* on unloading. These are shown schematically in Fig. 5.20. It will be noted that the strain is entirely recoverable, that is, there is no plastic flow involved.

Usually the characteristic time of anelasticity is short, the order of milliseconds, and the effect is observable as damping or *internal friction* in material subjected to vibrational stressing. The stress–strain curve depends on the frequency of the stress cycle relative to the characteristic time, see Fig. 5.21. At relatively high frequency the anelastic strain never develops and the stress–strain curve is a straight line with the Young's modulus due to the Hookean strain only (curve (*a*)). At relatively low frequency the anelastic component develops to its equilibrium value at each stress level and the stress–strain curve is again a straight line (curve (*c*)), but with a reduced, or relaxed, modulus, due to the presence of both Hookean and anelastic strain. At intermediate frequencies, when the period of the stress cycle is similar to the relaxation time, the anelastic strain is partially developed to values which depend on whether the stress is increasing or decreasing. The stress–strain curve now forms a hysteresis loop, representing energy converted into heat during a cycle (curve (*b*)). If the stress is varied harmonically, the strain is also harmonic but lagging by some phase angle ϕ.

The *damping capacity D* is defined as the fractional energy absorbed per cycle, that is, the area of the hysteresis loop divided by the maximum strain energy.

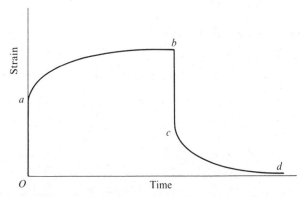

Fig. 5.20. Anelasticity. With a long relaxation time anelastic strain is observed as recoverable creep (*ab*) on loading and elastic after effect (*cd*) on unloading.

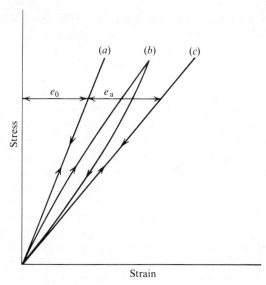

Fig. 5.21. Anelasticity appearing as a hysteresis loop when the stress is cycled at a critical rate: (a) $t \gg \tau$ (b) $t \approx \tau$ (c) $t \ll \tau$ (t = periodic time of stress cycle, τ = relaxation time of anelastic effect). *NB* Not to scale: maximum anelastic strain is typically $10^{-3} \times$ true elastic strain.

For anelastic phenomena, D is independent of the strain amplitude and varies with the angular frequency ω according to

$$D = \frac{\Delta E}{E} = 2\pi \frac{M_u - M_r}{\sqrt{M_u M_r}} \frac{\omega\tau}{1 + (\omega\tau)^2} \qquad (5.13)$$

where M_u is the unrelaxed elastic modulus and M_r is the relaxed elastic modulus (at relatively low straining rates). The variation of D versus $\ln \omega$ is plotted in Fig. 5.22. The curve is symmetrical about the peak damping, which occurs at $\omega\tau = 1$, and the maximum damping capacity is:

$$D_{max} = \pi \frac{M_u - M_r}{\sqrt{M_u M_r}}. \qquad (5.14)$$

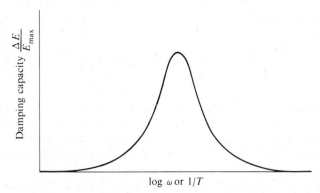

Fig. 5.22. Damping capacity versus frequency of stress cycle. The curve is independent of stress amplitude.

Several other parameters are also used to denote the damping or internal friction. They are related as follows (for models giving the time law given above, (5.12)):

$$D = \frac{\Delta E}{E} = 2\pi \tan \phi = 2\delta. \tag{5.15}$$

ϕ is the phase lag of the strain; δ is the *logarithmic decrement* (or decrement) which is the natural logarithm of the ratio of the amplitudes of two successive vibrations in the same direction, measured in a free vibration. The quantity $\tan \phi$ is often written as Q^{-1} by analogy with electrical systems.

Besides varying with frequency, anelastic damping varies with temperature. The relaxation time is dependent on atomic diffusion, which is a thermally activated process (see Chapter 2), and $\ln \tau$ is proportional to $1/T$. Hence, from (5.13), a plot of damping capacity versus $1/T$ will have the same form as damping capacity versus $\ln \omega$, that is, as shown in Fig. 5.22. Varying the temperature is experimentally easier than varying the frequency and this is the normal technique for investigating damping and the underlying atomic diffusion.

Typical values of the maximum damping due to anelasticity lie in the range 10^{-4} to 10^{-2}. Other mechanisms contribute a minimum background damping of around 10^{-5}.

Several physical phenomena are responsible for anelasticity. An important one is the shearing of the grain boundaries. Under stress two adjacent grains are able to slide past each other a small distance until the irregularities of the boundary interlock and the stress is carried by the crystalline material. In the Snoek effect, interstitial atoms in a bcc lattice diffuse to preferred positions in the tetragonal cells (formed by straining of the original cubic cells) and contribute a strain component. In the Zener effect, pairs of oversize atoms in substitutional solution align themselves in the direction of the applied stress. Magnetostriction in magnetic materials (see Chapter 15) can also cause anelasticity. Finally a macroscopic anelastic effect is produced in a flexed beam: the compressed material is heated and the extended material is cooled; isothermal conditions are restored by heat flow, producing a time dependent strain.

Another deviation from Hookean elasticity and a contribution to internal friction is static hysteresis. Unlike anelasticity, the damping is independent of frequency but very dependent on the maximum strain. It becomes significant at higher strains than for anelastic effects. In a quasi-static experiment, using strain gauges with a sensitivity of 10^{-6} to 10^{-7} to measure the strain as the stress is slowly cycled to successively higher stress levels, curves are obtained as shown in Fig. 5.23. At low maximum stress, the load and unload lines are indistinguishable (although it would be possible to detect some static hysteresis in damping type experiments). At some stress level, a hysteresis loop is detected (the damping capacity would then be about 0.01). As the maximum stress is increased, the loop area and maximum strain width increase rapidly but the loop continues to close (to the limits of the strain sensitivity) at zero stress and can be retraced any number of times. Finally, the loop fails to close on unloading and plastic flow is detected. The atomic mechanisms responsible for this type of internal friction are not elucidated, but one cause is the bowing of dislocation loops against the Peierls–Nabarro friction stress. The unpinning of dislocations is another. These phenomena are discussed in Chapter 9.

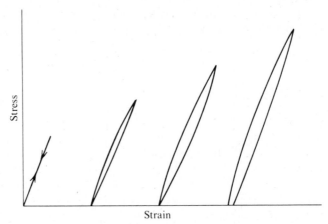

Fig. 5.23. Static hysteresis, which is independent of frequency, and the onset of microplasticity during stress cycles to increasing values of the maximum stress.

Another time dependent recoverable strain, of much greater magnitude, is found in linear high polymers, or thermoplastics. As mentioned briefly in § 5.1 these materials exhibit entropy elasticity due to rotation of the covalent bonds linking the atoms into chains and this strain often takes time to reach its equilibrium value. Discussion of this effect will be postponed until Chapter 8.

Whereas anelasticity and static hysteresis are departures from Hooke's Law but are still elastic phenomena, the next effect is concerned with the onset of plasticity, that is, permanent strain on unloading. With the development of strain gauges sensitive to 10^{-7}, it has been possible to investigate the onset of plasticity with increasing precision and recently (around 1960) a microyield stress has been defined as the stress for 10^{-6} permanent set on unloading. This plastic strain is three orders of magnitude smaller than is used in determining the conventional yield stress, but it remains an arbitrary point on the stress–plastic strain curve. The ratio of the microyield stress to the conventional yield stress is between about 10 and 90 per cent, depending on the specific material. The microyield stress is important in confirming the dislocation theory of plasticity (see Chapter 9) and in some practical applications, such as elastic hinges in precision instruments, where small deviations from elastic behaviour are critical.

5.10 Creep and stress rupture

Creep is the time dependent strain which sometimes follows the instantaneous strain obtained on stressing a material. Recoverable creep in metals has already been noted in the preceding discussion on anelasticity, and creep in high polymers will be discussed in Chapter 8. This section will consider briefly non-recoverable creep in metals. The subject is an extensive one, justifying a more expanded treatment, but for reasons of space and balance only an outline of the phenomenon will be included here.

A typical creep curve of strain versus time under constant stress is shown in Fig. 5.24. Several stages are recognised, which are due to different physical

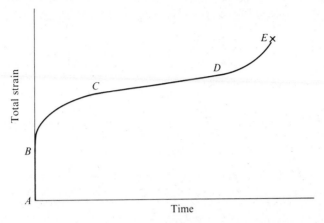

Fig. 5.24. Stages in a creep curve, leading to stress rupture. AB = initial elastic extension, BC = transient (primary) creep, CD = steady state (secondary) creep, DE = accelerating (tertiary) creep, E = stress rupture.

mechanisms. After the initial sudden extension AB, there is *primary* creep BC (or *transient* creep) in which the strain rate decreases with time. This is followed by *secondary* creep CD (or *quasi-viscous* creep) in which the creep rate is constant with time. Finally there may be *tertiary*, or accelerating, creep leading to fracture at E. The latter is called *creep rupture* or *stress rupture*. The total strain at rupture is typically only a small fraction of the elongation to fracture in a tensile test at the same temperature. If the specimen is unloaded at any point on the curve there is an immediate recovery of strain (according to Hooke's Law) but negligible subsequent recovery.

Creep is a function of stress and temperature for a given material. In a room temperature tensile test, creep is readily apparent at stresses approaching the tensile strength. At lower stresses the creep strain (in a given time) becomes less significant in relation to the instantaneous strain. As with the start of plasticity it is not possible to define an absolute limit below which creep is not present. In engineering design, arbitrary criteria have to be selected to define acceptable creep limits. One typical criterion, used on steels for boilers, is that the minimum creep rate shall not exceed 10^{-5} per hour; in turbine discs this is reduced to 10^{-7} per hour. Other criteria are also used involving the total creep strain in a given time.

The temperature parameter of importance is the *homologous* temperature T_{H} which is the ratio of the operating temperature to the melting point (on absolute scales), T/T_{m}. Different pure metals have similar creep characteristics when tested at the same homologous temperature. For example lead with a melting point of 326 °C exhibits similar creep phenomena at room temperature ($T_{\mathrm{H}} = 0.5$) as does nickel (m.p. 1455 °C) at 600 °C. Alloying will modify the creep behaviour at a given T_{H}, nevertheless the high melting point metals are the bases of creep resistant alloys. As the operating temperature is increased, the allowable stress, based on the yield stress at that temperature, will decrease but the creep stress, based on some criteria such as total creep strain or minimum creep rate, will decrease faster. Above T_{H} equal to about 0.4, the creep stress will be below the yield stress and it is necessary to design on a basis of creep properties rather than purely elastic ones.

The time laws of creep have been determined for pure metals, simple alloys and some ceramics. At T_H less than 0.3, transient creep is given by a logarithmic law:

$$e = \alpha \ln t + c'$$

where α and c' are constants (depending on the stress and temperature). For higher values of T_H the curve changes to a power law

$$e = \beta t^m + c'',$$

where β, m and c'' are constants. In many metals the index m equals $\frac{1}{3}$, originally found by Andrade in 1910. The secondary creep is given (*a priori*) by

$$e = \kappa t$$

where κ is a constant, which is very sensitive to stress and temperature.

The so-called instantaneous strain cannot be measured directly since, although the creep curve is nominally at constant stress, the load has to be applied initially at a finite rate and there is no discontinuity between the strain during loading and the transient creep at the full test load. The instantaneous strain is found (or really defined) either by fitting a time law to later stages of the creep curve and extrapolating back to zero time, or by measuring the strain at very low stresses and extrapolating linearly to the creep stress. (Quite different results are obtained by the two methods.)

Owing to the fact that a creep curve is the sum of several independent effects, obeying different time laws with parameters dependent on stress and temperature, analysis of experimental creep data is complex. It must be carried out systematically if misleading conclusions are to be avoided. Single point values of creep strain or creep rate cannot give the full picture of creep behaviour, although they have been, and still are, used by the unwary. The point is illustrated by the two creep curves in Fig. 5.25 which have different amounts of primary and secondary creep components.

Extrapolation of creep curves to longer times than the experimental data is a major preoccupation in technological creep work, both for the development of new alloys and for the provision of design data of established materials. Tests of only one year's duration are costly (and lengthy), yet many engineering components are intended for at least ten years life. The time laws of creep, which would provide a basis for extrapolation, have not, in general, been established for complex alloys and various empirical methods of extrapolation have been devised. It is usual to carry out creep tests over a limited period at either higher temperatures or higher stresses than it is expected will be eventually employed in service. The data are then plotted on axes chosen so that the points appear to form a straight line, when linear extrapolation over a decade is reasonably safe. The best method for a particular class of alloy under investigation is found empirically. Creep strain and creep rupture are separate phenomena and must be treated separately.

Closely linked with creep is the variation of the stress–strain curve with rate of straining and temperature, see Fig. 5.26. The yield curve obtained at high strain rate or low temperature will contain a smaller component of creep strain

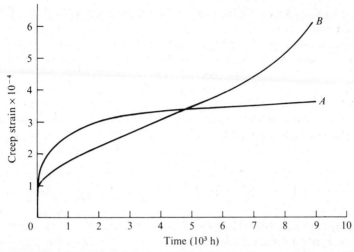

Fig. 5.25. Contrasting shapes of creep curves due to the independent stages of creep. Single para-meters, e.g. creep rate at 1000 h or creep strain at 1000 h, are not satisfactory measures of creep behaviour.

than the normal yield curve. Attempts have been made to rationalise the data obtained from tensile tests at various constant strain rates and temperatures and the data from creep tests. A mechanical equation of state has been proposed of the form:

$$F(\sigma, \varepsilon, \dot{\varepsilon}, T) = 0.$$

Unfortunately it has been conclusively shown by experiments involving step

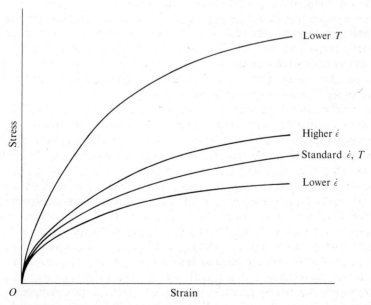

Fig. 5.26. Effect of strain rate and temperature on stress-strain curve. Except at low temperatures there is no mechanical equation of state, $F(\sigma, \varepsilon, \dot{\varepsilon}, T) = 0$.

changes of temperature or stress that no such equation exists, except at low stresses and low homologous temperatures, in the range where only the logarithmic transient creep is present.

The fundamental cause of all creep phenomena is the thermal energy of the atoms in a crystal. As explained in Chapter 2, the atoms in a solid are vibrating about their mean positions due to the thermal energy and the magnitude of the vibration of a particular atom or group of atoms is always fluctuating. Where an atomic rearrangement leading to plastic flow is prevented by some energy barrier, for example a dislocation held up by an obstacle in its slip plane, the random fluctuations of the local thermal energy level may in time cause the barrier to be overcome, that is, the dislocation to pass by the obstacle. Plastic deformation is thus due to the combined efforts of the applied stress and the thermal energy. However, there are many detailed atomic mechanisms leading to plastic flow and they are not equally sensitive to the applied stress and the thermal energy. This is because the thermal energy level can only fluctuate over a small volume whereas the applied stress acts throughout the whole material. As a consequence, the mechanisms of plastic flow, or their relative contributions to an overall strain, vary with strain rate and temperature.

Differences in the microstructure and dislocation arrangements can readily be observed in material strained to a given strain at high and low strain rates, for example, in a tensile test and a creep test. During secondary creep in metals the grain boundaries are found to develop cavities, which eventually join up and cause fracture. Thus stress rupture is intercrystalline, in contrast with normal plastic flow and fracture, which is transcrystalline. To eliminate grain boundary creep the filaments of electric light bulbs are often made of single crystals of tungsten. With different physical mechanisms at different strain rates it is unlikely that a mechanical equation of state can exist.

The preceding discussion has assumed that the material is in a stable condition and that no phase changes occur at the creep temperature in the absence of an applied stress. Many materials are used in a non-equilibrium condition to enhance their strength, for example, after cold rolling or precipitation hardening (see Chapter 11). In these, the ordinary creep behaviour may be complicated by phase changes such as recrystallisation or overageing.

A related effect which will only be noted here is *stress corrosion*. The combined action of stress and a mildly corrosive environment leads in some materials to rapid intergranular corrosion and cracking. The stress may be externally applied or be internal, for example as a result of non-uniform cold working. Brass cartridge cases are the classic example.

Turning briefly to materials for high temperatures, plain carbon and low alloy steels are adequate for use up to 550 °C, with creep being important above about 400 °C. These largely satisfied the requirements of steam and chemical plants up to the Second World War when the advent of gas turbines forced operating temperatures up to 800 °C. Austenitic stainless steels (Fe–18 % Cr–8 % Ni–0.5 % Mo) were used and later 80 % nickel–20 % chromium alloys (e.g. Nimonic, Hastelloy and Inconel). The latter contain titanium and aluminium which are precipitated as carbides to increase the creep strength. (Solution treatment at 1200 °C, and precipitation at 700 °C.) The addition of iron improves the high temperature fatigue strength. Cobalt based alloys (e.g.

Stellite and Vitallium) are favoured in American jet engines. For use above 1000 °C, four refractory metals are being developed: niobium (m.p. 1950 °C), molybdenum (m.p. 2622 °C), tantalum (m.p. 2996 °C) and tungsten (m.p. 3387 °C). Above 1500 °C ceramics, especially graphite, are used. Besides creep strength, resistance to oxidation is a critical property at such elevated temperatures.

5.11 Fatigue

The load on many structural components varies repeatedly and this can lead to fracture although the maximum stress is very much less than the tensile strength, even below the nominal yield stress. This type of failure is known as fatigue. It occurs in all classes of materials except glasses and is easily the most common cause of failure of engineering components in service. Only a limited discussion will be included here, from a phenomenological viewpoint.

In the simplest type of laboratory fatigue test a test piece is rotated continuously whilst supporting deadweight loads, thus subjecting the specimen to alternating bending moments. Various loading arrangements are shown in Fig. 5.27, giving pure bending moment or bending and shear force combined.

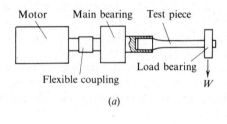

(a)

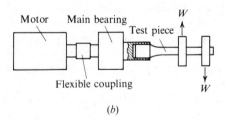

(b)

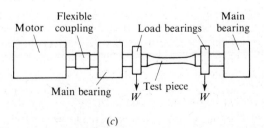

(c)

Fig. 5.27. Various methods of loading in rotating bending fatigue tests. (a) Single point loading; (b) two point loading; (c) beam loading. (After P. G. Forrest, *Fatigue of Metals*, Pergamon Press 1962.)

The methods was originated in 1858 by Wohler, the first to investigate fatigue. The maximum stress at any section occurs at the surface and fluctuates harmonically about zero, that is, between equal maximum tensile and compressive stresses. The specimen is slightly waisted to prevent fracture developing at the loading or support points, but very gentle changes of section must be used to avoid stress concentration which would seriously lower the observed fatigue strength. Alternating direct stress machines are also used and give slightly higher fatigue strengths.

A series of identical specimens are tested to fracture, starting at a high stress level and progressively reducing the stress on successive specimens. The stress amplitude S is plotted against the number of cycles to fracture N, using semi-log or log–log scales. Typical $S–N$ diagrams for non-ferrous and ferrous metals are given in Fig. 5.28 and Fig. 5.29.

The $S–N$ diagram indicates that the *fatigue strength*, or *endurance strength*, decreases with increasing number of cycles. For most materials, the endurance strength tends towards zero at very high cycles (Fig. 5.28), although data above about 10^9 cycles is very tedious to obtain. Where single values of fatigue strength are quoted these are the stress amplitude to cause failure in 10^7 or 10^8 cycles. The $S–N$ curve is sometimes divided into two regions in which different microscopic phenomena and laws are thought to hold: below $N = 10^4$ cycles the

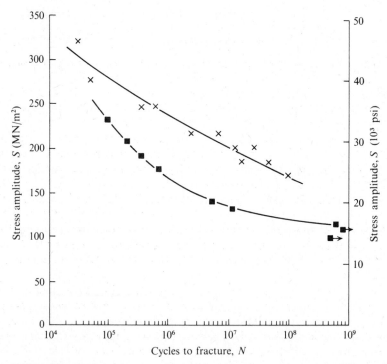

Fig. 5.28. $S–N$ diagram for two typical metals, neither showing a fatigue limit. ■→ indicates specimen was not fractured when the test was stopped. × Al–$4\frac{1}{4}$% Cu alloy (265) fully heat treated, ■ copper (99.95%) hard drawn.

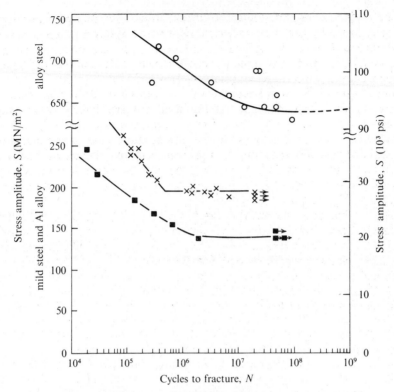

Fig. 5.29. *S–N* diagram for three metals which show a fatigue limit. ○ Ni–Cr–Mo alloy steel (TS = 1.2 GN/m²), × mild steel (0.2 % carbon), ■ Al–5 % Mg alloy.

effect is known as high stress, or low cycle, fatigue; above $N = 10^4$ cycles it is known as low stress, or high cycle, fatigue.

For ferrous metals and a very few others (e.g. Al–Mg, Ti alloys), the *S–N* curve approaches a finite stress amplitude, called the *fatigue limit*, below which fracture will not develop however great the number of cycles. It is important to plot data on a log (N) scale in determining any fatigue limit. In mild steel there is a distinct kink in the curve, around 10^7 cycles; in alloy steels the curve levels off more gradually, see Fig. 5.29. The reason for a finite fatigue limit in ferrous materials is not settled but it is probably connected with the strain-ageing effects. The ratio of the fatigue limit to the tensile strength for carbon steels is about 0.5, but for high strength alloy steels (>1 GN/m², 1.5×10^5 psi) it falls away to around 0.3.

For tests at a fixed stress amplitude there is a large scatter in the number of cycles to fracture, as shown in Figs. 5.28 and 5.29. The cause of scatter lies in the sensitivity of the fatigue mechanism to material and test parameters which can only be controlled to within finite tolerances. Static strength parameters are far more tolerant of small variations in the material and test conditions, at least in ductile metals. The large scatter has important considerations for design, and statistical methods must be used to obtain the probabilities of a desired life. Fortunately, where a fatigue limit exists and the *S–N* curve approaches zero slope, the scatter in the stress range (as distinct from the endurance) is small and it is fairly easy to determine the fatigue limit and design to keep the working stresses below it.

The *S–N* curve and the fatigue limit will be depressed towards the abscissa (*N*-axis) by several factors including larger size of specimen, stress concentration, corrosive environment, and stress fluctuation about a positive mean value (cf. zero, considered so far). If the stress amplitude is raised locally, due to some concentrator such as a notch, by a factor K_t (on the assumption that elastic theory holds), the fatigue strength will be reduced by some factor K_f, where $K_f \leqslant K_t$. The ratio K_f to K_t is normally given in terms of a *notch sensitivity factor*

$$q = \frac{K_f - 1}{K_t - 1}.$$

The factor q can vary between 0 (when notches have no effect on fatigue strength) and 1 (when the fatigue strength is reduced by the same factor as any stress concentrator raises the local stress). Unfortunately q is not purely a material constant but varies significantly with size, amongst other variables, as shown in Fig. 5.30 for two steels. The size effect in unnotched specimens is not so marked and is only significant below about one inch diameter: a 3 mm diameter specimen has a fatigue strength about 15 per cent higher than a 25 mm specimen.

Elastic stress concentration factors K_t can be as high as 3.5 before plastic flow relieves the stress and with the notch sensitivity factor around 0.7 gives a fatigue strength reduction factor of around 3. This explains the attention that should be given in the design of structures supporting fluctuating loads to the avoidance of sharp changes of cross-section, coarse surface scratches, too-small fillet radii, poor distribution of load between rivets or bolts, and weld flashes, to name a few of the frequent causes of fatigue failure. It is interesting to note

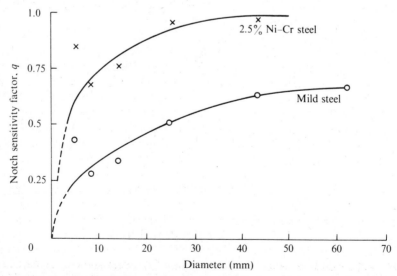

Fig. 5.30. Notch sensitivity factor q versus specimen size. Specimen had a transverse hole such that: hole dia./specimen dia. $= \frac{1}{6}$ ($K_t = 2.8$).

that higher strength alloy steels, besides having a lower fatigue limit to tensile strength ratio as compared with mild steel, have a higher notch sensitivity. For these reasons, where a mild steel component has failed by fatigue, it is unlikely to increase the endurance to make the component of high strength alloy steel.

Where a specimen or structure has previously corroded so that the surface is rough or pitted, the endurance under fatigue loading is reduced through the stress concentration factor, as already discussed. An even greater effect occurs when corrosion and fatigue loading are simultaneous. It has been shown for many metals that even dry air has an influence on the endurance, through the oxidising action, and the division between pure fatigue and corrosion fatigue is thus somewhat blurred. With more corrosive media, such as water or salt water, the S–N curve is sharply depressed towards the N-axis, especially at higher values of N and the fatigue limit of ferrous materials disappears. Some typical results in air and in salt spray are reproduced in Fig. 5.31. The corrosion

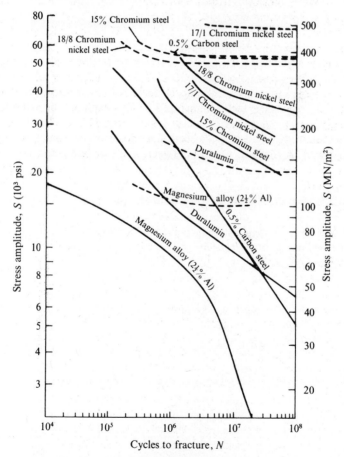

Fig. 5.31. S–N curves for corrosion fatigue (in salt spray) and normal fatigue (in air) for various metals: —— tested in salt spray; – – – tested in air. (From H. J. Gough and D. G. Sopwith, *J.I.S.I.* **127**, 301 (1933).)

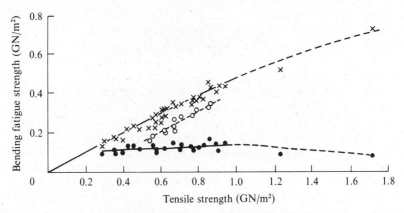

Fig. 5.32. Fatigue strength (bending, 20×10^6 cycles) in air and fresh water of various steels. Note the higher fatigue strength of the chromium steels in water. ● in fresh water (carbon and low alloy steels), ○ in fresh water (chromium steels), × in air (all steels).

resistance of a material is more important than its static tensile strength in determining the corrosion fatigue strength. For example, plain carbon steels show a marked reduction of fatigue strength in fresh water to values which are independent of the carbon content; on the other hand chromium steels are only slightly affected by water and the corrosion fatigue strength is proportional to the tensile strength, see Fig. 5.32.

An allied phenomenon is *fretting corrosion* (cf. stress corrosion, § 5.10). Fretting corrosion is found when two components are pressed against each other and repeated slight relative motion occurs. The effect is found, for example, in bolted or riveted joints, and in rolling bearings, either between the rings and the housing when running, or between the bearing balls and the raceways when the bearing is slightly vibrated when not running. Characteristic corrosion products are formed that in the case of steel appears as 'cocoa', a reddish brown Fe_2O_3, and in aluminium as black powder. The surface damage can originate a subsequent fatigue failure.

The appearance of a fatigue fracture has several characteristic features, at least in ductile material such as mild steel; in materials with less ductility, such as aluminium alloy, recognition of a fatigue fracture is not so easy. Unlike the tensile fracture, there is no plastic deformation adjacent to the fracture – at least not on a macroscopic scale. (In this respect the fracture may be described as brittle but this term is not usually employed in this context, see Chapter 10.) The actual fracture surface has two or three distinct zones, see Fig. 5.33. Around the point at which the crack started, usually on the surface, the surface is smooth with a few concentric lines like tide marks on a beach, called *arrest lines*. These are due to changes in the loading condition not necessarily to complete stoppage of the loading cycle. In this zone the crack has advanced very slowly and staining of the surface by oxidation or contamination will normally indicate the crack origin. Fatigue cracks can be detected at this stage by ultrasonic or magnetic crack detectors before they reach catastrophic lengths. Later the crack advances more rapidly and the surface is rougher. In the final zone the fracture is sudden since the cross-sectional area has been reduced to the point at which

Fig. 5.33. Fatigue fracture surface in mild steel, showing initiation point (dark region at top), **arrest** lines and crystalline (cleavage) area. (By courtesy of R. Weck and The Welding Institute.)

the maximum load cannot be carried statically. In mild steel this zone has frequently a bright crystalline appearance which led early workers to think that fatigue loading caused crystallisation of a metal. This is now clearly seen to be nonsense. Actually, the fracture is of the cleavage type, discussed in Chapter 10, and occurs when the fatigue crack reaches a critical length at which the energy for further growth can be obtained from the elastic energy of the surrounding metal.

The atomic mechanisms of fatigue will not be discussed here except to remark that some plastic deformation and the formation of slip lines appears to be an essential preliminary to the nucleation of a fatigue crack. Figure 5.34 shows one possible mechanism by which two intersecting slip planes, which are active in

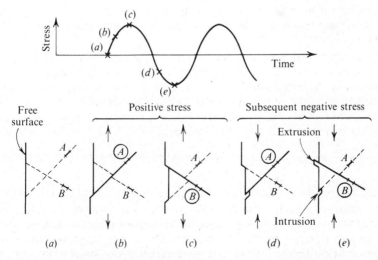

Fig. 5.34. Possible mechanism of nucleating a fatigue crack by two dislocation sources (*A* and *B*) which are alternately active during the stress cycle, as shown by the circle. (After A. H. Cottrell and D. Hall, *Proc. Roy. Soc.* **A242**, 211 (1957).)

succession, can produce extrusion and intrusion of material. The intrusion forms a crack nucleus and cycled plastic deformation at its root causes it to grow.

Designing against failure by fatigue, it can now be appreciated, is very much more complex and difficult than designing for static strength. There are a number of reasons for this, including: the pronounced size effect; the high sensitivity to local stress concentrations; the very large dependence on the environment, even if it is only mildly corrosive; and the wide statistical scatter. The size effect means that small specimens of representative material give different results in laboratory fatigue tests than do larger specimens or actual structures – and the larger the specimens the lower the fatigue strength, particularly when there are stress concentrators, as is normal. The sensitivity to stress concentrations, such as are caused by notches, bolt or rivet holes or surface scratches, means that so called 'design details' are often the controlling factor under fatigue loading and these are given far too little attention by designers. The influence of chemical action is complex and difficult to reduce to quantitative terms which can be utilised in design. Finally, the wide scatter of fatigue strength of a given material even under closely controlled test conditions adds to the difficulties of successful design against fatigue failure.

A consequence of these various problems is that fatigue is a subject in which structural engineers becomes as involved as materials scientists, and fatigue specialists are found in both disciplines, tackling the subject from the microscopic and phenomenological viewpoints.

QUESTIONS

1. (a) Sketch typical curves of (nominal) stress versus (linear) strain obtained in tensile tests of (i) ductile metal (ii) brittle ceramic (iii) mild steel.
   (b) Distinguish between: elastic and plastic strain; yield stress and yield point.
   (c) Define ductile and brittle materials.
2. The following figures were obtained in a tensile test of a specimen with 10 mm diameter and 50 mm gauge length:

Load (10^3 kg)	0.2	0.4	0.6	0.8	1.0	1.25	1.5	1.75	2.0	2.25	2.5	2.1
Extension (μm)	8.5	17	27	35	45	72	140	600	1500	3000	6900	10 000 (fracture)

   (a) Determine the allowable working stress if it must not exceed $0.25 \times$ TS or $0.8 \times$ YS (0.1 per cent proof).
   (b) From Barba's similarity law, what would be the value of the elongation (per cent) if the specimen diameter was 12.5 mm and the gauge length unchanged?
3. Derive an expression for the tensile strength (TS), given the strain-hardening curve can be represented by $\sigma_t = a\varepsilon^x$. Taking $a = 0.6$ kN/mm^2 and $x = 0.3$, what is the tensile strength and maximum uniform strain?
4. Using the tensile data in Fig. 5.2 determine the strain-hardening index and, assuming the power law holds to fracture, the true stress and local linear strain at fracture.
5. (a) How can the compressive yield curve be related to the tensile yield curve of a ductile metal?
   (b) For brittle material discuss the relationship between the fracture in tension, in compression and bending (flexural strength).

6. Describe briefly the three common tests of static indentation hardness. From the data in Question 2 (above) what would be the approximate hardness (DPN) of
   (*a*) the original annealed material, and
   (*b*) after fully strain hardening.
7. Derive an equation relating the tensile strength and the hardness (DPN).
8. (*a*) Distinguish between notch toughness and toughness.
   (*b*) Are all ductile metals notch tough?
   (*c*) What new factors are introduced into a notch toughness test (e.g. Izod) as compared with a tensile test? Which is the most significant?
9. Discuss: 'Plasticity is the most important mechanical property in structural materials.'
10. Which phenomena indicate that Hookean elasticity is only an idealised model for real materials?
11. (*a*) Distinguish between anelasticity and static hysteresis.
    (*b*) Define damping capacity.
    (*c*) How does the damping capacity vary with frequency or amplitude?
12. (*a*) Define creep and stress rupture.
    (*b*) Sketch a typical creep curve and identify the various stages.
    (*c*) Can the creep be correlated with the tensile test?
13. (*a*) Define homologous temperature and describe its significance to creep.
    (*b*) What is the fundamental physical cause of creep?
14. (*a*) What is an *S–N* diagram?
    (*b*) Distinguish between fatigue strength and fatigue limit.
    (*c*) Where does statistics enter the fatigue problem?
15. (*a*) Describe the appearance of a fatigue fracture, (i) in mild steel; (ii) in other materials.
    (*b*) What can be done to detect fatigue fractures before collapse?
16. Discuss some of the factors that depress the *S–N* curve to lower stresses.

FURTHER READING

General
D. McLean: *Mechanical Properties of Metals*. Wiley (1962).
R. E. Reed-Hill: *Physical Metallurgy Principles*. Van Nostrand (1964).

Plasticity
A. Nadai: *Theory of Flow and Fracture of Solids*, 2nd edition. McGraw-Hill (vol. 1, 1950; vol. 2, 1963).
P. W. Bridgman: *Large Plastic Flow and Fracture*. McGraw-Hill (1952).
Plasticity and Superplasticity. Institute of Metals Review Course (1961).

Hardness
D. Tabor: *The Hardness of Metals*. Clarendon Press (1951).
B. W. Mott: *Micro-indentation Hardness Testing*. Butterworth (1956).
H. O'Neill: *Hardness Measurement of Metals and Alloys*. Chapman and Hall (1967).

Notch toughness – see Chapter 10.

Anelasticity, damping capacity
C. Zener: *Elasticity and Anelasticity of Metals*. Chicago University Press (1948).
R. F. Hanstock: *The Non-Destructive Testing of Metals*. Institute of Metals (1951).

Yield point phenomena
E. O. Hall: *Yield Point Phenomena in Metals and Alloys*. Macmillan (1970).

Creep
A. H. Sully: *Metallic Creep*. Butterworth (1949).

F. Garofalo: *Fundamentals of Creep and Creep-Rupture in Metals.* Collier–Macmillan (1965).

A. I. Smith: Mechanical properties of materials at high temperatures. *Chart. Mech. Eng. (London)*, pp. 278–85 (1961).

Fatigue

Fatigue of Metals. Refresher Course, Institute of Metals (1955).

P. G. Forrest: *Fatigue of Metals.* Pergamon Press (1962).

6. Plasticity 1: macroscopic theory

6.1 Introduction

From the preceding chapter it will be clear that there are three separate phenomena to be distinguished when discussing the mechanical behaviour of solids:
elasticity, plasticity and fracture. Of these, plasticity is quite the most important
to an understanding of the 'strength of materials', at least as far as metals are
concerned, in spite of the fact that most books carrying this title are devoted to
the mathematical theory of elasticity. The useful strength of any material, or
the allowable stress as it is called in engineering design, is based on one or more
parameters which are functions of the plastic properties, such as the yield stress,
tensile strength or creep strength. Thus the yield stress is determined by the
onset of plastic flow and the tensile strength is dependent on the yield curve
(see § 5.3). Fracture is also important but, until recently, tends to be a secondary
factor: the fracture strength and notch toughness must be sufficiently high so
that the plasticity parameters are the controlling factors. Even fatigue strength
is closely tied to plastic flow to initiate the fracture. Much of the work in metallurgy is devoted to increasing the resistance to plastic flow by alloying or heat
treatment, as will be described in Chapter 11.

The theory of plasticity develops in three stages. In the first, macroscopic
theory, material is treated as a continuum. Given the observed behaviour of a
material undergoing plastic flow in simple tension, that is, its strain-hardening
curve, laws are developed which predict the onset of plasticity under combined
stresses (i.e. stresses along more than one axis) and the relations between stress
and strain in the plastic range. The theory finds application to various metal
working processes, such as rolling and extrusion, and to the design of structures
based on plastic collapse load, sometimes called *limit design*. (The more usual
procedure is to design against elastic failure of components.)

In the second stage, the crystal structure of materials is acknowledged and the
changes in atomic position during plastic flow are considered. As mentioned
briefly in § 2.8, the principal mode of plastic flow of crystalline materials is the
sliding of certain planes of atoms over each other in certain crystallographic
directions. The Miller indices of the slip planes and slip directions are called the
slip elements. The microscopic, or crystallographic, slip must be distinguished
from the macroscopic slip considered in continuum theory: the macroscopic
slip planes are not dependent on any particular physical mechanism of flow,
but in practice they represent the average of the crystallographic slip lines for
polycrystalline material. The crystallographic theory of slip is able to explain
why certain crystals are ductile and others brittle but, like the continuum theory,
it is unable to predict the value of the yield stress. Besides slip, plastic flow can
occur by a process called twinning. Details of slip and twinning in crystalline
materials will be given in Chapter 9.

The structural approach to the deformation of non-crystalline materials is much more complicated since there are no well defined slip planes on which sliding occurs. Some discussion of this type of plasticity will be given in Chapter 8 on high polymers, which are largely non-crystalline.

In the final, most important, stage of plasticity theory the mechanism of slip is considered. As outlined in § 2.8, slip on a slip plane does not develop homogeneously, like shearing a pack of playing cards, but occurs in an inhomogeneous manner, like the forward movement of a caterpillar. The 'hump' is a fault in the crystal lattice, called a *dislocation*, whose movement across a slip plane causes the atoms on opposite sides of the plane to be sheared relative to each other, (see Fig. 2.18). Dislocation theory, which will be given in Chapter 9, is able to explain and to some extent predict the yield stress, rate of shear hardening, and other plasticity effects. The influence of the microstructure, such as the presence of a second phase in various degrees of dispersion, on the yield stress and other plasticity parameters can also be rationalised.

This chapter covers the macroscopic, or continuum, theory of plasticity, commencing with the elements of elasticity theory to which it is closely allied. The plastic behaviour of real materials is very complex, and idealised models of materials have to be used for analysis to be possible. Nevertheless useful results are obtained. Problems involving strain in only two dimensions are more amenable to analysis and solutions are obtained by the method of slip-line fields, which will be described. The final section gives a very approximate method of analysing metal working processes, which gives the order of magnitudes of the forces and power involved.

6.2 Stress

The stress within a solid can be defined in two ways. True stress is the force divided by the instantaneous area over which the force is acting; nominal stress is the force divided by the original area (before loading). In a tensile test the cross-section area decreases as the specimen elongates, and the true stress is always larger than the nominal stress (see § 5.2). The reverse is true in compression. True stress is indicative of the actual stress the material is subjected to at a given moment, and therefore has a sound physical basis. Nominal stress is useful to engineers who are usually interested in the load that a given cross-section of a material will support. In elastic deformation strains are generally sufficiently small that it is unnecessary to distinguish between the two types of stress.

If the force is normal to the plane area over which it is acting, it produces a *normal* stress (Fig. 6.1(a)); if the force is parallel to the plane it produces a *shear* stress (Fig. 6.1(b)). A force acting at an arbitrary angle can be resolved into normal and shear components of stress (Fig. 6.1(c)).

In the study of both elasticity and plasticity, only deformation of a body is important; translation and rotation are restrained. To prevent translation, forces must occur as equal and opposite pairs acting on parallel planes. To prevent rotation, shearing forces on *perpendicular* planes must act in pairs in opposite directions such that their couples cancel; in Fig. 6.1(d), taking moments

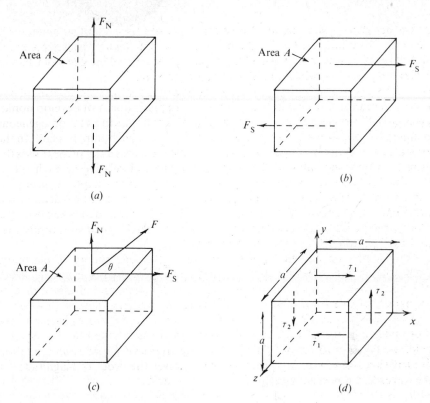

Fig. 6.1. Normal and shear stresses.
(a) Normal stress $\sigma = F_N/A$.
(b) Shear stress $\tau = F_S/A$.
(c) Resolving force F: $\sigma = F \sin \theta /A$, $\tau = F \cos \theta /A$.
(d) Complementary shear stresses: $\tau_1 = \tau_2$.

of the *forces* about the z-axis:

$$\tau_1 a^2 \times a = \tau_2 a^2 \times a$$

$$\tau_1 = \tau_2.$$

Such shear stresses are said to be *complementary*.

Figure 6.2 shows a small cube within a solid body subjected to external forces. The stress on each face of the cube is resolved into components which are parallel to the axes of the cube (X_1, X_2 and X_3). Each component is described as σ_{ij}, where the first index (i) indicates the face on which the stress acts and the second index (j) the direction. The index of the face (i) is that of the axis normal to it, the index of the direction (j) is that of the axis parallel to it. Thus σ_{31} acts on the cube face normal to the X_3-axis, in the X_1-direction. Only three faces are considered; from the preceding paragraph the stresses on the other three faces are equal and opposite. (This statement ignores second-order terms.) There are therefore nine components, and these are arranged to give the stress

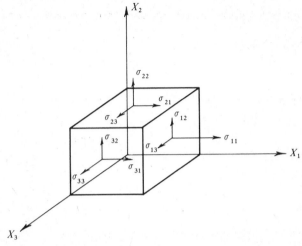

Fig. 6.2. Indexing of stresses: σ_{ij} indicates that the stress acts on the i-face (i.e. normal to the i-axis) in a direction parallel to the j-axis. σ_{ii} is therefore a normal stress; σ_{ij} ($i \neq j$) is a shear stress.

tensor

$$\begin{pmatrix} \sigma_{11} & \sigma_{12} & \sigma_{13} \\ \sigma_{21} & \sigma_{22} & \sigma_{23} \\ \sigma_{31} & \sigma_{32} & \sigma_{33} \end{pmatrix}.$$

The diagonal terms ($\sigma_{11}, \sigma_{22}, \sigma_{33}$) are all normal stresses. (A positive normal stress is called a tensile stress and a negative one is a compressive stress. The term tensile stress is sometimes used where normal stress is meant.) The off-diagonal terms are shear stresses. To prevent rotation the shear stresses must be complementary, that is $\sigma_{ij} = \sigma_{ji}$, and there are only six *independent* components of the stress tensor. The stress tensor completely defines the state of stress at a point in a body.

If the orientation of the cube axes is changed, the stress tensor will change. It can be shown that in some orientation the shearing stresses can always be reduced to zero, that is the stress tensor reduces to

$$\begin{pmatrix} \sigma_1 & 0 & 0 \\ 0 & \sigma_2 & 0 \\ 0 & 0 & \sigma_3 \end{pmatrix}.$$

The faces of the cube in this orientation are called the *principal (stress) planes* and the corresponding normal stresses are called the *principal stresses, σ_1, σ_2* and σ_3. These are defined such that $\sigma_1 > \sigma_2 > \sigma_3$. The principal planes are often obvious from the geometry of the system and this is one of the advantages of describing the system in terms of the principal stresses. There are still six variables to define the stress system: three for the principal stresses and three for the orientation of the principal axes. In an anisotropic material the direction of the axes in relation to the material property axes may be important. If $\sigma_1 = \sigma_2 = \sigma_3 = \sigma$ the system is said to be under *hydrostatic* stress, and $\sigma = -p$ where p is the hydrostatic pressure on the body.

The principal stress tensor may be separated into two parts:

$$\begin{pmatrix} \sigma_1 & 0 & 0 \\ 0 & \sigma_2 & 0 \\ 0 & 0 & \sigma_3 \end{pmatrix}$$

$$= \begin{pmatrix} \tfrac{1}{3}(\sigma_1 + \sigma_2 + \sigma_3) & 0 & 0 \\ 0 & \tfrac{1}{3}(\sigma_1 + \sigma_2 + \sigma_3) & 0 \\ 0 & 0 & \tfrac{1}{3}(\sigma_1 + \sigma_2 + \sigma_3) \end{pmatrix}$$

$$+ \begin{pmatrix} \sigma_1 - \tfrac{1}{3}(\sigma_1 + \sigma_2 + \sigma_3) & 0 & 0 \\ 0 & \sigma_2 - \tfrac{1}{3}(\sigma_1 + \sigma_2 + \sigma_3) & 0 \\ 0 & 0 & \sigma_3 - \tfrac{1}{3}(\sigma_1 + \sigma^2 + \sigma_3) \end{pmatrix}.$$

The first part is seen to represent a hydrostatic stress

$$\sigma_h = \tfrac{1}{3}(\sigma_1 + \sigma_2 + \sigma_3).$$

The second term is called the *deviatoric* stress (or *reduced* stress)

$$\sigma_{di} = \sigma_i - \sigma_h.$$

The hydrostatic component causes a change of volume (dilation) but no change of shape (distortion). This is because there are no shear stresses on planes of any orientation – as can be shown analytically or deduced from reasons of symmetry – every plane is identical and is therefore a principal plane. It has been found that hydrostatic stress does not cause plastic flow even at very high stress levels (see § 6.5). The deviatoric stress component, on the other hand, produces no dilation since their sum is always zero, but does produce shear stresses on other planes with consequent distortion and plastic flow.

Given the principal stress tensor, the stress tensor for any other orientation of the axes can be calculated by considering the equilibrium of forces acting on a small element of material. This will not be done here (see textbooks on theory of elasticity) but the following results will be quoted. For a plane with direction cosines l, m, n to the principal axes X_1, X_2, X_3, the normal stress is, see Fig. 6.3(a):

$$\sigma = l^2\sigma_1 + m^2\sigma_2 + n^2\sigma_3, \tag{6.1}$$

and the total shear stress is:

$$\tau^2 = l^2\sigma_1^2 + m^2\sigma_2^2 + n^2\sigma_3^2 - \sigma^2. \tag{6.2}$$

The maximum shear stress occurs on planes at 45° to the σ_1 and σ_3 directions (X_1, X_3), see Fig. 6.3(b). With $l = n = 1/\sqrt{2}$ and $m = 0$, the normal stress on these planes is $(\sigma_1 + \sigma_3)/2$ and the maximum shear stress is:

$$\tau_{max} = \frac{\sigma_1 - \sigma_3}{2} = \frac{\sigma_{d1} - \sigma_{d3}}{2} \tag{6.3}$$

(the subscript d indicates components of the deviatoric stress system). That is, the maximum shear stress is half the difference of the largest and smallest principal stress. This is an important result which should be remembered.

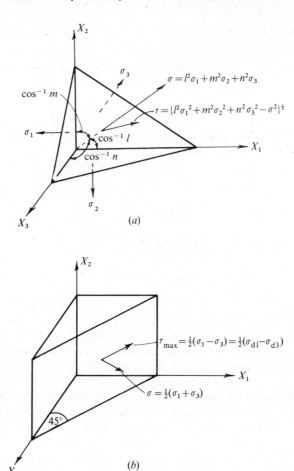

Fig. 6.3. (a) Normal stress σ and shear stress τ acting on a plane with direction cosines l, m, n, to the principal axes. (b) Plane of maximum shear stress.

6.3 Strain

Strain is a measure of the deformation of a body. There are, as for stress, two types of strain: normal (or direct) strain, and shear strain. Normal strain occurs in straightforward stretching, and is defined as the fractional change in length on deformation. Linear strain is the total change in length divided by the original length, see Fig. 6.4:

$$e = \int_{l_0}^{l_1} \frac{dl}{l_0} = \frac{l_1 - l_0}{l_0}.$$

Natural, or logarithmic, strain is the incremental change in length divided by the instantaneous length, $d\varepsilon = dl/l$, and the total natural strain resulting from a change in length from l_0 to l_1 is given by:

$$\varepsilon = \int_{l_0}^{l_1} \frac{dl}{l} = \ln \frac{l_1}{l_0}.$$

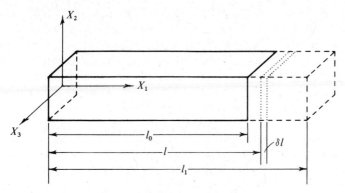

Fig. 6.4. Uniaxial normal strain.

$$\text{Linear strain } e = \int_{l_0}^{l_1} \frac{dl}{l_0} = \frac{l_1 - l_0}{l_0},$$

$$\text{Natural strain } \varepsilon = \int_{l_0}^{l_1} \frac{dl}{l} = \ln \frac{l_1}{l_0}.$$

Linear strain is larger than natural strain for the same extension in tension; the reverse is true in compression. For the small range of strains encountered in elastic deformation the difference between the two is negligible (as for nominal and true stress) and linear strain is generally used, as being mathematically simpler.

Shear strain occurs when parallel planes are moved relative to one another keeping the separation distance constant, such as occurs in the distortion of a rectangular block into a parallelepiped, see Fig. 6.5. The traditional shear

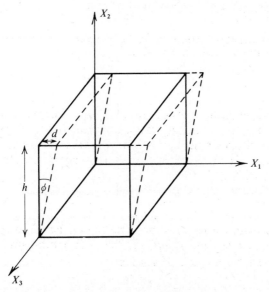

Fig. 6.5. Simple shear deformation. Shear strain $\sigma = d/h$.

strain is defined as the change of angle between two lines initially orthogonal to each other. Figure 6.6(*a*) shows shear deformation in the X_1X_2-plane. At the point O in the body, the lines OP and OQ are initially at right angles, and in the X_1 and X_2-direction; after deformation the lines have rotated to OP' and OQ' through angles e_{21} and e_{12} respectively (the reason for this symbolism will become apparent). The shear strain of the X_1X_2-plane in the X_1-direction, which equals that in the X_2-direction, is defined as

$$\gamma_{12} = \gamma_{21} = e_{21} + e_{12}. \tag{6.4}$$

(An alternative definition, sometimes used, is called the *rational* strain $\gamma'_{12} = (e_{21} + e_{12})/2$. This leads to complete identity between certain stress and strain equations. It will not be used here.)

If one line does not rotate, say $e_{21} = 0$, the deformation is called *simple shear*; the shear strain is then the rotation of one side only, i.e. ϕ in Fig. 6.5. If the angles are equal, $e_{21} = e_{12}$ (or in general terms $e_{ij} = e_{ji}$), it is called *pure shear*. Simple and pure shear are contrasted in Fig. 6.6(*b*) and (*c*). If both lines rotate by equal angles in the same direction, i.e. $e_{21} = -e_{12}$, solid body rotation has taken place without causing any straining.

It is always possible to divide the angular displacement (represented by e_{ij} and e_{ji} where $e_{ij} \neq e_{ji}$) into a pure shear component (given the symbol $\varepsilon_{ij} = \varepsilon_{ji}$) and a solid body rotative component (given the symbol $\omega_{ij} = -\omega_{ji}$).

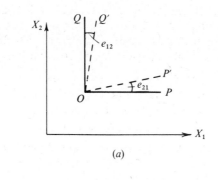

(*a*)

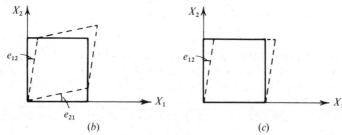

(*b*)   (*c*)

Fig. 6.6. Shear strain: pure and simple shear.
(*a*) General shear strain $\gamma_{12} = \gamma_{21}$ is defined as $e_{12} + e_{21}$. In general shear deformation, $e_{12} \neq e_{21}$.
(*b*) Pure shear $e_{12} = e_{21}$. $\gamma_{12} = \gamma_{21} = e_{12} + e_{21} = 2e_{12} = 2e_{21}$. (*NB* No rotation of principal axes.)
(*c*) Simple shear $e_{21} = 0$. $\gamma_{21} = \gamma_{12} = e_{21} + e_{12} = e_{12}$. (*NB* Equivalent to pure shear + rotation, see text.)

These are given by

$$\left.\begin{array}{ll}\text{pure shear component} & \varepsilon_{ij} = \tfrac{1}{2}(e_{ij} + e_{ji}); \\ \text{rotative component} & \omega_{ij} = \tfrac{1}{2}(e_{ij} - e_{ji}).\end{array}\right\} (6.5a)$$

By considering the effect of interchanging i and j it will be apparent that $\varepsilon_{ij} = \varepsilon_{ji}$ (i.e. ε is pure shear) and $\omega_{ij} = -\omega_{ji}$ (i.e. ω is solid body rotation). By adding,

$$\left.\begin{array}{l}e_{ij} = \varepsilon_{ij} + \omega_{ij}, \\ e_{ji} = \varepsilon_{ji} + \omega_{ji}.\end{array}\right\} (6.5b)$$

and by symmetry

Just as the stress tensor describes the state of stress at a point in a body, a strain tensor may be developed which describes the state of strain (i.e. the direct strains and the shear strains with respect to given axes). This is done by considering the displacement of points in the body relative to some fixed axes external to the body. The displacement (u_1, u_2, u_3) of a point is the difference between its coordinates before (x_1, x_2, x_3) and after (x'_1, x'_2, x'_3) deformation:

$$u_1 = x'_1 - x_1,$$
$$u_2 = x'_2 - x_2,$$
$$u_3 = x'_3 - x_3.$$

A constant displacement everywhere corresponds to a rigid displacement (translation), involving no straining, and is of no interest here. In a strained body the displacement must vary from point to point, and the u_is are functions of position (x_1, x_2, x_3). If the strain is non-uniform the function differs at different points, but for uniform strain the relation between displacement and position must be the same for all positions. Fortunately for mathematical simplicity, nearly all materials show a *linear* relationship between elastic displacement and position, given by:

$$\left.\begin{array}{ll}u_1 = e_{11}x_1 + e_{12}x_2 + e_{13}x_3 & (a) \\ u_2 = e_{21}x_1 + e_{22}x_2 + e_{23}x_3 & (b) \\ u_3 = e_{31}x_1 + e_{32}x_2 + e_{33}x_3 & (c)\end{array}\right\} (6.6)$$

where the e_{ij} are the *coefficients of proportionality*, and form the components of the strain tensor:

$$\begin{pmatrix} e_{11} & e_{12} & e_{13} \\ e_{21} & e_{22} & e_{23} \\ e_{31} & e_{32} & e_{33} \end{pmatrix}.$$

The physical meaning of these components can be illustrated by considering the deformation in uniaxial extension and simple shear. A bar (Fig. 6.4) aligned parallel to the X_1-axis has its length increased from l_0 to l_1, while being constrained so as to prevent any change in its other dimensions $(u_2 = u_3 = 0)$. The displacement of the free end of the bar in the X_1-direction is (the point $x_1 = 0$ being fixed):

$$u_1(l_0, x_2, x_3) = l_1 - l_0,$$

and of the mid-section:

$$u_1\left(\frac{l_0}{2}, x_2, x_3\right) = \frac{l_1}{2} - \frac{l_0}{2}.$$

From which it can be deduced that for a point x_1 along the bar the displacement is:

$$u_1(x_1, x_2, x_3) = \frac{x_1}{l_0}l_1 - \frac{x_1}{l_0}l_0 = x_1 e,$$

where e is the linear strain. As the deformation is uniform and only in the X_1-direction, u_1 depends only on x_1 but not on x_2 or x_3; that is, in (6.6a),

$$e_{12} = e_{13} = 0$$

and

$$u_1 = e_{11}x_1.$$

Thus e_{11} is the linear normal strain e in the X_1-direction.

Consider now simple shear of the rectangular block in Fig. 6.5; the $X_1 X_3$ planes (normal to X_2) are sheared in the X_1-direction. It can be seen that, again, displacements have occurred only in the X_1-direction ($u_2 = u_3 = 0$) and are given by:

$$u_1(x_1, x_2, x_3) = x_1' - x_1 = x_2 \phi.$$

This time the displacement depends upon x_2 alone, that is

$$e_{11} = e_{13} = 0$$

and

$$u_1 = e_{12}x_2 = \phi x_2.$$

Thus in simple shear e_{12} is the shear strain in the X_1-direction of planes normal to X_2. In general terms, e_{ij} represents the rotation of a line in the ij-plane initially parallel to the j-axis towards the i-axis direction, as indicated in Fig. 6.6(a).

From this it can be seen that the diagonal terms represent the normal strains in the three directions; the off-diagonal terms represent shearing. (Note the component $e_{ii} \neq \gamma_{ii}$, the shear strain in the ij-plane.)

When all components but e_{11} are equal to zero the body is in a state of uniaxial strain. This is not the same as simple tension (i.e. a tensile force acting along the X_1-axis alone) since this is accompanied by contraction in the X_2 and X_3-directions, due to the Poisson effect (see § 5.2). The strain tensor for simple tension is:

$$\begin{pmatrix} e_{11} & 0 & 0 \\ 0 & -ve_{11} & 0 \\ 0 & 0 & -ve_{11} \end{pmatrix}.$$

To obtain equivalence with the stress tensor it is necessary to separate the pure shear components and the solid body rotative components as indicated above in discussing the definition of shear strain, (6.5). That is, separate the e_{ij}-tensor into:

pure shear tensor components $\varepsilon_{ij} = \frac{1}{2}(e_{ij} + e_{ji})$

and   rotative tensor components $\omega_{ij} = \frac{1}{2}(e_{ij} - e_{ji})$.

The ω-tensor is then of no further interest. The pure shear strain tensor has six independent components, since $\varepsilon_{ij} = \varepsilon_{ji}$:

$$\begin{pmatrix} \varepsilon_{11} & \varepsilon_{12} & \varepsilon_{13} \\ \varepsilon_{21} & \varepsilon_{22} & \varepsilon_{23} \\ \varepsilon_{31} & \varepsilon_{32} & \varepsilon_{33} \end{pmatrix}.$$

Note now that $\varepsilon_{ij} = \varepsilon_{ji} = \gamma_{ij}/2$ (cf. e_{ij} above). The theory develops in a similar manner to that for stresses.

The strain tensor for any other orientation of the axes can be found from consideration of the compatibility of strains (cf. equilibrium of forces for the stress tensor). At some orientation the pure shear strain components are reduced to zero, giving a strain tensor

$$\begin{pmatrix} \varepsilon_1 & 0 & 0 \\ 0 & \varepsilon_2 & 0 \\ 0 & 0 & \varepsilon_3 \end{pmatrix}$$

where ε_1, ε_2 and ε_3 are called the principal strains. One of the advantages of working in terms of the pure shear tensor is that the directions of the principal strain axes do not alter during straining. In isotropic materials the principal strain axes coincide with the principal stress axes; in fact this can be used as a definition of isotropy. The maximum shear strains are at 45° to the principal axes and equal the difference of the principal strains (cf. (6.3), $\tau_{max} = (\sigma_1 - \sigma_3)/2$).

As with the stress tensor, the strain tensor may be divided into two parts:

$$\begin{pmatrix} \varepsilon_1 & 0 & 0 \\ 0 & \varepsilon_2 & 0 \\ 0 & 0 & \varepsilon_3 \end{pmatrix}$$

$$= \begin{pmatrix} \tfrac{1}{3}(\varepsilon_1 + \varepsilon_2 + \varepsilon_3) & 0 & 0 \\ 0 & \tfrac{1}{3}(\varepsilon_1 + \varepsilon_2 + \varepsilon_3) & 0 \\ 0 & 0 & \tfrac{1}{3}(\varepsilon_1 + \varepsilon_2 + \varepsilon_3) \end{pmatrix}$$

$$+ \begin{pmatrix} \varepsilon_1 - \tfrac{1}{3}(\varepsilon_1 + \varepsilon_2 + \varepsilon_3) & 0 & 0 \\ 0 & \varepsilon_2 - \tfrac{1}{3}(\varepsilon_1 + \varepsilon_2 + \varepsilon_3) & 0 \\ 0 & 0 & \varepsilon_3 - \tfrac{1}{3}(\varepsilon_1 + \varepsilon_2 + \varepsilon_3) \end{pmatrix}.$$

The first part with three equal strain components represents pure dilation, that is, change in volume but not in shape. (By symmetry there are no shear strains at any orientation of the axes.) The second part represents the deviation (distortion) strain and, as for the deviatoric stress, the sum of its three components is zero, that is, there is no volumetric change since

$$\frac{\Delta v}{v} = (1 + \varepsilon_1)(1 + \varepsilon_2)(1 + \varepsilon_3) - 1 = \varepsilon_1 + \varepsilon_2 + \varepsilon_3.$$

6.4 Elasticity

The mathematical theory of elasticity is based upon two laws. The first is Hooke's Law, which states that the stress in an elastic body is directly proportional to the strain, that is,

$$\sigma = \text{const. } \varepsilon.$$

The constant is called the elastic modulus. The second is the principle of superposition. This states that if a stress σ_A produces a strain ε_A, and a stress σ_B produces a strain ε_B, then the application of σ_A and σ_B together will produce a strain equal to $(\varepsilon_A + \varepsilon_B)$. This holds provided that $(\sigma_A + \sigma_B)$ does not exceed the limit of proportionality.

Three elastic moduli are recognised. Young's modulus E relates tensile stress to tensile strain; the shear or rigidity modulus G relates shear stress to shear strain; and the bulk modulus K relates hydrostatic stress to dilatational strain. The other constant characteristic of an elastic body is Poisson's ratio v, which relates longitudinal to lateral strain in simple tension. The value of v is 0.25–0.33 for most metals, and ~ 0.1 for solids with highly directional bonds, such as strongly covalent crystals. For there to be no change in volume on deformation, v should equal 0.5. The proof of this is left as an exercise at the end of the chapter.

Various relations can be shown to exist between the moduli:

$$K = \frac{E}{3(1 - 2v)}, \quad G = \frac{E}{2(1 + v)}, \quad \left(\text{or } E = \frac{3G}{1 + G/3K} \right).$$

Hence only two of the four constants need to be known for any material. Values of these constants for typical metals are given in Table 6.1.

For a solid whose elastic constants are isotropic, the general relationships between stresses and strains are fairly simple. Consider for example a body under principal stresses σ_1, σ_2 and σ_3 which have produced principal strains ε_1, ε_2 and ε_3. ε_1 is the strain in the X_1-direction due to σ_1 plus the Poisson strain in the X_1-direction due to σ_2 and σ_3, and is therefore given by

$$\varepsilon_1 = \frac{1}{E}\sigma_1 - \frac{v}{E}\sigma_2 - \frac{v}{E}\sigma_3;$$

similarly
$$\varepsilon_2 = \frac{1}{E}\sigma_2 - \frac{v}{E}\sigma_3 - \frac{v}{E}\sigma_1,$$

and
$$\varepsilon_3 = \frac{1}{E}\sigma_3 - \frac{v}{E}\sigma_1 - \frac{v}{E}\sigma_2.$$

$$\left. \right\} \quad (6.7a)$$

Eliminating E leads to the two equations:

$$\frac{\varepsilon_1}{\sigma_1 - v(\sigma_2 + \sigma_3)} = \frac{\varepsilon_2}{\sigma_2 - v(\sigma_3 + \sigma_1)} = \frac{\varepsilon_3}{\sigma_3 - v(\sigma_1 + \sigma_2)}. \quad (6.7b)$$

In most engineering materials the grain size is sufficiently small and randomly oriented for the material to be truly isotropic, and the above formulae may be used.

Table 6.1. *Room temperature elastic moduli for pure metals*

Metal	E GN/m² (10^6 psi)	G GN/m² (10^6 psi)	v
Al; Al alloys	72.5 (10.5)	27.6 (4.0)	0.31
Copper	110 (16.0)	41.5 (6.0)	0.33
Plain carbon steel (Fe–C alloy)	200 (29.0)	76.0 (11.0)	0.33
Stainless steel (18Cr–8Ni)	193 (28.0	65.5 (9.5)	0.28
Titanium	117 (17.0)	45.0 (6.5)	0.31
Tungsten	400 (58.0)	157 (22.8)	0.27

In single crystals and heavily cold-worked materials the elastic properties may be anisotropic, that is, vary in different directions. To account for this Hooke's Law must be written in completely generalised terms by relating each component of the strain tensor to the six components of the stress tensor. It can be seen that 36 coefficients are needed to do this. The equations may be written:

$$\varepsilon_{11} = S_{11}\sigma_{11} + S_{12}\sigma_{22} + S_{13}\sigma_{33} + S_{14}\sigma_{12} + S_{15}\sigma_{23} + S_{16}\sigma_{31}$$
$$\varepsilon_{22} = S_{21}\sigma_{11} + S_{22}\sigma_{22} + S_{23}\sigma_{33} + S_{24}\sigma_{12} + S_{25}\sigma_{23} + S_{26}\sigma_{31}$$
$$\varepsilon_{33} = S_{31}\sigma_{11} + \cdots$$
$$\varepsilon_{12} = S_{41}\sigma_{11} + \cdots \quad ; \quad \varepsilon_{23} = S_{51}\sigma_{11} + \cdots \quad ; \quad \varepsilon_{31} = S_{61}\sigma_{11} + \cdots$$

where the Ss are called moduli of compliance. A similar set of equations relates stress to strain:

$$\sigma_{11} = C_{11}\varepsilon_{11} + C_{12}\varepsilon_{22} + C_{13}\varepsilon_{33} + C_{14}\varepsilon_{12} + C_{15}\varepsilon_{23} + C_{16}\varepsilon_{31}$$

where the Cs are called the elastic coefficients (constants). Symmetry considerations reduce the number of coefficients from 36 to 21, as $S_{ij} = S_{ji}$ (or $C_{ij} = C_{ji}$). Thus, a crystal with the structure of lowest symmetry (triclinic) needs 21 coefficients to describe its elastic properties. Materials which crystallise in structures of higher symmetry need fewer coefficients, and in the cubic case:

$$S_{11} = S_{22} = S_{33},$$
$$S_{12} = S_{13} = S_{23},$$
$$S_{14} = S_{15} = S_{16} = S_{24} = S_{25} = S_{26} = S_{34} = S_{35} = S_{36}$$
$$= S_{45} = S_{46} = S_{56} = 0$$

and

$$S_{44} = S_{55} = S_{66}.$$

Thus only three independent coefficients are needed for a cubic material and:

$$\varepsilon_{11} = S_{11}\sigma_{11} + S_{12}(\sigma_{22} + \sigma_{33})$$
$$\varepsilon_{22} = S_{11}\sigma_{22} + S_{12}(\sigma_{33} + \sigma_{11})$$
$$\varepsilon_{33} = S_{11}\sigma_{33} + S_{12}(\sigma_{11} + \sigma_{22})$$
$$\varepsilon_{12} = S_{44}\sigma_{12}$$
$$\varepsilon_{23} = S_{44}\sigma_{23}$$
$$\varepsilon_{31} = S_{44}\sigma_{31}.$$

Comparing the above and (6.7a), it can be seen that:

$$S_{11} = \frac{1}{E}, \quad S_{12} = \frac{v}{E} \quad \text{and} \quad S_{44} = \frac{1}{2G}.$$

For an isotropic cubic material (a non-cubic material could not be isotropic) the number of independent constants reduces to two, since:

$$S_{44} = 2(S_{11} + S_{12}).$$

The dependence of the elastic constants upon interatomic forces has been demonstrated in Chapter 3. The elastic behaviour of a material is controlled by these forces and by the crystal structure, and is little influenced by microstructure.

6.5 Criterion of yielding for ductile materials

It has been shown (Chapter 5) that a ductile material yields in a simple tensile test at some stress level, say Y. This is the only principal stress present, the other two being zero. A yield criterion is required which will indicate when the same material will yield under combined stresses, that is, when there is more than one principal stress, for example, as in torsional loading.

Experiments have shown that the application of a pure hydrostatic stress system will not produce yielding. This is not unexpected since a hydrostatic stress produces only dilation whilst it is known that plastic deformation involves no volumetric change (§ 5.2). Thus any yielding criterion must be independent of the hydrostatic component of a stress system, or, in other words, yielding occurs when a function of the deviatoric stress components reaches a critical value, which depends on the particular material and its history. It should be noted that the concept of a yield criterion is not restricted only to the initial loading from the annealed state but also applies during strain hardening.

In Fig. 6.7 the principal stresses are plotted along coordinate axes and a point in the coordinate space thus defined represents a particular stress system, that is, one with principal stresses given by the coordinates of the point. For each material some points will represent stress systems which do not cause yielding, and some represent systems which cause the material to yield. A surface, called the *yield surface*, can be constructed so as to separate these two sets of points; points on this surface are those which just satisfy the yield criterion. Since a line making equal angles with the axes, that is $\sigma_1 = \sigma_2 = \sigma_3$, represents hydrostatic stress only, the yield surface will be centred on this line.

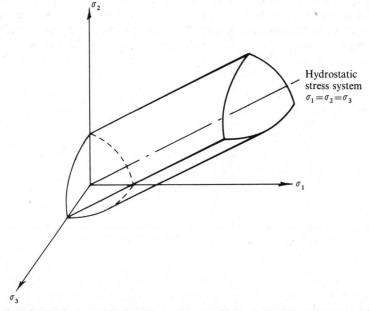

Fig. 6.7. Yield surface in principal stress space. Since yielding is independent of the hydrostatic stress component, the surface forms a prism on the axis $\sigma_1 = \sigma_2 = \sigma_3$.

Various shapes for the yield surface are possible, representing different criteria of yielding. For any isotropic material it must have threefold symmetry about the hydrostatic axis, and, if the yield stress in tension is equal to that in compression, the symmetry will be sixfold. This is assumed in the theory of plasticity.

The simplest yield criterion is Tresca's maximum shear theory of 1864, which states that yielding occurs when the maximum shear stress exceeds a critical value. In simple tension ($\sigma_1 = \sigma$, $\sigma_2 = \sigma_3 = 0$), the maximum shear stress is on planes inclined 45° to the line of action of σ, and is equal to $\sigma/2$ (see §6.2). When all three principal stresses have finite values, the maximum shear stress is equal to $\frac{1}{2}(\sigma_1 - \sigma_3)$, and acts on planes which are inclined at 45° to both σ_1 and σ_3 (see Fig. 6.3). The Tresca, or maximum shear stress, yield criterion is thus:

$$\tfrac{1}{2}(\sigma_1 - \sigma_3) \geqslant A, \tag{6.8a}$$

where A is a constant of the material. In terms of the deviatoric stresses ($\sigma_{di} \equiv \sigma_i - \frac{1}{3}(\sigma_1 + \sigma_2 + \sigma_3)$), the same equation holds:

$$\tfrac{1}{2}(\sigma_{d1} - \sigma_{d3}) \geqslant A.$$

The yield surface is a hexagonal prism with the line $\sigma_1 = \sigma_2 = \sigma_3$ as its axis. Its projection on a plane normal to this line is a hexagon, as shown in Fig. 6.8. In simple tension, σ_1 is equal to the tensile yield stress Y, and $\sigma_2 = \sigma_3 = 0$. The constant A thus has the value $Y/2$, and the Tresca criterion for general yielding becomes

$$\sigma_1 - \sigma_3 \geqslant Y. \tag{6.8b}$$

Consider a thin walled tube loaded in pure torsion (Fig. 6.9). If τ is the shear stress on the orthogonal section (τ = torque/wall section area × radius) then the principal stresses are $\sigma_1 = \tau$, $\sigma_2 = 0$ and $\sigma_3 = -\tau$. The maximum shear stress is the torsion shear stress (τ) on the orthogonal section – equal to the complementary shear stress on a longitudinal section. If yielding occurs when the shear stress exceeds K, that is, $\sigma_1 = K$ and $\sigma_3 = -K$, then by the Tresca

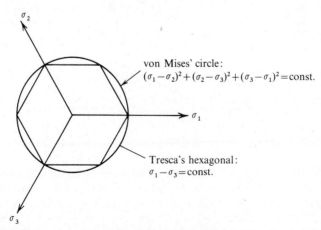

Fig. 6.8. Cross-section through the yield surfaces according to Tresca and von Mises theories. (Section is orthogonal to the hydrostatic stress line, see Fig. 6.7.)

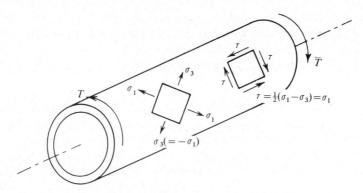

Fig. 6.9. Stress system for a thin walled tube in torsion.

yield criterion (6.8*b*) $2K = Y$. The tensile yield stress is twice the torsion shear yield stress.

An alternative criterion is due to von Mises (1913). The hexagonal surface is replaced by a cylindrical one (Fig. 6.8), represented by:

$$(\sigma_1 - \sigma_2)^2 + (\sigma_2 - \sigma_3)^2 + (\sigma_3 - \sigma_1)^2 = B \qquad (6.9a)$$

where B is a constant of the material. In terms of the deviatoric stresses the criterion becomes:

$$\sigma_{d1}^2 + \sigma_{d2}^2 + \sigma_{d3}^2 = B/3 \qquad (6.9b)$$

This criterion was chosen originally for mathematical convenience, but it applies to real materials rather better than the Tresca criterion; the only exception is the upper yield point of mild steel. Hencky showed in 1924 that the left-hand side of (6.9) represents the shear, or distortion, strain energy. It also represents the shear stress on the octahedral planes. (Put $l = m = n = 1/\sqrt{3}$ in (6.2).) To evaluate B, in simple tension at yield:

$$\sigma_1 = Y, \qquad \sigma_2 = \sigma_3 = 0$$

and
$$(Y - 0)^2 + (0 - 0)^2 + (0 - Y)^2 = 2Y^2 = B.$$

Considering again a thin walled cylinder subjected to pure torsion, at yield:

$$\sigma_1 = -\sigma_3 = K, \quad \text{and} \quad \sigma_2 = 0.$$

Hence
$$(K - 0)^2 + (0 - (-K))^2 + ((-K) - K)^2 = 6K^2 = B.$$

Thus, according to von Mises' criterion,

$$2Y^2 = 6K^2 \quad \text{or} \quad Y = \sqrt{3}K.$$

The ratio of the normal yield stress in tension $Y(=\sigma_1)$ to the shear yield stress in torsion K ($=\sigma_1 = -\sigma_3$) is 2.0 according to Tresca, and 1.73 according to von Mises. This provides a means of checking the theories by experiment. In some classic experiments by Taylor and Quinney in 1931, thin-walled tubes were loaded first to yield in tension (Y) and checked for isotropy by showing there was no change in volume of the tube. The tension load was then reduced to various levels and a torsion load applied up to yield. If the direct axial stress due to the tensile load is σ and the shear stress due to the torsion is τ, it can be

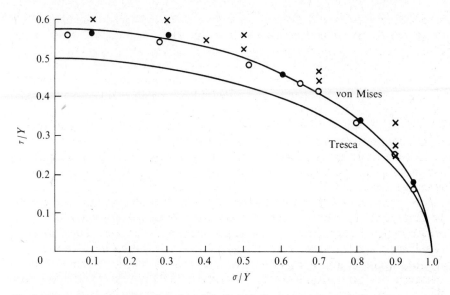

Fig. 6.10. Experimental confirmation of von Mises' yield criterion. ○ copper, ● aluminium, × steels. (From G. I. Taylor and H. Quinney, *Phil. Trans. Roy. Soc. Lond.* A**230**, 323 (1931).)

shown that the principal stresses are:

$$\sigma_1 = \tfrac{1}{2}\sigma + (\tfrac{1}{4}\sigma^2 + \tau^2)^{\frac{1}{2}}$$
$$\sigma_2 = 0$$
$$\sigma_3 = \tfrac{1}{2}\sigma - (\tfrac{1}{4}\sigma^2 + \tau^2)^{\frac{1}{2}}.$$

From Tresca's criterion

$$\sigma_1 - \sigma_3 = Y$$

i.e.

$$\left(\frac{\sigma}{Y}\right)^2 + \left(\frac{\tau}{Y/2}\right)^2 = 1.$$

From von Mises' criterion

$$(\sigma_1 - \sigma_2)^2 + (\sigma_2 - \sigma_3)^2 + (\sigma_3 - \sigma_1)^2 = 2Y^2$$

i.e.

$$\left(\frac{\sigma}{Y}\right)^2 + \left(\frac{\tau}{Y/\sqrt{3}}\right)^2 = 1.$$

Plotting the experimental data on σ/Y against τ/Y axes, ellipses should be obtained cutting the σ/Y axis at 1 and the τ/Y axis at 0.5 (Tresca) or 0.576 (von Mises). Figure 6.10 reproduces the results of Taylor and Quinney for several materials and it is apparent that von Mises' criterion closely fits the data, except for steels. Their experiments also showed that the principal strain axes correspond to the principal stress axes.

6.6 Continuum plasticity

Once the stress system satisfies the yield criterion, the laws of elasticity are no longer sufficient and a plasticity theory is required to relate the stresses and strains in the plastic regime. Both elastic (recoverable) and plastic (non-recoverable) strains will be present although the former can sometimes be neglected. As mentioned in the introduction (and Chapter 5) the behaviour of real materials in the plastic range is complex. The value of the initial yield stress, or the yield criterion, is a function of the material composition and its history. The rate at which it rises due to strain hardening is similarly dependent on the material and its history, and often obeys no simple analytical function. In addition, the plastic deformation is irreversible, that is, it is not dependent on the instantaneous value of the stress system – for example, the plastic strains do not become zero when the stresses are removed.

In order to develop a mathematical theory of plasticity, various simplifying assumptions have to be made. In many applications the plastic strains are very large compared with the elastic strains, which can be neglected. The material model is then called *rigid–plastic*. There are some problems where this cannot be done, notably where residual stresses are being investigated. These are important in various situations, for example, in increasing the elastic strength of gun barrels by prior plastic straining by internal pressure, called autofrettage, and in raising fatigue strength by 'shot peening' in which a surface layer in compression is formed. Another simplification which is often made is to ignore the strain hardening and assume the flow continues at constant stress. The material is then said to be *perfectly* (or *ideally*) *plastic*. In many high temperature metal working processes this is very reasonable, except that the value of the yield stress is dependent on the strain rate which may vary across the deforming body. Finally a combination of both approximations can be used, giving a material called *rigid–ideally plastic*. The various material models are illustrated in Fig. 6.11.

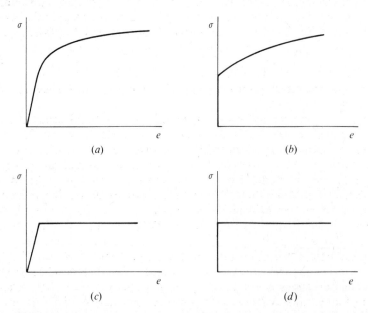

Fig. 6.11. Material models used in continuum plasticity theory. In the text only the rigid–ideally plastic model is used. (*a*) Real material, (*b*) rigid–plastic model, (*c*) elastic–ideally plastic model, (*d*) rigid–ideally plastic model.

After yielding, the total strains at any moment may bear no relation to the instantaneous stresses. For example a tubular specimen may be strained plastically in tension and then be further strained by internal pressure. The theory of plasticity postulates that *increments of plastic strain* always depend on the instantaneous values of the stresses. The constant of proportionality will have an instantaneous, non-negative, value. In a strain-hardening material, this varies throughout a straining programme and is assumed to depend on the total plastic work already done on the material. Just as in the elastic range, the strain increment in a particular principal direction depends not only on the stress in that direction but also on the orthogonal stresses. It is therefore possible to write equations similar to (6.7) for the plastic flow equations:

$$\delta\varepsilon_1^p = \lambda(\sigma_1 - v\sigma_2 - v\sigma_3),$$
$$\delta\varepsilon_2^p = \lambda(\sigma_2 - v\sigma_3 - v\sigma_1),$$
$$\delta\varepsilon_3^p = \lambda(\sigma_3 - v\sigma_1 - v\sigma_2).$$

Putting v equal to 0.5, since there is no volume change during plastic flow, and eliminating the constant of proportionality λ gives:

$$\frac{\delta\varepsilon_1^p}{\sigma_1 - 0.5(\sigma_2 + \sigma_3)} = \frac{\delta\varepsilon_2^p}{\sigma_2 - 0.5(\sigma_3 + \sigma_1)} = \frac{\delta\varepsilon_3^p}{\sigma_3 - 0.5(\sigma_1 + \sigma_2)}, \quad (6.10a)$$

or in terms of the deviatoric stresses the flow rule is

$$\frac{\delta\varepsilon_1^p}{\sigma_{d1}} = \frac{\delta\varepsilon_2^p}{\sigma_{d2}} = \frac{\delta\varepsilon_3^p}{\sigma_{d3}}. \quad (6.10b)$$

That is, the plastic strain increment in a principal direction is proportional to the deviatoric stress in that direction. Note that these equations give only the *ratios* of the plastic strain increments and their absolute magnitude has to be obtained otherwise. If the axis directions are not principal directions:

$$\frac{\delta\varepsilon_{11}^p}{\sigma_{d11}} = \frac{\delta\varepsilon_{22}^p}{\sigma_{d22}} = \frac{\delta\varepsilon_{33}^p}{\sigma_{d33}} = \frac{\delta\varepsilon_{12}^p}{2\sigma_{12}} = \frac{\delta\varepsilon_{23}^p}{2\sigma_{23}} = \frac{\delta\varepsilon_{31}^p}{2\sigma_{31}}. \quad (6.10c)$$

(In texts using the rational shear strain, half the conventional shear strain, the factor of two is eliminated.)

For an elastic–plastic material ε^p represents only part of the total strain, the other component being the elastic strain given by Hooke's Law. The stress–strain relationships for an elastic–plastic model are named after Reuss (in 1930) and Prandtl (in 1924). For large plastic flows, when elastic strains can be neglected and the rigid–plastic model is used, the only strains are those given by (6.10), and the superscript p can then be omitted. The equations are then named after Levy (in 1871) and von Mises (in 1913). Since the generalised theory is due to Reuss and all the other equations are special forms, (6.10) will be referred to here as the Reuss equations.

As an example of the application of these equations a thin walled cylinder subjected to an internal pressure p will be considered (Fig. 6.12). The hoop (circumferential) stress is evaluated by cutting the cylinder in half axially, as shown in Fig. 6.12(*b*). Balancing the pressure force against the wall force per unit length gives:

$$2rp = 2t\sigma_1, \quad \text{i.e.} \quad \sigma_1 = \frac{pr}{t}.$$

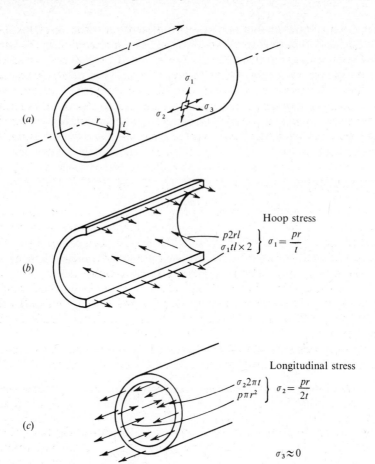

Fig. 6.12. Stress system for a thin walled tube under internal pressure.

Similarly for a radial section (Fig. 6.12(c)), the axial stress σ_2 is given by:

$$\pi r^2 p = 2\pi r t \sigma_2, \quad \text{i.e.} \quad \sigma_2 = \frac{pr}{2t} \quad (= \sigma_1/2).$$

The radial stress σ_3, provided the wall is thin, is effectively zero. According to von Mises the cylinder will yield when:

$$(2\sigma_2 - \sigma_2)^2 + \sigma_2^2 + (-2\sigma_2)^2 \geq 6K^2 \quad (\text{or } 2Y^2),$$

i.e.

$$\sigma_2 \geq K, \quad \text{or} \quad p = \frac{2tK}{r}.$$

The strain increments after yielding are given by Reuss's equation:

$$\frac{\delta\varepsilon_1}{2\sigma_2 - 0.5\sigma_2} = \frac{\delta\varepsilon_2}{\sigma_2 - \sigma_2} = \frac{\delta\varepsilon_3}{-0.5(2\sigma_2 + \sigma_2)}.$$

Hence

$$\delta\varepsilon_2 = 0 \quad \text{and} \quad \delta\varepsilon_1 = -\delta\varepsilon_3.$$

The cylinder does not strain plastically in the axial (σ_2) direction and the increase in diameter (and circumference) is compensated by a reduction in thickness to keep the volume of material constant.

This result has an interesting application to pneumatic tyres, which, since rubber has a Poisson's ratio of 0.5, behave elastically in the same way as metals in the plastic range (except that the strain is recoverable). If the cylinder is bent into a circle and the ends joined, a tyre configuration is obtained. Inflating the tyre causes no change in the major circumferential length and the tyre stays snugly on the wheel rim.

6.7 Plane plastic strain for rigid–perfectly plastic material

Many problems in plasticity involve only plane strain and solutions can be obtained more easily than in the case of general strain problems through the use of slip-line field theory (see § 6.8). Here all displacements are parallel to one plane and all dimensions perpendicular to that plane are unchanged. That is, one of the principal strains is zero and, from the convention that $\sigma_1 > \sigma_2 > \sigma_3$, this will be found to be ε_2. Examples include rolling, extrusion, machining and indenting. Another example is constraint due to a notch, which will be discussed later in this section. In the following, the material will be assumed to be rigid–plastic.

The strain increments just after yielding are given by the Reuss equation (6.10), and for plane strain with ε_2, and hence $\delta\varepsilon_2$, equal to zero:

$$\sigma_2 - \tfrac{1}{2}(\sigma_3 + \sigma_1) = 0,$$

i.e.
$$\sigma_2 = (\sigma_1 + \sigma_3)/2.$$

The stress in the no-strain direction is always the mean of the two principal stresses in the plane of strain. The principal stress tensor for plane strain can thus be divided into

$$\begin{pmatrix} \sigma_1 & & \\ & \sigma_2 \left(= \dfrac{\sigma_1 + \sigma_3}{2} \right) & \\ & & \sigma_3 \end{pmatrix}$$

$$= \begin{pmatrix} \sigma_2 \, (= -p) & & \\ & \sigma_2 & \\ & & \sigma_2 \end{pmatrix} + \begin{pmatrix} \dfrac{\sigma_1 - \sigma_3}{2} \, (= K) & & \\ & 0 & \\ & & \dfrac{\sigma_3 - \sigma_1}{2} \, (= -K) \end{pmatrix}$$

hydrostatic component   deviatoric component

It can be easily argued that the deviatoric stresses σ_{d1} and σ_{d3} are equal and opposite for plane plastic strain – if $\delta\varepsilon_2 = 0$ then $\sigma_{d2} = 0$ and σ_{d1} must equal $-\sigma_{d3}$ to give zero volume change. Whichever yield criterion applies, they can be shown, as follows, to equal $+K$ and $-K$ respectively, where K is the shear yield stress, making:

$$\sigma_1 = -p + K$$
$$\sigma_3 = -p - K$$
(6.11)

where p is the (compressive) hydrostatic stress component.

Since the Reuss equations involve only the *ratios* of strain increments, the magnitude of the increments can be vanishingly small. They are therefore valid at the start of the yield. Substituting the above value of $\sigma_2 = (\sigma_1 + \sigma_3)/2$ in von Mises' yield criterion (6.9) gives

$$\{\sigma_1 - \tfrac{1}{2}(\sigma_3 + \sigma_1)\}^2 + \{\tfrac{1}{2}(\sigma_3 + \sigma_1) - \sigma_3\}^2 + \{\sigma_3 - \sigma_1\}^2 = 6K^2.$$

Note that the constant is put equal to $6K^2$ where K is the yield shear stress of a tube in torsion because this represents a case of plane strain; $2Y^2$ would not be appropriate here. This reduces to

$$\sigma_1 - \sigma_3 = 2K$$
(6.12)

which is also the Tresca criterion of yield. So in plane strain the yield criterion is the same on both theories. The proof can also be done in terms of the deviatoric stresses, with $\sigma_{d2} = 0$. (N.B. This equivalence does not hold in an elastic–plastic material, since it is assumed above that $\sigma_2 = (\sigma_1 + \sigma_3)/2$, which is a consequence of the *plastic* strain ε_2 being equated with zero.)

Consider now a tensile test on a notched bar of rectangular cross section (Fig. 6.13). It appears at first sight that yield will occur at the reduced section at the notch but this is not always so – for reasons of plastic restraint this section may be stronger than the rest of the bar. If the bar yields locally to the notch the adjacent material (the shaded region in Fig. 6.13(b)), which is unstressed, prevents contraction in the X_2-direction, that is, perpendicular to the plane of the specimen, in the direction of the notch root. From (6.12), with $\sigma_3 = 0$, the notch section yields at $\sigma_1 = 2K$ (a uniform stress over the section is assumed). According to Tresca's yield criterion, $Y = 2K$ and thus yielding at the notch

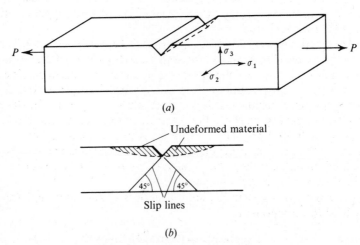

(a)

(b)

Fig. 6.13. Deformation of a notched bar.

occurs at the tensile yield stress of the parallel bar. This may be felt intuitively to be unlikely. However, given that von Mises' criterion is the correct one, $Y = K\sqrt{3}$ and the notch section yields at $\sigma_1 = 2Y/\sqrt{3} = 1.15Y$. Thus the yield stress is above that of the parallel bar, but allowance must be made for the reduced cross-sectional area in determining where yielding begins. If the notch area is less than $1/1.15 = 0.87$ of the bar area, yielding will begin at the notch; if the notch area is greater than this, yielding will begin in the bar.

6.8 Slip-line field theory

In the last example, the values of the principal stresses were assumed uniform throughout the material below the notch, and yielding occurred simultaneously across the specimen. This is not always so. The yield criterion may be satisfied in some regions, but only infinitesimal flow occurs due to the constraint by rigid material in which the yield criterion has not yet been met. Finite yielding can only occur when the criterion has been exceeded in a continuous path across the specimen.

Plastic flow at a point can be treated as shearing of the planes of maximum shear stress. In a uniform stress field these planes are parallel everywhere. In a non-uniform field their direction varies and the surfaces of maximum shear are no longer planes. However in plane strain conditions, the surfaces are always parallel to the axis of no displacement (i.e. X_2) and their intersections with the strain plane (i.e. defined by the X_1 and X_3-axes) are called *slip lines*. (Note that these are macro slip lines and must not be confused with micro slip lines to be discussed in Chapter 9.) The solution of problems involving varying principal stresses may be tackled by the theory of these slip lines – called *slip-line field theory*. The material model is still rigid–perfectly plastic.

In order to use the theory, a slip-line field, that is, an arrangement of slip lines, must be constructed for the particular problem. Herein lies the greatest difficulty. The choice of field must be made intuitively, with only experience as a guide. Once a field has been chosen, the solution of the problem is usually simple, and the achievement of a reasonable answer will indicate whether the initial choice of field was sensible.

Slip lines occur in pairs at right angles to one another and at 45° to σ_1 and σ_3 (see §6.2 on maximum shear stresses). The two sets of slip lines are called α-lines and β-lines. They are defined by the convention that in passing from an α-line to a β-line in an anti-clockwise direction, the σ_1-direction is crossed (Fig. 6.14(a)). Remember always that algebraically $\sigma_1 > \sigma_2 > \sigma_3$; the stress system may not be uniform throughout the material, with σ_1 changing direction from point to point (unlike the X and Y axes, which are fixed), but it is always algebraically the greatest principal stress. If the stresses do vary, then the slip lines will be curved, but they will always retain their identity, as defined above, and will always cross at right angles.

The principal stress system in plane strain was shown above (§6.7) to consist of a hydrostatic component:

$$\sigma_h = -p = \frac{\sigma_1 + \sigma_3}{2} \ (= \sigma_2),$$

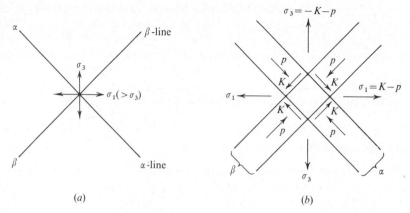

Fig. 6.14. Convention for labelling α and β-lines (*a*), and the associated stress system (*b*). (*NB* The sense of K is always towards the σ_1-axis.)

and a deviatoric component

$$\sigma_{d1} = K\left(=\frac{\sigma_1 - \sigma_3}{2}\right)$$

$$\sigma_{d2} = 0$$

$$\sigma_{d3} = -K.$$

K is a constant of the material (the maximum shear stress at yield in torsion) but p is a variable which must be established as detailed below.

The stresses acting on a slip line (i.e. the plane of maximum shear stress) will then consist of a normal compressive stress p and a shear stress K, see Fig. 6.14(*b*). The normal stress is due to the hydrostatic component, since this acts equally on planes of all orientation, and the deviatoric component contributes nothing, since the normal stress on planes at 45° to σ_{d1} and σ_{d3} is $(\sigma_{d1} + \sigma_{d2})/2$ (see §6.2), but $\sigma_{d1} = -\sigma_{d3}$. As regards the shear stress, this can only be caused by the deviatoric components of principal stress:

$$\frac{\sigma_{d1} - \sigma_{d3}}{2} = K.$$

At a free surface with no normal and shear stresses on it, the principal stresses run parallel and perpendicular to the surface and thus the slip lines are at 45°. At yield $\sigma_1 - \sigma_3 = 2K$ and at a free surface the principal stress parallel to the surface is thus always $\pm 2K$. Note that if it is positive (tension) it is labelled σ_1, and if it is negative (compression), σ_3. The hydrostatic component at the surface, $-p = (\sigma_1 + \sigma_3)/2$, is therefore $\pm K$. This provides a useful starting point in calculating the value of p at other points in a slip-line field.

The variation of p along a slip line is obtained as follows. Consider the simple field shown in Fig. 6.15(*a*) where the α-lines are all straight and the β-lines are arcs of circles; this is called a centred-fan field. Let the small angle between two α-lines be $d\phi$, measured *anti-clockwise*, and the hydrostatic pressures be p and $p + dp$ as indicated. Consider the equilibrium of forces acting on the sides of

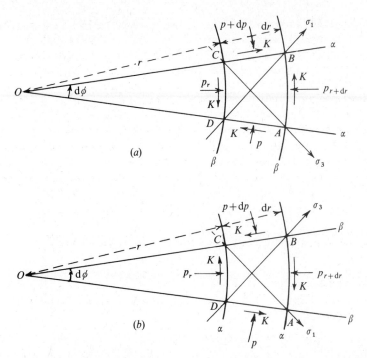

Fig. 6.15. Stresses in curved slip-line fields (centred-fan fields). (a) α straight, β curved; (b) β straight, α curved.

element *ABCD* (taking unit thickness). Taking moments about the fan centre *O* gives:

$$\circlearrowleft Kr\,d\phi\,dr + (p + dp)\,dr\left(r + \frac{dr}{2}\right) = \circlearrowright K(r + dr)\,d\phi(r + dr) + p\,dr\left(r + \frac{dr}{2}\right).$$

Normal stresses acting on *AB* and *CD* produce no couples. Ignoring second-order terms this reduces to:

$$dp - 2K\,d\phi = 0$$

along a β-line. If the α-lines are now curved and the β-lines are straight (Fig. 6.15(b)) the only change is the direction of the shear stresses. It is therefore merely necessary to reverse the sign of *K* in the above expression, to get

$$dp + 2K\,d\phi = 0$$

along an α-line. Integrating these two equations gives the Hencky relations of 1923:

$$p + 2K\phi = C_\alpha \quad \text{along an α-line,} \qquad (a)$$
$$p - 2K\phi = C_\beta \quad \text{along a β-line.} \qquad (b) \Bigg\} \ (6.13)$$

C_α and C_β are constants for any particular line and must be determined. Remember that ϕ is measured anti-clockwise about the origin of the radius of curvature. Though these relations have been deduced from rather special cases, they are true when both sets of slip lines are curved, and when the curvature is non-circular (though in the latter case it is then difficult to determine ϕ).

The speed and direction of flow of the material within a slip-line field is shown by means of a velocity diagram or *hodograph*. To draw this, certain rules have to be established. Since the normal stress acting on a slip line equals the hydrostatic stress component $(-p)$ the deviatoric stress is zero and, according to the Reuss flow rules, (6.10), there will be no linear strain rate or strain in the slip-line directions. In other words, slip lines behave as inextensible wires which are free to flex. For a curved slip line, see Fig. 6.16(a), consider the motion of two adjacent points, P_1 and P_2, on an α-line. Let the angles between the tangents to the slip line and the x-axis, and the velocity components be as indicated. Since the slip line is inextensible, the velocity components of P_1 and P_2 along $P_1 P_2$ must be the same; that is, resolving velocities in the physical plane:

$$v_\beta \sin \frac{\mathrm{d}\phi}{2} + v_\alpha \cos \frac{\mathrm{d}\phi}{2} = (v_\alpha + \mathrm{d}v_\alpha) \cos \frac{\mathrm{d}\phi}{2} - (v_\beta + \mathrm{d}v_\beta) \sin \frac{\mathrm{d}\phi}{2}.$$

Clearing up and ignoring a second-order term gives, for an α-line:

$$\mathrm{d}v_\alpha - v_\beta \, \mathrm{d}\phi = 0 \qquad (6.14a)$$

where $\mathrm{d}v_\alpha$ is the change of velocity in the α-direction and v_β is the velocity in the β-direction. Observe again that v_α is constant along a straight α-line.

Correspondingly, for a β-line

$$\mathrm{d}v_\beta + v_\alpha \, \mathrm{d}\phi = 0. \qquad (6.14b)$$

(The change of sign comes from the convention for labelling the lines, which makes the positive direction for a β-line point towards the negative x-direction, see Fig. 6.16. Note also that, unlike the Hencky relationships (6.13), the equation

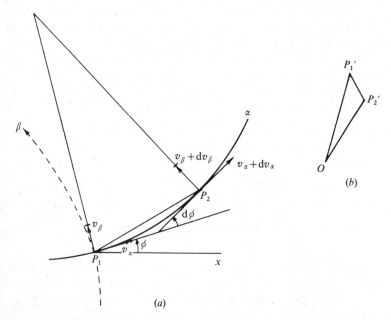

Fig. 6.16. Velocity diagram or hodograph. (a) Slip-line element $P_1 P_2$. (b) Image of $P_1 P_2$ in hodograph.

for the α-line has the negative sign.) These equations are due to Geiringer, in 1931.

The velocities of P_1 and P_2 are represented in the hodograph, Fig. 6.16(b), by the vectors OP'_1 and OP'_2 (O is the origin and represents zero velocity). The relative motion of P_2 with respect to P_1 is represented by the velocity vector $P'_1P'_2$ and, since P_1P_2 is inextensible, $P'_1P'_2$ in the hodograph is perpendicular to the slip line P_2P_1 in the physical plane. Thus each element of a slip line in the physical plane has a corresponding image in the velocity plane, the two being at right angles to each other. When the flow is pure translation, without rotation, the velocity image is reduced to a single point, e.g. P'_1 and P'_2 would coincide.

So far the velocity has been assumed continuous throughout the slip-line field, with no step changes. Lines of velocity discontinuity can occur, with the velocity step being tangential to the discontinuity line. The velocity component perpendicular to the discontinuity must be the same on both sides, since no volume change is possible. This is illustrated in Fig. 6.17(a); the small element A on side (1) of the discontinuity is transferred to B on side (2); for constant volume the heights of the parallelograms A and B must be equal, that is, the velocity components perpendicular to the boundary are equal. The tangential velocity discontinuity in passing from X to Y, just on opposite sides of the line, is v^*. If the velocity step is regarded as a velocity gradient over an infinitely small distance centred on the discontinuity line, it will be seen that the discontinuity line is one on which maximum shear straining is taking place and it is therefore always a line of maximum shear (6.10(c)), that is, a line of velocity discontinuity must be a slip line.

The velocity discontinuity across a slip line is the same all along its length. This follows from (6.14). Considering a point on an α-line, there will be a tangential velocity v_α on one side and v'_α on the other, v_β being the same. Further along the line both v_α and v'_α will have changed by $v_\beta \, d\phi$, $d\phi$ being rotation of the line which will be common to both sides. An element of the slip line P_1P_2 (Fig. 6.17) will now have two images in the hodograph, one for each side of the line. They will be separated by the velocity discontinuity and, since this is constant along the line, the two images are identical as shown in Fig. 6.17(b).

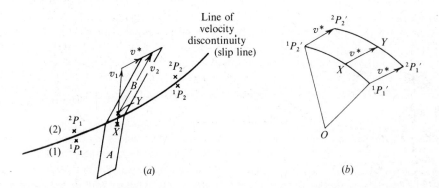

Fig. 6.17. Velocity discontinuity and hodograph. (a) Physical plane, (b) hodograph.

6.9 Applications of slip-line field theory

Applications of the slip-line field theory can now be considered, using in particular the relationships:

$$\sigma_1 = K - p,$$
$$\sigma_3 = -K - p,$$

$\left.\right\}$ [(6.11)]

and the Hencky equations:

$$p + 2K\phi = C_\alpha, \quad \text{along an } \alpha\text{-line}$$
$$p - 2K\phi = C_\beta, \quad \text{along a } \beta\text{-line}$$

$\left.\right\}$ [(6.13)]

and the Geiringer equations:

$$\mathrm{d}v_\alpha - v_\beta\, \mathrm{d}\phi = 0 \quad \text{along an } \alpha\text{-line}$$
$$\mathrm{d}v_\beta + v_\alpha\, \mathrm{d}\phi = 0 \quad \text{along a } \beta\text{-line}$$

$\left.\right\}$ [(6.14)]

where the angle ϕ is measured anti-clockwise.

One problem is that of the pressure required to sink a punch into a plastic material. This is obviously relevant to hardness testing, where it is desired to relate the measured hardness to the tensile yield stress of the material. For simplicity, the punch is considered to be a smooth rectangular bar which spans the whole width of the specimen (Fig. 6.18(a)); the thickness of the material under the punch is large compared with the punch width, that is, the material is semi-infinite. (Other solutions have been obtained allowing for wedge-shaped punches, finite thickness of material, and friction.) Plastic constraint is provided by the material away from the punch which does not deform, so that no strain occurs in the direction of the punch's long edges (i.e. plane strain deformation). Since friction between the punch and the material is assumed to be absent, the punch face must be a principal plane and the slip lines intersect it at 45°.

When a force is applied to the punch on a real material, regions of incipient yielding develop under both edges of the punch. In these regions the condition for yielding is satisfied, but general, finite, yielding cannot occur because the rest of the material under the punch remains elastic. As the pressure on the punch is increased, the size of these yielded regions increases until the two regions meet on the centre-line of the punch. General yielding of the material under the punch can now occur, and the punch begins to penetrate with the material being forced out to the edges. Hill in 1950 suggested a solution for a rigid–ideally plastic model based on the slip-line field shown in Fig. 6.18(b). Another slip-line field applicable to a rough punch was developed in 1920 by Prandtl and, curiously, gives the same punch stress (see Question 19).

The stress at which general yielding occurs can be calculated as follows. Consider the slip line $AFGHE$ which runs from the point A on the surface at the mid-point under the punch to the point E on a free surface (Fig. 6.18(b)). At E, $\sigma_1 = 0$, $\sigma_3 = -2K$ (remember $\sigma_1 > \sigma_3$ algebraically) and $\sigma_2 = -p = -K$. It should be intuitively clear that the surface stress is compressive (and is therefore σ_3) since the punch pressure squeezes the material out sideways (Poisson's effect). Disbelievers can try taking the stress positive and show that the punch/material interface stress comes out tensile, which is an obvious nonsense. By the rule given for labelling slip lines, AGE is seen to be a β-line

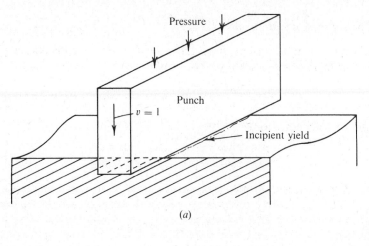

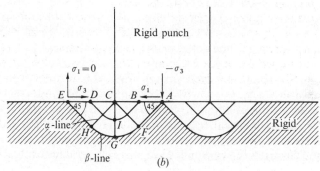

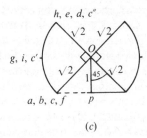

Fig. 6.18. Indentation using a flat smooth punch, showing the slip-line field at the start of finite yielding (b), and hodograph (c).

and hence the variation of p is given by (6.13b):

$$p - 2K\phi = C_\beta.$$

If ϕ is taken to be zero at E, then $C_\beta = p = K$, and the variation of p along this particular line is given by:

$$p = K(1 + 2\phi).$$

The angle turned from E to A is $\pi/2$, and therefore at A:

$$p = K(1 + \pi).$$

The compressive stress applied by the punch at A is:

$$-\sigma_3 = K + p = (2 + \pi)K.$$

The same arguments apply to all the slip lines, e.g. DIB, and the stress on the punch face is uniform. According to von Mises, $K = Y/\sqrt{3}$, and the stress required to force the punch into the material is:

$$\sigma = \frac{Y}{\sqrt{3}}(2 + \pi) = 2.97Y \simeq 3Y. \tag{6.15}$$

Thus the hardness of a material, as measured by a punch indentation test, should be approximately three times its yield stress in simple tension. This result has been obtained for a simple geometry, but is in fact remarkably accurate for circular, spherical and pyramid-shaped indentors. In a strain-hardening material, actual hardness values are three times the flow stress taken at some appropriate strain. Under a Vickers diamond pyramid this is 0.08, and the DPN hardness is found to be $\simeq 3 \times$ flow stress at a strain of 0.08. This has been discussed in § 5.7. It should be noted that in making hardness tests the indentor must be more than three times as strong as the material being tested since it lacks plastic constraint from adjacent material. This is one advantage of the diamond indentor.

Turning now to the hodograph, this is given in Fig. 6.18(c). The punch is assumed to have unit vertical velocity and is represented by the vector Op or, for short, the point p. Plastic material at A must move parallel to the interface with the rigid material AF and will have the same downward component of velocity as the punch. Hence the point a in the hodograph. Since the β-line AF is straight, there is no change in v_β (cf. (6.14b)) and the material at F (and throughout ACF) is moving with the same velocity represented by a. In the circular sector CFH the direction of flow swings round through 90°: the slip line FGH is represented by fgh in the hodograph, the velocity image being orthogonal to the slip line. The velocity of G is horizontal and represented by g. The velocity of I, represented by i, must be coincident with g: it can have no vertical velocity component (v_α) since IG is a straight α-slip line, and the horizontal velocity v_β must be the same as the velocity of B since, although $\Delta\phi = -45°$, $v_\alpha = 0$ (6.14). At CH all flow becomes parallel to HE and the image of CHE is a single point h. The plastic material at C is either marginally under the punch, with its velocity represented by c, or in the circular sector, represented by c', or marginally outside the punch and represented by c''. In this example the direction of flow coincides everywhere with the one of slip-line directions and there is no line of velocity discontinuity; it will be seen in the next example that this is not generally so.

Concluding this introduction to slip-line field theory, an example of extrusion of a sheet through a perfectly smooth wedge-shaped die will be discussed. The particular slip-line field and hodograph are shown in Fig. 6.19. Considering one half of the symmetric field, it consists of a 45° isosceles triangle BCD and a circular sector CAB. For the particular reduction chosen here AC intersects the axis of the die at 45° and the sector angle equals the die semi-angle θ. By

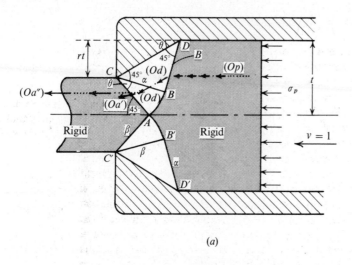

(a)

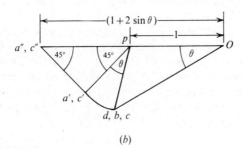

(b)

Fig. 6.19. Extrusion through a smooth wedge-shaped die with a reduction $r = 2 \sin \theta / 1 + 2 \sin \theta$, where θ is the die semi-angle. (a) Slip-line field, (b) hodograph.

geometry the fractional reduction of thickness $(t \to (1 - r)t)$ is:

$$r = \frac{2 \sin \theta}{1 + 2 \sin \theta}. \tag{6.16}$$

If the reduction is greater or less than is given by this formula the slip-line field is a modification of the one shown and will not be discussed further (see Johnson and Mellor).

To label the slip lines, consider the stresses acting in the axial direction on the extruded sheet bounded by CAC'; these must be zero since no external force exists. If CA is a β-line the stress in the axial direction is σ_3 (cf. Fig. 6.14(b)) and

$$\sigma_3 = -K - p,$$

i.e.

$$p = -K,$$

and

$$\sigma_1 = K - p = 2K.$$

That is, the stress acting orthogonal to the die axis is tensile. This is physically unrealistic. It can be concluded that CA is an α-line, so that the axial stress is

$$\sigma_1 = K - p = 0,$$

i.e.
$$p = K$$

and the orthogonal stress

$$\sigma_3 = -K - p = -2K.$$

The stresses on CA are thus established. From A to B, a β-line, p changes according to the Hencky Law

$$p - 2K\phi = C_\beta = K \qquad (\phi = 0 \text{ at } A).$$

At B
$$p = K(1 + 2\theta);$$

and this applies throughout BCD which has an orthogonal net of slip lines (not shown). The stress on the die face CD, which will be σ_3 (by the convention for α and β-lines), is:

$$\sigma_3 = -K - p = -2K(1 + \theta). \qquad (6.17)$$

The plunger stress σ_p can now be obtained by considering the equilibrium of (external) forces acting on the stock in the axial direction:

$$2K(1 + \theta)rt = \sigma_p t,$$

i.e.
$$\frac{\sigma_p}{2K} = \frac{2(1 + \theta)\sin\theta}{1 + 2\sin\theta}. \qquad (6.18)$$

$2K$ represents the yield stress in uniaxial tension according to Tresca. The variation of $\sigma_p/2K$ with die angle θ, together with the appropriate reduction for this particular slip-line field is:

θ	30	60	90
r	0.5	0.63	0.67
$\sigma_p/2K$	0.76	1.3	1.7

The hodograph is shown in Fig. 6.19(*b*). The plunger is assumed to be advancing with unit velocity and is represented by the point p (Op is the velocity in magnitude and direction). Material entering the plastic zone at D must travel parallel to CD, so that d lies somewhere on a line through O at θ to Op. Since the rigid material at D is travelling with the plunger and is represented by p there must be a velocity discontinuity, which, by the rules given above, must be in the direction of the slip line DB. Hence d also lies on a line through p drawn parallel to DB. This establishes d. All the material in BCD has the same velocity represented by d. From B to A the velocity image is the arc ba centred on p (since the velocity discontinuity is constant along a slip line). At A the material crosses a further velocity discontinuity into the rigid extruded material. Clearly the extruded material is travelling horizontally (as Op) and the discontinuity lies along the slip line AC. Hence a''. The material at C reaches the point with a velocity given by c, changes to c' momentarily on passing through the sector, and exits at c'' on joining the rigid extrusion.

The stream or flow line for the material is marked on the figure: on entry as rigid material its velocity is Op (in the plunger direction); within CBD it is Od (parallel to CD); within the sector ABC it swings round and increases in magnitude from Od to Oa'; at exit in the rigid extrusion its velocity is Oa'' (in the axial direction). By geometry of the hodograph the ratio of the exit to inlet velocities is $1 + 2 \sin \theta$; this could also be obtained from the reduction of thickness (6.16), since the volume flow is constant. It will be seen that here the flow is not along the slip-line directions.

6.10 Approximate analysis of metal working

Although the preceding analysis of plastic deformation is required to determine the stresses and strains to a fair degree of accuracy, a very elementary approach to plastic working will now be given which is capable of estimating the orders of magnitude of the forces and power involved in several metal working processes such as wire drawing, deep drawing and rolling.

There are two objectives in plastic working: to improve the metallurgical condition by dispersing the non-metallic inclusions and grain size refinement (see § 11.3); and to produce the desired shape and size of stock. This normally involves reducing one dimension very significantly and increasing another correspondingly, for example, in wire drawing the cross-section is reduced and the length increased. The former objective requires up to 25 per cent deformation but the latter may involve a factor of ten or more – steel ingots are regularly reduced by this amount in rolling down to plate. The cost of the power required is more than offset by the convenience of handling long continuous lengths.

The most efficient way of changing the dimensions of a bar is by plastic straining in tension. This involves no friction against dies and no wasted work in inhomogeneous shearing, see below. The work done in extending a bar from $A_1 \times l_1$ (cross-section $\times$ length) to $A_2 \times l_2$ is:

$$W_h = \int_1^2 \sigma A \, \mathrm{d}l,$$

where σ is the true yield stress. The work per unit volume ($= Al$) is

$$w_h = \int_1^2 \sigma \frac{\mathrm{d}l}{l} = \int_1^2 \sigma \, \mathrm{d}\varepsilon.$$

For an ideally plastic material:

$$w_h = Y \ln \frac{l_2}{l_1} = Y \ln \frac{A_1}{A_2} = Y\varepsilon \qquad (6.19a)$$

where ε is the logarithmic strain. For a work-hardening material the area under the true stress – logarithmic strain curve is the work done per unit volume. When the true stress changes only slowly an average value over the strain range, $\bar{Y}$, may be used, and the homogeneous work is:

$$w_h = \bar{Y} \ln \frac{A_1}{A_2} = \bar{Y}\varepsilon . \qquad (6.19b)$$

The actual work done in plastic working is always greater than the homogeneous work done, owing to two factors: the friction between the stock and the tool, and the inhomogeneous deformation that occurs. During its pass through the tool the stock suffers shear strains which do not contribute to the final shape change. This is shown schematically in Fig. 6.20. The actual work

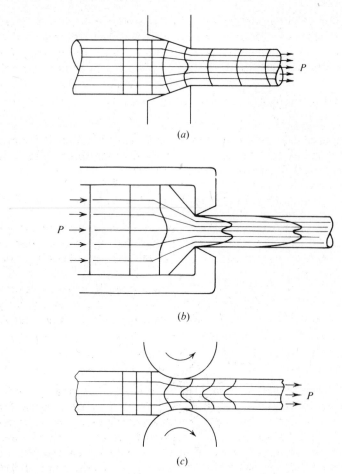

(a)

(b)

(c)

Fig. 6.20. Inhomogeneous deformation in various metal working processes. (a) Drawing through a fixed die; (b) extrusion; (c) drawing through free rollers (Stickel mill).

done is:

$$w_a = w_h + w_f + w_i,$$

where w_f is the work against friction and w_i work spent on inhomogeneous shearing. The efficiency of the process is defined as

$$\eta = w_h/w_a.$$

It has been found experimentally that the values of η for any particular type of process lie within fairly narrow bands:

$$\text{extrusion}\quad \eta = 45\text{--}55\%$$

$$\text{wire drawing}\ \eta = 50\text{--}65\%$$

$$\text{rolling}\quad\ \ \eta = 60\text{--}80\%.$$

Using these figures and knowing the yield stress of the material it is possible to calculate the force and power required.

Consider, for example, extrusion. If the pressure applied is P_1 over the initial area A_1, the work done as the plunger advances a distance l is $P_1 A_1 l$ and the work done per unit volume of material extruded is $P_1 A_1 l / A_1 l = P_1$. Similarly, for drawing, the drawing stress P_2 is work done per unit volume. Hence

$$P = w_a = \frac{w_h}{\eta} = \frac{\overline{Y}}{\eta} \ln \frac{A_1}{A_2}. \tag{6.20}$$

The force required is P times the relevant area and the power is P times the volume of material passed per second. A fuller drawing theory due to Siebel in 1947, widely used in industry, gives

$$P = \overline{Y} \left[(1 + \mu/2) \ln \frac{A_1}{A_2} + \frac{2}{3} \alpha \right], \tag{6.21}$$

where μ is the coefficient of friction and α is the semi-angle of the conical die. (μ is typically 0.1–0.5 and α in the range 3°–12°.) No rigorous solution of the problem has been developed, although there are slip-line field solutions to wedge and square dies.

Using (6.20) the maximum possible reduction in a single pass of a drawing operation can be determined, since the drawing stress cannot exceed the yield stress of the material at exit from the die (Y_e say). The maximum reduction of area $r_{max} (=(A_1 - A_2)/A_1)$ is then:

$$r_{max} = 1 - \exp \left(\frac{-\eta Y_e}{\overline{Y}} \right). \tag{6.22}$$

For fully strain-hardened material $Y_e = \overline{Y} = Y$, and taking $\eta = 0.5$, the maximum reduction is 40 per cent. Higher reductions are possible for an annealed material which strain hardens on passing through the die. It is possible to draw at stresses above the tensile strength because of the compressive stresses ($P_r = -\sigma_2 = -\sigma_3$) exerted by the die which lowers the yield curve from $Y(e)$ to $Y(e) - P_r$ (Tresca criterion) and suppresses the onset of necking (cf. § 5.3).

In deep drawing the stock is also drawn through a die but here the die is provided by the adjacent material. As shown in Fig. 6.21, material which starts on the periphery (radius R_1) with height $2\pi R_1$ is progressively compressed until it enters the cup (radius R_2) with height $2\pi R_2$. A blank holder is required to prevent the blank buckling under the compressive loads and if fixed firmly this will restrain the blank from thickening as it is drawn inwards, that is, plane straining occurs, with some increase in friction forces. By symmetry, no inhomogeneous shearing occurs during the compression phase. Consider the work done as the length of the cylindrical cup is increased by a small distance δx. An element at the periphery, width $R_2 \delta x / R_1$, can be regarded as being compressed all the way down from length $2\pi R_1$ to $2\pi R_2$. Of course in practice all the material in the blank suffers a slight compression. The homogeneous work done per unit volume is

$$w_h = \overline{Y}^* \ln \frac{R_1}{R_2}, \tag{6.23a}$$

where $\overline{Y}^*$ is the mean yield stress for plane strain (i.e. $Y^* = 1.15Y$ according to

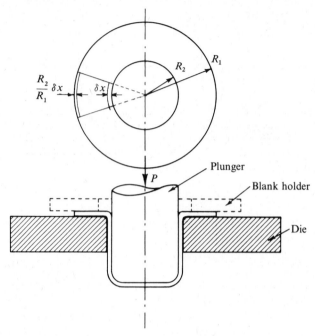

Fig. 6.21. Deep drawing. Note that the material acts as its own die in reducing the height from $2\pi R_1$ to $2\pi R_2$.

von Mises, $Y^* = Y$ according to Tresca). As for drawing and extrusion, the actual work done per unit volume is greater than the ideal figure and the drawing stress is given by

$$P = w_{\text{a}} = w_{\text{h}}/\eta = \frac{\bar{Y}}{\eta} \ln \frac{R_1}{R_2}. \tag{6.23b}$$

The value of P decreases as the drawing proceeds. For a fully strain-hardened material the maximum drawing ratio is when $P = \bar{Y}$ and, for $\eta = 0.5$, $R_1 = 1.65R_2$. In reality the process is more complex than this simple theory suggests due to bending under tension at entry to the cup and squeezing between the plunger and die.

Another important example of plastic working is rolling, and here again a simple approach gives the order of magnitude for the roll force, which is important for the roll bending strength and the main bearings, and for the roll power, which governs the motor size.

In Fig. 6.22, consider the projected length of the arc of contact (and hence the projected contact area per unit length of roll); by simple geometry:

$$\frac{x}{L} = \frac{L}{2R}$$

i.e.

$$\frac{L^2}{2R} = x = \frac{h_1 - h_2}{2}$$

i.e.

$$L = \sqrt{R(h_1 - h_2)}, \tag{6.24}$$

where h_1 is the plate thickness on entry to, and h_2 at exit from, the rolls.

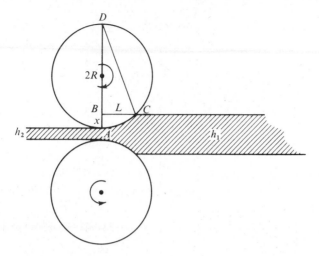

Fig. 6.22. Arc of contact on rolling mill. Projected length of arc of contact, $L = \sqrt{R(h_1 - h_2)}$.

The roll force per unit length is then:

$$f_R = \bar{Y}^* \sqrt{R(h_1 - h_2)}, \tag{6.25}$$

where $\bar{Y}^*$ is the mean yield stress for plane strain compression since friction prevents the stock from expanding sideways along the rolls.

Several effects are responsible for this value being too small by a factor of two: friction restraining expansion in the rolling direction (friction hill); elastic deformation of the rolls, called roll flattening; and a plastic restraint effect. Figure 6.23 shows the friction hill effect in a specimen compressed between rough parallel plates. Consider the equilibrium of the small element (unit thickness into the paper):

$$h \, d\sigma_3 = 2\mu\sigma_1 \, dx,$$

where σ_1 and σ_3 vary with x. According to Tresca

$$\sigma_1 - \sigma_3 = \text{const.},$$

i.e.

$$d\sigma_1 = d\sigma_3.$$

Hence

$$\frac{d\sigma_1}{\sigma_1} = \frac{2\mu}{h} \, dx.$$

Integrating

$$\sigma_1 = Y^* e^{2\mu x/h} = Y^* \left(1 + \frac{2\mu x}{h}\right),$$

where Y^* is the yield stress for plane strain at the edge, $x = 0$. The normal stress σ_1 increases towards the centre as shown in Fig. 6.23. The average stress for a stock width L is

$$\bar{\sigma}_1 = Y^* \left(1 + \frac{\mu L}{2h}\right).$$

Hence the roll force per unit length is

$$f_R \simeq Y^* L \left(1 + \frac{\mu L}{h_1 + h_2}\right).$$

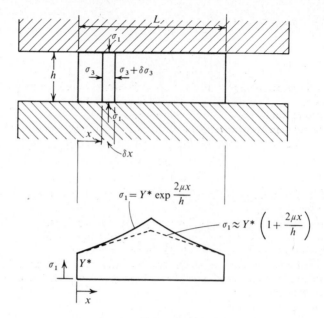

Fig. 6.23. Friction hill (drawn for $L = 3h$, $\mu = 0.3$).

The theory breaks down at high values of μ since it assumes that the normal stress σ_1 is a principal stress. More exact solutions by slip-line field theory have also been determined.

The value of L has been given in (6.24) but, due to roll flattening, this also is an under-estimate and the roll force is increased by a factor similar to that due to the friction hill effect. Flattening is reduced by using small rolls, but to prevent bending, a series of back-up rolls is necessary.

The third factor increasing the roll force is the plastic restraint exerted by the material adjacent to the roll gap. The strengthening effect is more marked the smaller the arc of contact in relation to the stock thickness. The net result of these various factors is that the simple formula, (6.19), for the roll force per unit length needs to be multiplied by a factor of about two.

The roll torque per unit length is approximately the roll force times the lever arm to the roll centre, $L/2$. For *two* working rolls

$$t_{2R} = f_R L = \alpha \bar{Y}^* R(h_1 - h_2). \tag{6.26}$$

where the numerical constant α is about 2.

QUESTIONS

(Y = yield stress in uniaxial tension, K = yield shear stress in torsion.)
   1. Distinguish between
       (a) macroscopic and microscopic theory of plasticity,
       (b) macro slip lines and micro slip lines,
       (c) deformation, dilation and distortion.

2. The stress tensor at a point is:

$$\begin{matrix} \sigma_{11} & \sigma_{12} & \sigma_{13} \\ \sigma_{21} & \sigma_{22} & \sigma_{23} \\ \sigma_{31} & \sigma_{32} & \sigma_{33} \end{matrix}$$

(a) Show the stress system on the faces of a cube with its edges parallel to the X_1, X_2, X_3 axes.

(b) What is meant by *complementary shear stress*?

3. (a) Define principal stresses. For the principal stress tensor (in MN/m^2):

$$\begin{matrix} 70 & 0 & 0 \\ 0 & 33 & 0 \\ 0 & 0 & 20 \end{matrix}$$

(b) Write down the corresponding stress tensor when the axes are orientated for maximum shear stress.

(c) Write down the hydrostatic and deviatoric stress components.

(d) What is the shear strain energy ($E = 200 \, GN/m^2$ and $v = 0.3$)?

4. (a) Define linear normal strain and logarithmic normal strain.

(b) Define shear strain, and distinguish between simple shear and pure shear.

(c) Define Poisson's ratio and show that in plastic flow its value is 0.5.

5. For a general strain tensor:

$$\begin{matrix} e_{11} & e_{12} & e_{13} \\ e_{21} & e_{22} & 0 \\ e_{31} & 0 & e_{33} \end{matrix}$$

(a) Show schematically the deformation of a cube initially with its edges parallel to the X_1, X_2 and X_3 axes.

(b) What is the corresponding pure shear strain tensor?

(c) What are the shear strains: γ_{12}, γ_{13} and γ_{23}?

6. (a) Define principal strain.

(b) If the strain tensor is (per cent):

$$\begin{matrix} 1 & 0 & 0 \\ 0 & 0.05 & 0 \\ 0 & 0 & 0.03 \end{matrix}$$

what is the dilation tensor and the distortion tensor?

7. (a) Give Tresca's criterion of yielding in terms of the principal stresses and of the deviatoric stresses.

(b) Give von Mises' criterion of yielding in terms of the principal stresses and of the deviatoric stresses.

(c) If yield occurs in simple tension at stress Y and in torsion of a thin walled tube at shear stress K find the relationship between Y and K according to Tresca and von Mises.

8. (a) Describe briefly the experiments by Taylor and Quinney to investigate the various yield criteria.

(b) If in these experiments the tensile load produced an axial stress σ and the torsion load a shear stress τ, calculate the values of the principal stresses.

(c) Hence show that at yield

$$\left(\frac{\sigma}{Y}\right)^2 + \left(\frac{\alpha\tau}{Y}\right)^2 = 1$$

where α is 2 according to Tresca and $\sqrt{3}$ according to von Mises.

9. A thin walled tube was loaded axially in tension until yielding began. The load was

then halved and an internal pressure applied until an axial strain of 5 per cent was recorded. What were the diametral and wall thickness strains?

10. (a) Define plane strain.

    (b) Determine σ_2 as a function of σ_1 and σ_3 (the principal stresses in the plane of strain) in the elastic and rigid–plastic regimes.

    (c) Show that for a rigid–plastic material the Tresca and von Mises' criteria of yield are identical.

11. Discuss the simplifying assumptions made for various material models and in continuum plasticity theory and distinguish between : (i) elastic–plastic, (ii) rigid–plastic, and (iii) rigid–ideally plastic.

12. State the basic law of plastic flow in terms of

    (a) the principal stresses,

    (b) the deviatoric stresses.

13. (a) Explain the reason for assuming that plane strain conditions exist at the notch section of a notched bar tensile test piece.

    (b) Hence show that if von Mises' criterion holds, a notched bar is not weakened in tension as long as the notch section is not less than 0.87 times the full section.

    (c) How is the argument in (b) affected by using Tresca's criterion?

14. (a) Define a *slip line* in the context of slip-line field theory.

    (b) What material model is used?

    (c) Give the convention for labelling α and β-lines.

    (d) Sketch the stress system (in terms of K, p, σ_1, σ_3).

15. (a) Prove one of the Hencky relations :

    for an α-line

$$p + 2K\rho = C_\alpha,$$

    for a β-line

$$p - 2K\phi = C_\beta.$$

    (Consider the equilibrium of a centred-fan field with circular arcs.) Define the positive direction for measuring ϕ.

    (b) What is the use of the Hencky relations?

16. (a) What is a *hodograph*?

    (b) Demonstrate that slip lines are not extending (or contracting) during plastic flow.

    (c) Hence prove for a curved α-line :

$$dv_\alpha - v_\beta \, d\phi = 0;$$

    or, for a β-line :

$$dv_\beta + v_\alpha \, d\phi = 0.$$

17. (a) Draw the slip-line field under a smooth flat die pressed into a semi-infinite plastic material.

    (b) Calculate the stress on the punch for finite yielding.

    (c) What are the stresses under the punch parallel to the surface?

18. (a) Draw a simple slip-line field for extrusion through a 90° wedge-shaped die. Indicate the rigid material.

    (b) What is the reduction for which this field applies?

    (c) Calculate the stress on the die face and on the plunger.

    (d) Draw the hodograph.

19. The slip-line field for a rough flat punch pressing into a semi-infinite rigid–plastic material is (due to Prandtl) :

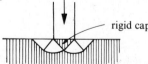

rigid cap

(*a*) Indicate the α and β-lines.

(*b*) Show that the indentor stress (load/normal area) at finite yielding is $(2 + \pi)K$.

(*c*) Draw the hodograph.

20. A compression test of annealed copper gave:

nominal stress (MN/m^2)	100	200	250	300	325
linear strain (per cent)	3.5	12.5	20	33	60

Draw the curve of true yield stress versus logarithmic strain and determine the homogeneous work for 66 per cent reduction of height.

21. Wire was drawn from the above copper. After annealing, two draws were made, each with 20 per cent reduction of area, to give a 2 mm diameter wire. Assuming 50 per cent efficiency determine the drawing stress and force in the final pass. What was the motor power in this pass for a delivery speed of 10 m/s?

22. A cup is to be deep drawn, 50 mm diameter × 50 mm deep, from mild steel 1.5 mm thick. Determine the diameter of stock required and the maximum drawing force. The yield stress can be taken as 250 MN/m^2.

23. The yield stress of a medium carbon steel at 1000 °C and 1000 per cent per second compression rate is 150 MN/m^2. A slab of 50 mm thickness and 300 mm wide is rolled down to 35 mm in a mill. Calculate the approximate value of the roll force and roll torque (using a factor of 2 on the simple estimate). What is the maximum normal stress allowing for the friction hill effect ($\mu = 0.6$)?

24. A 60/40 brass rod is extruded at 700 °C at which the yield stress is 140 MN/m^2. The diameter of the extrusion plunger is 200 mm and two 10 mm rods are extruded simultaneously. Assuming 60 per cent efficiency, calculate the plunger force and energy consumed per Mg (tonne) of product.

FURTHER READING

A. Nadai: *Theory of Flow and Fracture of Solids*, 2nd edition. McGraw-Hill (vol. 1, 1950; vol. 2, 1963).

W. Prager: *An Introduction to Plasticity*. Addison-Wesley (1959).

R. Hill: *Mathematical Theory of Plasticity*. Oxford University Press (1960).

W. Johnson and P. B. Mellor: *Plasticity for Mechanical Engineers*. Van Nostrand (1962).

P. Feltham: *Deformation and Strength of Materials*. Butterworth (1966).

G. C. Spencer: *Introduction to Plasticity*. Chapman and Hall (1968).

7. Ceramics and glasses

7.1 Introduction

Ceramics may be defined as inorganic materials with ionic and covalent atomic bindings. The word comes from Cerami, a district of ancient Athens where potters manufactured their wares (keramos = burnt stuff). Traditional ceramics cover the raw materials and products of the pottery industry such as earthenware, china, porcelain, tiles and bricks. The principal raw material is natural clay, which consists mainly of small crystals of hydrated alumino-silicates, that is, compounds containing Al_2O_3, SiO_2 and H_2O in various proportions. After shaping by plastic deformation, the ceramic body is fired at a high temperature to drive off the water and to cause various reactions. The final microstructure consists of refractory crystalline components embedded in a glassy (non-crystalline) matrix. The system is complex, heterogeneous, non-equilibrium and not tightly controlled : cheapness is a major requirement. Art and empiricism are still important factors and the solid state physicist has hardly begun to explain the properties in terms of the individual phases.

The term ceramics now covers a very much wider field. Besides the traditional ceramics based on clay, it embraces the new ceramics, such as very pure and dense single oxides, carbides and nitrides; graphite; cement and concrete; cermets (ceramics in a metal matrix); glass and glass ceramics. In some definitions of ceramics, concrete and glass are excluded because they do not involve high temperature processing and irreversible reactions, which are common to the other groups.

The ceramic industry, covering pottery, house and refractory bricks, cement and glass, has for long been a large one and vital to important sections of engineering. Nevertheless there has been a tendency to regard ceramics as beyond the range of 'engineering materials'. Since World War II, many new requirements of mechanical, electronic and nuclear engineering have appeared which could not be met by 'conventional' engineering materials, notably metals. Only high grade ceramics could possibly satisfy the arduous specifications and consequently many of the developments in materials have been in the field of ceramics, which are now being rightly recognised as an important class of technological materials, along with metals and high polymers.

The 'new' ceramics, in a limited sense, cover various pure compounds such as oxides, carbides and nitrides. In impure grit form they have been used as abrasives, for example, alumina (corundum) and silicon carbide (carborundum). By improved manufacturing techniques they are now being produced in an increasingly pure state, in fully crystalline solids, with no glass phase and nearly theoretical density. The porosity is typically less than 0.5 per cent, compared with 5 to 15 per cent in many high quality traditional ceramics.

Cement consists of complex compounds of lime (CaO), silica (SiO_2) and alumina (Al_2O_3), which combine with water during setting and hardening.

It is made by firing limestone with some clay and grinding the clinker to a fine powder. The cement powder is then either mixed with fine sand and water to form 'cement' or with sand, crushed rock (aggregate) and water to form concrete. Cement and concrete are very complex ceramic systems and, although their chemistry is now fairly well understood, their physics has progressed little. Nevertheless they are the most widely used engineering material, with a total annual consumption of concrete exceeding one tonne (10^3 kg) per head, five to ten times that of steel. The applications are widespread, ranging from domestic building works to pressure vessels for nuclear reactors.

Cermets are particles of crystalline ceramic held in a matrix of metal, typically 5 to 15 per cent by weight. They are often referred to loosely as ceramics, although there is the proportion of metal present which greatly affects the properties. The object is to overcome the brittleness of the pure ceramic. Their exceptional hardness combined with limited toughness, maintained to high temperatures, has led to their use as cutting tools for many years. In recent years, attempts have been made to develop cermets with much higher proportions of metal for application as high temperature constructional material, in particular, for gas turbine blades. However the necessary properties have not been achieved.

Glasses may be considered alongside ceramics for several reasons. They are oxides or mixtures of oxides, but unlike ceramics they are amorphous rather than largely crystalline. As already noted, traditional ceramics contain a good proportion of glass phase surrounding the crystalline components. Lastly, the properties of glass and ceramic have many similarities, such as brittleness and low thermal and electrical conductivity.

A recent development is glass-ceramic, in which a glass is caused to crystallise (devitrify) in a controlled manner to produce a very fine grained ceramic, free of porosity. Apart from the excellent range of properties, the manufacturing techniques are advantageous.

The following section treats the physical and mechanical properties of ceramics in general terms. Later sections describe the structure and mechanical properties of specific types of ceramics. The chapter concludes with two sections on the structure and properties of glasses. The electrical and magnetic properties of ceramics are discussed in more detail in Chapters 14 and 15.

7.2 Physical and mechanical properties

The physical and mechanical properties of ceramics stem from their atomic binding and crystal structure. The binding is ionic or intermediate between ionic and covalent, as described in Chapter 3. In both, there are energy gaps between the Brillouin zones and the outermost zone is completely filled with electrons. The absence of free electrons is responsible for making ceramics poor conductors of electricity and heat. Hence their important applications as thermal and electrical insulators, and, where the energy gap is small, as semi-conductors. For the same reason ceramics are transparent in the fully dense state, although normal production methods leave sufficient residual porosity to make them translucent or, more often, opaque.

The bonds are highly stable and ceramics have very high melting points and high chemical stability. The melting points of some ceramics are given in Table 7.1; hafnium carbide at 4150 °C has the highest melting point of any known material. Consequently ceramics are widely used as furnace linings and containers for high temperature reactions.

The electronic properties of ceramics are extensive and will only be touched on here. They will be considered further in Chapters 14 and 15. Ionic materials have separated electric charges, from the nature of their bindings, and the dipoles respond to an applied electric field (see § 3.10). They are widely used in electrical engineering as dielectrics. In some ceramics, e.g. barium titanate $BaTiO_3$, an interaction exists between the dipoles leading to their spontaneous

Table 7.1. *Melting points of ceramics contrasted with refractory metals*

Ceramics		Melting point (°C)	Refractory metals
Hafnium carbide	HfC	~4150	
Tantalum carbide	TaC	3850	
Graphite	C	3800*	
Zirconium carbide	ZrC	3520	
Niobium carbide	NbC	3500	
		3370	W Tungsten
Tantalum nitride	TaN	3350*(?)	
Hafnium boride	HfB_2	3250	
Titanium carbide	TiC	3120	
Thoria	ThO_2	3110	
Zirconium boride	ZrB_2	3060	
Tantalum boride	TaB_2	3000	
		2996	Ta Tantalum
Titanium boride	TiB_2	2980	
Tungsten carbide	WC	~2850	
Magnesia	MgO	2798	
Zirconia	ZrO_2	2770	
Boron nitride	BN	2730*	
		2622	Mo Molybdenum
Beryllia	BeO	2570	
Silicon carbide	SiC	2500	
Zircon	$ZrO.SiO_2$	2495	
		2468	Nb(Cb) Niobium
Boron carbide	B_4C	2450	
Alumina	Al_2O_3	2050	
Chromium oxide	Cr_2O_3	1990	
Torsterite	$2MgO.SiO_2$	1830	Cr Chromium
Mullite	$3Al_2O_3.2SiO_2$	1810	
		1800	Ti Titanium
		1772	Pt Platinum
Silica (crystobalite)	SiO_2	1715	
Titania	TiO_2	1605	
		1554	Pd Palladium
		1527	Fe Iron
		1490	Co Cobalt
		1455	Ni Nickel

* Sublimes.

alignment and the phenomenon of ferroelectricity (see § 14.13). A parallel effect is the interaction between magnetic dipoles, arising from electron spin, and in the class of ceramics called ferrites this leads to the phenomenon of ferri-magnetism (see § 15.1). As non-conducting magnetic materials they find use in many high frequency devices and as computer memory units.

Some ceramics have energy levels in the band gap produced by impurities, leading to extrinsic semiconduction (see Chapter 14). They can absorb incident radiation which is re-emitted in a coherent beam (i.e. radiation is in phase and in one direction). These materials form the active element in lasers (light amplification by stimulated emission of radiation) and masers (microwave amplification), see § 14.16.

The crystal structures of ceramics are many and varied (see § 2.7). They range from the cubic structure of MgO, through the layer structure of mica (see Fig. 7.3) to the linear structure of asbestos (hydrous magnesium silicate). The structures are often of low symmetry, which may lead to the effect of *piezoelectricity*, the formation of a static charge upon elastic deformation. Quartz (one crystalline form of SiO_2) is an example, and is used in electronic oscillators.

Turning to the mechanical properties, these are characterised by the lack of plasticity, high hardness and stiffness, and good compressive strength. The tensile strength is low due to influence of microcracks (see Chapter 10). Thus the mechanical behaviour is in marked contrast with that of metals. The mechanical and physical properties of some typical ceramics are listed in Table 7.2.

The absence of plasticity in crystalline ceramics can be traced to the nature of their bindings. It was briefly noted in § 2.8 (Fig. 2.18) that the mechanism of plastic flow is the movement of line faults in the crystal structure, called dislocations, over certain crystal planes, called the slip planes. Although full discussion of dislocations will be taken later (Chapter 9), it is sufficient to note here that they can move at a low stress in metals because the non-directional metallic bond leads to a wide dislocation (§ 9.11). In covalent ceramics the bond is between specific atoms, leading to a narrow dislocation and a high resistance to movement. In terms of the macroscopic parameters, it can be shown that the dislocation width is given by a function of the ratio G/K where G is the shear modulus and K the bulk modulus. Now the ratio is given by

$$\frac{G}{K} = \frac{3(1 - 2v)}{2(1 + 2v)} \tag{7.1}$$

where v is Poisson's ratio. For metals v is about 0.3 and $G/K = 0.37$, whereas in ceramics $v = 0.1$ and $G/K = 1$. Consequently the yield stress, at which dislocations move, is much higher for ceramics than for metals, and approaches the fracture stress. Thus, covalent ceramics are brittle in the single crystal and polycrystalline state.

On the other hand, in ionic ceramics the bond is not directed, being due to the overall electrostatic attraction between interlocking arrays of positive and negative ions. Dislocations are wide and resistance to movement is low. Plastic flow in single crystals can occur on a considerable scale. Figure 7.1(*a*) shows the stress–strain curve obtained in compression with a NaCl crystal; this has

Table 7.2. *Properties of some ceramics (at room temperature)*

Material and approximate composition (by weight)	Porosity (per cent)	Melting point (°C)	Specific gravity	Coefficient of expansion (deg^{-1} × 10^6)	Thermal conductivity (W m^{-1} deg^{-1})
Concrete: hydrated cement (CaO, SiO$_2$, Al$_2$O$_3$.H$_2$O), sand (SiO$_2$), ballast	~20		2.4	10–14	1.8
Building brick: 40–60% SiO$_2$, 30–20% Al$_2$O$_3$	~20		1.7	5–6	0.7
Alumina fire brick: 10–40% SiO$_2$, 50–90% Al$_2$O$_3$	~20		2.2	4–5	1.3
Electrical porcelain (high voltage): 45% SiO$_2$, 45% Al$_2$O$_3$, 10% K$_2$O	<0.1 w		2.3	3	1.0
Steatite: 60% SiO$_2$, 5% Al$_2$O$_3$, 30% MgO	<0.05 w		2.65	8.3	3.3
Alumina: single crystal: 99.9% Al$_2$O$_3$	0	2050	3.98		42
translucent: 99% Al$_2$O$_3$, 0.1% MgO	<0.5		3.98	5.6	30
dense (tool): >90% Al$_2$O$_3$	1.2	2000	3.93	8.8	
electrical: 95% Al$_2$O$_3$	<0.05 w		3.65		
Beryllia: translucent BeO	<0.5	2570	3.01	9.0	200–300
Silica (SiO$_2$): quartz single crystal	0	1600	2.65	$\begin{cases} 7.5(a) \\ 13.7(c) \end{cases}$	$\begin{cases} 5.4(a) \\ 9.2(c) \end{cases}$
quartz polycrystalline	0	1600	2.65	16.9	7.1
glass (non-crystalline SiO$_2$)	0		2.22	0.55	1.2
Silicon carbide: 95% SiC	0.5	2500	3.1 (~3.4 theor.)	5.5	100
bonded 26% Si$_3$N$_4$, 72% SiC	8		2.89	4.4	17
Diamond (single crystal, tetrahedral) C	0	1500d	3.52	1.1	665
Commercial graphite: 98% C(hex)	20–30	3800s	1.5–1.8	1–3	120–180
Pyrolitic graphite: C(hex)	<1	3800s	2.2	$\begin{cases} \sim10(c) \\ \sim1(a) \end{cases}$	$\begin{cases} 2.5(c) \\ 420(a) \end{cases}$
Graphite fibre: C(hex)	<1	3800s	2.0 (2.27 theor.)		
Silicon nitride: Si$_3$N$_4$ hot pressed	<1	1910s	3.18 (3.2 theor.)	2.9	~25
sintered	20	1910s	2.4	2.6	10
Tungsten carbide: WC		2700	15.5 (15.7 theor.)	6.2	2.9
Cermet: 90% WC, 10% Co	0	1400	14.9	5.3	70

w water absorption   s sublimes   d decomposes   (a) and (c) ∥ and ⊥ to basal planes

many similarities to those obtained with metals (see Chapter 9). Figures 7.1(*b*) and (*c*) show the increase in density of dislocations intersecting a cross-section, cut orthogonal to the specimen after compression; slip began on two orthogonal systems, the bands in one system increasing in number and broadening with further compression. In polycrystalline material, however, the adjacent grains are compelled to change their shape in the same manner, if voids are not to be formed at the boundaries. Von Mises showed in 1928 that this required five independent slip systems. A slip system is the combination of a slip plane and slip direction and an independent system is one that produces a deformation that cannot be produced by a combination of the others. Conversely there cannot therefore be more than five *independent* systems. Whereas metals have many slips systems, ionic crystals are only able to slip on a limited number of slip planes, due to the restriction that similarly charged ions must not be forced into near-neighbour positions such that cohesion would be lost. Consequently,

Electrical data (at 1 MHz)		Mechanical data			
Dielectric constant	Loss tangent ($\times 10^3$)	Young's modulus (GN/m²)	Median TS (MN/m²)	Compressive strength (MN/m²)	Remarks
		15–35	2.5	25–40	Effect of time and water content see Fig. 7.7
		15	5	25	
		40	8	15	High alumina raises working temperature (1400–1600 °C)
6.1	5–10	70	45	350	
6.1	~1	100	70	600	
	<0.1	360–460	500–700	2000	Sapphire or corundum. Hexagonal crystal
		410	700		⎱ Properties are very dependent
		400	400–600*	3000–3500	⎰ on density and grain size.
8.9	3	~260	200	2000	
7.0	0.4	360–390	150–200*	~1500	
4.5		54			Hexagonal crystals a-axis parallel basal plane
4.5		54			
4.1	0.1	72	100–120		
	<0.1	350–450	250*	650	Refractory; abrasive grit
			40*	140	Refractory
		950	4–8000		(cf. Cu thermal conductivity 400)
		3.5–15	5–25*	35–80	Hexagonal crystallites in amorphous matrix. Wide variations.
				360(c)	c-axis (⊥ to basal planes) is deposited
		40–70(a)	100–250(a)	330(a)	⊥ to substrate
		450–700	2000		Imperfect polycrystalline with c-axis ⊥ to fibre axis
		250	110*	600	
		160	150*	350	
		700	350	550	
		600	1600	4500	Cutting tool material – not a true ceramic

* flexural strength ($\approx$ 2TS)

polycrystalline ionic material is brittle with cracks forming at the grain boundaries.

This lack of plasticity is responsible for the low tensile strength of most ceramics, well below their theoretical strength. It will be shown in Chapter 10 that, according to Griffith theory, fracture of brittle materials occurs not at the theoretical fracture stress but at some lower stress at which small cracks, in the crystals or on the surface, are able to propagate. The problem in strengthening ceramic is to raise the fracture stress nearer to its theoretical value, either by introducing some plasticity, possibly by alloying, so that microcracks do not lead to premature fracture, or by eliminating the microcracks. In carefully made glass and ceramic fibres the latter is achieved and the strength approaches the theoretical value. To maintain this strength in practical environments the fibre must be protected by mounting in a metallic or plastics matrix. The subject of fibre composites will be discussed in Chapter 11.

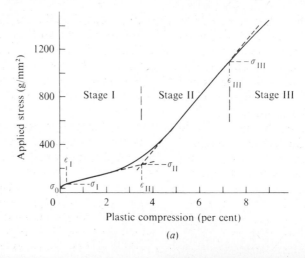

(a)

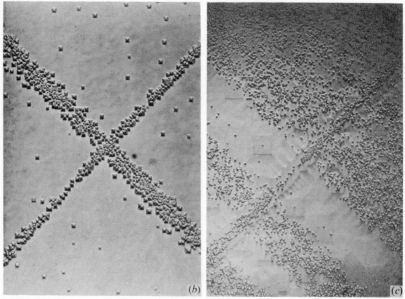

Fig. 7.1. Plastic flow in an ionic crystal (NaCl). (a) Stress–strain curve in compression. (b) Dislocation etch pits observed on a section cleaved perpendicular to the compression axis at the start of yielding; two active slip systems. (c) As (b), but 0.1 per cent strain: one of the two slip systems has broadened. ((b) and (c) × 500.) (From R. W. Davidge and P. L. Pratt, *Phys. Stat. Sol.* **6**, 759 (1964).)

As many ceramics are used at high temperatures, their ability to withstand thermal shock is of very great importance. Thermal stresses arise from several causes. Non-cubic materials may have anisotropic coefficients of thermal expansion; change in temperature will produce thermal stresses between the grains which may lead to fracture. A rapid change in temperature leads to thermal gradients and differential thermal expansion; this effect is aggravated by the low thermal conductivities of ceramics. Rapid cooling is usually more disastrous than a rapid heating as the former puts the surface into tension, where small flaws then grow into cracks. Lastly, differential effects also arise from non-uniform composition. For example, a refractory used as a furnace lining may react with the contents of the furnace and change its composition at the surface. Hence the chemical stability of ceramics is of great importance to thermal shock resistance.

According to the thermoelastic theory of thermal stress resistance the relevant parameter is the ratio of the elastic strain at fracture to some thermal strain. For an imposed temperature gradient the appropriate thermal strain equals the coefficient of thermal expansion α and the resistance is $\sigma_f/E\alpha$. But if the temperature gradient arises from a sudden rise in the external temperature or an imposed heat flow, the appropriate parameter is $\sigma_f k/E\alpha$. Here σ_f is the fracture stress, k the thermal conductivity, E Young's modulus and α the coefficient of thermal expansion. These quantities are given in Table 7.3 for various ceramics. Metals show good resistance to thermal shock because k is large and E is small, but in ceramics the reverse is often true.

An alternative approach to thermal shock resistance has been developed by Hasselman and others (1963 onwards). Instead of considering only the stress to initiate fracture, consideration is given to whether such a crack on formation is able to propagate catastrophically by the rapid conversion of the elastic strain energy of the material into the surface energy of fracture, that is, Griffith criterion of brittle fracture. For high thermal shock resistance, low fracture strength and high elastic modulus are required together with a high fracture

Table 7.3. *Thermal shock resistances of ceramics and glasses.* σ_f/E_α *applies when temperature gradient is imposed;* $\sigma_f k/E_\alpha$ *applies when a heat flow rate is imposed*

Material	Tensile fracture stress σ_f (MN/m²)	Elastic modulus E (GN/m²)	Thermal expansion α (10^{-6} deg⁻¹)	Thermal conductivity (k) at 100 °C (Wm⁻¹deg⁻¹)	$\dfrac{\sigma_f}{E\alpha}$	$\dfrac{\sigma_f k}{E\alpha} \times 10^{-3}$
Al_2O_3	145	350	8.8	30	47	1.4
MgO	96	210	13.5	36	34	1.2
BeO	145	300	9.0	220	53	11.5
Mullite $3Al_2O_3.2SiO_2$	82	145	5.3	6.3	107	0.67
Cermet WC–6% Co	1600	600	5.2	60 (?)	510	30
Cermet TiC–Co	1100	410	9.0	35	300	10.5
Hardened tool steel	2500	200	10	46	1250	57
Fused silica	105	725	0.55	1.5	2640	4.0
Soda-lime glass	70	69	9.0	1.0	113	0.1
Hard drawn Cu wire	410	117	14.0	380	250	95
Commercial graphite	20	10	3.0	150	670	100

energy, in contrast with the parameters of the thermoelastic theory. Good correlation with experiment is claimed. Reference should be made to the literature for this radically new approach to the problem of thermal stress resistance in brittle materials (see Hasselman's 1969 paper).

7.3 Traditional ceramics

Classic ceramics are made from three components: clay, flint and feldspar. The clay consists mainly of compound oxides of Al_2O_3, SiO_2 and H_2O, and provides workability before firing, when the H_2O component is driven off. The flint is a crystalline form of SiO_2, also called quartz, and is a cheap refractory component. Potash feldspar is a compound of K_2O, Al_2O_3 and SiO_2, and is a low melting component which forms a glass during firing and binds the refractory crystalline components together. For structural clay products, such as bricks and tiles, natural clay containing all three components, and many other impurities, is used directly. For whiteware products, such as china and porcelain, the composition is more closely controlled by compounding the purer clays, flint and feldspar.

Quartz, the refractory component of traditional ceramics, is one of three common crystalline forms of silica out of about twenty known; the other two are *tridymite* and *cristobalite*. Each of these minerals has high and low temperature modifications. In addition silica forms a glass (i.e. amorphous state, see § 7.8), which is often erroneously referred to as quartz glass – quartz is a specific crystalline form of silica. The ratio of the atomic radius of silicon to that of oxygen atoms is 0.37 and the four oxygen atoms always arrange themselves into a tetrahedron around the silicon atom. In the different forms of silica, all four corners are shared in a variety of arrangements, too complex to describe here. In the various silicates (clays) the tetrahedral groups are joined by one, two or three of the corner oxygen atoms into various layer structures, described below.

The various crystalline forms of silica undergo two types of transformation, as found in metals (cf. § 11.9). In displacive, or martensitic, transformation, the change from one phase to another occurs rapidly by slight shear distortions of the crystal structure. In reconstructive transformation, or conversion, the change proceeds slowly because activation energy is required to break bonds, promote diffusion and nucleate a new structure. Figure 7.2 shows the free energy of silica in its various crystalline forms. Within each crystalline species (quartz, tridymite or crystobalite) there are two or more modifications; inversions between modifications are martensitic transformations and rapid. The α–β quartz inversion at 573 °C is particularly significant to ceramic technology; it involves a volume change which produces cracking. Between the species the transformations are reconstructive and proceed very sluggishly; indeed, the different phases of silica can coexist side by side indefinitely.

Clay consists of very fine grains of hydrated aluminium silicate, tertiary compounds of Al_2O_3, SiO_2 and H_2O. Present in smaller quantities are CaO, Fe_2O_3, FeO, Na_2O, K_2O, MgO and TiO_2. In the study of ceramics the metallic oxides (minerals) are considered as the basic units, or end members, in the same way as Fe_3C (cementite) is taken as a component of mild steel (cf. § 4.8). The

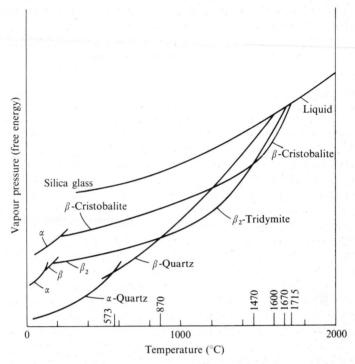

Fig. 7.2. Phase diagram for silica. Note that displacive transformations between modifications $(\alpha \rightleftharpoons \beta)$ are rapid but reconstructive transformations; (quartz $\rightleftharpoons$ cristobalite $\rightleftharpoons$ liquid) are very sluggish.

names and compositions of the minerals commonly encountered in ceramics technology are listed in Table 7.4. The composition of natural clays covers a wide range. One of the purest is English china clay, extensively mined in Cornwall; its principal component is the mineral *kaolinite* with the composition $Al_2O_3.2SiO_2.2H_2O$.

The crystal structures of the common clay minerals are based on various combinations of layers of Si_2O_5, which is made from SiO_4 tetrahedra joined at their corners, and $AlO(OH)$ layers. In kaolinite, the oxygen atoms projecting from the Si_2O_5 layer are built into the $AlO(OH)$ layer to give the composition $Al_2O_3.2SiO_2.2H_2O$, see Fig. 7.3(a). Another important clay mineral is *montmorillonite* $Na_2O.2MgO.5Al_2O_3.4SiO_2.nH_2O$. Here the $AlO(OH)$ sheet is placed between two Si_2O_5 layers, see Fig. 7.3(b). The multilayer spacing is larger in montmorillonite than in kaolinite, 15 Å against 7.2 Å, and this is responsible for the ability of montmorillonite to absorb large quantities of water in interstitial solution. In the mineral *muscovite* $K_2O.3Al_2O_3.6SiO_2.2H_2O$ (the most common of the mica group of minerals), the layers are weakly held together by potassium ions (Fig. 7.3(c)) and it is possible to cleave mica to give almost atomically smooth surfaces, the potassium ions adhering about equally to each face. In many of the minerals, other elements, or oxides such as CaO, K_2O and TiO_2, are taken into substitutional solid solution.

Table 7.4. *Ideal compositions of various minerals*

Silica	
Quartz	
Tridymite	Common crystalline phases of SiO_2
Cristobalite	
Alumina silicate	
Kaolinite (china clay)	$Al_2O_3.2SiO_2.2H_2O$
Pyrophyllite	$Al_2O_3.4SiO_2.H_2O$
Metakaolinite	$Al_2O_3.2SiO_2$
Sillimanite	$Al_2O_3.SiO_2$
Mullite	$3Al_2O_3.2SiO_2$
Alkali alumina silicate	
Potash feldspar	$K_2O.Al_2O_3.6SiO_2$
Soda feldspar	$Na_2O.Al_2O_3.6SiO_2$
(Muscovite) mica	$K_2O.3Al_2O_3.6SiO_2.2H_2O$
Montmorillonite	$Na_2O.2MgO.5Al_2O_3.24SiO_2.(6 + n)H_2O$
Leucite	$K_2O.Al_2O_3.4SiO_2$
Magnesium silicate	
Cordierite	$2MgO.5SiO_2.2Al_2O_3$
Steatite	$3MgO.4SiO_2$
Talc	$3MgO.4SiO_2.H_2O$
Chrysotile (asbestos)	$3MgO.2SiO_2.2H_2O$
Forsterite	$2MgO.SiO_2$
Magnesium alumina	
(Magnesium aluminate) spinel	$MgO.Al_2O_3$
Spinel group	$M^{2+}O.M_2^{3+}O_3$
Alumina	
Corundum	Al_2O_3
Sapphire	Al_2O_3 (very pure and perfect, often single crystal)
Ruby	Al_2O_3 + trace Cr_2O_3
Emery	Al_2O_3 + some (SiO_2 + Fe_2O_3)
Bauxite	$Al_2O_3.2H_2O$
Gibbsite	$Al_2O_3.3H_2O$
Zinc sulphide	
Zinc blende	
Wurtzite	Crystalline phases of ZnS

Clay particles are very fine and usually plate-shaped (arising from the crystal structure). The particles of kaolinite are flat hexagonals about 1–5 μm across and 0.05 μm thick (cf. 25 μm grains in fine grained steel and fine sand). Other clays, known as ball clays, are even finer, below 1 μm, with some rod shaped and some irregular. Different names are given to minerals with the same basic composition when the particles have different size or shape, or when the impurity elements are different. China and ball clay are often mixed to allow dense packing of the particles.

Clay forms a colloidal suspension with water, which is responsible for the clay's plasticity during shaping and 'throwing' prior to firing. Electric charges play a vital role in the formation of the water film, aided by the fine lamellar habit of the clay particles. At certain water contents clays are *Bingham* solids, that is, they are elastic at low stresses but above some yield stress they deform plastically at a rate proportional to the excess stress. The plasticity of clays,

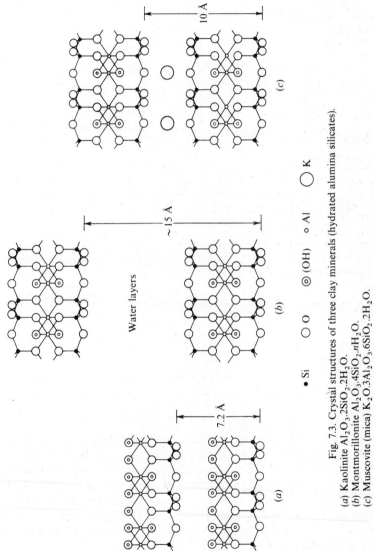

Fig. 7.3. Crystal structures of three clay minerals (hydrated alumina silicates).

(a) Kaolinite $Al_2O_3.2SiO_2.2H_2O$.
(b) Montmorillonite $Al_2O_3.4SiO_2.nH_2O$.
(c) Muscovite (mica) $K_2O.3Al_2O_3.6SiO_2.2H_2O$.

since it is so vital, has been extensively studied, but will not be considered further here.

The third component of traditional ceramics is feldspar. Feldspar is alkali alumina silicate, a mixture of: potash feldspar $K_2O.Al_2O_3.6SiO_2$, soda feldspar $Na_2O.Al_2O_3.6SiO_2$ and lime feldspar $CaO.Al_2O_3.2SiO_2$. Each feldspar has several crystalline forms. Although the smallest component, feldspar exerts a large influence on the properties of the final product since during firing it forms the low melting point flux, which binds the other components together.

A typical composition by weight for a porcelain before firing is:

25 per cent quartz (flint)	SiO_2
25 per cent ball clay (fine)	
25 per cent china clay (coarse)	$Al_2O_3.2SiO_2.H_2O$
25 per cent potash feldspar	$Al_2O_3.6SiO_2.K_2O$

After firing the overall composition becomes about: 65 weight per cent SiO_2; 30 weight per cent Al_2O_3 and 5 weight per cent K_2O. This is marked in the composition diagram, Fig. 7.4. The composition is adjusted to suit various applications. Increased amount of feldspar leads to greater densification at lower temperatures and translucency. Increased clay content makes forming much easier, but requires higher firing temperature, and leads to superior mechanical and electrical properties.

The raw materials are ground into a wet slurry with up to 30 per cent water and then formed to shape. The ware is first dried by gentle heating. Initially there is a film of water separating the grains and over the outside surface of the body and the evaporation rate is independent of the water content. When the particles begin to touch each other, around 20 per cent water content, the rate drops off.

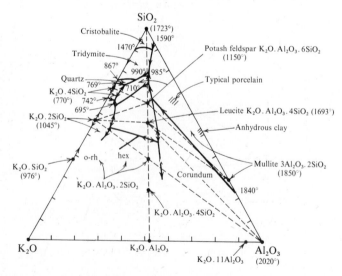

Fig. 7.4. SiO_2-Al_2O_3-K_2O systems showing the compositions of various complex oxides and of typical porcelains.

Shrinkage up to 25 per cent by volume occurs, the majority during the constant rate stage, and care must be taken to avoid warping and cracking due to uneven drying.

After drying, the body is fired at 1200 °C, or higher for high clay content. Details of the reaction are complex, vary with composition, and do not proceed to equilibrium. Nevertheless the trend of the reactions is indicated by the equilibrium diagrams, see Fig. 7.4 for the ternary system K_2O, Al_2O_3 and SiO_2, and Fig. 7.5 for two isopleths (vertical sections): (*a*) between leucite

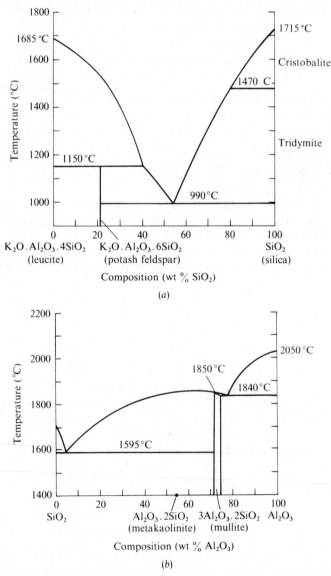

Fig. 7.5. Isopleths (vertical sections) through the SiO_2–Al_2O_3–K_2O diagram (Fig. 7.4.). (*a*) $K_2O.Al_2O_3.4SiO_2$–SiO_2; (*b*) SiO_2–Al_2O_3.

($K_2O.Al_2O_3.4SiO_2$) and SiO_2, which includes the potash feldspar composition ($K_2O.Al_2O_3.6SiO_2$); and (*b*) between alumina and silica, which includes mullite ($3Al_2O_3.2SiO_2$). Marked on the ternary equilibrium diagram are the various compounds which can be formed and the primary phase in each field, that is, the first phase to crystallise on cooling a melt of any given composition. The boundary lines separating the primary phase fields indicate where two phases are formed and, where they join, the composition of ternary eutectics. Note that the point representing the potash feldspar composition lies outside its own primary phase field on account of its incongruent melting by peritectic breakdown at 1150 °C (see § 4.9). Cross reference between the three figures will assist in their interpretation.

In outline, the reactions of a whiteware composition during firing are as follows. Over a narrow temperature range around 450 °C the combined (bound) water in the clay is expelled, leaving the quasi-compound metakaolinite $Al_2O_3.2SiO_2$ in a reactive amorphous state. The quartz undergoes the α–β inversion at 573 °C (and the reverse on cooling). Between 1000 and 1200 °C the metakaolinite breaks down to *mullite* $3Al_2O_3.2SiO_2$ and amorphous silica in two or three stages. (Intermediate phases are probably defect spinel A_2S_3 and a mullite-like phase AS.) The mullite is precipitated as very fine crystals in a matrix of amorphous silica. About 1100 °C the feldspar melts sluggishly. Note that the binary eutectic between feldspar and silica melts at 990 °C and the tertiary eutectic with a slight addition of mullite (or metakaoline) at 985 °C, but these compositions do not have time to form during firing. They do however limit the use of this composition for refractories. The liquid feldspar flows into the interstices under the pull of surface tension and there is a contraction of up to 40 per cent by volume. Reaction now occurs between the feldspar and the clay phases. Alumina diffuses into the feldspar phase and mullite crystals grow from the boundaries. Alkali diffuses into the amorphous silica to form a homogeneous glass. Some dissolving of the quartz grains may also occur.

On cooling to room temperature the microstructure consists of the original quartz grains, with some rounding of the corners and possibly a glassy surface film, the relict feldspar grains now consisting of glass criss-crossed with mullite needles, and a glass matrix containing very fine mullite crystals, the residue of the clay. A typical micrograph of an electrical porcelain is given in Fig. 7.6. The structure has been likened to concrete in which the cement and sand (glass and mullite) bind the coarse aggregate (quartz grains) together. Considerable porosity remains, up to 20 per cent, and many microcracks.

7.4 Cement and concrete

Cement, which consists of several compounds of lime CaO, alumina Al_2O_3 and silica SiO_2, is mixed with sand, coarse aggregate (e.g. crushed rock) and water to form concrete, which is the principal material of civil engineering. The cement and water form a paste, which initially makes the mixture into a plastic mass and later, after a complex setting reaction, binds it together into a hard brittle material with good compressive strength. Although the use of mortar and cement has a long history, their scientific study is still in its infancy. The chemical processes involved are now fairly well understood but the relationship between

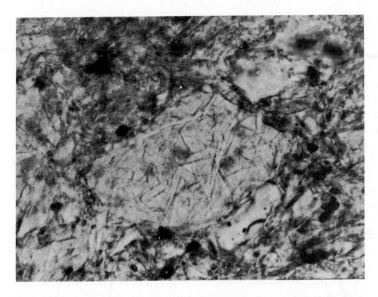

Fig. 7.6. Electrical porcelain microstructure, showing feldspar relict criss-crossed by mullite needles (large central area), quartz (light areas), unresolved matrix (dark) and some pores (black) $\times 900$. (By courtesy of The British Ceramic Research Association.)

the engineering properties and the physical structure is still little known. This section gives a short outline of the large amount of knowledge on cement. Further details are given in monographs on this subject.

Lime mortar, the forerunner of cement, has been known and used for over 5000 years. Quicklime (CaO) is first produced by heating limestone or chalk (forms of $CaCO_3$) to drive off carbon dioxide (the term 'quick' indicates that the lime is finely divided and hence reactive). The quicklime is then reacted with water to give hydrated, or slaked, lime ($Ca(OH)_2$). Because the reaction causes a 20 per cent volumetric expansion, a very fine powder is formed with a particle size around $2 \mu m$; excess water gives a highly dispersed white paste. Much heat is also evolved. If the paste, or paste and sand mixture (mortar), is allowed to dry out, a coherent solid of low strength is obtained as a consequence of the fine particle size. Hardening occurs subsequently by reaction with carbon dioxide from the air, reforming calcium carbonate. The reactions are:

$$CaCO_3 \rightarrow CaO + CO_2 \quad \text{(lime burning, calcining)},$$

$$CaO + H_2O \rightarrow Ca(OH)_2 \quad \text{(lime slaking)},$$

$$Ca(OH)_2 + CO_2 \rightarrow CaCO_3 + H_2O \quad \text{(carbonisation, hardening)}.$$

The hardening takes place slowly from the surface and in massive sections never penetrates to the centre; it cannot occur under water.

Hydraulic lime, that is, one that hardens by reaction with water and remains hard under water, is obtained by mixing slaked lime with volcanic ash (*pozzolana*) which contains very finely divided and reactive silica. Powdered tiles or bricks have a similar effect. The Romans used hydraulic limes extensively. Dicalcium silicate ($2CaO.SiO_2$) is formed and slowly hardens by hydration, as described below for Portland cement. The burning of limestone containing some clay gives a hydraulic lime without the use of pozzolana material, as discovered by Smeaton during the rebuilding of the Eddystone Rock lighthouse in 1756. This material is the forerunner of modern cements but contains free quicklime which on slaking breaks up the burnt product, without the need for grinding. Modern cements also contain a high proportion of tricalcium silicate ($3CaO.SiO_2$) due to burning at a higher temperature with fluxes and this is responsible for the much quicker hardening of cement.

Portland cements are mainly calcium silicates and calcium aluminate, a typical composition (in terms of the oxides) being 65 per cent CaO, 20 per cent SiO_2, 5 per cent Al_2O_3, 10 per cent balance (including Fe_2O_3). It is customary to write this: 65C, 20S, 5A, 10 % balance. They take their name from Portland Bill on the south coast of England where the limestone is somewhat similar in appearance. Their invention is attributed to Aspdin in 1824. In their manufacture, lime-bearing materials such as limestone, chalk, or shell, are mixed with clay, which provides the silica and alumina; ferric oxide (Fe_2O_3) is also present or is added. The raw materials are ground together and fired in a rotary kiln at around 1500 °C. A molten phase is formed by the alumina and ferric oxide in which the lime and silica can dissolve and react together to give di- and tricalcium silicates, which then crystallise out. The presence of this molten phase is essential to the process as the rate of reaction in the solid state (as occurs in firing clay-bearing limestone) is quite inadequate. It is important that no free lime is left after burning as this will eventually also hydrate in the hardened cement and disintegrate it. The calcium carbonate content is closely controlled (± 0.1 per cent) to avoid free lime yet promote the production of C_3S. The reaction products are, in quantity order (with typical weight percentages for ordinary Portland cement):

tricalcium silicate	C_3S or $3CaO.SiO_2$	(45 per cent)
dicalcium silicate	C_2S or $2CaO.SiO_2$	(30 per cent)
tricalcium aluminate	C_3A or $3CaO.Al_2O_3$	(10 per cent)
tetracalcium aluminoferrite	C_4AF or $4CaO.Al_2O_3.Fe_2O_3$	(10 per cent)
balance		(5 per cent)

The proportions of these compounds, called the mineralogical composition, are more important than the oxide composition in determining the properties of the cement such as the rate of hardening and heat evolution, or the resistance to attack by sulphate-bearing water. Standard compositions are discussed below. The burnt product from the kiln consists of dense hard lumps called clinker (cf. soft porous burnt product of hydraulic lime). A few per cent of gypsum ($CaSO_4.2H_2O$) is added and the clinker ground to a very fine powder, cement. The purpose of adding gypsum is to limit the rate of hydration of the tricalcium aluminate, which if not suppressed leads to rapid 'flash setting'.

The hydration reactions which occur on mixing cement particles and water are very complex. About two parts by weight of cement to one of water are necessary to complete hydration and achieve full strength. It should be noted that the hardening of concrete is not a process of evaporation as is sometimes believed. Indeed precautions are sometimes required to prevent surface evaporation and the entry of air before hydration has been completed. On the other hand excess water, which makes the mixing and placing of cement or concrete much easier, is also undesirable, leading to porosity and loss of strength. The end products of hydration are thought to be as shown in Table 7.5, though the details of various intermediate reactions are too complicated and obscure to present here. The most important reaction product is tricalcium disilicate hydrate ($C_3S_2H_3$), better known as *tobermorite gel*; this is formed by both the di- and tricalcium silicates. The name comes from Tobermory in the Isle of Skye where a natural mineral has an almost identical structure. A gel is a special type of material involving particles in the range 1–100 nm. In tobermorite the particle diameter is around 10 nm, compared with about 10 μm for the grains in the dry cement powder. In addition hardened Portland cement contains crystals of calcium hydroxide and some unreacted cement grains which become sealed off from the water by surface layers of insoluble hydration products. If a hardened cement is ground to fine powder it will undergo a second, weaker, setting reaction on this account.

The tobermorite gel is responsible for binding the other constituents of concrete into a solid mass, through a process of adhesion. An ion at the surface of an ionic solid is surrounded by fewer oppositely charged ions than a similar ion within the body of the crystal. It tries to correct for this by attracting other atoms or molecules, called adsorption. It is also possible for two particles to be attracted to one another, known as adhesion. The specific forces of adhesion are extremely strong between particles of tobermorite and other substances; and in the finely divided state of a gel, the specific surface (area per unit mass) is very large, 3×10^5 m^2/kg. The unique combination in tobermorite gel of high specific adhesive force and large surface area gives exceptionally high adhesive forces.

Table 7.5. *The hydration of cement powder. Only the initial and end products are given, not various complex intermediate reactions*

C[†]	Cement powder	Formulae	Hydrated cement
45%	Tricalcium silicate C_3S	$2(3CaO.SiO_2) + 6H_2O \rightarrow$ $3CaO.2SiO_2.3H_2O + 3Ca(OH)_2$	*Calcium silicate hydrate + calcium hydroxide
30%	Dicalcium silicate C_2S	$2(2CaO.SiO_2) + 4H_2O \rightarrow$ $3CaO.2SiO_2.3H_2O + Ca(OH)_2$	*Calcium silicate hydrate + calcium hydroxide
10%	Tricalcium aluminate C_3A	$3CaO.Al_2O_3 + 12H_2O + Ca(OH)_2 \rightarrow$ $2CaO.Al_2O_3.Ca(OH)_2.12H_2O$	Tetracalcium aluminate hydrate
		$3CaO.Al_2O_3 + 30H_2O + 3(CaSO_4.2H_2O) \rightarrow$ $3CaO.Al_2O_3.3CaSO_4.36H_2O$	Calcium mono-sulphoaluminate hydrate
10%	Tetracalcium aluminoferrite C_4AF	$4CaO.Al_2O_3.Fe_2O_3 + 10H_2O + 2Ca(OH)_2 \rightarrow$ $6CaO.Al_2O_3.Fe_2O_3.12H_2O$	Calcium alumino-ferrite hydrate

† Composition of ordinary Portland cement.   * Tobermorite gel.

Although the changes in structure and properties commence directly the cement is mixed with water and continue by a series of overlapping reactions, it is useful and customary to distinguish certain stages. The workability of a cement/water paste is almost unaffected at first, or can be maintained by continuous working (an effect called *thixotropy*, also found in some household paints). After a time the paste is evidently stiffer and a simple test (Vicat) involving the penetration of a standard needle is used to define the start and finish of *setting*. In ordinary Portland cement these are respectively not less than 30 minutes and not longer than 10 hours. During this time the paste loses its plasticity and becomes a brittle solid. Subsequent reactions lead to *strengthening* or *hardening*, which is normally measured up to 28 days but continues for many years. Different reactions predominate in the early and later stages of the overall reaction: calcium aluminate (C_3A) hydrates very rapidly (and, indeed, has to be controlled by the addition of gypsum); this is followed by tricalcium silicate (C_3S) and over a long period by C_2S. It will be apparent that a quick setting cement is not synonymous with rapid hardening.

The composition of Portland cements can be varied to give a range of properties to suit specific requirements. As already mentioned, the mineralogical (compound) composition is the more important parameter than the oxide composition. Typical compositions of cements for different requirements are given in Table 7.6; these are covered by appropriate specifications. When rapid hardening is required, that is, nearly the full strength in three days rather than twenty-eight or more, the proportion of C_3S is increased, through higher lime to silica ratio and finer grinding and mixing of the raw materials. The clinker is also ground to a finer state. In large concrete structures, such as dams, the heat evolved during hydration can cause troublesome temperature gradients and cracking. Low heat cement is obtained by reducing the amount of C_3A and C_3S in relation to the C_2S, which is slower to hydrate and has lower heat of reaction. The heat of evolution values are respectively: 207, 120, 62 cal/g.

Table 7.6. *Average compositions of various types of Portland cement. (From F. M. Lea and C. H. Desch,* The Chemistry of Cement and Concrete, *Edward Arnold Limited)*

Type	Oxide composition (per cent)					Mineralogical composition (per cent)				
						C_3S	C_2S	C_3A	C_4AF 4CaO.	
						3CaO.	2CaO.	3CaO.	Al_2O_3.	Free
	CaO	SiO_2	Al_2O_3	Fe_2O_3	Balance	SiO_2	SiO_2	Al_2O_3	Fe_2O_3	CaO
Ordinary	64.6	21.3	6.1	2.6	5.4	43	29	11	8	1.4
Rapid hardening	64.3	20.8	5.2	3.0	6.7	50	21	9	9	1.8
Sulphate resisting	62.3	21.2	4.0	5.6	6.9	49	24	1.5	17	1.9
Low heat	61.9	25.4	4.6	2.1	6.0	16	61	9	6	0.8

Hardened cement is chemically attacked by sulphates, for example by calcium sulphate often present in ground water. Reaction occurs with the calcium hydroxide crystals and the hydrated calcium aluminate, forming products with over twice the volume. Sulphate resistant cement therefore contains little C_3A and increased C_4AF.

The compressive strength of concrete is dependent on many factors, such as the properties and proportions of the constituents: cement, sand, coarse aggregate and water; the way they are mixed together; the temperature, the humidity and, of course, the curing time. The tensile strength is too low to be of any value. In carrying out comparative strength tests all these factors must be clearly specified. The water/cement ratio is of the greatest practical importance as the strength falls off rapidly with increasing water content, see Fig. 7.7. This is attributed to porosity. On the other hand the lower the water content the more difficult it is to mix and place the concrete. A standard *slump test* is carried out on site to ensure that too much water is not added, with consequent low strength at a later date. For full hydration the water/cement ratio by weight should be 0.4, of which 0.25 goes into chemical combination and 0.15 into adsorption on the tobermorite.

The aggregate is graded in size to pack in closely and give a dense material with the minimum amount of expensive cement. To this end a range of particle sizes is required, typically from 50 mm down to about 0.25 mm. (Above 5 mm it is arbitrarily defined as coarse aggregate.) The amount of cement must be sufficient to coat all the particles and bind them together. A common mix is $1:2:4$ by volume of cement, sand, and aggregate.

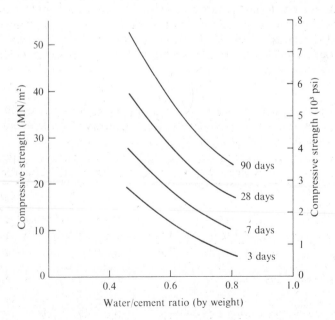

Fig. 7.7. Compressive strength of concrete as a function of water/cement ratio and time.

7.5 Cermets

Cermets consist of small particles of crystalline ceramic held together by a matrix of metal, typically tungsten carbide in cobalt. Although the term 'cermet' is to be preferred, these materials are frequently referred to as *cemented carbides* or *hard metals* (and on occasions, simply, ceramics!). Cermets were developed in the 1920s as materials for cutting tools. Since the particular requirements for cutting-tool material are not simply related to the mechanical tests discussed in Chapter 5, they will be briefly discussed before proceeding to the structure of cermets.

The properties required of a cutting-tool material are complex and material selection has normally been made on the basis of machine shop experience. For a single point cutting tool for use on a lathe, the loading conditions which have to be supported have been enumerated by Trent in 1968:

    (i) High compressive stress, especially near the cutting edge.

   (ii) Tensile stress at points removed from the cutting edge.

  (iii) Local stress concentrations near the cutting edge.

  (iv) High temperature. (Strength and oxidation resistance must be maintained.)

   (v) Rapid temperature fluctuations and steep temperature gradients.

  (vi) Abrasion, leading to wear. (Tool indentation hardness is only an approximate measure of resistance to abrasive wear.)

 (vii) Diffusion and reaction between the tool and workpiece materials.

(viii) Built up edge, due to the adhesion of the swarf or chip to the rake face of the tool, causing high localised stresses.

Which of these factors will be the most critical will depend on the particular application, and the best material composition will vary accordingly. High compressive strength and hardness at temperature, which are often considered as the most important criteria in a tool material, are only part of the requirements except in special circumstances, such as fine wire drawing and fine drilling, where stresses are low.

Although a full discussion of tool materials will not be given in this book, a brief mention of their development leading up to cermet cutting tools is appropriate here. In the modern era, the original tool materials described by Huntsman in 1740 were plain carbon steels with high carbon content (0.6–1.4 per cent C) used in the quenched condition (see Chapter 12). They are still the most common materials for the less arduous applications requiring wear resistance and the ability to maintain a sharp edge, for example, for dies. For metal cutting they have several drawbacks; their hardness drops rapidly with temperature above 200 °C; their wear resistance is limited, although it increases with carbon content at the expense of toughness; finally, they are water hardening, which leads to distortion and cracking troubles in heat treatment. They were succeeded by alloy tool steels (or non-deforming tool steels) containing tungsten and manganese (Mushet in 1868) and later by 'high speed steels' (Taylor and White in 1898). The latter were responsible for a sharp improvement in cutting properties as they actually increase their hot hardness during tempering up to 600 °C. Cutting speeds increased from below 1 m per minute to more than 30 m per minute. The basic grade of modern high speed steel contains 18% W–4% Cr–1% V and 0.7% C (known as 18–4–1). Subsequent developments have been in non-ferrous materials: cobalt based alloys ('Stellite'), tungsten carbide cermets (around 1925) and, on a limited scale, alumina (around 1950, see § 7.6). Cutting speeds with WC–Co cermets can be as high as 300 m per minute, with good wear resistance and toughness. At the same time the machine tools have been radically improved as regards stiffness and power to keep pace with the new cutting tool materials.

The principal cermet used for high speed metal cutting tools and mining drills is composed of tungsten carbide WC and 6–20 wt per cent of cobalt Co. The tungsten carbide is a hard and brittle ceramic with a simple hexagonal crystal structure (see Fig. 7.8) and cobalt is a ductile metal with a cph structure which is able to wet the carbide phase and form a strong adhesive bond with it. Most other hard compounds cannot be bonded successfully. Although the cermet remains brittle, the plastic deformation and energy absorbed before fracture is increased by at least an order of magnitude over values for pure WC. Since the development around 1925 of straight WC–Co cermets for cutting tools, many multicarbide compositions have been introduced in which the tungsten carbide is partially or wholly replaced by other carbides, notably TiC, TaC and MoC. These give better performance when cutting steel at high speeds although the original compositions are better for cutting cast iron and non-ferrous metals.

In the manufacture of the cermet, a fine powder of mixed WC and Co is either compacted and sintered, or hot pressed (sintered under pressure). The powder is around 1 μm in size and will contain some impurities unavoidably collected during processing, for example, from the balls used in the wet milling. The sintering is carried out at about 1400 °C, which is below the melting points of WC (2600 °C – a peritectic reaction) and Co (1492 °C), nevertheless a liquid phase is formed due to alloying of the components Co, WC and free C. The changes which take place during sintering are complex but a simplified version gives an adequate model for explaining the final microstructure. The cermet is

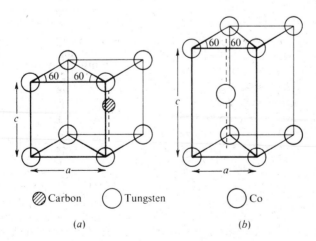

Fig. 7.8. Crystal structure of (a) tungsten monocarbide (simple hexagonal), $a = 0.29$ nm, $c = 0.28$ nm; (b) cobalt (close-packed hexagonal), $a = 0.25$ nm, $c = 0.41$ nm.

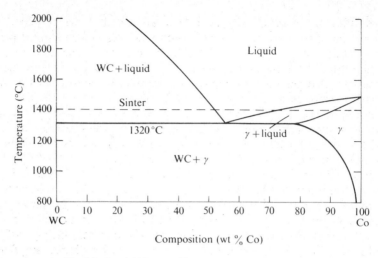

Fig. 7.9. Quasi-binary equilibrium diagram for WC–Co.

regarded as a binary alloy with an equilibrium diagram as shown in Fig. 7.9. At the sintering temperature, the tungsten carbide diffuses into the solid cobalt which becomes fully liquid at 30 per cent WC. The liquid is able to wet the remaining WC particles, and is drawn into the interstices, dissolving further WC up to the equilibrium concentration of 50 per cent. During sintering the component shrinks 50 per cent by volume. Some recrystallisation and growth of the WC particles also occurs. On cooling, the liquid phase deposits carbide and solidifies by eutectic reaction at about 1320 °C. Further carbide is precipitated in the solid state as the cermet cools to room temperature at which WC is nearly insoluble in Co.

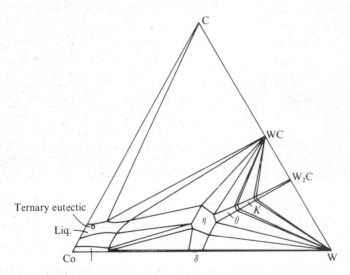

Fig. 7.10. W–Co–C ternary system: isothermal section at 1400 °C.

Some idea of the complexity of the various processes which occur during sintering (but giving the end product shown above) can be gauged from the ternary equilibrium diagram for W–Co–C, from which an isotherm at 1400 °C is given in Fig. 7.10. It must be appreciated that equilibrium is seldom achieved but that the changes on sintering are determined by the kinetics of various alternative solid state reactions. Further the composition is not usually stoichiometric WC, but will contain free graphite or WC_2. During sintering the ternary eutectic forms some liquid phase at as low a temperature as 1280 °C and the binary Co–C eutectic occurs at 1315 °C. η, θ and K-phases may also be formed. The full details will not be considered here.

Typical microstructures of tungsten carbide cermets are shown in Fig. 7.11. The white areas are the cobalt binder; (a) is a fine grain WC with 6 per cent Co, as used for cutting cast iron and non-ferrous metals, (b) has rather coarser grains of WC bound with 11 per cent Co and is extensively used for mining, (c) is an example of a multicarbide composition for steel cutting: 70 % WC–9 % TiC–12 % TaC–9 % Co. Three phases are present here: angular WC

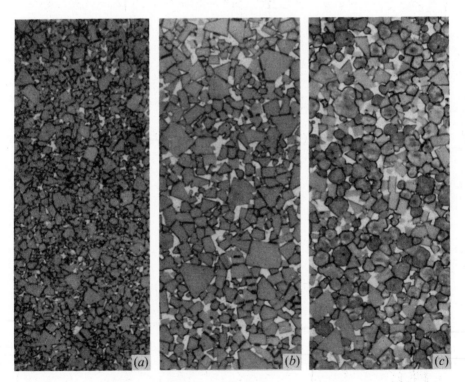

Fig. 7.11. Tungsten carbide and cobalt cermet microstructures. Etched in Murikama's reagent. × 1500. (By courtesy of Wickman Wimet Limited.)

(a) WC–6 % Co: fine WC grains bonded by light etching cobalt. Typical grade for cutting cast iron and non-ferrous metals.

(b) WC–11 % Co: medium-coarse grains of WC bonded by light etching cobalt. Typical grade for rock drilling.

(c) 70 % WC, 9 % TiC, 12 % TaC–9 % Co: light grey angular particles are WC, darker small particles are TiC–TaC–WC solid solution and white areas are cobalt. Typical grade for steel cutting.

particles, darker irregular particles of solid solution TiC–TaC–WC, and the white cobalt matrix.

There is considerable discussion as to whether at low cobalt content (6 per cent) the carbide particles are separated by a thin film of cobalt or whether they form a continuous three-dimensional skeleton. A thin cement layer would largely explain the great increase of the transverse rupture modulus from 500 MN/m^2 (75 000 psi) for sintered pure WC to 1.7 GN/m^2 (250 000 psi) for 6 per cent Co–WC. The thin film is restrained from plastic deformation by the adjacent hard carbide yet prevents cracks traversing through a continuous carbide phase. At higher cobalt content, the WC particles are clearly embedded in a cobalt matrix.

The most important development in cutting tools during the 1960 decade has been in the use of throw-away cermet tips. These are supplied by the manufacturer with several cutting edges, prepared to a very high standard, for holding mechanically in a tool-holder. Previous practice was to braze the tip to a shank and preparation of the cutting edge was the responsibility of the individual tool room. As a rival to the cermet tool, alumina is beginning to compete (see § 7.6). Given the right machine tool this is able to machine some materials faster and for longer than cermets. However greater care is needed to avoid accidental fracture. The physical properties of the rival cutting tool materials are contrasted in Table 7.7.

The development of cermets as high temperature structural materials has been given much effort in the last twenty years, especially for gas turbine blades. Here the thermal shock resistance, creep and impact strength at temperature are more important than hardness and wear resistance. The proportion of ceramic to metal is accordingly moved towards a majority of the metallic phase (30–80 per cent). So far the development of the necessary properties has proved elusive. Attention initially focussed on oxide based cermets, especially Al_2O_3–Cr and Al_2O_3–Cr–Mo, on account of their superior oxidation resistance. Of the carbides, only the TiC based cermets have comparable oxidation resistance and their mechanical properties are possibly superior, if still inadequate. Several oxide and carbide cermets and a nickel alloy are compared for this type of application in Table 7.8.

Table 7.7. *Comparison of the physical properties of various tool materials (after A. G. King and W. M. Wheildon:* Ceramics in machining processes, *Academic Press* (1966))

	18–4–1 high speed steel Fe–18W–4Cr– 1V–0.7C	WC–Co cermet	Al_2O_3 <1% MgO	Al_2O_3 10% TiO
Grain size (μm)	10	2	3	2
Bend strength (MN/m^2)	3500	1600	600	600
Compressive strength (MN/m^2)	4100	4500	3000	3500
Hardness, Knoop (100g)	740	1800	1570	1660
Max. working temp. (°C)	600	900	1000	1000

Table 7.8. *Comparison of cermets, ceramic and metal for high temperature structures (Data from A. G. Thomas, J. B. Huffadine and N. C. Moore, Metall. Rev. **8** no. 32, 461 (1963).)*

| Material | Ceramic Al_2O_3 | Cermets | | | | | Metal Nimonic 105 |
		LT1	LT1B	LT2	K163B	40N	
Composition % Al_2O_3	99	23	18	15			
% TiC					57	54	
% Cr		77	60	25			10
% Ni					30	40	55
% W				60			
% Mo			20		5		5
Balance			$2TiO_2$		8	6	20Co
Specific gravity	3.9	5.9	6.0	8.8	6.0	6.1	8.0
Expansion (deg^{-1} × 10^6)	7.2	8.9	8.5	8.4			19.7
Conductivity (Wm^{-1} deg^{-1})	30	17			34		13
Modulus of 20 °C	310	310	390	510	1120	950	1250
rupture (MN/m²) 1000 °C							125
1150 °C			125	200	310		
100 h Stress 900 °C		240			250		210
rupture (MN/m²) 1000 °C		220			135		80
Impact resistance* (J/cm²)	0.34	0.25	0.37		13.5	14.2	24
Oxidation: Wt gain (mg/cm²) 1000 °C, 100 h	0	1.66	0.80		10.4	28.9	1.43

* Measured on a Hounsfield plastics impact tester.

7.6 New ceramics: alumina

The term 'new' ceramics is limited here to describing simple compounds such as oxides, carbides, and nitrides which are now being produced in the pure crystalline state with very low or nil porosity. The term has also been used to describe all ceramics not based on clay. Very close quality control is exercised to achieve a very well defined product, in contrast with traditional ceramics. Sintering or hot pressing of dry powder is the normal manufacturing process. The annual production is insignificant in volume but their importance lies in their ability to fulfil various vital functions in advanced technology, on account of their special electronic, chemical, and mechanical properties. Examples of the new ceramics include: beryllia and uranium oxide, used in nuclear reactors; boron carbide, the hardest known material, which is used in lightweight armour and gas bearings; silicon carbide, long used for heater elements, refractories and abrasives; silicon nitride, in experimental gas turbine blades and bearings; and barium titanate which has a dielectric constant of over 1000, or its modifications which have dielectric constants approaching 10 000. It is not possible to discuss significantly each in turn and only two ceramics, ferrite and alumina, will be considered here.

Ferrites are a class of ceramic magnetic materials, which are derived from the naturally occurring *magnetite*, Fe_3O_4 or double oxide $FeO.Fe_2O_3$. These will be discussed in detail in Chapter 15, but a brief description will be included now to emphasise that many of the new ceramics are developed for their electrical and magnetic properties. In brief, magnetism arises from electron spin, which is in two senses, supporting or opposing an applied magnetic field. In some atoms, for example iron, the spins are not fully paired off so that each atom is a permanent magnetic dipole. Depending on the interatomic distance these dipoles interact with each other, either supporting or opposing their neighbour. The former leads to ferromagnetism, found in α-iron (§ 4.8), and the latter to antiferromagnetism and, when the opposing groups are unequal as in the ferrites, ferrimagnetism. The crystal structure of magnetite is spinel, see Fig. 15.5; that is, the atoms are in the same geometrical arrangement as the mineral spinel $MgO.Al_2O_3$ but the Mg and Al cations are replaced by Fe^{2+} and Fe^{3+} ions respectively. The creation of two groups, with opposing magnetic dipoles, is due to the two types of interstitial sites in which the iron atoms are located between the closely packed cubic lattice of oxygen atoms: one set has a coordination number of 4 and the other 8. In synthetic ferrites, developed since 1946, the divalent ferrous ion is replaced chemically by other metallic ions such as Mg^{2+}, Zn^{2+}, Cu^{2+}, although they may be physically located in different interstitial sites, and the magnetic properties can be tailored to requirement. The soft ferrites have minimum hysteresis loop and respond linearly to an applied magnetic field. They are used as high frequency ($> 100 \, kHz$) transformer cores, where their high resistance eliminates eddy currents, a cause of overheating in soft iron cores. Other ferrites, containing manganese and magnesium, are designed to give square hysteresis loops for use in computer memory stores and other logic circuits. Another group of ferrites are excellent permanent magnets, for example, $BaFe_{12}O_{19}$ (Magnadur). The development of synthetic ferrites in the last 25 years has been vital to the development of many electronic devices, particularly computers. The theory of magnetic materials is considered further in Chapter 15.

Alumina is another ceramic on which a lot of development has been carried out in order to purify the material and to improve the processing techniques so as to obtain a fine grained structure, free of porosity. Alumina has several crystalline forms but the common one is $\alpha\text{-}Al_2O_3$, called *corundum* or, particularly as single crystal, *sapphire*. At least six other crystal forms have been reported but their structures are not established and they are frequently lumped together as $\gamma\text{-}Al_2O_3$ to distinguish them from the common $\alpha\text{-}Al_2O_3$. The crystal structure of corundum consists of oxygen atoms in an approximately close-packed hexagonal array with aluminium atoms occupying the octahedral interstitial sites (coordination number = 6). To obtain the correct ratio of aluminium to oxygen atoms one site in three is unoccupied, so the aluminium atoms form sheets of open hexagons, see Fig. 7.12. Although the aluminium atoms are lying directly above each other in successive sheets, the presence of the empty centres of the hexagonals means that the repeating unit contains 6 layers of oxygen atoms and 6 of aluminium atoms; this can be represented as $A \, C_1 \, B \, C_2 \, A \, C_3 \, B \, C_1 \, A \, C_2 \, B \, C_3$, where A and B are close-packed hexagonal sheets of oxygen atoms and C_1, C_2 and C_3 are intervening sheets of aluminium atoms in open hexagonal array. Only half this unit is shown in Fig. 7.12.

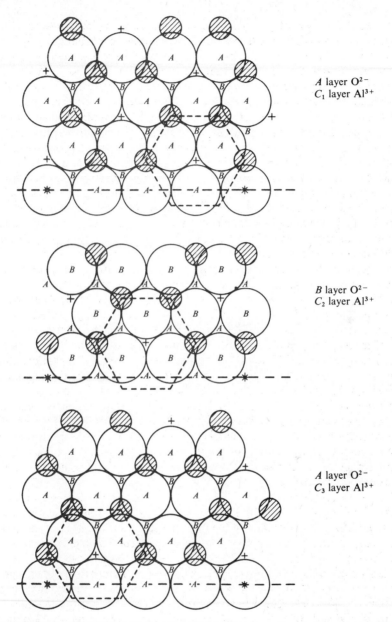

A layer O^{2-}
C$_1$ layer Al^{3+}

B layer O^{2-}
C$_2$ layer Al^{3+}

A layer O^{2-}
C$_3$ layer Al^{3+}

Fig. 7.12. Crystal structure of α-alumina (corundum). Oxygen atoms are in approximately cph array with smaller aluminium atoms in octahedral interstices, with one site in three vacant. There are thus three types of Al layer depending on the vacancy position, and the repeat unit consists of twelve layers: $AC_1BC_2AC_3BC_1AC_2BC_3$.

Extremely pure corundum is white or translucent but there are usually traces of other oxides which give attractive colours in gem stones: TiO_2 is responsible for blue sapphire; Cr_2O_3 for red sapphire, or ruby; and MgO gives yellow sapphire. Emery, which is widely used for abrasives, is also natural α-alumina containing a large proportion of iron oxide.

High density commercially pure alumina (>99 per cent Al_2O_3) and high alumina ceramics (>85 per cent Al_2O_3) have been increasingly used since World War II on account of their combination of electrical, mechanical and chemical properties. The purity, grain size, and porosity have been continuously refined until a fine grained fully dense translucent alumina was produced in 1959 (containing some MgO). Electrically, they possess high resistance, good dielectric strength and a very low loss factor at high frequencies: mechanically, they have high hardness, dimensional stability and resistance to abrasive wear; chemically, oxides are stable in most environments. These characteristics are maintained to high temperatures: the melting point of alumina is 2050 °C and the working temperature can be over 1000 °C. In addition they are available commercially at reasonable prices, can be metallised (coated with a metal) and diamond ground to fine tolerances. The many applications include supporting wafers for electronic components or thin films, microwave valve envelopes, radomes, lightweight armour, high speed cutting tools, thread guides and gland seals.

The microstructures of several grades of alumina are given in Fig. 7.13: (a) is a debased alumina containing 95 per cent Al_2O_3 and alkaline earth silicates.

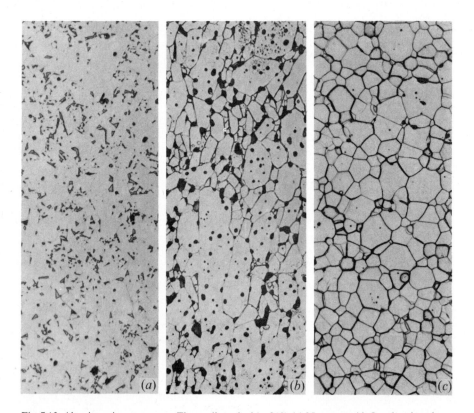

Fig. 7.13. Alumina microstructures. Thermally etched (× 210). (a) 95 per cent Al_2O_3, showing glassy phase between grains. (b) 99 per cent Al_2O_3, showing porosity. (c) 99.7 per cent Al_2O_3–0.1 per cent MgO, translucent alumina. (By courtesy of The British Ceramic Research Association.)

The crystals of alumina are embedded in a glassy matrix. These debased aluminas are cheaper to produce but generally have degraded properties. (b) is 99 per cent Al_2O_3 with no glass phase but considerable porosity. (c) is 99.7 per cent Al_2O_3 translucent alumina with no glass or porosity.

   The difficulty of producing high purity, fine grained, fully dense alumina is illustrated in Fig. 7.14, which plots the grain size versus porosity obtained from 99.9 + per cent Al_2O_3 (containing only a few ppm of Si, Cu and Mg) by varying the pressure and temperature during sintering. Below 2 per cent porosity the grain size increased rapidly. In 1959 Coble found that the addition of 0.1 per cent MgO prevented grain growth during sintering so that a fully dense, fine grained (about 2 μm) alumina was obtained. With no residual porosity the material becomes transparent in thin sections, translucent in thicker ones. The mechanical and electrical properties are significantly improved. Since alumina is highly resistant to the alkali metals, the first experimental application of translucent alumina was to sodium vapour lamps operating at 1000 °C, which give an attractive light at very high efficiency. It may also be used in magneto

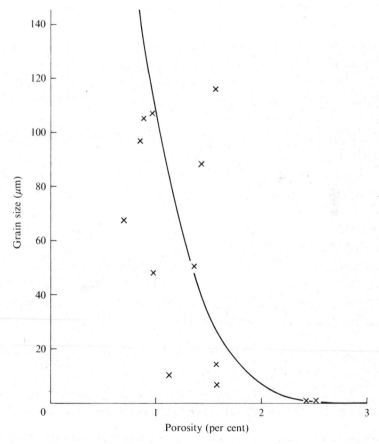

Fig. 7.14. Grain size versus porosity of very high purity alumina (99.9+ per cent). (Data from Spraggs and Vasilos, *J. Am. Ceram. Soc.* **46**, 224 (1963).)

hydrodynamic generators to contain gases containing potassium at 2000 °C. More immediately it is being used in envelopes and windows of high powered radar valves on account of its very low dielectric loss at high frequency, combined with good mechanical properties.

In experiments to determine the variation of properties with microstructure the interdependence of grain size and porosity, discussed above, has to be considered. Figure 7.15 shows the variation of zero porosity bend strength with grain size for 99.9+ per cent Al_2O_3, obtained by Spriggs and Vasilos. The bend strengths (S) of several specimens with a constant grain size but different porosities (P) were measured and the zero porosity strength (S_0) obtained by extrapolation according to an exponential law:

$$S = S_0 \exp{(-bP)}$$

where the constant b was adjusted for each grain size. The zero porosity strength was found to be given by the relation

$$S_0 = \text{const.}\ G^{-\frac{1}{3}}$$

where G is the grain size. This law indicates less dependence on grain size than is given by Petch's law ($\propto G^{-\frac{1}{2}}$, see § 10.4) for the fracture of metals. It should be remarked that in commercially pure aluminas, except of the highest purity, there is no simple correlation of strength with alumina content or density due to the wide variation in microstructure.

In the use of alumina for cutting tools its high hot hardness and resistance to oxidation at over 1000 °C and its poor wettability by metals, which leads to low wear rates, are valuable properties and in the right conditions more than compensate for the lower rupture and compressive strength as compared with rival materials. The relative merits of several cutting tool materials were

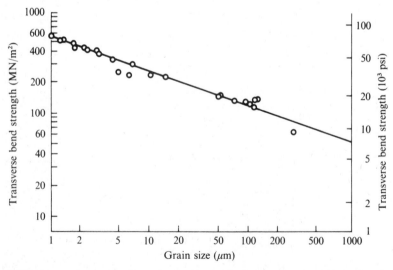

Fig. 7.15. Transverse bend strength (extrapolated to zero porosity) versus grain size for alumina. (After Spraggs and Vasilos, *J. Am. Ceram. Soc.* **46**, 224 (1963).)

discussed in the previous section (see Table 7.7). Many proprietary grades have become available since the 1950s, when they were introduced in the USSR; their compositions and processing treatment are not revealed. They may be 'pure' alumina (>99 per cent) or contain up to 10 per cent of other oxides which are reputed to increase the toughness or make manufacture cheaper.

Turning now to the electrical properties of aluminas, they have high resistivity, moderate dielectric constant (5–10), high dielectric strength (50–200 kV/cm) and very low dielectric loss (tan $\delta < 10^{-3}$). It is for the last reason, combined with mechanical strength at high temperature, that alumina is invaluable for microwave valve envelopes, windows and radomes. The dielectric constant of 21 commercial aluminas is plotted against the density in Fig. 7.16. This shows that there is a systematic increase in the dielectric constant with increasing density, irrespective of the detailed composition and microstructure, although greater scatter appears at the lower densities, which correspond with lower alumina constants (<95 per cent Al_2O_3). The loss tangent tan δ (see § 3.10) is plotted against temperature in Fig. 7.17 for some representative compositions. This demonstrates, on the other hand, that the dielectric loss is very sensitive to the microstructure and cannot be simply related to the alumina content. It is only roughly true that tan δ increases with impurity content and temperature.

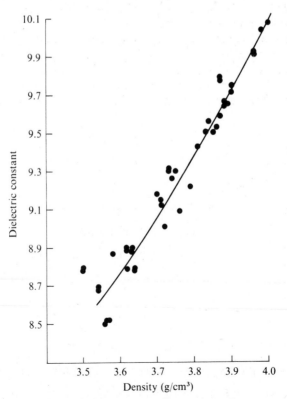

Fig. 7.16. Dielectric constants of 21 commercial aluminas (88 per cent Al_2O_3) at 9368 MHz and 20 °C. (After W. George and P. Popper, *Proc. Br. Ceram. Soc.* **10**, 63 (1968).)

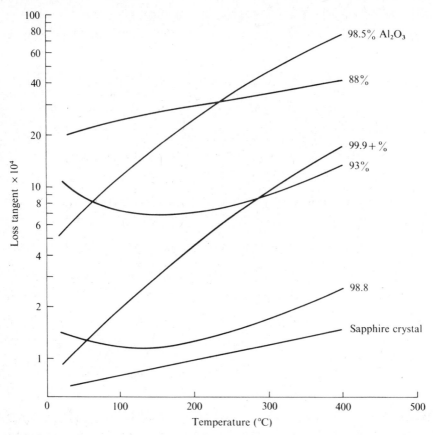

Fig. 7.17. Loss tangent versus temperature of some representative aluminas at 9368 MHz. (From W. George and P. Popper, *Proc. Br. Ceram. Soc.* **10**, 63 (1968).)

High alumina ceramics are used for high duty sparking plugs where the requirements are much tougher than for most insulators: dielectric strength at elevated temperatures, resistance to repeated thermal and mechanical shock, and corrosion resistance to hot combustion products at 2500 °C including lead antiknock compounds. Early spark plug porcelains were based on the classic components of feldspar, quartz and clays, but the quartz gave poor thermal shock resistance, with its phase change at 573 °C, and the alkali feldspar caused electrical leakage at elevated temperatures. The quartz was subsequently replaced by calcined sillimanite $Al_2O_3.SiO_2$ which breaks down to mullite $(3Al_2O_3.2SiO_2)$ and silica; and the feldspar by talc $3MgO.4SiO_2.H_2O$. Firing temperatures were increased to 1425 °C. Since about 1935 high alumina ceramics have been employed on account of their superior thermal stability, heat conductivity and insulation capacity. Some 5 per cent of silica is retained to ease manufacture and reduce the thermal expansion.

Another form of alumina which deserves a brief note here is that formed on the surface of aluminium metal, either naturally in air or as a result of electrolytic action, called *anodising*. The natural film is very thin, under 10 nm (cf. 0.3 nm

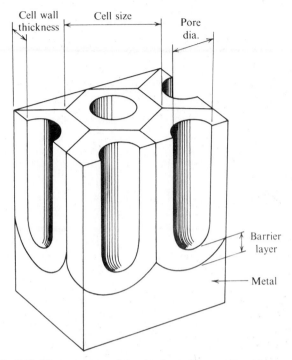

Fig. 7.18. Microstructure of alumina formed by anodic oxidation.

aluminium atom diameter), when further reaction is prevented by self sealing. The film can be thickened several hundred times by anodising. The metal is made the anode in an electrolytic cell with, for example, sulphuric acid or oxalic acid as the electrolyte. Passing of a current draws the oxygen ions to the aluminium where they react to form oxide, with some modification due to the electrolyte. If the alumina is insoluble in the electrolyte the oxide is a dense film which rapidly insulates the underlying metallic electrode. This is called a *barrier layer* and has a thickness roughly proportional to the applied volts (1.5 nm/V). However with an electrolyte in which alumina is sparingly soluble a very much thicker layer is formed although this is not dense. Its cellular microstructure is shown in Fig. 7.18: the thin dense layer is still called the barrier layer and the remainder is termed the porous layer. The size of the pores increases with concentration and temperature of the electrolyte and the current density but the thickness of the pore wall and barrier layer are proportional to the voltage. Thus the microstructure can be infinitely varied. Subsequently the pores can be sealed by hydration of the oxide. A typical film thickness is in the range 10–40 μm (0.0004–0.0015 in). In *hard anodising* the process is basically the same but the parameters are optimised to give an abrasion resistant film; a typical film is then in the range 25–250 μm. As expected on consideration of their relative microstructures there are considerable differences in the properties of the aluminas produced by anodising and powder metallurgy. This large subject of anodising will not be discussed here.

Large single crystals of ceramics are manufactured by growth from the vapour, by precipitation from a molten flux, or, if the material does not decompose, by the Verneuill technique. In the latter, finely powdered material is fed through an electric arc, or an oxy-hydrogen flame, onto the surface of a seed crystal which is slowly withdrawn from the flame as the fused powder builds up. Alumina crystals up to 100 mm diameter have been grown by this method. Single crystals of many ceramics are commercially available. Examples are quartz crystals for precision electronic oscillators – the crystal is piezo-electric, that is, an electric field causes a mechanical strain; synthetic sapphire for low wear bearings in watches and instruments; synthetic ruby for lasers; alkaline earth crystals (e.g. CaF, LiF) for prisms and windows in optical equipment.

7.7 Graphite

In one of its crystal forms, carbon has a hexagonal layer structure, as shown in Fig. 7.19; this is called graphite (cf. diamond with tetrahedral structure). It will be seen that the structure repeats every alternate layer. The atoms within a layer are 0.14 nm (1.4 Å) apart and bound by strong covalent bonds but between layers the separation is 0.33 nm (3.3 Å) and there are only weak van der Waals forces. The anisotropy of the structure is repeated in the physical and mechanical properties. Strength, stiffness, and thermal and electrical conductivity are high in directions parallel to the hexagonal planes (a-directions) and low in the perpendicular direction (c-axis). The layers are easily sheared and this was long considered the source of the low coefficient of friction, but it now appears that this is also dependent on the presence of an adsorbed film, such as water. Thus the graphite brushes in aircraft DC generators will wear excessively

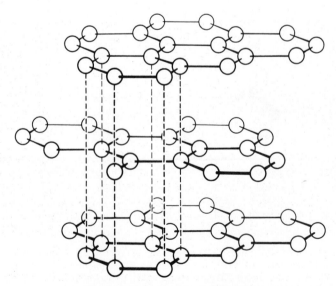

Fig. 7.19. Crystal structure of graphite.

at high altitudes (10 000 m) because of the dry air and low partial pressure of oxygen. In recent years ($\sim$ 1958) graphite whiskers have been grown. These are near-perfect single crystals with high length/diameter ratios and the *c*-axis orientated perpendicular to the whisker length. The strength is about 7 GN/m^2 (1×10^6 psi) and the stiffness around 700 GN/m^2 (100×10^6 psi). Since the specific gravity is only 2.2 the specific strength and stiffness are exceptionally high.

Polycrystalline graphite covers a wide range of microstructures and properties, including conventional, pyrolitic and fibre material. Conventional commercial graphite is manufactured from petroleum coke, a by-product of oil refineries, and a binder such as pitch. Finely crushed coke is mixed with the binder and shaped by extrusion or rolling. The volatiles are then driven off by baking at 750–950 °C for up to 3 days. This is followed by the conversion to graphite by heating to 2600–3000 °C in an electric furnace, where the charge is itself the heating element. Prior to graphitisation, the microstructure contains coke particles which consist of very imperfect small crystals in which the hexagonal layers are more widely spaced than theoretical and incorrectly orientated to each other (*turbostatic* structure). In between the coke particles, there are even smaller crystallites arising from the binder. The material is more amorphous than crystalline and has many voids. Graphitisation causes the crystallites to become more perfect and the larger ones to grow, mainly in the *c*-direction. Considerable porosity remains, up to 30 per cent, and there are many very small imperfect crystallites. By varying the raw materials and processing, the crystal size, the degree of preferred orientation (which is clearly very important with such an anisotropic crystal structure), the proportion of small crystallites, and the voidage can be controlled over very wide limits. The porosity can be reduced to almost zero by subsequent impregnation with furfuryl alcohol and further graphitisation. This is done for the cans of fuel elements in some nuclear reactors. A high quality fine-grained microstructure is shown in Fig. 7.20(*a*).

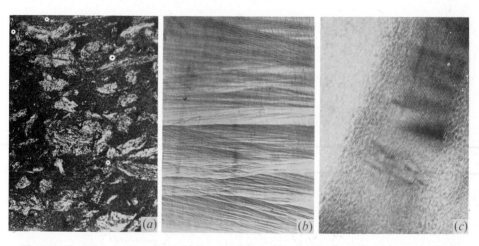

Fig. 7.20. Graphite microstructures: (*a*) Commercial graphite, showing fine coke grains, carbonised lamp black and pitch mix. × 125. (By courtesy of W. Watt.) (*b*) Pyrolytic graphite. × 125. (By courtesy of W. Watt.) (*c*) Carbon fibre, showing basal planes within fibril. × 1.225 × 10^6. (By courtesy of D. Crawford and D. J. Johnson.)

The properties of commercial graphite vary widely in line with the variability in microstructure. In general the material is noted for its strength at elevated temperature (the strength actually increases up to 2500 °C due to increased plasticity), low density, good resistance to thermal shock and low nuclear absorption. The disadvantages are its inhomogeneity, variability, brittleness and poor oxidation resistance. Strenuous efforts are being made to overcome these limitations. Typical properties are listed in Table 7.2.

Pyrolytic graphite is prepared by vapour deposition and leads to almost theoretically dense graphite with highly preferred orientation. The properties are markedly different from conventional commercial graphite. The technique of chemical vapour deposition has been applied to metals for many years and to ceramics more recently. In the formation of pyrolytic graphite, methane (CH_4) decomposes as it passes over a mandrel at 1000–3000 °C; elemental carbon is deposited and hydrogen flows out. The crystals have their c-axis orientated perpendicular to the mandrel surface. Although there is no amorphous carbon, the crystals are not perfect; the layer planes tend to be kinked and successive layers not correctly positioned. The microstructure is shown in Fig. 7.20(*b*). The material is used in rocket nozzles.

Graphite is also produced in the form of very fine fibres, 10 μm diameter. The fibre is polcrystalline, each crystal being an elongated fibril lying about parallel to the fibre axis, with many fibrils in a fibre cross-section. The c-axis is perpendicular to the fibril and fibre axis so that tensile load is in directions parallel to the hexagonal layers. The basal layers are visible in Fig. 7.20(*c*). The crystal size and preferred orientation are dependent on the details of the production process and the properties of fibres vary accordingly. For example, fibres made by pyrolysis of rayon and hot stretched at 2700 °C have a tensile strength of 1.5 GN/m^2 (2 $\times$ 10^5 psi) and a Young's modulus of 460 GN/m^2 (70 $\times$ 10^6 psi), whilst for fibres made from polyacrylonitrile the values are 2.0 GN/m^2 (3 $\times$ 10^5 psi) and 260 GN/m^2 (40 $\times$ 10^6 psi) respectively. There is still some way to go before the strength values for whiskers are achieved, and this may not be possible in full since the near perfection of the structure of whiskers will be difficult to match in polycrystalline fibres. Nevertheless carbon fibre has a high specific strength, equal to that of glass fibre (§ 7.9) and a very high Young's modulus, an order of magnitude greater than that of glass. For this reason it is superior to glass fibre in composite materials where the high strain associated with the high allowable stress causes undesirable distortion in structures (see § 11.12). At present the price of carbon fibre is limiting its use.

7.8 Glasses

Glasses are a class of inorganic materials whose physical state is intermediate between a liquid and a solid. They are mainly metallic oxides. Organic materials (see Chapter 8) can also exist in the glassy state but they are not usually referred to as glasses. In contrast with true solids, which are crystalline, glasses are non-crystalline or amorphous (vitreous). On the other hand a glass can be distinguished from a supercooled liquid: in the latter the nearest neighbour atoms (first coordinates) are random but in a glass the nearest neighbours are regular, and the second and subsequent coordinations are random. In other words there is a short range order, but no long range order (periodicity) as found in crystals. The structure of glasses is explained in more detail below. On cooling a liquid to the glassy state there is no crystallisation and no discontinuous change in specific volume and mechanical properties. However a transition, at the *glass transition temperature*, does occur in some parameters, notably the coefficient of thermal expansion (i.e. the gradient of the specific volume – temperature curve), which in the glassy state is very much less than for a liquid.

Rigidity in a glass is achieved by a steady increase of viscosity with falling temperature. Typical viscosity–temperature curves are shown in Fig. 7.21. Several reference temperatures corresponding to certain viscosities are defined as follows:

(*a*) Working point: viscosity equals 10^3 Ns/m^2 (10^4 poises). Many glass fabrication methods, such as pressing and drawing, are carried out around this temperature. (N.B. An alternative working point corresponds to 10^4 Ns/m^2.)

(*b*) Softening point: viscosity equals $\sim 10^7$ Ns/m^2 ($\sim 10^8$ P). This is the temperature at which glass flows appreciably under its own weight (see ASTM C338); the exact viscosity is dependent on the glass density. Hand fabrication is done around this temperature.

(*c*) Annealing point: viscosity equals 10^{12} Ns/m^2 (10^{13} P). Internal stresses are relieved at this temperature.

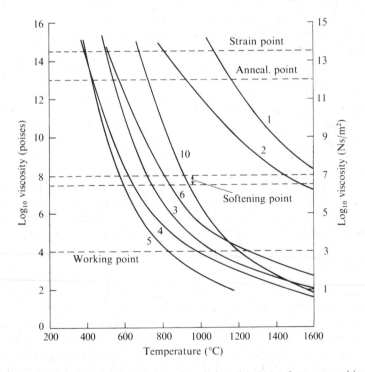

Fig. 7.21. Viscosity–temperature curves for various glasses. Numbers refer to compositions given in Table 7.9. (Data from E. B. Shand, *Engineering Glass*, Modern Materials Vol. 6, p. 247. Academic Press (1968).)

(d) Strain point: viscosity equals 10^{13} Ns/m^2 (10^{14} P). Below this temperature a glass can be cooled rapidly without introducing undue internal stresses.

The glassy state is a non-equilibrium one and is only achieved because crystallisation is prevented by rapid cooling through a temperature range just below the thermodynamic freezing point, that is, the temperature at which crystalline phases would start to form if time were allowed for equilibrium. The room temperature structure and properties of a glass are affected to some extent by the rate of cooling through the temperature range. Over very long periods glasses will devitrify, or become crystalline – hundreds of years at ambient temperature. In glass-ceramics (see § 7.10) crystallisation is promoted by special techniques.

The composition of glass formers is limited in the main to three acidic oxides, SiO_2, B_2O_3 and P_2O_5 (i.e. a closely knit group in the Periodic Table). The oxides of arsenic and germanium, and a few non-oxides, such as sulphur, selenium and lead fluoride, can also form glasses. Besides the glass formers, there are a number of other oxides which can be incorporated in a glass, known as network modifiers. They are the basic oxides of the alkali metals (Na_2O, K_2O) and of the alkaline earths (MgO, CaO). Forming a third category are some oxides which are not themselves capable of forming a glass network but can join in a network already existing; they are known as intermediate oxides. Examples are alumina and beryllia. The distinction between network-modifying oxides and intermediate oxides will be made clearer below in the discussion of the microstructure of glasses. The composition and properties of the principal types of commercial glass are listed in Tables 7.9 and 7.10. The majority are silicates and borosilicates.

The basic structural unit of the glasses is a tetrahedron with the small cation surrounded by four oxygen anions, for example SiO_4 (Pauling, § 2.7). Boron oxide, however, is built up of triangular units. In the glassy state adjacent tetrahedra share corner oxygen atoms to form an irregular three-dimensional network, as shown schematically in two dimensions in Fig. 7.22(a) The same

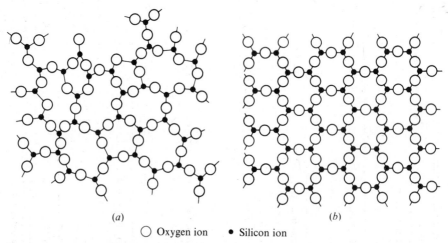

(a)   (b)

○ Oxygen ion   ● Silicon ion

Fig. 7.22. Silica as (a) a glass and (b) a crystalline solid. This two-dimensional representation shows only three of the normal four oxygen atoms around each silicon atom.

Table 7.9. *Composition of some glasses*

Glass	SiO_2	Na_2O	K_2O	CaO	MgO	BaO	PbO	B_2O_3	Al_2O_3	Remarks
1 (Fused) silica	99.5+									Difficult to melt and fabricate but usable to 1000 °C. Very low expansion and high thermal shock resistance.
2 96% silica	96.3	<0.2	<0.2					2.9	0.4	Fabricate from relatively soft borosilicate glass; heat to separate SiO_2 and B_2O_3 phases; acid leach B_2O_3 phase; heat to consolidate pores.
3 Soda-lime: plate glass	71–73	12–14		10–12	1–4				0.5 –1.5	Easily fabricated. Widely used in slightly varying grades, for windows, containers and electric bulbs.
4 Lead silicate: electrical	63		6	0.3	0.2		21	0.2	0.6	Readily melted and fabricated with good electrical properties. High lead absorbs X-rays: high refractive used in achromatic lens. Decorative crystal glass.
5 high lead	35		7.2				58			
6 Borosilicate: low expansion	80.5	3.8	0.4					12.9	2.2	Low expansion, good thermal shock resistance and chemical stability. Widely used in chemical industry.
7 low electrical loss	70.0		0.5				1.2	28.0	1.1	Low dielectric loss.
8 Aluminoborosilicate: standard (apparatus)	74.7	6.4	0.5	0.9		2.2		9.6	5.6	Increased alumina, lower boric oxide improves chemical durability.
9 low alkali (E-glass)	54.5	0.5		22				8.5	14.5	Widely used for fibres in glass resin composites.
10 Aluminosilicate	57	1.0		5.5	12			4	20.5	High temperature strength, low expansion.

11 Glass-ceramic	SiO_2	Al_2O_3	MgO	TiO_2	Crystalline ceramic made by devitrifying glass. Easy fabrication (as glass), good properties. Various glasses and catalysts.
	40–70	10–35	10–30	7–15	

Table 7.10. *Properties of some glasses*

		Viscosity data				Physical data	
Glass	Strain point (°C) $\mu = 10^{13}$ Ns/m^2	Annealing point (°C) $\mu = 10^{12}$ Ns/m^2	Softening point (°C) $\mu = 10^{7}$ Ns/m^2	Working point (°C) $\mu = 10^{3}$ Ns/m^2	Specific gravity	Coeff. of expansion (deg^{-1} × 10^7) (0–300 °C)	Thermal conductivity (Wm^{-1} deg^{-1}) (at 0 °C)
1 (Fused) silica	1050	1120	1600		2.2	5.5	1.45
2 96 % silica	820	910	1500		2.18	8.0	1.25
3 Soda-lime plate	510	553	735	1070	2.47	87	1.0
4 Lead silicate: electrical	395	435	625	985	2.86	93	0.7
5 high lead	395	430	580	820	4.28	91	
6 Borosilicate: low expansion	520	565	820	1245	2.23	33	1.17
7 low electrical loss	455	495		1070	2.13	32	
8 Aluminoborosilicate: standard (apparatus)	540	580	795		2.36	49	
9 low alkali (E-glass) fibre					2.55		
10 Aluminosilicate	670	715	915	1190	2.53	42	
11 Glass-ceramic (Pyroceram 9606)			1250		2.6	57	3.6

units can also combine through sharing of corner atoms into regular arrays, or crystalline solid, see Fig. 7.22(*b*). The addition of a network-modifying oxide changes the structure to that shown schematically in Fig. 7.23(*a*). The large metallic ions do not join in the network but are held interstitially by ionic-type bonds. The additional oxygen atoms are incorporated in the network, but owing to the excess of negative ions some of the links are broken, weakening the strength of the network and consequently lowering the viscosity. For example, the addition of (15 per cent Na_2O, 10 per cent CaO) to silica lowers the softening point from 1600 °C to 700 °C. This is the basis of soda-lime glass, widely used for bottles, windows and light bulbs. The ease of glass manufacture and subsequent fabrication gives a cheap product. The probable effect on the structure of adding an intermediate oxide is shown in Fig. 7.23(*b*). Here the cation is small and can form a tetrahedral group, e.g. AlO_4, which replaces an SiO_4 group in the network; because the valency is 3 instead of 4, alkali metal or alkaline earth ions are required to ensure electroneutrality.

The ability of oxides to form glasses is dependent on the following rules, according to Zachariasen (1932):

1. The oxygen atom should be linked to not more than two metal cations.

2. The number of oxygen atoms around a cation should be as small as possible, four or less.

3. Polyhedra (formed by the oxygen atoms around the cation) should share corners with one another, but not edges or faces.

Mechanical data					Electrical data			
Refractive index (sodium D-line)	Young's modulus (GN/m²)	Poisson's ratio	Hardness DPN (100 g load)	Median tensile strength (MN/m²)	Log₁₀ (volume resistivity) (ohm cm) 25 °C 250 °C		Dielectric constant	Loss tangent (at 1 MHz, 20 °C)
1.45	72	0.17	700–750	100–120		11.8	3.8	0.0001
1.46	69	0.17	650	100–120	17	9.7	3.8	0.0005
1.51	69	0.21	540–580	80–100	12	6.5–7.0	7.0–7.6	0.004 −0.011
	61	0.21	420–470			8.9	6.7	0.0016
1.97	52	0.23	290–340			11.8	9.5	0.0009
1.47	62	0.20	550–600	80–100	15	8.1	4.6	0.005
1.47	51	0.22			17	11.2	4.1	0.0006
						6.9	5.6	0.01
	72		freshly drawn: 3800 in service: 2000					
1.53	87.5	0.25	580–630			11.4	7.2	0.0038
	120	0.24	620–640	185		10	5.6	0.002

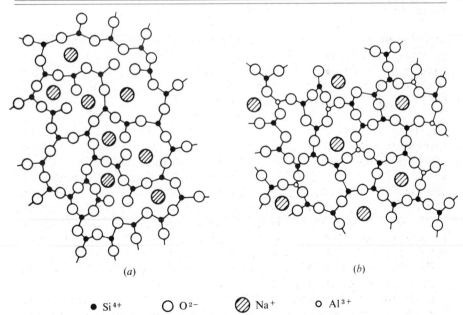

(a)   (b)

● Si⁴⁺   ○ O²⁻   ⊘ Na⁺   ○ Al³⁺

Fig. 7.23. (*a*) Network modified glass, e.g. soda–silica glass (note that large metallic ions do not form part of the network). (*b*) Intermediate oxide glass, e.g. alumina–silica glass (note that small metallic ions form part of the network).

4. At least three corners of a polyhedron must be shared to ensure a three-dimensional network.

Various exceptions to these rules can be found and an alternative theory of glass-formation considers it as a rate phenomenon. The relevant factors are the liquidus temperature, activation energy for nucleation of crystals, viscosity and the number of defects (to act as nucleation centres). All glass formers have a high viscosity in the liquid state, thus restraining the atoms or molecules from moving into a crystalline array. The value of the viscosity is around $10\ \mathrm{Ns/m^2}$ (10^2 P) compared with $10^{-3}\ \mathrm{Ns/m^2}$ (10^{-2} P) for normal liquids. This can be due either to high molecular weight, as in high polymers, or to intermolecular binding. The crystalline state of glass formers always has an open structure with low coordination number (number of nearest neighbours). The density and free energy of the glassy and crystalline states are therefore similar and there is little incentive for crystallisation to occur.

7.9 Properties of glasses

The most important property of glass is its transparency to visible light, though the addition of transition metal oxides will produce colouring or even make it opaque. The transparency arises because glass, being cooled from the liquid state, does not normally contain voids or flaws with dimensions close to the wavelength of light, the source of scattering in non-metallic crystals formed by sintering. The refractive index can be controlled by suitable additions: flint glass containing lead oxide has a high refractive index and is used in achromatic lenses.

Electrically, glass is an insulator, though some conduction can occur by the diffusion of ions such as sodium through the open network: the conductivity rises rapidly with temperature. The dielectric constant is generally low, dependent on the modifier; lead in silica raises the value from 4 to 10, see Table 7.10.

Thermal conductivity is several orders of magnitude lower than for crystalline ceramics because the passage of thermal vibrations is severely restricted by the disordered structure. The thermal expansion of silica glass is very low (5.5×10^{-7} per degree cf. metals around 10^{-5} per degree) due to the open network. It increases rapidly with the addition of a modifier, the value for soda-lime glass being 8.7×10^{-6} per degree. The borosilicate glasses (Pyrex) have a rather lower coefficient of expansion (3×10^{-6}), which makes them suitable for applications involving thermal shock. In extreme cases, however, fused silica must be used but this is much more expensive.

As regards mechanical properties, glass is a Hookean solid if loaded quickly and a Newtonian liquid under slow rates of deformation. In a Newtonian liquid a shear stress causes adjacent layers of liquid to move relative to one another, see Fig. 7.24. The shear stress σ' is proportional to the velocity gradient:

$$\sigma' = \eta \frac{\mathrm{d}v}{\mathrm{d}x}$$

where η is the viscosity.

Now

$$\frac{\mathrm{d}v}{\mathrm{d}x} = \frac{\mathrm{d}}{\mathrm{d}x}\left(\frac{\mathrm{d}y}{\mathrm{d}t}\right) = \frac{\mathrm{d}}{\mathrm{d}t}\left(\frac{\mathrm{d}y}{\mathrm{d}x}\right)$$

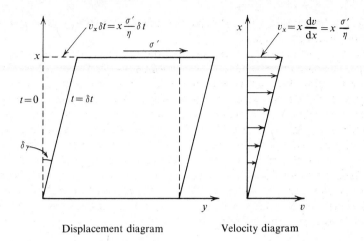

Displacement diagram   Velocity diagram

Fig. 7.24. Shear of a Newtonian fluid.

and $\mathrm{d}y/\mathrm{d}x = \gamma$, the shear strain.

Hence

$$\sigma' = \eta \frac{\mathrm{d}\gamma}{\mathrm{d}t}. \tag{7.2}$$

The stress is proportional to the strain rate, in contrast with a Hookean solid for which the stress is proportional to the strain.

The relative importance of the elastic and viscous components of strain in a visco-elastic solid such as glass can be expressed in terms of a *relaxation time* τ which is defined by the equation

$$\tau \frac{\mathrm{d}\gamma}{\mathrm{d}t} = \frac{\sigma'}{G} \tag{7.3}$$

where G is the instantaneous (elastic) shear modulus in the absence of viscous flow. In other words, τ is the time required for the viscous strain to equal the elastic strain. Combining (7.2) and (7.3) gives

$$\tau = \frac{\eta}{G}. \tag{7.4}$$

Whether a glass is behaving predominantly elastically or viscously depends on the duration of the loading in relation to the relaxation time. The value of G for glass is about $30\,\mathrm{GN/m^2}$ (4×10^6 psi) and at room temperature η is $10^{13}\,\mathrm{Ns/m^2}$ (10^{14} P) or greater, making τ equal to 330 seconds. Under typical short term loading to fracture at room temperature glass is therefore elastic. At the hand working temperature (400–600 °C) the viscosity is reduced to $10^7\,\mathrm{Ns/m^2}$ (10^8 P), making τ equal to 3.3×10^{-4} seconds, and glass now behaves as a viscous fluid at normal rates of straining.

An interesting and important consequence of its Newtonian viscosity is that glass can be drawn down without necking (cf. § 5.3). Consider a rod of glass of cross-section area A with an applied force F. Since the shear strain rate $\dot{\gamma}$ is proportional to the rate of extension $\dot{\varepsilon}$, (7.2) gives the tensile stress

$$\sigma = \frac{F}{A} = \kappa\eta\dot{\varepsilon}.$$

At constant volume

$$d\varepsilon = -\frac{dA}{A}, \quad \text{and} \quad \dot{\varepsilon} = -\frac{\dot{A}}{A}.$$

Substituting in the above gives

$$\dot{A} = -\frac{F}{\kappa\eta}.$$

Thus the rate of reduction of area is a function only of the applied force, and not the local value of the stress. Should an incipient neck develop, it does not become unstable as happens in a normally plastic material. On the contrary, the higher cooling rate increases η and helps stabilise the drawing process, allowing glass fibres to be drawn down without the use of dies.

When glass is behaving as a Hookean solid its strength is limited by the onset of fracture. The strength of glass in tension is normally very low, around $100\ \text{MN/m}^2$ ($\sim 10^4$ psi), with a wide scatter about the median value. Very much higher stresses, $\sim 10\ \text{GN/m}^2$ (10^6 psi), can be supported in compression, about the theoretical strength value $E/10$ (see § 10.2). These figures are fairly independent of the glass composition. The low tensile strength is due to the presence of surface cracks, which act as local stress raisers. Because of the inability of the glass to deform plastically, the high local stress at the tip of the crack causes it to propagate at a very rapid rate, the small energy of the fracture surfaces being obtained entirely from the elastic energy of the surrounding material. The classic experiments of Griffiths in 1920 into the fracture of glass, which laid the foundation of the fracture theory of brittle materials, will be described in detail in Chapter 10. It is found that freshly drawn glass fibres have a high tensile strength, $4\ \text{GN/m}^2$ (0.6×10^6 psi) (cf. the theoretical value 7–$10\ \text{GN/m}^2$), but in air it decreases with time to an equilibrium value which depends on the fibre size: the larger the specimen the lower the strength. This is attributed to the formation of surface cracks by interaction with water vapour, dust particles or, simply, handling.

Another factor in assessing the strength of glass is the duration of the stressing. A longer test lowers the fracture stress: after one month under stress it may be one third of the value obtained in a three second loading cycle. This effect is called *static fatigue* (N.B. there is no alternating stress involved, cf. § 5.11). It is quite independent of any viscous deformation that may develop during the longer test period. It is thought that water vapour is responsible and tests *in vacuo* only show a much reduced effect.

On account of these various factors, the allowable working stress in tension of glass components is extremely low, typically $10\ \text{MN/m}^2$ (10^3 psi) or ten per cent of the median short term tensile strength. Even so, great care must still be taken to avoid local stress concentrations around holes, sharp corners, joints and supports. Surface damage must also be avoided.

There are several methods of significantly increasing the value of the tensile fracture stress (but only to an order of magnitude below the theoretical strength). These rely on introducing a high residual compressive stress into the surface

Table 7.11. *Allowable working stresses in glasses with various surface treatments. (From E. B. Shand, loc. cit.)*

	Untempered glass	Thermally tempered glass	Chemically tempered glass	Glass-ceramic
Median TS (MN/m^2)	100	210	390	185
2% probability TS (MN/m^2)	37	160	300	125
Static fatigue factor	0.4	0.75	0.75	0.75
Safety factor	2.5	2.5	2.5	2.5
Working stress (MN/m^2)	6	48	90	37
% of median TS	5.8	22.5	23	20

of the glass, which prevents the propagation of cracks until the net local stress (applied tensile stress minus residual compressive stress) is positive. The median fracture stress can be as high as 400 MN/m^2; at the same time the scatter of results and also the static fatigue effect are reduced. A working stress of 100 MN/m^2 is then possible, a factor of ten improvement over untreated glass. Some typical values are given in Table 7.11.

The oldest technique of toughening is *thermal tempering* which consists of heating the glass to near its softening point and cooling rapidly by means of water jets. The surface layer cools more rapidly than the core so it is not able to contract and extends by viscous flow. When a uniform temperature is restored the surface layer, which has become overlength, is forced into compression, typically 200 MN/m^2. In high-temperature chemical strengthening, a surface layer is formed by ion exchanging with some medium such that it has a lower coefficient of thermal expansion than the bulk glass. On cooling to room temperature the surface layer contracts less and is again forced into compression, typically 500 MN/m^2. Finally, in low-temperature chemical strengthening, there is again an exchange of ions but those entering the glass are oversize and create a compressive surface stress directly. Chemical strengthening tends to be more costly but it can produce a higher compressive stress and can be used on thinner sections.

Commercial glass fibres are typically 15 μm (0.6 × 10^{-3} in) in diameter. They have a fracture strength of about 3.5 GN/m^2 (5 × 10^5 psi) when drawn but, in spite of the application of a size to prevent surface damage, the strength in service is reduced to half the value, 1.5–2.0 GN/m^2 (2.5 × 10^5 psi). This high specific strength (cf. Table 5.1) is utilised in glass fibre plastic composites, which will be discussed in Chapter 11.

7.10 Glass-ceramics

Glass-ceramics are polycrystalline ceramics prepared by controlled crystallisation of a glass. They are a new class of material, developed around 1960, which covers a range of physical and mechanical properties superior to those of glass and traditional ceramics. In addition, the production methods, based on glass forming techniques, have various advantages over those used for normal ceramics. Exploitation of the unique characteristics of glass-ceramics is only beginning but a brief review of this special type of material will be given here in view of its scientific interest and potential importance for special applications.

Glass-ceramics were developed from photosensitive glasses in which small amounts of copper, silver and gold are precipitated by irradiation with ultra-violet light and subsequent heat treatment. It was discovered by Stookey that heating above the normal temperature caused the glass to crystallise on the nuclei provided by these metallic crystals. Because of the large number and uniform distribution of the nuclei, the resultant ceramic was very fine grained, uniform in composition and free of any porosity. Mechanical strength and electrical resistance were high. Subsequent researches have found a number of glass compositions and various catalysts which will crystallise without the need for ultra-violet irradiation. For example, the platinum group of metals will form metallic nuclei when present in amounts between 10^{-3} and 10^{-1} per cent. Alternatively various oxide nuclei can be used; for example 2–20 per cent titania can be precipitated out of a cooling silicate glass.

The heat treatment process following fabrication by conventional glass-shaping processes is two stage. In the first, a large number of nuclei are produced, and in the second, the temperature is raised to cause crystal growth until little glass phase remains (2 per cent). Some care must be taken to avoid cracking, distortion or oversize crystals. A typical crystal size is 1 μm, but values down to 0.02 μm can be obtained (cf. hot pressed alumina at 2–20 μm).

During heat treatment the properties change. The glass becomes opaque due to scattering at the interfaces between the new crystals and remaining glass phase, which have different refractive indices. Some surface roughness develops and the volume changes up to 3 per cent (cf. 50 per cent in the firing of traditional ceramics). The coefficient of thermal expansion usually decreases, but can be controlled over a wide range, typically between 10^{-7} and 10^{-5} per deg, to match that of metallic components. It can even be made negative. Thermal conductivity increases sharply during the change from glass but it is never quite as high as in the translucent oxides. The softening temperature is raised from around 500 °C to around 1000 °C. Generally speaking the electrical properties are enhanced, with greater resistivity and lower dielectric loss.

Perhaps most important of all, the median mechanical strength increases by at least a factor of two, from about 100 MN/m^2 (10^4 psi) to 200 MN/m^2 (2×10^4 psi) or higher, possibly due to improved resistance to surface damage. The allowable working stress can be increased by a greater factor because the scatter is less. Resistance to thermal shock is sharply increased due to improvement in the several parameters involved. The indentation hardness is not high, around 700 DPN, but the resistance to scratching and abrasion is exceptional, close to that of sapphire (pure dense alumina). Some typical values of the physical and mechanical constants are included in Table 7.10.

Glass-ceramics are currently being applied to cooking and tableware, on account of their mechanical strength, resistance to thermal shock and the possibility of mass-production techniques; to bearing surfaces and pump impellers on account of their resistance to abrasion, good surface finish and ability to bond to metals; to insulators on account of their high dielectric strength and ability to match the coefficient of thermal expansion of metal parts; to substrates for printed circuits and other electronic components. Other potential applications are being investigated.

QUESTIONS

1. (a) Define ceramics and contrast the structure and properties with metals.
   (b) List and briefly describe the principal divisions of ceramics and give some examples.
2. Outline in general terms some of the electrical, mechanical and chemical characteristics of new ceramics.
3. (a) Account for the plasticity of ionic single crystals and the brittleness of ionic poly-crystals.
   (b) Why are covalent ceramics brittle as single crystals and polycrystals?
   (c) Why have ceramics very low tensile strength (cf. theoretical value) but very much higher compressive strength?
4. (a) What are the causes of thermal stress?
   (b) What parameters are considered to represent the resistance to thermal stress according to the thermoelastic theory of failure?
   (c) What new consideration is introduced in the unified theory of thermal shock resistance?
5. (a) What are the chemical formulae of silica, lime, alumina?
   (b) Name the three classes of materials in which each is the major oxide.
6. (a) Distinguish between silica, quartz, tridymite and fused silica.
   (b) Distinguish between reconstructive and displacive transformations.
   (c) Describe the structure of a typical clay mineral.
7. (a) What are the components of a traditional ceramic and what is the main function of each?
   (b) Outline the processing of a typical whiteware and the changes on firing.
   (c) Describe the microstructure of porcelain. Give an approximate oxide composition.
8. (a) Write down the three chemical equations describing the production and hardening of lime mortar.
   (b) Is hardening of lime mortar a process of hydration?
9. (a) What are the two materials for making Portland cement and give an approximate oxide composition of cement powder (at least put them in quantity order).
   (b) Why is about 5 per cent Fe_2O_3 a small but vital component of the raw materials?
   (c) Give an approximate mineralogical composition of cement powder.
10. (a) What is the principal reaction product on hydrating cement powder?
    (b) Discuss the role of tobermorite gel in the cohesion of cement.
    (c) Distinguish between the start of the hydration reactions, setting and hardening.
11. (a) What are the four components required to make concrete and their approximate proportions?
    (b) How and why is the water content of concrete limited during mixing?
    (c) How does engineering design allow for the particular strength characteristics of concrete?
12. (a) Describe the microstructure of a cermet. Give two alternative names.
    (b) What is the main property advantage of introducing the metallic phase (cf. fully crystalline ceramic)?
    (c) Distinguish between single carbide and multicarbide cermets.
13. (a) Discuss some of the requirements for a single-point cutting tool.

(b) Outline briefly the development of cutting tool materials giving the advantages achieved in successive stages.

14. (a) Give three characteristics of new ceramics.
    (b) Draw the crystal structure of α-Al_2O_3.
    (c) Distinguish between alumina, corundum, sapphire, ruby, debased alumina, high density alumina, fully dense (translucent) alumina, emery and anodised aluminium.

15. (a) Discuss the influence of porosity, grain size and purity of alumina on the mechanical properties (e.g. bend strength) and electrical properties (e.g. dielectric constant and loss tangent).
    (b) Why does production of very high purity, fine grained alumina not appear to be feasible?

16. (a) Draw the crystal structure of graphite.
    (b) Relate this to the anisotropy in strength and conductivity.
    (c) Distinguish the microstructures of commercial graphite, pyrolitic graphite, carbon fibre and carbon whiskers.

17. (a) Distinguish between crystalline ceramics, glass and glass-ceramics.
    (b) Distinguish between liquid, supercooled liquid, glassy state and a crystalline solid. Define glass transition temperature.
    (c) Which oxides are the three main glass formers.

18. (a) Draw schematically the structure of glass formers, network modifiers and intermediate oxides.
    (b) Explain the action of network modifiers in lowering the viscosity at any given temperature.
    (c) What are the common network modifiers, e.g. in glass for bottles and windows? Why are they added to silica?

19. (a) Distinguish between a Hookean solid, a strain-hardening material, a Newtonian viscous fluid, a visco-elastic material (define relaxation time).
    (b) Which of them is a glass?
    (c) Show analytically why a glass can be drawn at elevated temperature.

20. (a) Why has normal glass a low tensile strength?
    (b) How is this affected by various factors, e.g. time since drawing, environment, size, duration of load?
    (c) Distinguish between static fatigue and fatigue.

21. Outline three methods of increasing the strength of glass.

22. (a) When and why does a normal glass devitrify?
    (b) How can the process be accelerated?
    (c) Contrast the manufacture and microstructure of crystalline ceramics (e.g. alumina) and glass-ceramics.

FURTHER READING

General

P. Schwarzkopf and R. Kieffer: *Refractory Hard Metals*. Macmillan (1947).

W. D. Kingery: *Introduction to Ceramics*. Wiley (1960).

W. W. Kriegel and H. Palmour (eds.): *Mechanical Properties of Engineering Ceramics*. Interscience (1961).

P. W. Lee: *Ceramics*. Reinhold (1961).

E. M. Levin, C. R. Robbins and H. F. McMurdie: Phase diagrams for ceramicists. *Am. Ceram. Soc.* (1964).

J. E. Hove and W. C. Riley (eds.): *Ceramics for Advanced Technologists*. Wiley (1965).

J. E. Hove and W. C. Riley (eds.): *Modern Ceramics*. Wiley (1965).

J. H. Brunton: Diamonds as an engineering material. *Chart. Mech. Eng.* (*London*) (January 1966).

R. F. Hilton: Refractories and other ceramics. *Chart. Mech. Eng.* (*London*) (February 1966).

L. W. Marrison: *Crystals, Diamonds and Transistors.* Penguin (1966).

L. W. Marrison: Mechanical properties of non-metallic crystals and polycrystals. *Proc. Br. Ceram. Soc.* **6** (June 1966).

L. W. Marrison: Electrical and magnetic ceramics. *Proc. Br. Ceram. Soc.* **10** (March 1968).

R. M. Fulrath and J. A. Pask (eds.): Ceramic microstructures: their analysis, significance and production. *3rd Berkeley International Materials Conference 1966.* Wiley (1968).

E. C. Henry: *Electronic Ceramics.* Doubleday (1969).

D. P. H. Hasselman: Unified theory of thermal shock fracture initiation and crack propagation in brittle ceramics. *J. Am. Ceram. Soc.* **52**, 600 (1969).

D. W. Budworth: *An Introduction to Ceramic Science.* Pergamon (1970).

Cement and concrete

F. M. Lea and C. H. Desch: *The Chemistry of Cement and Concrete.* Arnold (1956).

W. Czernin: *Cement Chemistry and Physics for Civil Engineers.* Crosby Lockwood (London) (1962).

Cermets and ceramic tool materials

P. Schwarzkopf and R. Kieffer: *Cemented Carbides.* Macmillan (1953).

A. G. Thomas, J. B. Hazzadine and N. C. Moore: Preparation, properties and application of metal/ceramic mixtures. *Metall. Rev.* **8**, no. 32, 461 (1963).

A. G. King and W. M. Wheildon: *Ceramics in Machining Processes.* Academic Press (1966).

E. M. Trent: Cutting-tool materials. *Metall. Rev.* no. 127 (Oct. 1968).

Alumina

E. Ryshkewitch: *Oxide Ceramics: Physical Chemistry and Technology.* Academic Press (1960).

D. B. Binns and P. Popper: Mechanical properties of some commercial alumina ceramics. *Proc. Br. Ceram. Soc.* **6**, 71 (1966).

Anodic Oxidation of Aluminium and its Alloys. Information bulletin No. 14, Aluminium Federation Ldn (1966).

W. George and P. Popper: The dielectric properties of some commercial alumina materials. *Proc. Br. Ceram. Soc.* **10**, 63 (1968).

Graphite and carbon fibres

W. Watt: Production and properties of high modulus carbon fibres. *Proc. Roy. Soc. Lond.* A **319**, 5 (1970).

Glass and glass-ceramics

G. O. Jones: *Glass,* 2nd edition. Chapman and Hall (1971).

P. W. McMillan: *Glass-ceramics.* Academic Press (1964).

E. Venis: Glass for engineers. *Chart. Mech. Eng.* (*London*) (September 1967).

E. B. Shand: Engineering glass. *Modern Materials* **6**, 247. Academic Press (1968).

8. High polymers

8.1 Introduction

High polymers are enormous molecules which are formed by the repeated joining together by covalent bonds of some small atomic group, known as a monomer unit. The word 'high' was introduced to distinguish this class of materials, with molecular weights of around 10^5 to 10^7, from polymers composed of only a few monomers; in this chapter polymer should be taken as synonymous with high polymer. The connecting links in many high polymers are carbon atoms; for example, in polyethylene (or 'polythene') the monomer, ethylene

$$
\begin{array}{cc}
H & H \\
| & | \\
C & = C \\
| & | \\
H & H
\end{array}
$$

is joined many thousands of times into the polymeric chain

$$
\begin{array}{cccccccc}
H & H & H & H & H & H & H & H \\
| & | & | & | & | & | & | & | \\
-C-&C-&C-&C-&C-&C-&C-&C- \\
| & | & | & | & | & | & | & | \\
H & H & H & H & H & H & H & H
\end{array}
$$

which may be written more simply as

$$
\left[
\begin{array}{cc}
H & H \\
| & | \\
C & - C \\
| & | \\
H & H
\end{array}
\right]_n
$$

The — symbol indicates a covalent bond, that is, two electrons are shared by the connected atoms (see Chapter 3). The monomer unit in this example has a double bond ($=$), one bond of which is broken to form the covalent bonds with adjacent monomers. The chain begins and ends with two other atomic groups, the terminal groups, which may or may not be identical to each other. With long chains they have little influence on the properties (see § 8.2).

The macromolecules can be divided into four types according to their structure, see Fig. 8.1. They are linear, branching, (lightly) cross-linked and (close) network. The divisions are not clear cut in practice since a nominally linear polymer may have some short branches, as in low density polyethylene, and the cross-linked polymer merges into the network polymer.

The linear polymers, which constitute the largest and most important group, consist of monomers joined end to end in a line or chain. Because the binding between the molecular chains is due to van der Waals forces, which are weak, this type of polymeric material soon softens and becomes liquid as the tempera-

272

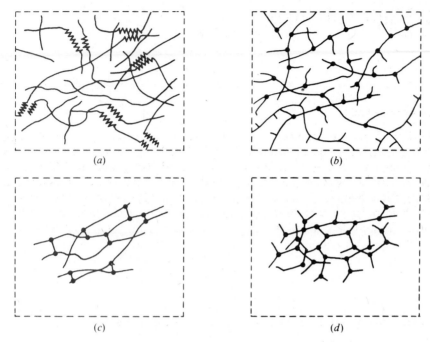

Fig. 8.1. The principal types of polymer structure: (a) linear, with some crystallites shown schematically; (b) branching; (c) lightly cross-linked; (d) close network.

ture is raised. They are the basis of *thermoplastics* and can be worked into required shapes at temperatures typically in the range 300–400 °C.

At room temperature the structure of linear polymers is either fully amorphous, with the chains twisted and interlocked in a random manner, or partially crystalline, with some regions of the polymer in which the molecular chains are in a regular array and others where the chains are random. Fully crystalline linear polymers have not been produced so far. The degree of crystallinity is dependent on the material and the processing procedure, for example a high cooling rate limits the time available for forming regular crystalline arrays (see § 8.5).

Examples of linear polymers are polythene, already described, and the vinyl compounds which are obtained by replacing one of the hydrogen atoms in the ethylene monomer:

$$\ldots -\underset{\underset{H}{|}}{\overset{\overset{H}{|}}{C}}-\underset{\underset{H}{|}}{\overset{\overset{X}{|}}{C}}\left[\underset{\underset{H}{|}}{\overset{\overset{H}{|}}{C}}-\underset{\underset{H}{|}}{\overset{\overset{X}{|}}{C}}\right]_{n}\underset{\underset{H}{|}}{\overset{\overset{H}{|}}{C}}-\underset{\underset{H}{|}}{\overset{\overset{X}{|}}{C}}- \ldots$$

Many polyvinyl compounds are household words: polyvinyl chloride (PVC) has X = chlorine, polypropylene has X = CH_3 and polystyrene has X = C_6H_5, a phenyl group, which is a benzene ring with one hydrogen atom removed. Note that the benzene ring, which occurs frequently in organic chemistry (aromatic compounds, cf. aliphatic) is often shown in a simplified manner:

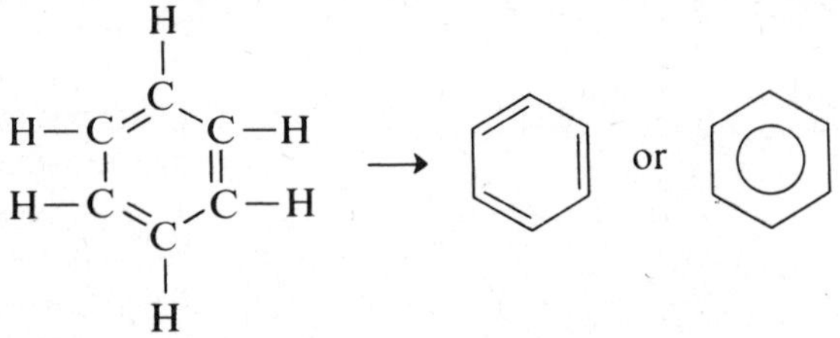

$$
\begin{array}{c}
H \\
| \\
\text{H}-\text{C} \overset{C}{=} \text{C}-\text{H} \\
| \quad || \\
\text{H}-\text{C} =_{\text{C}} \text{C}-\text{H} \\
| \\
H
\end{array}
\quad \rightarrow \quad \bigcirc \quad or \quad \hexagon
$$

The latter version, which is becoming more popular, indicates correctly that the double bonds are shared, that is, they are not static but move around the ring of carbon atoms. Some authors (or their printers) omit the circle but this can be misleading.

Substitution of both hydrogen atoms on the same carbon atom (1,1-disubstituted ethylene) leads to the vinylidene compounds, a well known example of which is polymethyl methacrylate ('Perspex', 'Plexiglas'):

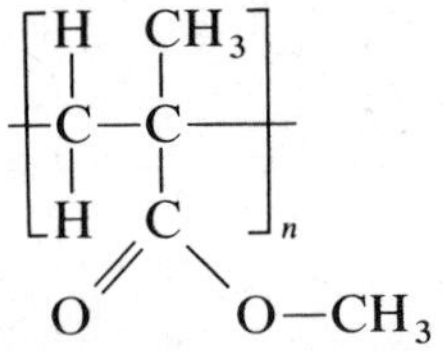

$$
\begin{bmatrix}
H & CH_3 \\
| & | \\
-C-C- \\
| & | \\
H & C
\end{bmatrix}_n
$$
$$
O \quad O-CH_3
$$

On account of their chain structure, linear polymers can easily be drawn into high strength fibres: nylon (polyamides) and Terylene (polyethylene terephthalate) are well known examples. These and many other important linear high polymers, whose structures are given in Table 8.1, will be referred to in the following text, which discusses in more detail the influence of the number of monomer units in a chain (degree of polymerisation) and the spatial arrangement of the side groups (molecular architecture).

At the other limit of structural types are the close-network polymers, in which the macromolecules are formed from monomer units having more than two active bonds, with the result that a two or three-dimensional molecule is created. Primary covalent bonds link the whole structure so these materials do not soften appreciably on heating and decompose before melting. They are the basis of *thermosetting plastics* or *thermosets*. They do not have quite the versatility of form and properties of the other types of polymer and will not be given as much consideration here.

Some important examples of thermosets are given in Table 8.2; the structures shown are two-dimensional representations of three-dimensional molecules. Included are three thermosets involving formaldehyde: phenol-, urea-, and melamine-formaldehyde. Phenolformaldehyde was the first synthetic high polymer to achieve commercial application in 1916 (the original 'Bakelite'), although other plastics based on naturally occurring cellulose predate it. The formaldehyde group

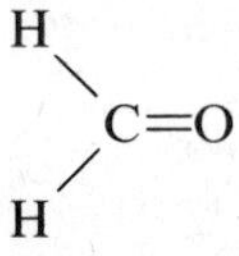

$$
\begin{array}{c}
H \\
\diagdown \\
\quad C=O \\
\diagup \\
H
\end{array}
$$

Table 8.1. *Structure of various linear high polymers (thermoplastics) (see also Table 8.4 for elastomers). N.B. In reality, the carbon bonds are directed towards the corners of a tetrahedran and rotation about single bonds occurs*

Chemical name	Repeat unit	Notes
Olefines C_nH_{2n}:		
polyethylene (polythene)	H H $\|$ $\|$ —C—C— $\|$ $\|$ H H	Available in low density and high density forms. Original low density due to some short side chains which are avoided in the high density form.
Vinyl compounds $CH_2.CHR$ (*monosubstituted ethylene*):		
polyvinyl chloride (PVC)	H Cl $\|$ $\|$ —C—C— $\|$ $\|$ H H	Stiff and brittle in pure form. Usually plasticised.
polypropylene	H CH_3 $\|$ $\|$ —C—C— $\|$ $\|$ H H	First stereospecific commercial polymer (using Natta–Ziegler catalysts).
polystyrene	H C_6H_5 $\|$ $\|$ —C—C— $\|$ $\|$ H H	Phenyl group C_6H_5 is $T_G = 100\,°C$
polyvinyl acetate	H O—C $\|$ $\|$ $\diagdown$ —C—C— O $\diagup$ $\|$ $\|$ CH_3 H H	Adhesive. $T_G = 29\,°C$

Table 8.1 (*continued*)

Chemical name	Repeat unit	Notes
Vinylidene compounds $CH_2:CRS$ (*1,1-disubstituted ethylene*)		
polyvinylidene chloride (PVDC)		Low permeability to gases and water vapour. Copolymerised with PVC, it is used for food packaging. 'Saran'.
polyisobutylene (PIB)		
polymethyl methacrylate (PMMA)		'Perspex', 'Plexiglas' transparency. $T_G = 72\,°C$.
Polyamides		
nylon-6		All nylons contain amide linkages $-NH.CO-$ but properties vary significantly with the number of methylene ($-CH_2-$) groups. Nylon-6 has medium strength, high impact and easy processing.
nylon-6,6		Highest m.p., strongest and most rigid of the nylons.

Chemical name	Repeat unit	Notes
nylon-6,10	$-\!\overset{\displaystyle H}{N}\!(CH_2)_6\overset{\displaystyle H}{N}\!-\!\overset{\displaystyle O}{C}\!(CH_2)_8\overset{\displaystyle O}{C}\!-$	Best chemical resistance and least water absorption of the nylons.
Various		
polytetrafluoroethylene (PTFE)	$-\!\underset{F}{\overset{F}{C}}\!-\!\underset{F}{\overset{F}{C}}\!-$	Low friction. High temperature. High chemical resistance. 'Fluon', 'Teflon'.
polyacetal (polyformaldehyde)	$-\!\underset{H}{\overset{H}{C}}\!-\!O-$	Highly crystalline. m.p. 180 °C. 'Delrin' (1960).
polyurethane	$-O-R-O-\overset{\displaystyle O}{C}\!-\!\overset{\displaystyle H}{N}\!-R'-\overset{\displaystyle H}{N}\!-\!\overset{\displaystyle O}{C}\!-$	Various groups at R and R' give a wide range of urethanes, containing $-NH.CO.O-$ group.
polyethylene terephthalate (PET)	(terephthalate ester repeat unit with —CO.O— links and benzene ring, $-O-CH_2-CH_2-O-$)	A polyester (containing $-CO.O-$ link). Good fibre and sheet former. 'Terylene', 'Dacron'. $T_G = 70\,°C$.
polycarbonate	(bisphenol-A carbonate repeat unit with two benzene rings, $C(CH_3)_2$ bridge and $-O-\overset{O}{C}-$ carbonate group)	High impact strength transparency. Crystalline. m.p. 230 °C.

Table 8.1 (*continued*)

Chemical name	Repeat unit	Notes
polyimides		Exceptional high temperature stability (300 °C and above). $T_G = 70$ °C. **R** stands for one of several aromatic compounds (see p. 324).
cellulose		The basis of plants and raw material for rayon, celluloid and cellophane. $T_G = 70$ °C.
Copolymers (random)		
polyvinyl chloride–vinyl acetate		Widely used for gramophone records.
styrene acrylonitrile (SAN)		A toughened form of polystyrene.

Chemical name	Repeat unit	Notes
Copolymers (graft)		
styrene–butadiene (HIPS)	$\begin{array}{cc} H & C_6H_5 \\ -C-C- \\ H & H \end{array}$, $\begin{array}{cccc} H & H & H & H \\ -C-C=C-C- \\ H & & & H \end{array}$	Particles of styrene–butadiene rubber in styrene matrix. High impact polystyrene (see § 8.7).
acrylonitrile–butadiene–styrene (ABS)	$\begin{array}{cc} H & C\equiv N \\ -C-C- \\ H & H \end{array}$, $\begin{array}{cccc} H & H & H & H \\ -C-C=C-C- \\ H & & & H \end{array}$, $\begin{array}{cc} H & C_6H_5 \\ -C-C- \\ H & H \end{array}$	Particles of acrylonitrile–butadiene rubber (nitrile) in styrene acrylonitrile matrix (SAN). Very tough and strong. Can be metal-plated.

always links the other group involved by dropping its oxygen atom to make two bonds available; the mating bonds in the other groups are obtained by removing hydrogen atoms. Water molecules are thus a by-product of the reaction (i.e. polycondensation) and must be removed. In contrast, formaldehyde can also undergo addition polymerisation to polyacetal with no by-product, see Table 8.1. Once fully reacted thermosets cannot be hot worked (hence the name).

The branching and cross-linked polymers are intermediate in structure and properties to the limiting types just described.

Since polymers have been defined as large molecules with covalent bonds, many inorganic materials can be included in the class. For example, the silicate

Table 8.2. *Structure of various network high polymers (thermosetting plastics).*
N.B. Two-dimensional representation of three-dimensional molecules

(PF thermoset) + mH$_2$O

(a) *Phenolformaldehyde, PF (close-network)*. Schematic two-dimensional representation of complex condensation reaction. N.B. By convention the C and H atoms around the benzene ring are often not marked. This is the original thermoset (Dr Baekeland, 1909; 'Bakelite', 1916). A slight reaction between phenol and air prevents a pure white plastic.

(UF thermoset) + mH$_2$O

(b) *Urea formaldehyde, UF (close-network)*. Condensation reaction between formaldehyde and hydrogen atoms in the urea creates a close network polymer, with water.

clays have an SiO_4 monomer unit repeated many times to form layer molecules, as described in § 7.3. Graphite and diamond are also covalent solids with respectively two and three-dimensional crystal structures (§ 2.3 and § 7.7). However, the term high polymer is normally restricted to organic molecules (i.e. based on carbon) although, since chains of silicon and oxygen atoms (called polysilicones or polysiloxanes)

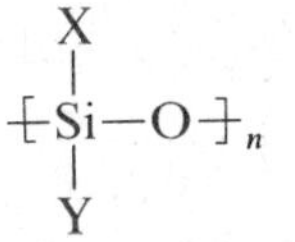

$$\begin{matrix} & X & \\ & | & \\ \{-Si&-O\}_n & \\ & | & \\ & Y & \end{matrix}$$

are now included, the dividing line is becoming diffuse.

Table 8.2 (*continued*)

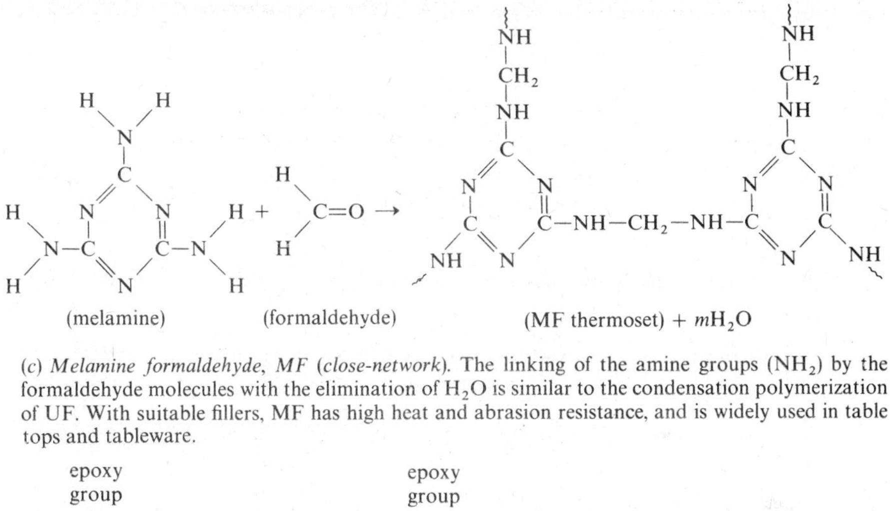

(melamine)   (formaldehyde)   (MF thermoset) + mH$_2$O

(c) *Melamine formaldehyde, MF (close-network)*. The linking of the amine groups (NH$_2$) by the formaldehyde molecules with the elimination of H$_2$O is similar to the condensation polymerization of UF. With suitable fillers, MF has high heat and abrasion resistance, and is widely used in table tops and tableware.

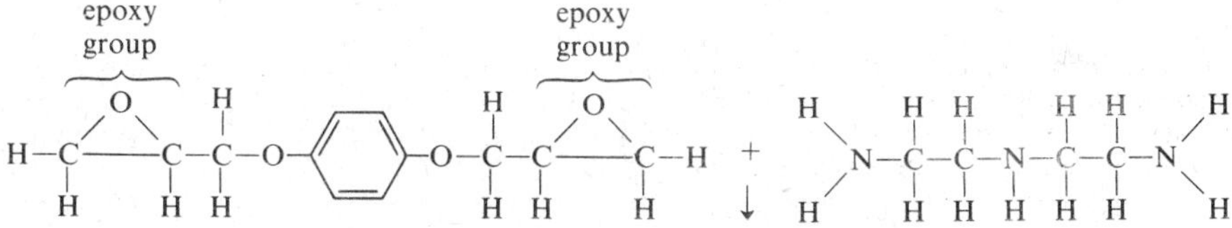

(diglycidyl ether of bisphenol A)   (diethylene triamine)

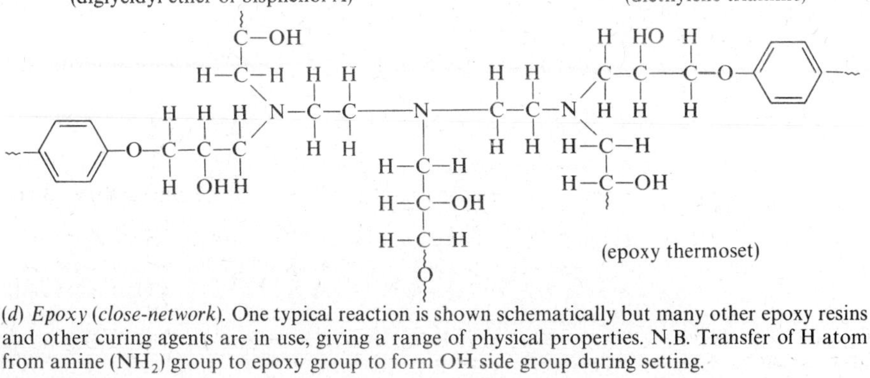

(epoxy thermoset)

(d) *Epoxy (close-network)*. One typical reaction is shown schematically but many other epoxy resins and other curing agents are in use, giving a range of physical properties. N.B. Transfer of H atom from amine (NH$_2$) group to epoxy group to form OH side group during setting.

Stemming from the acceptance in the 1930s of the nature of high polymers (some fifty years after the idea of macromolecules was first put forward) and the advances in understanding of the relationship between structure and properties, very many new processes for synthesising polymers have been developed, especially since the Second World War. Besides providing the basis of plants and animals, high polymers are now responsible for at least four major technological products: paints, rubbers (or elastomers), fibres and plastics (including adhesives). Thermoplastics and elastomers will be the major consideration in this volume.

Plastics materials get their name from the fact that many of them can be formed into their final shape by plastic deformation at elevated temperature, that is, they are linear polymers. (Note that the 's' is present in the adjectival form). At room temperature, however, many high polymers in the pure state show little ductility and fail by brittle fracture (Chapter 10). To overcome this brittleness and improve the properties in other ways, or simply to reduce the

Table 8.2 (*continued*)

(unsaturated dibasic acid)   (dihydric alcohol e.g. ethylene glycol)

$$OH-\overset{\overset{O}{\|}}{C}-\overset{\overset{H}{|}}{C}=\overset{\overset{H}{|}}{C}\{CH_2\}_n\overset{\overset{}{}}{\underset{\|}{C}}-OH + OH-\overset{\overset{H}{|}}{\underset{\underset{H}{|}}{C}}-\overset{\overset{H}{|}}{\underset{\underset{H}{|}}{C}}-OH$$

$$\sim\!\!-\overset{\overset{H}{|}}{\underset{\underset{H}{|}}{C}}-\overset{\overset{H}{|}}{\underset{\underset{H}{|}}{C}}-O-\overset{\overset{O}{\|}}{C}-\overset{\overset{H}{|}}{C}=\overset{\overset{H}{|}}{C}\{CH_2\}_n\overset{}{\underset{\|}{C}}-O-\overset{\overset{H}{|}}{\underset{\underset{H}{|}}{C}}-\overset{\overset{H}{|}}{\underset{\underset{H}{|}}{C}}-O-\overset{\overset{O}{\|}}{C}-\!\!\sim + mH_2O$$

$$\underbrace{\qquad}_{\text{ester group}}\qquad\underbrace{\qquad}_{\text{ester group}}\qquad\underbrace{\qquad}_{\text{ester group}}$$

Stage 1: Polymerisation of unsaturated linear polymers involving the formation of ester groups.

unsaturated
linear polyester $+ \overset{\|}{M}$ $\left(\text{unsaturated monomer,} \atop \text{e.g. styrene}\right)$ $\overset{\overset{H}{|}}{\underset{\underset{H}{|}}{C}}=\overset{\overset{}{}}{\underset{\underset{H}{|}}{C}}$ ⬡

$$\sim\!\!-\overset{|}{\underset{|}{C}}-\!\!\sim$$

$$-O-\overset{\overset{O}{\|}}{C}-\overset{\overset{H}{|}}{\underset{\underset{M}{|}}{C}}-\overset{\overset{M}{|}}{\underset{\underset{H}{|}}{C}}\{CH_2\}_n\overset{}{\underset{\|}{C}}-O-\overset{\overset{H}{|}}{\underset{\underset{H}{|}}{C}}-\overset{\overset{H}{|}}{\underset{\underset{H}{|}}{C}}-O-\!\!\sim$$

$$\sim\!\!-C\{CH_2\}_n\overset{\overset{}{}}{\underset{\underset{M}{|}}{C}}-\overset{\overset{H}{|}}{\underset{\underset{H}{|}}{C}}-\overset{\overset{}{}}{\underset{\underset{O}{|}}{C}}-O-\!\!\sim \qquad \text{(thermoset)}$$

$$\}$$

Stage 2: Cross-linking of unsaturated ester polymers by means of an unsaturated monomer, e.g. styrene.

(e) *Polyester* (*cross-linked*). A typical reaction is shown schematically (two dimensions). Many other reactions are used to form the linear polymer containing the ester groups and other unsaturated monomers to create the cross-links. N.B. The absence of by-products in the cross-linking which is important in forming fibre-glass composites.

price, a plastic usually contains a considerable proportion of additives, such as plasticisers, fillers, lustrants, etc. The modification of polymers by such additions is discussed in § 8.7. Here it should be noted that the presence of these additives introduces considerable complications into the employment of plastics in engineering components since each manufacturer tends to devise his own additives, and although the polymer is normally specified (but not necessarily the degree of polymerisation) the manufacturer rarely discloses the nature of the additives. Since these additions affect many of the properties, data on mechanical and physical properties become specific to a particular manufacturer and brand name.

Nevertheless there is an increasing use of plastics in technology on account of their advantageous overall price, including the economic manufacturing methods, and their specific properties. The thermoplastics currently of importance are: polyvinyl chloride (PVC), polyamides (nylons), polyethylene, polypropylene, polyethylene terephthalate (PET) polycarbonates, acetals, polystyrene, acrylonitrile butadiene styrene (ABS) and fluorocarbons. Data on these materials are given in Table 8.3.

It is important that the distinctive behaviour of plastics as compared with the more conventional metallic materials should be appreciated by designers. As will be discussed later, plastics do not show Hookean elasticity, although various elastic moduli are definable. Many plastics exhibit *cold flow*, which is creep at ambient temperatures, and the environment (e.g. water vapour) often affects their properties. As with other new technologies, initial mistakes have been made which gave plastics a bad name, and it is only now that reliable engineering type data are being collected and a rational design approach being developed. These points will be discussed in § 8.10.

8.2 Polymerisation reactions

The process by which monomer units are combined again and again into a giant molecule is known as polymerisation. A monomer must be multifunctional, that is, capable of producing at least two active chemical bonds for it to be able to polymerise. With one active bond it can only combine with one other monomer to form a dimer. With two active bonds it can react with two other monomers, each of which can then react with another monomer to form by repetition a long chain, or linear, polymer. With more than two active bonds per monomer, reactions can take place in several directions to build up a two or three-dimensional molecule.

The number of active bonds is called the *functionality* of the monomer. Thus a monomer which utilises two bonds to form a linear polymer is bifunctional. Monomers may not be able to use all their possible active bonds due to space limitations. For example, phenol C_6H_5OH has five such bonds (by removing the five hydrogen atoms) and might be expected to be pentafunctional, but it is, in fact, only trifunctional because of spatial interference (steric hindrance) – see, for example, phenolformaldehyde, Table 8.2.

The chemical reactions which produce synthetic polymers are of two types: *addition* polymerisation and *condensation* polymerisation (or *polycondensation*). Addition polymerisation occurs between monomers with an unsaturated, or

Table 8.3. *Properties of various high polymers (at 20 °C) N.B. Spread of values about average is ±10 per cent except when marked ~, where spread is ±20 per cent. (Based on G. R. Palin: Plastics for Engineers, Pergamon Press 1967)*

Material (for structure see Tables 8.1, 8.2)	Physical data					Mechanical data				Electrical data		
	Specific gravity	Coeff. of expansion (deg⁻¹ ($\times 10^3$))	Thermal conductivity ($Wm^{-1} deg^{-1}$)	Softening point (Vicat) (°C)	Glass transition (°C)	Initial modulus (GN/m²)	Tensile strength (MN/m²)	Impact strength	Apparent modulus 1000 h (GN/m²)	Dielectric constant	Loss tangent ($\times 10^3$)	Dielectric strength (V/μm)
										(at 1 MHz)		
Olefines												
polyethylene: low density	0.91–0.94	16–22	~0.3	~75	<50	0.12–0.24	7–17	excellent		2.3	<0.2	18–27
high density	0.95–0.97	11–16	0.5	125	<50	0.55–1.0	21–38	excellent		2.3	<0.2	>30
Vinyl compounds												
polyvinyl chloride (unplasticised)	1.4	5–8	0.15	85	75	~2.4	55	poor		3.0	15	15–20
PVC (low plasticiser content)	~1.3	5–25	~0.17	75		~4.0	35	medium		3.5	70	>16
PVC (high plasticiser content)	~1.2	5–25	~0.17				14–28	good		4.5	140	>12
PVC–vinyl acetate copolymer	~1.4	7–8	0.15			~3.0	55		0.4	3.3	~12.5	>40
polypropylene	0.9	~11	~18	150	0	1.2	35	good		2.2	<0.5	>30
polystyrene	1.1	~7	0.08–0.2	90	100	2.4–4.1	35–62	poor		2.6	~0.07	20–28
toughened with SBR (HIPS)	1.1	~7	0.04–0.16	90		1.7–3.1	17–45	good		2.6	<2	12–24
Vinylidene compounds												
polyvinylidene chloride (PVDC)	1.7	19	0.13	75		~0.4	28	medium		3.5	50–80	~20
polymethylmethacrylate (PMMA)	1.2	5–8	0.20	100	~85	~3.0	~60	poor		2.5–3.5	15–30	20
Polyamides												
nylon-6: dry	1.3	~11	0.29	125–175†		~2.5	~75	good		3–7	20–130	20
nylon-6,10: dry	1.08	9–15	0.25	155†		~2.0	~60	good		3–4	20–30	>10
nylon-6,6: dry	1.13	~12.5	0.25	100–200†	45	~3.0	~85	good	2.3	3	60	14
saturated (8.5% H₂O)	1.17					~1.0	~35	excellent	0.5	7	~200	
30% glass, dry	1.4	~3	0.25			~7.5	~170	medium	6.5	3.5	20	
Miscellaneous												
polytetrafluoroethylene (PTFE)	2.2	~11	0.25	120†		0.35–0.62	17–28	good	2.1	2	<0.3	~20
polycarbonate	1.2	6.5	~0.22	220		2.3	62	excellent	1.4	2.6	~12.5	16
polyacetal	1.4	8.3	0.23	170		3.0	65	good	1.5	3.7	4	20
ABS	1.05	6–13	0.04–0.3	90		1.4–3.5	17–60	excellent		~3	7–20	~14
phenolformaldehyde (unfilled)	1.27	2.5–6	0.12–0.24			9	35–55	poor		~5	15–30	~14
Pf (wood floor filled)	1.4	3–4.5	0.17–0.3			7	~50	medium		~5	30–70	8–16
balsa wood ⎱15% water content	0.2					4	27			1.4	12	
Douglas fir ⎰along grain	0.5		0.15			10	70			2.25	25	
hickory	0.75	~50	0.25			15	140					

† Heat distortion temperature.

double, bond. One half of the bond opens to provide two half bonds able to unite with similar bonds in two adjacent monomers. No product other than the polymer is formed. Some examples have already been given; another is the polymerisation of propylene:

$$n\begin{bmatrix} H & CH_3 \\ | & | \\ C{=}C \\ | & | \\ H & H \end{bmatrix} \rightarrow \begin{bmatrix} H & CH_3 \\ | & | \\ C{-}C \\ | & | \\ H & H \end{bmatrix}_n$$

This has better heat and chemical resistance than polythene from which it is derived.

Addition polymerisation can also occur between different monomers, a process known as *copolymerisation*. An example is the copolymer formed from vinyl chloride and vinyl acetate

$$n\begin{bmatrix} H & Cl \\ | & | \\ C{=}C \\ | & | \\ H & H \end{bmatrix} + m\begin{bmatrix} H & COOCH_3 \\ | & | \\ C{=}C \\ | & | \\ H & H \end{bmatrix} \rightarrow$$

$$-CH_2-\overset{\overset{\displaystyle Cl}{|}}{C}H\!\!+\!\!CH_2-\overset{\overset{\displaystyle COOCH_3}{|}}{C}H\!\!\Big]_x\!\!\Big[CH_2-\overset{\overset{\displaystyle Cl}{|}}{C}H\!\!\Big]_y\!\!CH_2-\overset{\overset{\displaystyle COOCH_3}{|}}{C}H-$$

Usually the comonomers are randomly arranged along the chain:

.A.A.B.A.B.B.A.A.A..

This particular copolymer is widely used, for example in gramophone records, with about one vinyl acetate monomer to ten vinyl chloride. The presence of the occasional bulky side group keeps the chains apart and increases the flexibility and toughness as compared with pure PVC (in other words, lowers the glass transition temperature, see below). It is also possible to group each monomer into separate sequences, either along the main chain (*block* copolymers) or with one type in the main chain and the other confined to branches (*graft* copolymer). This arrangement can lead to two phase polymers, for example, in high impact polystyrene (see § 8.7).

Other important copolymers are found in the synthetic rubbers (elastomers, § 8.9). SBR rubber, which has replaced natural rubber for many applications, is a copolymer of styrene and butadiene:

$$\begin{bmatrix} H & \\ | & \\ C{-}C \\ | & | \\ H & H \end{bmatrix} + \begin{bmatrix} H & H & H & H \\ | & | & | & | \\ C{-}C{=}C{-}C \\ | & & & | \\ H & & & H \end{bmatrix}$$

The proportion is about one styrene unit to three butadiene. In nitrile rubber (NBR) butadiene is copolymerised with acrylonitrile:

$$-\overset{\overset{\displaystyle H}{|}}{C}-\overset{\overset{\displaystyle CN}{|}}{C}-$$
$$\overset{}{\underset{\displaystyle H}{|}}\ \overset{}{\underset{\displaystyle H}{|}}$$

The copolymerisation of three monomers: acrylonitrile, butadiene and styrene has produced an important two-phase plastic known as ABS (see § 8.7). By varying the proportion and arrangement it is possible to design copolymers with a wide range of properties. Only a small addition of a second monomer can play an important role, for example by providing reaction centres for the dyeing of fibres.

Condensation polymerisation does not require an unsaturated bond in the monomer. At each point of reaction a molecular product is formed, independent of the polymer. An example already given is the formation of 'Bakelite' from phenol C_6H_5OH and formaldehyde CH_2O, with water as the by-product. The reaction must be carried out so that the by-product is removed from the system.

Polyamides, or nylons, are a group of linear high polymers formed by both condensation and addition polymerisation. The common feature of the many diverse types of nylon is the presence of amide groups $(-NH.CO.-)$ in a chain of methylene groups $(-CH_2-)$. For example nylon-6,6 has the repeat unit

$$\begin{array}{cc} \text{diamine} & \text{dibasic} \\ \text{residue} & \text{acid} \\ & \text{residue} \end{array}$$

This is polycondensed from hexamethylene diamine $NH_2(CH_2)_6NH_2$ and adipic acid $COOH(CH_2)_4COOH$ with the elimination of water. The numbers 6,6 refer to the number of carbon atoms in the diamine and dibasic acid respectively. Thus nylon-6,10 is formed from hexamethylene diamine and sebacic acid $COOH(CH_2)_8COOH$. Another nylon group is made by addition polymerization of lactams. For example, on heating ε-capro-lactam, the ring structure opens and polymerises to nylon-6, a process known as ring-scission polymerisation:

Here the single number indicates the number of carbons in each repeat unit.

Every nylon contains the linking amide group $-\overset{\overset{\text{O}}{\|}}{\text{C}}-\underset{\underset{\text{H}}{|}}{\text{N}}-$, but each type must be regarded as a distinct material with its own individual properties. These can be traced to the affinity between the protruding oxygen atoms of the carbonyl groups (CO) and the protruding hydrogen atoms of the amino groups (NH) in adjacent polymers. In nylon-6,6 the frequency of these groups leads to maximum interaction between polymers, with a high degree of crystallisation and a strong, rigid, material. All the nylons absorb water from the atmosphere, probably through hydrogen bonding with the amino group, with a marked effect on the properties.

Another important group of polymers are the polyesters, also formed by condensation polymerisation. Esters are the reaction products of organic acids

(containing $-C\overset{\displaystyle O}{\underset{\displaystyle OH}{\diagdown}}$ groups) and alcohols (containing $-\overset{\displaystyle H}{\underset{\displaystyle H}{C}}-OH$ groups) and

they contain the ester linkage $-C\overset{\displaystyle O}{-}O-$. Depending on the functionality of the monomers, the polymer is saturated (i.e. no double bonds) and thermoplastic, or unsaturated and can be cross-linked (thermoset). The most noted example of a linear saturated polymer is polyethylene terephthalate, better known as 'Terylene' or 'Dacron'. This is the product of the reaction between terephthalic acid ($HOOC.C_6H_4.COOH$) and ethylene glycol HOC_2H_4OH (a glycol is a dihydric alcohol, i.e. with two OH groups). It has the repeating structure:

terephthalate acid residue   glycol residue

In practice, dimethylterephthalate ($CH_3OOC.C_6H_4.COOCH_3$) is reacted with the ethylene glycol and methyl alcohol is eliminated. The polymer is excellent for forming fibres and films.

In the thermosetting polyesters an unsaturated acid such as fumaric acid $HOOCH{=}CHCOOH$, together with a proportion of saturated acid to control the density of subsequent cross linking, is reacted with a glycol, see Table 8.2. The cross-linking may be developed directly between double bonds in adjacent chains, or more frequently, through an unsaturated monomer such as styrene $CH_2{=}CHC_6H_5$. The fact that this process occurs without the application of heat or pressure, and with no by-product, makes polyester the principal resin for use with glass fibre reinforcement. The more expensive epoxy resins may also be used.

Addition polymerisation is a complex reaction involving three distinct stages: initiation, propagation and termination. By contrast, polycondensation is a straightforward stepwise chemical reaction, with no specific initiation and termination process, and the polymer can always react with further monomer. The monomer unit in addition polymerisation is normally a fairly stable molecule and the reaction must be started by attacking it with a free radical or an ion which is able to break the double bond. The initiator is then attached to one side of the broken bond and the other is free to react with another monomer – the propagation stage. Propagation will continue (at a decreasing rate as the supply of monomer unit decreases) until another process intervenes which removes the reactivity, for example, the chain end may react with a group having only one active bond, known as a terminal group. Alternatively two chain ends may combine and preclude further reaction with monomers. The chain structures will then be respectively [initiator—n(monomer)—terminator] or [initiator—$(n + m)$(monomer)—initiator]. As long as there are many monomers in the chain (typically 10^4–10^5), the nature of the end groups is unimportant to the physical properties, but often critical to the chemical properties.

Several techniques of initiation have been discovered and much of the development of polymers is due to progress in this area. Besides starting the reaction, the initiator affects the propagation stage, determining the molecular architecture, and also the termination. The subject is complex and extensive, and will be only touched on here. One important class of initiators consists of free radicals, for example, formed by the breakdown of hydrogen peroxide H_2O_2. The other principal classes are cationic, anionic and heterogeneous. The latter is an important new discovery (1953) by Ziegler and Natta, which won them a Nobel Prize in 1963. These catalysts provide close steric control over the attachment of the monomers and require only low temperature and pressure. For example, in the polymerisation of ethylene, the original ICI process (1937) involved a pressure around $150 \, MN/m^2$ (20 000 psi) and a temperature of 200–400 °C. Using a Ziegler–Natta catalyst, polymerisation proceeds at $3.5 \, MN/m^2$ (500 psi) and the polymer is truly linear. High pressure polyethylene contains a certain amount of branching – branches up to 5 carbons in length occur one to three times every 100 carbon atoms in the main chain. Linear polyethylene is able to crystallise, leading to high density and improved properties.

Termination reactions depend on the type of initiator. Free radical initiation is not very specific and various chain terminations are possible. The propagating chain may terminate by reacting with a hydrogen atom extracted from a monomer, initiator, solvent, or specially added terminal, or transfer, agent. The latter reaction is

$$-CH_2CXY \cdot + HR \rightarrow -CH_2CHXY + R\cdot.$$

The free radical $R\cdot$ which is left is able to initiate a new chain, hence the term transfer agent. Alternatively, the hydrogen atom may be taken from a polymer chain already formed, providing a reaction centre for the growth of a branch. The length of chains is determined by the relative rates of initiation, propagation and termination and these factors must be adjusted to achieve the required degree of polymerisation.

8.3 Degree of polymerisation

The properties of polymeric materials, whether natural or synthetic, are sensitive to the size of the polymer molecules, that is, to the number of monomer units comprising each molecule. This is called the degree of polymerisation, and is equal to the polymer molecular weight divided by the monomer weight. The degree of polymerisation for commercial polymers usually lies in the range 10^3–10^5 monomers per molecule.

In any sample of a polymer, not all the molecules will have undergone the same degree of polymerisation, and the measured molecular weight will be an average value. There are several ways of averaging the molecular weight and specific properties are dependent upon a particular average. Conversely, the molecular weight determined experimentally will depend upon the method of measurement.

All measurements are made on dilute solutions of the polymer in a suitable solvent. The number of molecules present in a dilute solution can be determined by measurement of one of several related properties, known as the *colligative* properties: lowering of the vapour pressure, freezing point and osmotic pressure; raising of the boiling point. These are all directly proportional to the mole fraction, that is, the number of molecules, N, per unit weight of solution:

$$N = \frac{W}{M},$$

where W is the weight concentration and M is the molecular weight. For a dispersion of molecular weights, say, X_1 molecules with molecular weight M_1, X_2 with molecular weight $M_2, \ldots$ and X_i molecules with molecular weight M_i, then:

$$N = \sum_i X_i$$

and

$$W = \sum_i X_i M_i.$$

Thus the measurement of the colligative properties gives the *number average molecular weight* $\overline{M}_n$:

$$\overline{M}_n = \frac{W}{N} = \frac{\sum_i X_i M_i}{\sum_i X_i}.$$

The mechanical properties of a polymer are largely dependent on the numbers of molecules in each molecular weight range, and the number average molecular weight is therefore an important quantity. A typical variation of properties with molecular weight is shown in Fig. 8.2.

Light scattering of a dilute solution is another common method of measuring molecular weight. The intensity of light scattering by particles is proportional to the square of the mass of the particle and leads to the *weight average molecular weight* defined by:

$$\overline{M}_w = \frac{\sum_i X_i M_i^2}{\sum_i X_i M_i}.$$

Sedimentation methods yield the *z-average molecular weight*:

$$\overline{M}_z = \frac{\sum_i X_i M_i^3}{\sum_i X_i M_i^2}.$$

If all the molecules have the same weight, then $\overline{M}_n = \overline{M}_w = \overline{M}_z$. If there is a distribution of molecular weights then $\overline{M}_n < \overline{M}_w < \overline{M}_z$ because of the larger number of molecules in the low weight fraction. $\overline{M}_w$ is typically double $\overline{M}_n$.

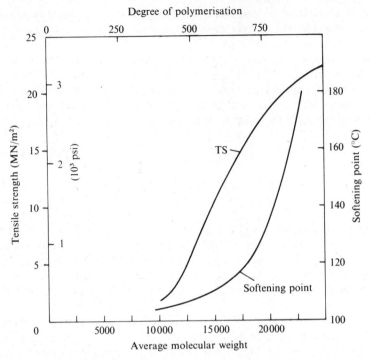

Fig. 8.2. Properties of polyethylene against molecular length.

Another simple method for determining molecular weight is by measuring the apparent viscosity of a dilute solution. In theory the viscosity should depend on the weight average, but empirically it is found to be related to the molecular weight of linear polymers by:

$$\eta = KM^a$$

where K and a are constants; K varies from 0.5 to 5×10^{-4} and a from 0.5 to 1. Viscosity measurements lead to the *viscosity average molecular weight*:

$$\overline{M}_v = \left(\frac{\sum\limits_i X_i M_i^{1+a}}{\sum\limits_i X_i M_i} \right)^{1/a}.$$

If $a = 1$, then $\overline{M}_v = \overline{M}_w$. More usually, $\overline{M}_n < \overline{M}_v < \overline{M}_w$.

Thus it will be seen that there are many different ways of averaging the distributed molecular weights of a polymer and it is necessary to specify the test method when quoting the average value.

8.4 Molecular architecture

The term molecular architecture refers to the spatial, or steric, arrangement of the monomer groups in a polymer molecule. This plays a dominant role in determining the structure and properties of the polymeric solid. Of particular importance are the size and position of the various substitute atoms or side groups along the chain. As mentioned above, these can be controlled by means of the polymerisation technique.

Linear polymers such as polyethylene have been shown previously as straight with all the carbon and hydrogen atoms lying in plane. This has been merely for typographical convenience. The bonds of a carbon atom are directed towards the corners of a tetrahedron, the included angle being about 109°. When fully extended the polyethylene chain can be represented by

$$
\begin{array}{ccccccc}
\odot & & \odot & & \odot & \\
| & & | & & | & \\
C & & C & & C & \\
/ \;\; \backslash & / \;\; \backslash & / \;\; \backslash \\
& C & & C & \\
& | & & | & \\
& \odot & & \odot &
\end{array}
$$

where $\odot$ represents two hydrogen atoms, one above and one below the plane of the carbon zigzag. If the ends of the polymer are not constrained, all the atoms are free to rotate about the bonds, although the bond lengths and relative directions remain fixed. However, rotation does not occur about any double bonds that are present. The polymer now takes up a random three-dimensional conformation, see Fig. 8.3.

The random walk of the chain links is limited by interactions between side groups. These may be repulsive due to steric interference or attractive if the side groups are polar. When there is a regular arrangement of attractive side atoms or groups along a chain, the conformation with the minimum internal energy (U) is a regular helix which arranges the attractive groups close to each. However, for minimum free energy ($U - TS$) (cf. § 2.9) the equilibrium arrangement consists of a more random and kinked coil, increasing the entropy (S). The change of

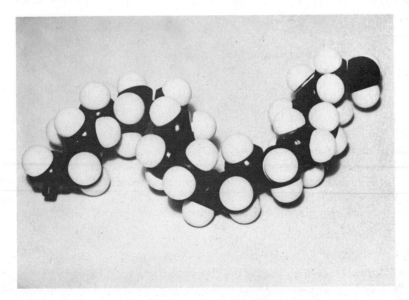

Fig. 8.3. Polymer model, showing random conformation due to rotation about single bonds.

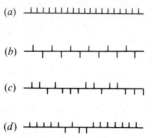

Fig. 8.4. Tacticity in polymers – schematic representation of the arrangement of side groups. (a) Isotactic; (b) syndiotactic; (c) atactic; (d) partially tactic.

polymer length due to bond rotation is an important feature of this class of material and is responsible for high elasticity. These effects are considered later (§ 8.9).

There are three ways of arranging side groups along a polymer chain, shown schematically in Fig. 8.4. In the isotactic configuration, the side groups are all on the same side. For example, an isotactic polymer chain of a vinyl compound would be (when extended and viewed in the plane of the carbon zigzag)

$$
\begin{array}{ccccccc}
\text{H} & \text{R} & \text{H} & \text{R} & \text{H} & \text{R} \\
| & | & | & | & | & | \\
-\text{C}-\text{C}-\text{C}-\text{C}-\text{C}-\text{C}- \\
| & | & | & | & | & | \\
\text{H} & \text{H} & \text{H} & \text{H} & \text{H} & \text{H}
\end{array}
$$

In the syndiotactic polymer the side groups alternate regularly from side to side

$$
\begin{array}{ccccccc}
\text{H} & \text{R} & \text{H} & \text{H} & \text{H} & \text{R} \\
| & | & | & | & | & | \\
-\text{C}-\text{C}-\text{C}-\text{C}-\text{C}-\text{C}- \\
| & | & | & | & | & | \\
\text{H} & \text{H} & \text{H} & \text{R} & \text{H} & \text{H}
\end{array}
$$

Regular arrangements are called tactic. In atactic polymers the arrangement is completely random

$$
\begin{array}{cccccccc}
\text{H} & \text{R} & \text{H} & \text{R} & \text{H} & \text{H} & \text{H} & \text{H} \\
| & | & | & | & | & | & | & | \\
-\text{C}-\text{C}-\text{C}-\text{C}-\text{C}-\text{C}-\text{C}-\text{C}- \\
| & | & | & | & | & | & | & | \\
\text{H} & \text{H} & \text{H} & \text{H} & \text{H} & \text{R} & \text{H} & \text{R}
\end{array}
$$

Many polymers show partial tacticity, that is, they are stereoregular over limited lengths of the chain. Tactic polymers are able to crystallise whereas atactic ones do not, the difference being more marked the larger the side groups. The presence of substituted atoms in the ethylene monomer means that it now has a head and a tail. As shown above, head to tail polymerisation is the normal sequence but a few irregularities are to be expected.

Tacticity is an example of steric isomerism, that is, polymers with the same composition having different structures (cf. polymorphism Chapter 2).

Another source of isomerism is the double bond found in some polymers. Polyisoprene, for which the monomer unit is $-\text{CH}_2.\text{CCH}_3=\text{CH}.\text{CH}_2-$ (see Table 8.5), is an example. (The dots in this formula are a method of indicating bonds along the main chain.) In the *cis*-isomer, which occurs in natural rubber, the bonds are arranged thus:

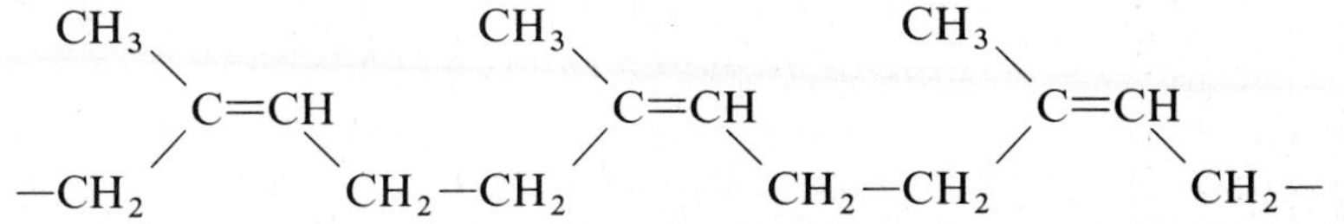

$$
\begin{array}{ccc}
CH_3 & CH_3 & CH_3 \\
\diagdown & \diagdown & \diagdown \\
C{=}CH & C{=}CH & C{=}CH \\
\diagup \quad \diagdown & \diagup \quad \diagdown & \diagup \quad \diagdown \\
-CH_2 \qquad CH_2{-}CH_2 & CH_2{-}CH_2 & CH_2-
\end{array}
$$

Here the double bonds lie in the direction of the chain (and on one side when drawn out straight). In the *trans*-form, which occurs in gutta-percha, the bond arrangement is:

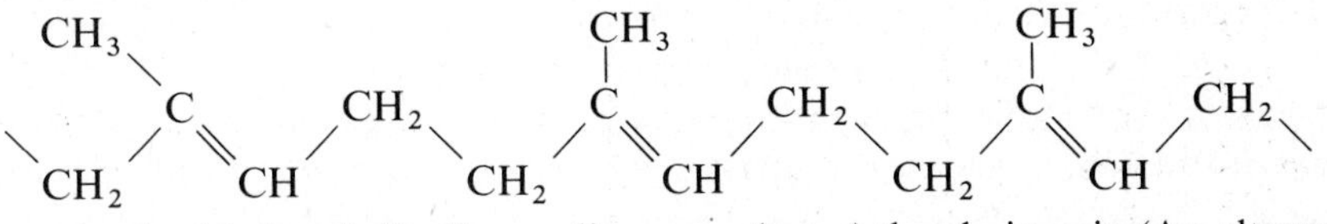

$$
\begin{array}{cccccc}
CH_3 & & CH_3 & & CH_3 & \\
\diagdown & & | & & | & \\
C & CH_2 & C & CH_2 & C & CH_2 \\
\diagdown \;\Vert\; \diagup \quad \diagdown & \diagup \;\Vert\; \diagdown & \diagup \quad \diagdown & \diagup \;\Vert\; \diagdown & \diagup & \diagdown \\
CH_2 \quad CH & CH_2 \quad CH & CH_2 \quad CH &
\end{array}
$$

Here the double bonds lie diagonally across (*trans*) the chain axis. (An alternative way of looking at the *cis*- and *trans*-forms is to consider the positions of the chain-connecting CH_2 groups in relation to the plane of the double bond.) The *trans*-isomer molecule is symmetrical and easily crystallises into a rigid material. The *cis*-isomer is unsymmetrical and unable to crystallise (except under stress) on account of the unbalanced forces associated with the double bond. It forms a rubbery material. The use of Ziegler catalysts has in recent years allowed the synthesis of *cis*-polyisoprene, exactly equivalent to natural rubber, where previously a random mixture of *cis*- and *trans*-orientations was obtained.

Another important structural factor is cross-linking, the formation of chemical bonds between one linear molecule and another. This is most conveniently brought about subsequent to polymerisation. It usually requires an unsaturated bond in the main chain or side group and the presence of impurities such as oxygen or sulphur to provide the connecting link. The cross-linking of unsaturated polyesters has already been described (§ 8.2, Table 8.2). The original and still important example is the vulcanisation of rubber by means of sulphur, first carried out by Goodyear in 1839. The reaction is complex but the final outcome is the opening of a few of the double bonds and the formation of sulphur cross-links:

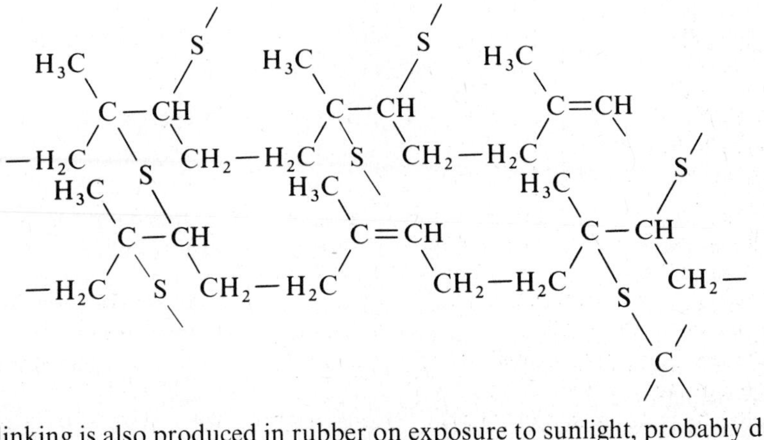

Cross-linking is also produced in rubber on exposure to sunlight, probably due to removal of hydrogen atoms, leaving free radicals which then attack double bonds to form cross-links. Cross-linking can also be achieved by the introduction of a trifunctional copolymer at a low concentration. The effect of cross-linking is to prevent the chains sliding past each other, suppressing viscous flow (see § 8.6). Excessive cross-linking will lead to brittle behaviour, as found in rubber aged by exposure to sunlight.

Finally, branching is a major characteristic of nominally linear polymers. This is an easier process than cross-linking since the reaction takes place at one site instead of two. Branches may form from unsaturated side groups or, as mentioned, as a result of the removal of a hydrogen atom from a back-bone carbon atom during termination of some other chain, leaving a free radical to attack a monomer. The length of the side chain may be as long as the parent, but more often for steric reasons it is relatively short. An example is low density polythene:

$$H_3C\diagdown$$
$$\qquad\diagup CH_2$$
$$H_2C\diagup$$
$$\quad CH\ \ CH_2[CH_2]^nCH_2CH_2\ \ CH_2\ \ CH_2[CH_2]^nCH_2CH_2\ \ CH\ \ CH_2$$
$$\quad CH_2\ CH_2\ CH_2\ CH_2\ CH\ \ \ CH_2\ CH_2\ CH_2\ CH_2\ CH_2\ CH_2\ CH_2\ CH_2$$

$$H_3C\diagdown$$
$$\qquad\diagup [CH_2]^4$$
$$H_2C\diagup$$

$$\qquad\qquad\qquad H_2C\diagup CH_2$$
$$\qquad\qquad\qquad\qquad\diagdown CH_2$$
$$\qquad\qquad\qquad H_2C\diagdown$$
$$\qquad\qquad\qquad\qquad\quad CH_3$$

Branches of up to five carbon atoms length are formed about three times per hundred carbon atoms of the main chain. Branching affects the properties for the same reason as do side groups: by keeping the chain apart crystallisation is prevented; on the other hand, sliding of the chains past each other in viscous flow is also obstructed. The detailed result of these conflicting changes will depend on the specific polymer.

8.5 Glass transition and crystallinity

The solidification of a linear polymer from the liquid state gives either an amorphous solid or a partially crystalline one, as mentioned briefly in § 8.1. With a fully amorphous polymer there is no sharp temperature at which various physical parameters change discretely, as occurs with a crystalline solid. Instead there is a progressive alteration of the properties towards the characteristics of a solid. However, as with inorganic glasses (§ 7.8), there is a second-order transition at which there is a change in the rate of change with temperature of various parameters such as specific volume, specific heat, damping and viscosity. A typical specific volume–temperature curve is given at Fig. 8.5(*a*). The change point is called the *glass transition temperature* (or, simply, glass temperature, symbol T_G). Above T_G, segments of the polymer chain are free to move due to thermal agitation; below T_G, the thermal energy is insufficient to activate movement of chain segments. Segment motion is responsible for an important mechanical property of polymers, called high elasticity, which will be discussed in the next section. It should be noted that a polymer is effectively a solid at

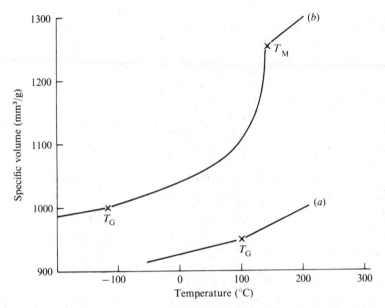

Fig. 8.5. Transition temperatures shown by measuring specific volume against temperature: (a) polystyrene (amorphous); (b) high density polyethylene (crystalline).

temperatures well above ($\sim 100\,^{\circ}\mathrm{C}$) its glass transition temperature. In fact, the useful temperature range tends to be slightly above T_G; at lower temperatures the polymer is brittle, and at higher temperatures creep becomes excessive. Elastomers should have their T_G about $100\,^{\circ}\mathrm{C}$ below the operating temperature.

The dependence of the glass transition on the molecular architecture has been extensively studied. Particularly important is the nature of any side groups. If stiff and bulky, they prevent chain rotation and raise T_G; but if flexible they keep the chain apart and assist movement, so lowering T_G. The presence of polar groups will raise the glass temperature by increasing the intermolecular attraction. Polymers with low molecular weight have a lower transition than similar polymers of high molecular weight. This can be ascribed to the ends of the chain producing free volume. A similar effect is responsible for the lower transition of branched chains as compared with linear chains of the same molecular weight. Finally the addition of plasticisers which are relatively low molecular weight substances lowers the glass temperature (see § 8.7).

A partially crystalline polymer has a crystalline melting point T_M and a glass transition temperature T_G, as shown by Fig. 8.5(b). T_M is typically around $200\,^{\circ}\mathrm{C}$ above T_G. Crystallisation begins on cooling past T_M and continues over a temperature range. Any amorphous polymer still present when the temperature reaches T_G undergoes the second order transition just discussed. There is no discrete step in the physical properties at T_M, because of the way individual polymer molecules are partially in the crystalline regions and partly in the surrounding amorphous matrix. Clearly crystallinity has a marked effect on properties, especially between T_G and T_M. It may be likened in some respects to cross linking since the chains become locked together in the crystallites.

A crystalline region (crystallite of *spherulite*) is about 0.1 μm across and the atoms of adjacent chains are arranged in a regular three-dimensional array with a definite space lattice (Chapter 2) so that X-ray diffraction patterns can be obtained. In the amorphous regions, which normally form the matrix around the crystallites, the polymer chains are randomly arranged and intertwined, though during drawing or rolling a preferred direction (orientation) of the chains (and any crystallites) will develop. A particular polymer chain, which is typically 1–50 μm long when stretched out, will traverse several crystalline and amorphous regions. Details of the arrangement of the chains within a crystallite are discussed later in this section.

The degree of crystallinity (θ) is defined as the weight fraction of the polymer which is in the crystalline regions. (The crystalline/amorphous weight ratio, sometimes quoted, is then $\theta/1 - \theta$.) It ranges from 0 to 80 per cent. Several methods are available for determining the fraction, such as X-rays, specific volume, and heat of fusion; they will not be detailed here. The various methods give results which differ by ± 5 per cent, due to the particular way a given method apportions the indeterminate boundary regions around each crystallite – there is not a sharp transition from crystallite to amorphous region.

Three physical factors favour crystallinity: low viscosity in the liquid state, regularity of the polymer molecules (tacticity) and strong intermolecular forces. A low viscosity allows the necessary molecular rearrangement to take place in the liquid prior to crystallisation Large and irregular polymers, particularly those with short branches, besides producing high liquid viscosity, are unlikely to pack into a regular crystalline array. Highly polarisable groups or bonds will provide large van der Waals binding forces and increase the tendency to crystallise.

As with inorganic glasses, the rate of cooling from the liquid affects the degree to which crystallisation develops. Slow cooling around the crystallisation temperature allows time for the necessary molecular alignments. Crystallisation is also induced when high polymers are drawn into fibres or films – the chains becoming aligned in the direction of extension. A measure of the combined effects of crystallisation and chain orientation can be gauged from the tensile strength of nylon: in bulk form, 10^8 N/m² (145×10^3 psi); as fibre, 10^9 N/m² (1.45×10^6 psi).

Polyethylene is a non-polar polymer which nevertheless crystallises strongly on account of its regularity, particularly the high density form produced by low temperature polymerisation by the Ziegler technique. Another is the recently introduced acetal plastics (1960), successfully polymerised from formaldehyde CH_2O after earlier techniques had led to unstable brittle polymers.

Single crystals of several high polymers have been studied in recent years following the production in 1957 of small single crystals of linear polyethylene by three independent workers (Till, Keller, Fisher). Grown from dilute solution, the crystals are lozenge-shaped, or trapezoidal, plates up to 100 μm across and 0.1 μm thick. A typical electronmicrograph is shown in Fig. 8.6. Each crystal consists of several 0.01 μm thick layers, which can be seen as growth spirals on the crystal face. The details alter with temperature, solution concentration and average molecular weight. Keller showed that the polymer chains lie folded between the faces of each platelet as shown in Fig. 8.7. The chain axes are, surprisingly, perpendicular to the crystal plates: the folds are very sharp with only three carbon atoms in each 180° bend.

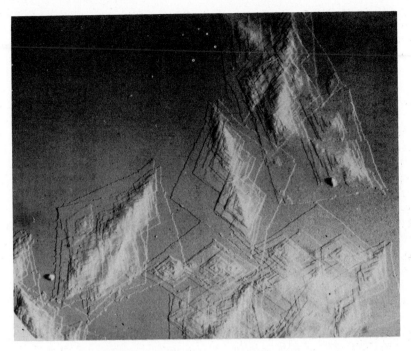

Fig. 8.6. Single crystals of polyethylene (× 16 000). (Courtesy of R. J. Seward.)

In bulk polymer, the ordered regions are called crystallites on account of their small size, the order of 0.01 to 0.1 μm (cf. metallic grains $\sim 50\,\mu$m). In the original *fringed micelle* theory (1932–42), the crystallites were thought to consist of straight and parallel polymer chains which were randomly arranged in an amorphous matrix, see Fig. 8.8. A single polymer molecule ran through several crystalline and amorphous regions. It has now been shown that the crystallites are organised into larger groups with approximately spherical outline, called spherulites. These can be observed as Maltese crosses under an

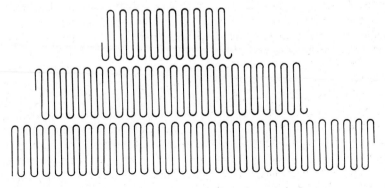

Fig. 8.7. Folded polymer chains in layers of a single crystal.

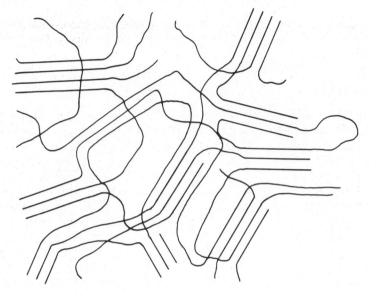

Fig. 8.8. Fringed micelle theory of crystallinity (now discarded): crystallites are randomly orientated.

optical microscope using polarised light, see Fig. 8.9. Keller has shown that a spherulite develops from a fibre-shaped nucleus (fibril) which as it grows lengthwise, branches and twists at regular intervals. The role of the single crystal platelets in this mechanism is not certain but it is probable that a fibril consists of twisted ribbons of folded molecules. Some modification of the single crystal structure must be expected as the long molecules become involved in

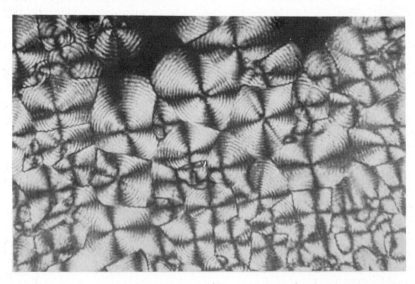

Fig. 8.9. Spherulites in polyethylene appearing as Maltese crosses under polarised light (× 55). (Courtesy of R. J. Seward.)

several different folding exercises. The size of the spherulites depends on the conditions of crystallisation, the maximum size being obtained by isothermal crystallisation a few degrees below the melting temperature.

8.6 Physical states

The physical state of linear, branched and lightly cross-linked high polymers depends on the temperature and it undergoes discrete changes at one or more transition temperatures. The fundamental cause of the changes of state is the thermal energy, which vibrates the molecular chains. Close network polymers (thermosets) do not change with temperature in the same way and only exhibit gradual variation of properties up to the temperature at which they decompose; the breakdown of polymers will be discussed in § 8.11. Before defining the various states in terms of their characteristic mechanical behaviour, it is necessary to consider a new type of elasticity, peculiar to high polymers.

Three modes of deformation occur in linear high polymers. Two are well known, being found in other solids and liquids: normal elasticity, due to the displacement of atoms from their positions of minimum potential energy, in other words, due to the stretching of the interatomic bonds (cf. § 3.1); and viscous flow, due to the slow diffusion of molecules past one another (cf. § 7.9).

The third mode is called *entropy*, or *high*, *elasticity*. The quantitative theory will be given later (§ 8.9). It suffices to note here that it is due to the ability of a molecular chain to rotate about its single bonds to produce many kinked and twisted shapes, known as *conformations*. The end to end length is infinitely variable. The most probable length, which corresponds to the condition of maximum entropy (see § 2.9) is that which occurs in the greatest number of completely random conformations of the chain. A tensile stress tends to straighten and lengthen the chains, decreasing the entropy, and at equilibrium the entropy stress F_S is given by:

$$F_S = -T\frac{dS}{dL},$$

where dS/dL is the rate of entropy change with length. This equation is derived in § 8.9. Increasing the temperature increases the entropy stress proportionately at a given strain. Very large elastic strains and very low elastic moduli can be produced by this effect. Although found to some extent in all polymers high elasticity is particularly marked in a class of polymers known as elastomers (or rubbers).

Entropy elasticity is a thermally activated phenomenon. Bond rotation may not occur immediately a stress is applied because of interaction between atoms in the same chain or in adjacent chains. In time, fluctuation of the local thermal energy will enable the bonds to rotate and the equilibrium chain conformations to be achieved. When this is large compared with the loading time, the entropy elasticity can be observed developing, and the phenomenon is then known as *retarded high elasticity*. At higher temperatures the thermal energy is greater and the entropy elasticity develops effectively simultaneously with the stressing, and it is then known as *instantaneous high elasticity*.

The relative importance of the various deformation modes changes with the temperature, and it is convenient to distinguish four physical states of high polymers, although in practice the divisions are not always clear cut. Starting at the low temperature end, the states are:

(1) glassy state,
(2) retarded highly elastic state (leathery),
(3) instantaneous highly elastic state (rubbery),
(4) viscous state.

The corresponding mechanical characteristics are shown in Fig. 8.10.

In the glassy state, there is ideally only normal elasticity, that is, there is an immediate elastic strain on the application of a stress, which then remains constant. Removal of the stress immediately restores the strain to zero. Young's modulus is about $10 \, \text{GN/m}^2$ (10^6 psi) for all polymers. In accord with the structure, this is below the moduli of metals (around $100 \, \text{GN/m}^2$) and fully covalent materials (diamond: $1000 \, \text{GN/m}^2$) but above that of a van der Waals

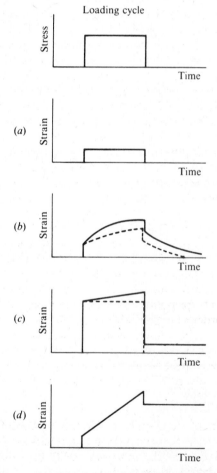

Fig. 8.10. Four physical states of a polymer characterised by its mechanical behaviour. (*a*) Glassy state; (*b*) leathery state (retarded high elasticity); (*c*) rubbery state (high elasticity); (*d*) viscous state. —— linear polymers, – – – cross-linked polymers.

solid (argon: 2–5 GN/m^2). In reality, polymers in the glassy state are not ideally elastic and some creep will be found, given sensitive gauges. This is very much less than occurs in the higher temperature states, where it is readily observed, but it is sufficient for it to be a factor in engineering design (see § 8.10).

About some temperature corresponding to the glass transition temperature discussed in the previous section, the glassy state gives way rapidly to the leathery state. Thermal agitation of the polymer chains is now sufficient to promote retarded high elasticity and, in non-cross-linked polymers, viscous flow. On stressing, the small instantaneous elastic strain is followed by a much larger time dependent elastic strain (primary creep) and viscous flow (secondary creep) (see Fig. 8.10(b)). Primary creep strain reaches an equilibrium value in time ranging from seconds to very many hours, whereas secondary creep continues indefinitely at a constant rate. On removing the stress, the original instantaneous strain is recovered at once and the primary creep strain is recovered over the same time period as was required for its development. The viscous flow strain is not recovered. The apparent elastic modulus in the leathery state becomes a function of the strain and time (see § 8.10).

About 30 °C above the glass transition the leathery state merges into the rubbery state. The activation energy for movement of the chain segments is now small compared with the thermal energy kT and the entropy strain develops simultaneously with the internal energy strain (Fig. 8.10(c)). Viscous flow in non-cross-linked polymers will also occur at an increased rate. The elastic modulus for the combined elastic modes drops by a factor of 10^3 to between 1 and 10 MN/m^2 (1.5×10^2–1.5×10^3 psi).

Finally, at still higher temperatures the viscous flow rate increases to the point at which it dominates the other deformation modes. This is the viscous state. The temperature at which this occurs is raised due to branched chains, crystallinity and cross-linking, and many cross-linked polymers will degrade before viscous flow takes place.

Only the glass transition is well marked, the other divisions being somewhat indeterminate due to overlapping of the deformation modes. Plastics are normally used at temperatures which put them into the glassy or leathery state; elastomers are used in the rubbery state. Many tests have been devised to determine the temperature at which a plastic flows, softens or distorts. Each test gives a single temperature point with no fundamental physical significance. For a given test an order of magnitude comparison of different plastics can be made and production quality controlled. They will not be detailed here. Data required for engineering design which involves a series of tests over a range of temperatures and stresses will be discussed later (§ 8.10).

8.7 Modification of polymers

There is a wide range of techniques, in addition to controlling the degree of crystallisation, whereby a polymer may be modified to obtain special properties such as increased strength, improved ability to be fabricated by plastic flow processes, colouration, or reduced cost.

The strength of a polymer is usually increased by promoting cross-links between individual polymer molecules. Cross-linking raises the glass transition and various flow temperatures of a polymer by making it more difficult or impossible, if the density of cross-links is high enough, for the molecules to slide past one another. Cross-links may be induced by the addition of active chemicals such as oxygen or sulphur. The vulcanisation of rubber by the addition of sulphur has already been mentioned (§ 8.4). Another method, used experimentally, is irradiation with high voltage electrons, X-rays, or nuclear particles. Bonds in side groups are broken, and when they reform due to the proximity of other molecules, there is a large probability of bonding on to another molecule thus forming a cross-link. Excessive irradiation however, will cause degradation by breaking links in the main chain.

The addition of a 'plasticiser' is made to lower the glass transition temperature, thus increasing the flexibility and toughness of a plastics at ambient temperature; it also reduces the temperature required for fabrication. Plasticisers are low vapour pressure liquids, with molecular weights in the range of 100–1000, which form a highly concentrated solution with the polymer. The polymer chains in the amorphous regions are forced apart by the plasticiser, allowing them to slip more readily over one another, thus reducing the viscosity. Up to 35 weight per cent of plasticiser may be added, although in some systems there may be a solubility limit which should not be exceeded. Plasticisers lower the mechanical strength of a polymer and care must be taken that this is not sacrificed for improved toughness and formability.

Other additions to polymers are called 'fillers'. Fillers are inert materials which may be added as pigments to give colouration, or as hard material to improve strength and abrasion resistance (e.g. glass fibre in nylon). Cheap filler may be added, especially to thermosets, to reduce the volume of an expensive polymer required. Fillers are also added to polymers used as adhesives so as to match their coefficient of thermal expansion to that of the materials to be joined.

The properties of a polymer may also be modified by the introduction of particles of a second polymer phase. This may be introduced as a mechanical mixture known as *polyblend*, or chemically combined as a special type of copolymer. Blending is achieved by milling together the previously polymerised individual constituents and heating to above the softening temperature of the major constituent. On cooling, a matrix is formed in which the minor constituent is dispersed. An example is styrene–butadiene dispersed in a styrene matrix. In the pure form, polystyrene is cheap and strong, but because its glass transition is around 100 °C it fails in a brittle manner at ambient temperatures, especially under impact. By introducing small particles of styrene–butadiene copolymer, which is a rubber (SBR, see § 8.9), crack propagation is made more difficult and the composite material has increased fracture toughness.

The two-phase copolymer is produced by simultaneous polymerisation of two constituents and there is now a chemical bond between the phases. Copolymers are ordinarily random, that is, each monomer is randomly distributed along the chain. Polymerisation techniques are available which produce *block copolymers* and *graft copolymers*. In block copolymers, the main chain contains a long sequence of one monomer followed by a long sequence of the other monomer. In graft copolymers, side chains composed entirely of one monomer are grafted onto a backbone of the other monomer. If the physical properties of the constituent polymers are very different, then on solidification they separate into two phases. The chemical bond between the two phases in a copolymer renders its mechanical properties superior to those of a polyblend of the same constituents.

For example, in high impact polystyrene (HIPS), styrene and about 5 per cent of styrene–butadiene random copolymer are polymerised together. This results in the formation of three polymers (a terpolymer): homogeneous polystyrene, homogeneous styrene–butadiene (SBR) and a graft copolymer consisting of SBR main chains with styrene side chains. The resulting microstructure, see Fig. 8.11, consists of a matrix of polystyrene in which large 'rubber' particles are dispersed. Although referred to as rubber particles, they really consist largely of polystyrene sub-particles held together by a thin continuous film of rubber. Any advancing crack which encounters a large particle has its energy dissipated in the tortuous paths of SBR between the sub-particles. This is very much more effective than the homogeneous particles of rubber found in the mechanical polyblend described above. The fracture toughness is an order of magnitude higher than unmodified polystyrene.

ABS (acrylonitrile butadiene styrene) is another two phase polymer which is obtained by grafting random copolymers of styrene–acrylonitrile (known as SAN) on to main chains of acrylonitrile–butadiene random copolymer (nitrile rubber). The microstructure consists of particles of nitrile rubber, 20–35 per cent by volume, in a matrix of SAN. It is finer than that of HIPS, and the properties

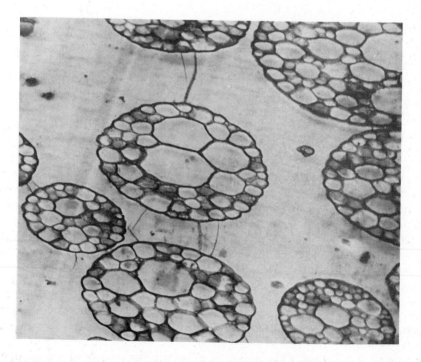

Fig. 8.11. Two-phase polymer: high impact polystyrene. Continuous phase of polystyrene contains spheroids composed of many small polystyrene particles bonded by styrene–butadiene rubber ($\times 11\,000$). (Courtesy of R. J. Seward.)

are correspondingly improved. ABS has a very desirable combination of chemical, electrical, mechanical and thermal properties, is easy to process and moderately priced. It is widely used in automobiles (interior trim), household appliances, pipes, packaging, telephones and other electrical devices.

As the properties of two phase polymers can, to a large extent, be tailored by the appropriate choice of constituents and microstructure, further developments are to be expected.

8.8 Wood and cellulose

An important constructional material and also a raw material for industry, such as papermaking and plastics, is wood. The annual world consumption, excluding firewood, is some 1000 million tons compared with 400 million tons for steel.

Dry wood consists of 50–60 per cent of a linear polymer called *cellulose*, about 25 per cent of related chemicals (*hemicelluloses*) and some 25 per cent of highly viscous fluid called *lignin*. Fresh wood also contains a lot of water, up to 100 per cent of the dry weight. The same polymer occurs in all plants, but in slightly different arrangements, that is, the super-molecular structure varies; in cotton fibre, the purest natural form of cellulose, the proportion of cellulose is 95 per cent. The molecule consists of a chain of up to 5000 glucose rings, linked by oxygen atoms:

The hydroxyl groups and hydrogen atoms are positioned above and below the plane of the glucose rings. In a tree the glucose monomer is first synthesised in the leaves from CO_2 and H_2O by the action of sunlight in the presence of chlorophyll. It then moves in the sap water to the growing cells just below the bark, where polycondensation of two hydroxyl groups leads to the formation of the oxygen linkage and the elimination of water. It is interesting to note that this oxygen bond provides a weak link in the chain, chemically, and it is attacked by enzymes in the stomachs of animals and by bleaches in the hands of a laundry.

The polar nature of the hydroxyl groups is responsible for strong attraction between cellulose molecules and the formation of small fibres (*fibrils*) with up to 100 molecules and forty per cent crystallinity. The hydroxyl groups in the amorphous regions strongly attract water molecules, which force the chains apart, an important feature of timber. The crystallites prevent the chains going into solution.

The microstructure of wood consists of a large number of long, closed, tubular cells (*tracheids*), see Fig. 8.12(*a*); their hollow centres (*lumens*) are partly full of water or sap, see Fig. 8.12. (The name cellulose is a combination of *cell*, and

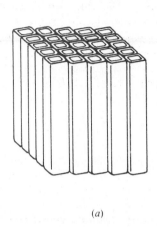

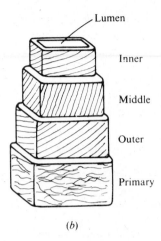

(a)   (b)

Fig. 8.12. Structure of wood: (a) assembly of cells (tracheids) up to 1 mm long by 10 μm across; (b) layers in cells walls, showing differing orientation of cellulose fibrils.

ose which is the chemical termination of sugars.) The fibrils in the cell walls are lying approximately helically, although the details of the arrangement are complex and vary with the type of wood or plant. Distinct layers with different pitch angles of the fibrils can often be identified within the wall, see Fig. 8.12(b). The wall is impregnated with up to 30 per cent of lignin. During the period of rapid growth, the cell walls are thin and the central passage large. As the growth slows down during the summer the cell walls are thicker and the core smaller. This is responsible for the characteristic rings and grain in wood.

Differences in the strength and stiffness of various woods can be traced to the variation of their densities, which will depend on the ratio of the wall thickness to core diameter. The specific gravity of pure cellulose is 1.5 and that of wood varies between 1.3 for lignum vitae, the densest and hardest wood, and 0.15 for balsa. Since the tensile strength in the grain direction is due to the cellulose chains, it is approximately proportional to the dry density. The tensile strength/weight ratio has a value of about 2.5×10^4 m (10^6 inch) which compares favourably with the equivalent figure for mild steel of 6×10^3 m (2.5×10^5 inch). It is apparent that the tensile strength of the cellulose polymer is excellent. Unfortunately, apart from the many natural faults which are a major concern in the selection of timber, the compressive strength is weaker, in contrast with most materials. This is because the adhesion between cells is low, around 10^7 N/m^2 (1000 psi) shear strength, and under compressive loads the tubular cells are able to buckle individually, rather than as a coherent assembly, see Fig. 8.13. The compressive strength/weight ratio is about 7.5×10^3 m (3×10^5 inch). The modulus of rupture (or maximum stress in bending) is correspondingly lower than the tensile strength. The strength of some typical woods is given in Table 8.4.

When a tree is felled the wood contains a large amount of water, as much as 100 per cent of the dry weight. Most of this is the hollow core of the cell, or lumen, but about 25 per cent is adsorbed on the hydroxyl groups in the cell wall. After cutting, the water evaporates slowly until equilibrium with the surrounding

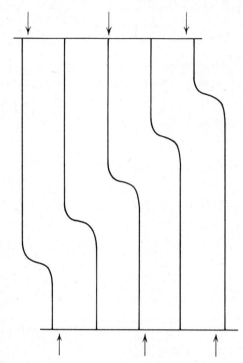

Fig. 8.13. Compression failure in wood due to buckling of the individual cells.

air is reached, a process called *seasoning*. The equilibrium value will depend on the relative humidity of the air, ranging from around 5 to 23 per cent (of the dry weight) in dry and moist air respectively. Accompanying the drop in water content is a shrinkage perpendicular to the cells or cellulose chains; there is little longitudinal change. This occurs only in the final stage as the water adsorbed in the cell walls is decreased. A one per cent change of adsorbed water corresponds to about one half a per cent change in dimension lateral to the grain,

Table 8.4. *Mechanical properties of wood*

Species	Water content (per cent)	Specific gravity (SG)	Young's modulus (GN/m^2)	Elastic limit (MN/m^2)	Modulus of rupture (MR) (MN/m^2)	Compressive strength (MN/m^2)	MR/SG
Ebony	12	1.01	17.7	98	189	92	185
Hickory	9.6	0.73	15.1	86	154	72	210
Oak (European)	12	0.69	10.1	56	97	52	140
Beech	13.1	0.62	12.6	15	102	44.5	165
Ash (white)	10.8	0.58	12.3	68	108	52.5	185
Fir (Douglas)	12	0.50	10.5	50	91	48	180
Spruce (Sitka)	12	0.38	8.1	40	67	36	175
Pine (Yellow)	12	0.35	5.5	40	53	30	150
Balsa	15	0.15	4.0	14	25	10	165

so a maximum strain of over ten per cent is possible. It is advisable for this to have taken place before the wood is used! Nevertheless the humidity of the air is variable and wooden components are dimensionally unstable. Natural seasoning with the wood in an open stack takes one or two years, depending on the size. Attempts to hurry the process have to be treated cautiously as too large a gradient of moisture content will lead to warping and cracking.

Apart from dimensional instability, the water content also affects the strength. There is approximately a factor of three increase of strength between freshly cut timber and the dry state.

8.9 Rubbers (elastomers) and high elasticity

An important group of high polymers are the rubbers. They may be formed by nature or synthesised in a chemical plant; the latter are nowadays called *elastomers*. The chief characteristics of this class of plastics material are the very large elastic strain and low elastic modulus. In most crystalline materials the elastic range is limited to about 0.1 per cent extension. Most polymeric materials extend to about 1.0 per cent, but the elastomers can achieve extensions of up to 1000 per cent. As mentioned briefly in § 8.6, this exceptional elasticity is due to rotation about the bonds of the polymer chain, and is called entropy, or high, elasticity.

A typical stress–strain curve for a rubber in tension is shown in Fig. 8.14. The following features of the mechanical properties should be noted:

(1) The large elastic range under tension (500–1000 per cent).

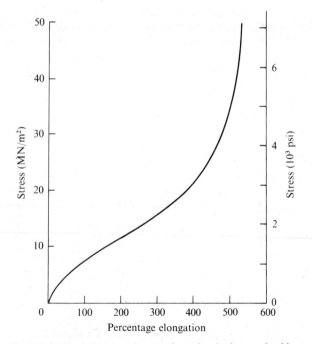

Fig. 8.14. Typical stress–strain curve for vulcanised natural rubber.

(2) The low elastic modulus (~ 10 MN/m^2, 1.5×10^3 psi). The material is non-Hookean and the stress–strain curve is S-shaped. At high stress the modulus increases rapidly to a value typical of normal polymers (~ 10 GN/m^2, 1.5×10^6 psi).

(3) The strain is rapidly recovered on removing the applied stress with little hysteresis and permanent set. In practice, recovery is not entirely instantaneous or complete: at high frequency the effective modulus is increased and considerable conversion of work into internal heating occurs; under long duration loads creep takes place.

(4) The bulk modulus of rubber is high (~ 1 GN/m^2, 145×10^3 psi) and Poisson's ratio nearly 0.5. In other words rubber is virtually incompressible.

The original explanation of high elasticity was that the polymer chains were naturally coiled in the unstressed state and gradually uncoiled under an increasing stress. The internal energy was assumed to increase as the chains were straightened. The increase of modulus at higher strains occurs when the chains are straight, and further extension is due to the stretching of the interatomic and intermolecular bonds. This latter stage corresponds to the elastic process in normal solids.

Three observations however could not be explained by this theory:

(1) The elastic modulus in the initial, elastomeric, range increases with increasing temperature.

(2) An elastomer increases its temperature when stretched rapidly and cools when suddenly released. (This can be checked by placing a rubber band against the upper lip.)

(3) The coefficient of thermal expansion of an unstretched elastomer is positive but that of a stretched elastomer is negative.

All three effects are contrary to the behaviour of other elastic solids.

These facts led Meyer in 1935 to propose an explanation of elastomeric behaviour in terms of kinetic theory. The polymer chains are considered free to rotate about single bonds with no change of internal energy, contrary to the preceding theory. Variation of the polymer length with applied stress is then a consequence of entropy considerations.

Applying the First and Second Laws of Thermodynamics to a material a tensile load F gives:

$$F \, dL = dU - T \, dS,$$

where U is the internal energy, S the entropy and L the specimen length. Hence:

$$F = \left(\frac{\partial U}{\partial L}\right)_T - T\left(\frac{\partial S}{\partial L}\right)_T. \tag{8.1}$$

This shows that an elastic force will arise if the internal energy or the entropy of a body is changing with length. The first term is important in normal solids, the second in gases, and both terms in elastomers.

Consider now only the entropy elasticity

$$F_S = -T\left(\frac{\partial S}{\partial L}\right)_T. \tag{8.2}$$

Due to its thermal energy a polymer chain in the unstressed state can be likened to an angry eel twisting and curling rapidly. There is a constantly changing distance between the head and tail, but there will be a most probable length corresponding to maximum entropy (see § 2.9). As the chain length is increased the number of possible conformations decreases, that is, the entropy is decreased, and the entropy force F_S has a finite positive value.

Extensive theory has been developed which considers the statistics of chains in several stages of approximation to real elastomers: ideal freely jointed chains; chains with fixed bond angles; restricted rotation of bonds; bonds of finite volume; and cross-linked networks. It has been shown that the decrease of entropy of a cross-linked network as it is extended from its unstressed length L_0 to length L is given by

$$S_0 - S = \frac{1}{2} N_0 k \left[\left(\frac{L}{L_0} \right)^2 + 2\frac{L_0}{L} - 3 \right], \tag{8.3}$$

where N_0 is the total number of sub-chains, a sub-chain being the portion of a polymer between cross links. The cross-links act as anchors and only the sub-chains between them are free to assume any conformation.

Differentiating (8.3) with respect to L and substituting (8.2) gives

$$F_S = \frac{N_0 kT}{L_0} \left[\frac{L}{L_0} - \left(\frac{L_0}{L} \right)^2 \right].$$

Converting to true tensile stress by dividing by the cross-sectional area yields

$$\sigma = nkT \left[\left(\frac{L}{L_0} \right)^2 - \left(\frac{L_0}{L} \right) \right], \tag{8.4}$$

where n is the number of sub-chains per unit volume (N_0/AL).

The isothermal modulus of elasticity is defined as

$$\left(\frac{\partial \sigma}{\partial e} \right)_T = L \left(\frac{\partial \sigma}{\partial L} \right)_T$$

and is obtained by differentiating (8.4):

$$E = L \left(\frac{\partial \sigma}{\partial L} \right)_T = nkT \left[\frac{L_0}{L} + 2 \left(\frac{L}{L_0} \right)^2 \right], \tag{8.5(a)}$$

which at zero stress, reduces to

$$E = 3nkT. \tag{8.5(b)}$$

Thus the initial elastic modulus is shown to be proportional to the absolute temperature, in agreement with experiment. At very large strains the chains are completely straight and no further contribution to the elasticity from entropy consideration is possible. All strain is then due to change of internal energy by stretching of the bonds.

The elastic modulus is also shown in (8.5) to be proportional to the number of segments, or cross-links, per unit volume. However, without any cross-links the chains flow past each other under stress and although entropy elasticity is a maximum it is accompanied by large non-recoverable deformation. This occurs, for example, in natural rubber before vulcanisation (latex). In the process of vulcanisation, some cross-links are formed by reacting the natural rubber with, for example, about 2 per cent of sulphur, as described in § 8.4, eliminating viscous flow. Increasing the number of cross-links progressively raises the modulus by inhibiting entropy elasticity. Vulcanisation with 25–30 per cent sulphur increases the modulus of high elasticity to the point at which a hard and glassy material called *vulcanite* is formed.

Whilst all linear polymers show some degree of high elasticity over a temperature range (see § 8.6), certain polymers are exceptional in this respect. The requirement is clearly for rotation about single bonds to be subjected to minimum constraint, in other words, the energy barriers opposing rotation must be small compared with the available thermal energy. Thus the glass transition temperature, at which segment mobility commences, must be at least $100\,^{\circ}C$ below the operating temperature. The value of T_G is low in the absence of bulky and highly polar side groups or chains. At the same time crystallisation must not occur, which effectively raises the number of cross-links. This is particularly liable to occur at high stress and strain when the polymer chains become aligned. A *cis*-configuration, giving an asymmetric structure, as found in natural rubber, helps to prevent this. Finally the presence of double bonds makes for ease of rotation of adjacent single bonds since the atoms at each end of the double bond have less side atoms. It also allows cross-linking to take place.

Natural rubber is a high molecular weight *cis-polyisoprene* (as mentioned in § 8.4):

$$\begin{array}{c} CH_3 \\ \diagdown \\ C{=}CH \\ \diagup \qquad \diagdown \\ {+}CH_2 \qquad\quad CH_2{+}_n \end{array}$$

It has a number average molecular weight of $\sim 5 \times 10^5$ with a broad distribution. Though formed in many plants, the major source is the tree *Hevea brasiliensis*: hence the alternative name of *hevea rubber*. As its name indicates, the tree originated in Brazil, but it is now cultivated extensively in Malaysia, whence most of the world's natural rubber is obtained. The rubber exists in the form of *latex*, an aqueous dispersion of the polymer, stabilised by some proteins and fatty acids. After tapping from a cut in the bark of the tree, the rubber is coagulated by adding acetic acid, rolled into sheets and dried. This is followed by severe mechanical working, called mastication, which breaks down the molecular chains, and facilitates subsequent compounding with vulcaniser, (Goodyear, in 1839), carbon black, inert fillers, plasticisers, antioxidants and dyes. These improve the ease of processing and the properties of the final product; they form an important part of rubber technology but will not be considered here.

Early attempts to synthesise polyisoprene gave a mixture of *cis* and *trans*-forms (see § 8.4), which was not satisfactory. In 1955 the use of Ziegler catalyst led to the synthesis of entirely *cis*-polyisoprene which is essentially identical to natural rubber.

The first rubber to be synthesised was polybutadiene, in Germany early in this century:

$$+CH_2{-}CH{=}CH{-}CH_2{+}_n$$

This differs from polyisoprene in that it has no CH_3 side group. Alkali metals were used as a catalyst and gave a mixture of *cis* and *trans*-configurations but the properties were adequate. It was called Buna rubber. Recently the use of stereo specific catalysts of the Ziegler type has enabled pure *cis*-polybutadiene to be polymerised with even greater elasticity than natural rubber, due to the absence of the methyl group.

Prior to the Second World War, a copolymer of butadiene and styrene:

$$+C\overset{\displaystyle H}{\underset{\displaystyle H}{\mid}}\!\!\underset{}{\diagdown}\!\!C+$$

was produced, known now as SBR (formerly Buna S in Germany and GR-S in America). A proportion of three butadiene to one styrene is usual. The presence of styrene give a much tougher and stronger rubber and the phenyl side groups (C_6H_5) scattered randomly along the copolymer chain oppose a tendency to crystallise at high stresses. SBR is at least equal to natural rubber in many applications, such as light duty tyres (the heat generation due to hysteresis is higher but the wear resistance better), and superior for many moulded applications. It accounts for much the largest part of the world production of synthetic elastomer.

Nitrile rubbers (NBR; Buna N) are copolymers of butadiene and acrylonitrile

$$+\underset{\displaystyle H}{\overset{\displaystyle H}{\mid}}C-\underset{\displaystyle H}{\overset{\displaystyle CN}{\mid}}C+$$

Like SBR which they resemble in most respects they are random copolymers and the proportion is typically 2–3 parts butadiene to 1 part acrylonitrile. The substitution of the polar nitrile group CN results in greatly improved resistance to oil and other organic liquids without greatly increasing the intermolecular forces which would reduce elasticity. The nitrile rubbers are used for containers, pipes and gaskets for handling such liquids.

Butyl rubbers are copolymers of isobutylene:

$$+\underset{\displaystyle H}{\overset{\displaystyle H}{\mid}}C-\underset{\displaystyle CH_3}{\overset{\displaystyle CH_3}{\mid}}C+$$

with a small amount of isoprene. Polyisobutylene itself is saturated (no double bonds) but otherwise shows elastomeric behaviour. The isoprene is added to allow cross-linking and suppression of the viscous flow. The butyl rubbers are very impermeable to air as compared with natural rubber and are therefore used for tyre inner tubes. They resist attack by chemicals, including ozone. They are also used in high grip outer covers for tyres on account of the high hysteresis at ambient temperatures due to some hardening of the bond rotation. This disappears at about 60 °C. (One source of friction is the energy lost as asperities in the road deform the rubber.)

The neoprene rubbers are polychloroprene:

$$\underset{\displaystyle +CH_2 \qquad\quad CH_2\!\!\!+_{\!n}}{\overset{\displaystyle Cl}{\underset{\diagup}{\overset{\diagdown}{C=CH}}}}$$

They are similar to polyisoprene except that the methyl group has been replaced by a chlorine atom. They have good resistance to ageing and oil, especially at high temperatures; at low temperatures they lose their elasticity.

Finally in this short survey of the principal rubbers or elastomers, the silicone rubbers will be mentioned. The silicones (polysilicones, polysiloxanes) are high polymers with the structure

$$+\!\!\underset{\displaystyle Y}{\overset{\displaystyle X}{\underset{|}{\overset{|}{Si}}}}\!\!-\!\!O\!+_{\!n}$$

where X and Y represent atoms such as hydrogen, or groups such as methyl or phenyl. They may also be another siloxane group ($-\overset{|}{\underset{|}{Si}}-O-$) when a cross-linked or network polymer is formed. Of the many silicone elastomers the commonest, polydimethyl siloxane, has $X = Y = CH_3$, with n in the range 5–10 000. Cross-linking is obtained by adding a suitable initiator to the gum, which reacts to two methyl groups with the elimination of H_2. Alternatively some unsaturated side groups can be included in the chain which are subsequently cross linked. Silica dust is included to improve the strength. Silicone elastomers have an exceptionally large temperature range, from about $-80\,°C$ to $+250\,°C$.

The structure and properties of various commercial rubbers are given in Tables 8.5 and 8.6.

8.10 Mechanical properties of thermoplastics

The glass transition of thermoplastics is never far removed from room temperature and plastics are usually used either just in the glassy state or just in the leathery state. In either event it is necessary to consider the three components of strain: instantaneous elastic (internal energy), delayed elastic (entropy) and viscous. Because time is always an important factor in their mechanical behaviour thermoplastics are said to be visco-elastic. The term must not be taken to mean that the behaviour is a simple summation of the behaviour of a Newtonian viscous fluid (stress $\propto$ strain rate) and of a Hookean elastic solid (stress $\propto$ strain).

Although the importance of time and temperature to mechanical behaviour has been recognised for some time, there are no well established methods of testing and data presentation, nor design procedures. It must be remembered that in the past the majority of designs in plastics, such as haircombs and domestic buckets, have been based on rule of thumb procedures concerned with the

Table 8.5. *Common elastomeric polymers (rubbers). The useful temperature range refers to the compounded and vulcanised rubber; the glass transition to the pure polymer.*

Chemical name	Repeat unit(s)	Useful temperature range (glass transition) (deg)
(*Cis*-)polyisoprene (natural or synthetic rubber)	CH_3 $C=CH$ $-CH_2$ CH_2-	$-50, +120 (-75)$
(*Cis*-)polybutadiene (Buna)	$CH=CH$ $-CH_2$ CH_2-	(-108)
Styrene–butadiene copolymer (SBR, Buna S, GRS)	C_6H_5 $-CH_2-CH-$ (33%) and $CH=CH$ $-CH_2$ CH_2- (67%)	$-50, +140$
Nitrile–butadiene copolymer (nitrile rubber NBR)	CN $-CH_2-CH-$ (25%) and $CH=CH$ $-CH_2$ CH_2- (75%)	$-35, +175$
Polydimethylsiloxane (silicone rubber)	CH_3 $-Si-O-$ CH_3	$-70, +275$ (-123)
Butylene–isoprene copolymer (butyl rubber)	CH_3 $-CH_2-C-$ CH_3 and CH_3 $C=CH$ $-CH_2$ CH_2- (trace)	$-50, +150$
Polychloroprene (neoprene, CR)	Cl $C=CH$ $-CH_2$ CH_2-	$-35, +130$
Polysulphide (Thiokol)	S $\parallel$ $-CH_2-S-S-CH_2-$ $\parallel$ S	$-40, +120$
Vinylidene fluoride– perfluoropropylene copolymer (fluoro rubber, 'Viton')	$H\ F$ $-C-C-$ $H\ F$ (70%) and $F\ CF_3$ $-C-C-$ $F\ F$ (30%)	

manufacturing methods, economics and aesthetics. This was quite sufficient. There is now a desire to use plastics for more critical components where behaviour under load in service must be accurately predicted without too much trial and error. For this, rational design procedures are necessary, but so far they have proved difficult to evolve. It is to be hoped that the subject will crystallise in the near future.

Table 8.6. *Qualitative comparison of the properties of various elastomers (rubbers). (Based on F. T. Bowden: Chartered Mechanical Engineer, July 1964, by permission of the Council of the Institution of Mechanical Engineers.)*

Type of rubber	Natural rubber (cis-poly-isoprene)	Buna (poly-butadiene)	SBR (styrene butadiene)	Nitrile rubbers (acrilo-nitrile butadiene)	Silicone rubbers	Butyl rubber (isoprene-isobuty-lene)	Neoprene (chloro-prene)	Polysul-phide rubbers	Fluoro rubbers	Polyur-ethane	EPR ethylene-propylene
Elasticity	α	α+	α−	β	γ+	?	β	?	γ+	β	β
Tensile strength	α	β	α−	β	?	β	α−	?	β−	α+	α−
Tear strength: cold	α	α+	β−	β	β	α	α	α	β−	α+	α
hot	α	α+	β	?	?	β	α	?	β	α+	α+
Compression set	α−	α	α−	β+	β	β	α	?	β+	α	α
Atmospheric ageing	β−	β−	β	α−	α	α−	α−	α+	α+	α	α+
Sunlight resistance	?	β−	β−	β−	α	β+	α−	α	α+	α	α
Ozone resistance	β	β	?	β	α	?	α−	α	α+	α+	α
Flame resistance	?	?	?	β	?	?	β	?	α	γ	γ
Oil resistance: mineral	?	?	?	β+	α	?	α+	α+	α+	β+	γ
anim. and veg.	β−	β−	β−	α	β	α	β	α	α	α−	γ
Acid resistance: dilute	α	α	α−	α	α	α+	α+	α	α+	β−	α
concentrated	β−	β	β−	β−	β	β+	β+	?	α+	β−	β
Water resistance	α	α	β+	β	β	α−	β	β	α	γ	α
Impermeability to gases	β	β−	β	α−	β	α+	β+	α−	α+	β+	β−
Temperature resistance: lower limit	α	α	α−	β	α+	α	β	α−	β	α	α−
upper limit	β−	β−	β	β+	α+	β+	β+	?	α	β−	α−

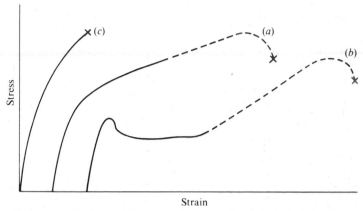

Fig. 8.15. Three types of yield curve found in polymers. (*a*) Gradual yield (e.g. polyethylene, nylon-6,6). (*b*) Sharp yield point (e.g. nylon-6). (*c*) Brittle (e.g. polymethylmethacrylate). (Note: broken curves indicate reduced strain scale.)

Three possible forms of the tensile stress–strain curve are shown in Fig. 8.15. A plastic may yield gradually as in (*a*), or suddenly as in (*b*) (called cold drawing), or fracture without appreciable flow as in (*c*). The same plastic can change the type of curve with variation in the strain rate, temperature or moisture content. In all three curves there is little linear portion to the curve and the definition of the tensile elastic modulus and proportional limit need to be reconsidered. Various methods of specifying the modulus are used. The initial modulus is taken as the slope of the tangent to the stress–strain curve at the origin; this is rather difficult to determine unequivocally. The secant modulus is defined as the slope of a line from the origin to some point on the curve, either at 1.0 or 1.5 per cent strain. Finally, the tangent modulus may be determined from the slope of the yield curve at some particular strain. The various alternatives are illustrated in Fig. 8.16. The secant modulus is currently favoured but it should be noted that it is sensitive to strain rate and temperature. The very low value (1 GN/m², 145 × 10³ psi) as compared with metals (∼100 GN/m², 15 × 10⁶

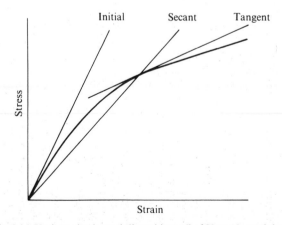

Fig. 8.16. Various elastic moduli used instead of Young's modulus.

psi) can present problems in design on account of the unacceptable deflections and errors introduced into the normal stress analysis theory.

In place of the proportional limit, a modulus accuracy limit (here referred to as MAL) has been defined as the strain at which the secant modulus has decreased to 0.85 times the initial modulus. The strain at the MAL varies for different plastics between 0.75 and 2.0 per cent and is independent of the temperature and strain rate used in the tensile test, although the corresponding stress changes. It is claimed that, if the initial strains on loading are within the MAL, obtained by putting the initial modulus into the normal equations of elastic theory, the strains and deflections at any later time (or stresses and loads if there is relaxation type of loading) can be estimated simply by inserting an 'apparent modulus' into the equations. This will be defined later in this section. Beyond the MAL the secant modulus decreases rapidly: for example, nylon (2.5 per cent water) has the MAL at 0.85 per cent strain and by 2 per cent strain the secant modulus has dropped to 55 per cent of the initial modulus.

The yield stress of plastics is defined (ASTM D638) as the lowest stress at which the stress–strain curve has zero slope. Thus it may refer to a yield point in sharply yielding plastics (curve (*b*) in Fig. 8.15) or to the tensile strength if yielding is gradual (curve (*a*)). An alternative definition of the yield, used in theoretical analysis, is the stress at which the strain-hardening rate becomes constant, as shown by the linear portion in a plot of true stress against logarithmic strain. Yield is then normally at 10 to 20 per cent strain. Variation with temperature of the tensile strength and initial modulus of nylon-6,6 is shown in Fig. 8.17 and data on several other plastics in Table 8.3.

For rational design, data on creep at constant load, or relaxation at constant strain, are absolutely essential. As has been emphasised for metals in the creep

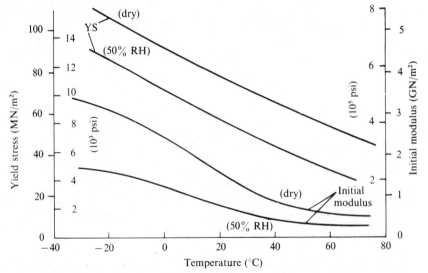

Fig. 8.17. The variation of initial modulus and tensile strength with temperature of nylon-6,6, dry and in equilibrium at 50 per cent relative humidity. (Note: at 50 per cent relative humidity the water content of nylon is 2.5 per cent by weight. The yield curve of nylon is of type (*a*) (Fig. 8.15) and hence the tensile strength and yield strength (ASTM D638) are coincident.)

range (§ 5.10), the stress–strain curve obtained in a normal tensile test gives little indication of the long term response to stress or strain. With many plastics creep occurs even at room temperature, that is cold flow. The creep curves of polymers have the same general form as metals, though the comparison must not be carried far owing to the profound differences in their microstructures. There is an initial sudden extension merging into a decreasing strain rate (primary creep), to be followed by steady state creep (secondary) and possibly accelerating creep (tertiary) and stress rupture. Separation of the initial strain can be done by taking the initial modulus at very low stresses and the same temperature and multiplying by the stress for the creep test. It is often not clear from quoted results whether this procedure has been followed or whether the total strain is being plotted; the latter is the normal practice for plastics and should be assumed where there is ambiguity. Figure 8.18 shows the creep strain of dry nylon-6,6 at 20 °C, typical of creep curves for plastics. It is apparent that creep is an important factor at stresses very far below the tensile strength (80 MN/m², 12×10^3 psi). Unlike metals the primary creep is mainly due to entropy elasticity and is recovered over a period when the load is removed. This can be important where the load is applied intermittently. Some typical creep recovery curves are given in Fig. 8.19, for polypropylene; about 75 per cent of the creep strain reached after creep tests lasting 10^5 seconds was recovered in about the same time scale.

There are three other methods of plotting the data from constant load creep tests besides the direct one of strain versus time (or log time) and they are more convenient for designers. The data can be presented as isochronous (constant time) curves of stress versus strain or as isometric (constant strain)

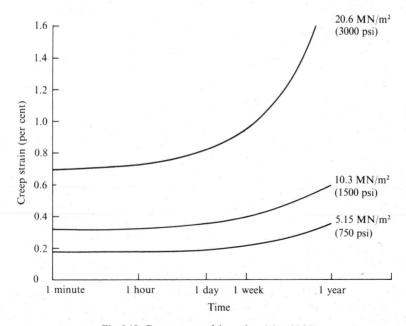

Fig. 8.18. Creep curves of dry nylon-6,6 at 20 °C.

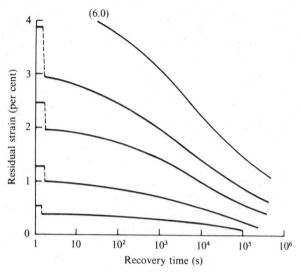

Fig. 8.19. Recovery creep of polypropylene at 60 °C following creep tests for 10^5 s at various stresses. The strain just prior to unloading is indicated. (Data from S. Turner.)

curves of stress versus time (or log time). The method of deriving them from the creep curve is illustrated in Fig. 8.20, which is based on actual data for poly-propylene. The third method, widely used, is to plot isometric curves using an ordinate scale of stress divided by strain, called the *creep modulus*, and an ab-scissa of log(time). It will be seen (Fig. 8.20(d)) that the creep modulus increases with decreasing strain level. This may be confusing until it is recalled that to obtain the creep stress the modulus has to be multiplied by the strain.

With some amorphous plastics the creep modulus curves at the lower strains get closer to each other and approach an asymptote. Following Baer (Baer, Knox, Linton and Mair, 1960) this will be called the *apparent modulus*. In the literature a single curve for creep modulus versus log(time) is sometimes found with no indication of the strain and this probably refers to the limiting isometric curve at low strains. However, it can be misleading since the creep modulus as generally understood is a function of the strain value, *inter alia*. The use of the term apparent modulus would avoid this ambiguity. (Some authors are guilty of using this term for the secant modulus which is clearly a function of strain since polymer elastic range is non-linear.) It has been proposed by Baer that for design purposes the apparent modulus curve can be taken as the creep modulus as long as the initial stresses or strains on loading are below the MAL, in other words the creep strain at a given time is proportional to the applied stress. This can only be an approximation, although adequate for design. (At short times the creep modulus is equivalent to the secant modulus at the same strain; at the MAL this is 15 per cent less than the initial modulus, the equivalent of the apparent modulus.) It is also claimed by Baer that the apparent modulus curve is the same whether derived from constant load creep data, as considered here, or from constant deformation relaxation data. That is, there is a mechanical equation of state (see § 5.10). Apparent modulus data on several plastics are given in Fig. 8.21.

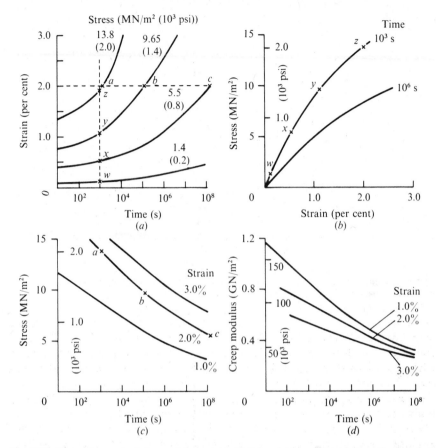

Fig. 8.20. Alternative presentations of creep data. (*b*), (*c*) and (*d*) are derived from the data in (*a*). (*a*, Constant stress creep curves; (*b*) isochronous stress versus strain; (*c*) isometric stress versus time; (*d*) isometric creep modulus versus time. (From ICI data on polypropylene, 'Propathene', at 20 °C.)

Materials in which the instantaneous and time dependent strains are proportional to the stress at any time are called *linear visco-elastic*. Over the last fifty years an extensive theory of linear visco-elasticity has been developed by mathematicians, paralleling elastic and plastic theory. Models are postulated consisting of elastic springs and viscous dashpots. By adding more and more components closer approximations to actual materials are achieved. Attempts to incorporate non-linearity into the formal theory have also been made. So far the theory has not been applied to engineering design but it is the only formal and rigorous theory available which might be used to describe the mechanical behaviour of some linear polymers. It will not be treated here.

A simple approximate method of designing for plastics has been developed by Baer *et al.* (1960 paper), known as the du Pont procedure. The strains on loading are calculated from the normal equations of elastic theory using the initial modulus. The maximum strain should be within the MAL, defined above. Deflections, or loads, at any subsequent time are then obtained by using the

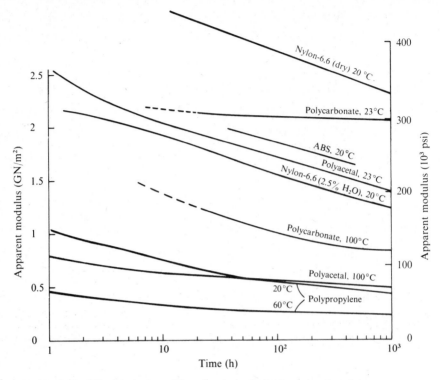

Fig. 8.21. Apparent modulus of various plastics (stress/total strain).

apparent modulus at that time. Creep will be entirely recoverable on unloading. Poisson's ratio is taken as independent of time (0.4). Figure 8.22 shows a good correlation between theory and experiment obtained with a beam in three point loading. Nevertheless, the rough and ready nature of this method must be appreciated. There appears to be little data so far to support the basic hypothesis that there is a finite strain range over which the creep modulus is effectively constant (and equal to the limiting value at zero strain, the apparent modulus). The above agreement may be fortuitous due to taking the creep modulus at 0.5 per cent strain as the apparent modulus for insertion in the equations. The restriction of the method to very low strains ensures that stress rupture is avoided but at the expense of rather understating the capabilities of plastics. Possible improvements to this pseudo-elastic approach have been suggested by Ratcliffe and Turner (1966 paper) with the object of extending the maximum allowable strain.

Extrapolation of creep data on thermoplastics to longer times (but by not more than one decade) is done by plotting log(strain) against log(time), when the curves are usually almost linear. An alternative method for amorphous polymers above their glass transition is to make use of the *time–temperature superposition* principle (associated with the name of Ferry). It was found experimentally and has now been given theoretical justification that, if the data from a series of creep tests at one stress but at various temperatures are plotted as strain versus log(time), the curves have identical shape but are displaced along the time axis. In other words, they can be made to coincide by shifting

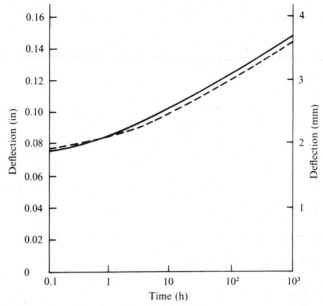

Fig. 8.22. Theoretical and experimental deflections of a polyformaldehyde (Delrin) beam in three-point loading, 22.8 °C. Theory uses the du Pont design method. —— actual experimental curve, − − − predicted curve. (From E. Baer, J. R. Knox, T. J. Linton and R. E. Maier, *S.P.E. Journal*, **16** (no. 4, April 1960).)

them along the time axis. Results at a given stress obtained over a limited period at elevated temperature (say, 20 degree increase) can thus be used to predict results for the same stress over a longer period at a lower temperature. The shift factor can be found by curve fitting if some short-duration test results are available at the lower working temperature, or on theoretical grounds. The method, which also applies to relaxation data, is not satisfactory with crystalline plastics or with amorphous plastics below their glass transition temperature.

To interpolate creep curves at stresses intermediate to experimental results, the experimental points should first be plotted on a log(strain)–log(time) basis. The position of an intermediate creep curve can be obtained by constructing two isochronous stress–strain curves from the data (at, say, 100 s and 100 h) and transferring the strain values at 100 s and 100 h for the required stress level back to the log(strain)–log(time) graph. The creep curve can then be drawn through them, parallel to the existing curves.

8.11 Other polymer properties

The physical properties of high polymers follow from their predominantly covalent binding within the chains. The polymer structure is very important, whereas the molecular weight has relatively little effect. This is because physical behaviour, excluding mechanical, involves the cooperative movement of no more than small groups of atoms. Mechanical properties, on the other hand, involve movement of the whole molecular chain, and are very dependent upon molecular weight.

Polymers are electrical insulators with resistivities in the range 10^{11}–10^{17} Ωm, and as such are considered in greater detail in Chapter 14. Polymers can be given some conductivity by the inclusion of particles of finely divided conducting material such as powdered graphite or carbon black. Conjugated bonds (alternate single and double bonds) along the main chain can give some conductivity, or at least a semiconductor-like energy gap. Irradiation or partial thermal degradation can produce broken bonds which act as donor or acceptor sites (see Chapter 14). Resistivities of ~ 1 Ωm, comparable with that of germanium, can be produced in this way. Still lower values of resistivity have been produced by adding donor or acceptor radicals along the molecular chain. The chain ending groups may also contribute to conductivity, as will any ions present, such as catalyst residues.

The dielectric constant of polymers is usually in the range 1 to 10. Dielectric losses are low; some plastics (see Table 8.3) have a loss tangent (§ 3.10) of $\sim 10^{-4}$ at 1 MHz, which is difficult to measure and is effectively zero. Losses rise rapidly with absorbed water and an intrinsic low loss is only of value if there is little tendency to absorb moisture. Dielectric properties are dependent on the polar atoms or groups present and whether they are disposed symmetrically or not. Symmetrical polymers, such as polyethylene or PTFE, have dielectric constants in the range 1–3, and low dielectric losses at high frequencies. PTFE may be regarded as being produced from polyethylene by the replacement of the four hydrogen atoms in the monomer unit by fluorine:

$$
\left[\begin{array}{c} \text{H} \quad \text{H} \\ | \quad\quad | \\ \text{C}-\text{C} \\ | \quad\quad | \\ \text{H} \quad \text{H} \end{array}\right]_n \rightarrow \left[\begin{array}{c} \text{F} \quad \text{F} \\ | \quad\quad | \\ \text{C}-\text{C} \\ | \quad\quad | \\ \text{F} \quad \text{F} \end{array}\right]_n
$$

If only 1, 2 or 3 of the hydrogen atoms are replaced, an asymmetric polymer is formed, with much higher dielectric constant and losses. For example, poly-(difluoroethylene) has the loss increased by a factor of 100. Another example is chlorinated polystyrene. In poly(2,5-dichlorostyrene) two opposite hydrogen atoms in the benzene ring are replaced by chlorine, and the losses are low, whereas in poly(3,4-dichlorostyrene) an asymmetric structure is formed with high losses:

2,5-dichlorostyrene
symmetric, low loss

3,4-dichlorostyrene
asymmetric, high loss.

Dielectric losses arise from the dissipative motion of polar groups, which are trying to follow the changing electric field, and are a function of frequency, showing a maximum when the natural frequency (*relaxation* frequency) of the particular motion coincides with the field frequency. At frequencies above the relaxation frequency the polar groups cannot keep pace with the electric field; for one half cycle the *polarisation* (see Chapter 14) of the polymer leads the field, and for one half cycle it lags behind, and the time-averaged tan δ is small. At lower frequencies the polarisation is always in step with the field, and again tan δ is small. At the relaxation frequency, the polarisation is always 90° behind the field. A *loss peak* is found as the frequency of the applied field is swept, its position and magnitude being affected by any factor which influences molecular or atomic motion in the polymer. Since the relaxation processes are thermally activated, the peaks move to higher frequencies as the temperature is increased. Conversely the peaks may also be found at fixed frequencies by varying the temperature. Generally one peak (the α-peak) is associated with the glass-transition temperature, and up to three more peaks, due to side group and chain segment movement and rotation, may be found at lower temperatures. (These peaks may also be detected mechanically, by internal friction measurements.) Plasticisers tend to increase losses, crystallisation and cross-linking to decrease them.

Polymers are widely used in the electrical industry as insulators, and as dielectrics. Wire conductors may be coated with a monomer solution, which is then polymerised *in situ*. Rigid polymers, particularly thermosets such as PF impregnated fabric, are used as insulating supports, and as formers for coils. As dielectrics they are used in capacitors, coaxial cables and transmission lines. Polyethylene has been described as the polymer that won the war; its excellent dielectric properties being essential for the success of radar, with its short wavelength, high frequency (10^2–10^3 MHz) radiation.

Optically, polymers may be transparent or opaque. Amorphous polymers are transparent, approaching the transmission of glass, in the pure state, but can be easily coloured by the addition of pigments or *lustrants* (usually TiO_2). Crystalline polymers are opaque in the pure state, due to scattering by the spherulites. These have a different density, and hence refractive index, to the surrounding amorphous polymer and are of the same size as the wavelength of light, especially the ultra-violet. Opacity increases as the spherulite size increases. For example, PTFE is nearly fully crystalline and appears milky white. In thin films, however, the size of the spherulites is limited and crystalline films may be quite transparent.

Polymers are also thermal insulators, with thermal conductivities typically of the order of 0.1 W m^{-1} deg^{-1}. Their use is normally confined to temperatures below 100 °C, but high temperature resistant materials, such as the polyimides, are used in fire-resistant protective clothing. Expanded polystyrene (styro foam) in which a foam, entrapping many small bubbles of air, is created before polymerisation is complete, is now a very common domestic insulator.

The structure of a polymer may change slowly over a period due to a variety of causes, quite small chemical changes producing a marked degradation of properties. Besides excessive temperature, the main causes are ultra-violet light, ionising radiation and chemical attack by the environment such as oxygen, ozone and water. Degradation may involve breaking of chains, chemical modification of side groups and cross-linking. Chain breaking may be random or the original constituents may be produced in appreciable amounts. Thus

Bakelite (phenol formaldehyde), formed by condensation polymerisation with water as a by-product, is de-polymerised on exposure to superheated steam.

The stability of many plastics is limited by the carbon–hydrogen bond and it can be increased by replacing the hydrogen with an atom or groups of atoms with greater combining strength. For example the substitution of fluorine atoms in place of the hydrogen atoms in the ethylene monomer to give polytetra-fluorethylene raises the useful temperature from 150 °C up to 250 °C. Still higher temperatures are possible if the carbon–carbon backbone is replaced by more strongly bonded silicon and oxygen atoms as in polysilicones. For example, the elastomer polydimethyl siloxane:

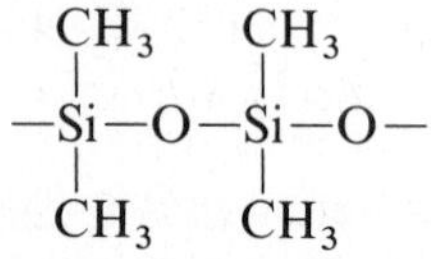

$$
\begin{array}{ccc}
\text{CH}_3 & \text{CH}_3 \\
| & | \\
-\text{Si}-\text{O}-\text{Si}-\text{O}- \\
| & | \\
\text{CH}_3 & \text{CH}_3
\end{array}
$$

can be used over the temperature range $-70\,°C$ to $300\,°C$. Silicone polymers can be produced which correspond to many of the carbon-based polymers, and in each case they have a wider working temperature and greater resistance to chemical attack.

Thermal stability is also conferred by the incorporation of aromatic (benzene) rings into the polymer chain. The extra bonds provided by the ring structure make the polymer more resistant to thermal degradation. Examples are the aromatic polyamides, in which aromatic rings are incorporated into a nylon polymer chain. For example, Nomex, with the structure:

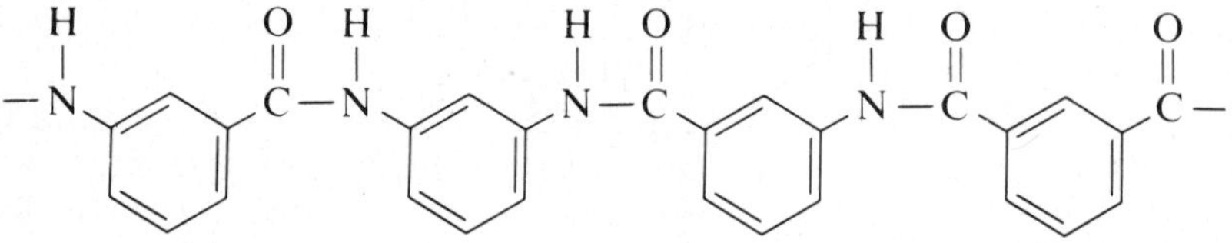

is highly crystalline and melts at 300 °C, whereas nylon melts at 225 °C. Aromatic polyamide fibres are ignited only after long exposure to direct flame, and are quenched as soon as the flame is removed. They are used in the manufacture of fire-resistant protective clothing.

Even higher temperature performance can be attained with aromatic hetero-cyclic polymers. In these polymers, aromatic rings alternate with ring-shaped molecules of the parent compound, to produce an extremely strong and rigid chain. For example, polyimides are formed by the condensation polymerisation of pyromellitic dianhydride with any diamine:

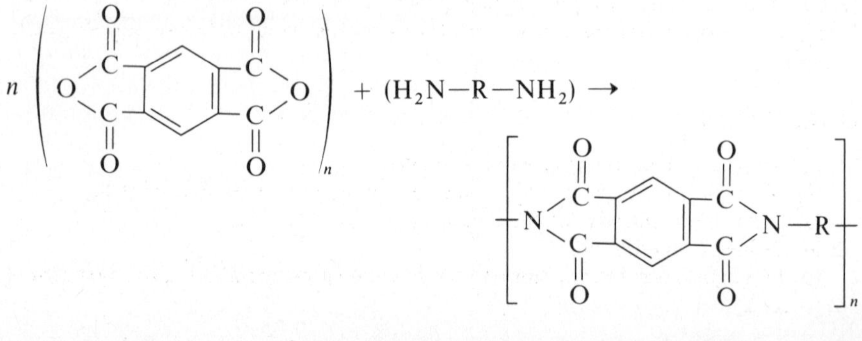

Various aromatic diamines, such as *m*-phenylene diamine (H_2N⟨⟩NH_2),

benzidene (H_2N⟨⟩⟨⟩NH_2) or diamine diphenyl ether

(H_2N⟨⟩O⟨⟩NH_2), produce polyimides with a very high

degree of oxidative and thermal stability. Melting points are above 600 °C and they may be used continuously at temperatures above 300 °C, and intermittently above 400 °C. Current applications include their use as insulation in high-temperature electric motors.

   Ladder polymers, with even better high temperature properties, are under development. They have structures which may be schematised thus:

and it can be appreciated that at least two opposite bonds must be severed before the polymer is degraded. Ladder polymers include polyquinoxaline:

$$\left[\begin{array}{c}\text{quinoxaline structure}\end{array}\right]_n$$

and mixed quinoxaline and thiazine:

$$\left[\begin{array}{c}\text{mixed quinoxaline and thiazine structure}\end{array}\right]_n$$

The latter is completely stable to above 500 °C. Such materials are serious rivals to many aluminium and copper alloys for medium temperature applications, just as many ordinary polymers have replaced metals at ambient temperatures. Further developments are expected to produce polymers with even higher useful temperatures.

QUESTIONS

1. (a) Define a high polymer; give two examples.
   (b) Distinguish the four types of polymer structure; give an example of each.
   (c) Define thermoplastic and thermoset.
2. (a) Define polymerisation.
   (b) Distinguish between addition and condensation polymerisation; give an example of each.

(c) What is an essential requirement of a monomer to undergo addition polymerisation?

3. (a) Describe three stages of addition polymerisation.

(b) Is condensation polymerisation as complex?

(c) What is special about polymers produced with Ziegler type catalysts?

4. (a) Define a copolymer. Give an example.

(b) Distinguish between random, block and graft copolymers.

(c) Describe the structure of high impact polystyrene.

5. (a) The structure of polyethylene is frequently written as $+CH_2.CH_2\frac{}{}_n$. How does this fail to indicate the whole truth about the structure?

(b) Define polyvinyl and polyvinylidene polymers. Give an example of each.

6. (a) What is the amide linkage, as in nylons?

(b) Distinguish between: nylon-6,6, nylon-6,10, nylon-6.

(c) What is the significance of the various structures?

7. (a) What is the ester linkage?

(b) Distinguish between linear and cross-linked polyester.

8. (a) Distinguish between degree of polymerisation and molecular weight.

(b) Discuss the different ways of averaging the molecular weight.

(c) What is the order of magnitude of the weight average molecular weight?

9. (a) What is meant by molecular architecture?

(b) Show diagrammatically the difference between: syndiotactic, isotactic, atactic.

(c) Distinguish between *cis* and *trans*-polyisoprene ($-CH_2.CCH_3=CH.CH_2-$).

(d) What is the significance of (c)?

10. (a) Define glass transition temperature T_G in terms of molecular structure and macro-scopic properties.

(b) Define crystalline melting point T_M.

(c) Sketch a typical curve of specific volume versus temperature for amorphous and partially crystalline polymers.

(d) What factors tend to raise the glass transition?

11. (a) What factors promote crystallinity in a polymer?

(b) Describe the structure of a polymer single crystal.

(c) Describe the structure of a partially crystalline polymer.

12. (a) Describe the four physical states of linear polymers.

(b) How are they modified by cross-linking?

13. (a) Describe the structure of wood.

(b) How is the strength related to this structure?

(c) Why season?

14. (a) Briefly describe the difference between a pure linear high polymer and a commercial thermoplastic.

(b) What is a plasticiser and its purpose?

15. (a) Define a rubber (elastomer).

(b) Distinguish between normal (internal energy) elasticity and high (entropy) elasticity.

(c) Contrast briefly the mechanical behaviour of an elastomer and a crystalline solid.

16. (a) Discuss in qualitative terms the kinetic theory of high elasticity.

(b) Given the variation of entropy with chain length is

$$S_0 - S = \frac{1}{2}N_0k\left[\left(\frac{L}{L_0}\right)^2 + 2\frac{L_0}{L} - 3\right],$$

show that the isothermal elastic modulus is $E = 3nkT$.

(c) Discuss the structural requirements for a polymer to be a good elastomer.

17. (a) Discuss: 'Thermoplastics are visco-elastic materials.'

(b) What replaces Young's modulus and the proportional limit (as defined for metals)?

(c) Sketch three types of tensile stress–strain curves for plastics.

18. What features give a plastic stability at high temperature?

FURTHER READING

K. Hutton: *Chemistry: The Conquest of Materials*. Pelican Books (1957).

M. Morton (ed.): *Introduction to Rubber Technology*. Reinhold (1959).

E. Baer, J. R. Knox, T. J. Linton and R. E. Maier: Structural design of plastics. *S.P.E.J.* **16**, 396 (1960).

C. C. Winding and G. D. Hyatt: *Polymeric Materials*. McGraw-Hill (1961).

F. W. Billmeyer Jr.: *Textbook of Polymer Science*. Wiley (1962).

F. Bueche: *Physical Properties of Polymers*. Interscience (1962).

L. E. Nielsen: *Mechanical Properties of Polymers*. Reinhold (1962).

J. A. Brydson: *Plastics Materials*, 2nd edition, Iliffe (1964).

D. W. Saunders and J. M. Stuart: *Plastics for the Engineer*. Reinhold (1964).

F. T. Bowden: Rubber as an engineering material. *Chart. Mech. Eng.*, p. 374 (July 1964).

Various authors: Symposium on 'Plastics and the Mechanical Engineer', *I. Mech.E.* (*London*), **179**, part 3B (1964).

P. Meares: *Polymers: Structure and Bulk Properties*. Van Nostrand (1965).

P. D. Ritchie (ed.): *Physics of Plastics*. Plastics Institute, London (1965).

K. Ziegler: Organometallic compounds in macromolecular chemistry. *Trans. J. Plastics Inst. (London)*, **33**, 1 (1965).

W. F. Ratcliffe and S. Turner: Engineering design: data required for plastics materials. *Trans. J. Plastics Inst. (London)*, **34**, 137 (1966).

Various authors: *Plastics Application Series*. Reinhold.

E. Baer (ed.): *Engineering Design for Plastics*. (Soc. Plastics Eng. Inc.) Reinhold (1964).

H. J. Stern: *Rubber: Natural and Synthetic*, 2nd edition. Maclaren (1967).

G. R. Palin: *Plastics for Engineers*. Pergamon Press (1967).

C. T. Greenwood and E. A. Milne: *Natural High Polymers*. Oliver and Boyd (1968).

A. H. Frazer: *High Temperature Resistant Polymers*. Interscience (1968).

Plastics Materials Guide. Issued annually with *Europlastics Monthly*.

9. Plasticity 2: physics of plasticity

9.1 Introduction

The continuum theory of plasticity, which has been presented in Chapter 6, is concerned with the effect of combined stress on the onset of yielding and the magnitudes of the subsequent plastic strains, especially under certain constraints, such as plane strain, which are found in many metal working processes. It does not, however, give any physical insight into the nature of the deformation process. For this the structure of material must be taken into account and the movement of atoms during deformation investigated.

The physics of plasticity is studied at two levels. The phenomenological study, which is described in the first part of this chapter, is concerned with the crystallography of the deformation process, which is largely based on experimental observation using single crystals. The principal mode of deformation is shown to be slip on certain close-packed crystallographic planes in close-packed directions; the deformation is analysed in terms of the resolved shear stress and strain. Yielding is found to commence at a critical value (for a given material) of the resolved shear stress, irrespective of the other stresses present, and continues at an increasing stress, the phenomenon of strain hardening. The theory is unable to predict values for the yield stress or the subsequent strain hardening, for which a new body of theory is required, that of crystal dislocations.

The theory of dislocations is given in the second part of this chapter. The primary problem is the calculation from first principles of the yield stress of pure metals. The answer obtained assuming homogeneous slip over a slip plane is found to be several orders of magnitude too large. By postulating the existence of faults in the crystal lattice, the dislocations, slip can occur inhomogeneously at a very much lower stress. In pure metals there is no clear cut dominant factor controlling this stress and quantitative prediction of the yield stress is difficult. However, in many alloys the controlling factor can be identified and dislocation theory is able to make quantitative explanations of the phenomena, for example, in precipitation and dispersion-hardened alloys (§ 11.7, § 11.8) and in low carbon steels with a sharp yield point (§ 11.5). Dislocations are also involved in the mechanism of fracture (see § 10.15).

Dislocations are also found in non-metals but, due to the more complicated crystal structures, they are generally fewer in number and much less mobile than in metals. This accounts for the lack of ductility of most non-metallic crystalline solids. Dislocations in these materials are discussed briefly in § 9.12.

CRYSTALLOGRAPHY OF PLASTIC DEFORMATION

9.2 Tensile test of single crystals: slip lines

Many investigations of the physics of plastic flow have been made, and are continuing, using single crystals measuring several millimetres thick by several

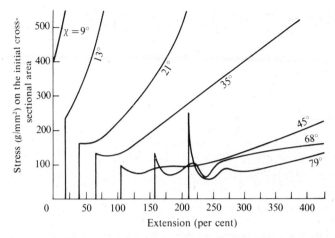

Fig. 9.1. Tensile tests of cadmium single crystals at different orientations. Angles marked are between the basal plane and the tensile axis. (From E. Schmid and W. Boas, *Krystallplastizität*, Springer-Verlag (Berlin 1935); English edition, F. A. Hughes & Co. Ltd. (London 1950).)

centimetres long. From a tensile test of a pure metal crystal, the curve of stress (load/initial orthogonal area) versus strain (extension/initial length) is not unique but varies with orientation of the crystal lattice relative to the tensile axis. This is most marked in hexagonal metals and a typical set of curves for cadmium at various orientations is shown in Fig. 9.1. The curves start with a nominally proportional and elastic range in which Young's modulus varies slightly with the crystallographic direction. Then yielding begins, rather more sharply than in polycrystalline metal, at a stress which depends on the orientation, as do the subsequent curves to fracture.

At yield, parallel lines develop on the surface of the crystal forming loops around it parallel to a principal crystallographic plane, see Fig. 9.2. Each loop is the step on the surface produced when two parts of the crystal slide over each

Fig. 9.2. Slip steps on the surface of an aluminium–4 per cent copper alloy single crystal, aged 48 hours at 130 °C (× 7).

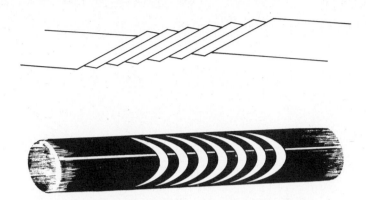

Fig. 9.3. Schematic representation of crystal undergoing single slip.

other on a crystallographic plane by a distance of between ten and a thousand nanometres (100–10 000 Å). Such surface marks are called slip lines,[†] or slip bands, since in fcc metals several slip lines are found grouped together at higher strains. The direction of slip is the same on all slip planes along the length of the specimen, parallel to a close-packed crystal direction. The deformation is thus superficially like the shearing of a pack of playing cards, see Fig. 9.3, with shear occurring on widely spaced planes and the material in-between remaining crystalline. Slip lines were first observed (1900) in polycrystalline metal by Ewing and Rosenhain under a light microscope; studies using an electron microscope were started in 1948 by Heidenreich and Shockley.

As the plastic strain is increased the shear in the slip bands and their number increase. In bcc and fcc metals slip also occurs on equivalent crystallographic planes and directions, intersecting the original lines. This is called double or multiple glide. Finally, fracture occurs either by parting of a slip plane, following single slip, or by necking, following multiple slip.

9.3 Critical resolved shear stress: shear-hardening curve

Once it is realised that the plastic deformation in a crystal is due to slip on a parallel set of planes in a definite direction, it is possible to analyse the tensile stress–strain curves of Fig. 9.1 in terms of shear of the slip system, and obtain a shear-hardening curve.

In Fig. 9.4 the crystal under test is arranged in a system of rectangular co-ordinates with one end at the origin O and the other end at a point P, coordinates $x_0 y_0 z_0$, such that the slip planes are parallel to the xy-plane and the slip direction parallel to Oy. χ_0 and λ_0 are the initial angles between the specimen axis and the slip plane and slip direction respectively. When the axial tensile load is applied, assuming end O is fixed at the origin, end P moves in the slip direction, that is, parallel to Oy. The initial elastic strain is small compared with the plastic strain and is ignored. When the free end of the specimen is at any point (xyz), the initial

---

[†] The term 'glide' is also used, although some authors distinguish between slip and glide. They must be distinguished from the slip lines defined in the continuum theory of plasticity (Chapter 6).

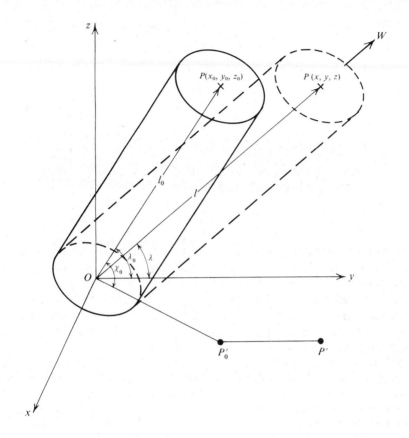

Fig. 9.4. Coordinate system for the extension of a crystal by slip on a single system.

and instantaneous lengths, l_0 and l respectively, are related by

$$\left(\frac{l}{l_0}\right)^2 = \frac{x^2 + y^2 + z^2}{x_0^2 + y_0^2 + z_0^2}.$$

The shear strain γ is defined as the relative displacement in the slip direction of two planes parallel to the slip planes and with unit separation. Hence

$$y = y_0 + \gamma z.$$

Also

$$x = x_0$$

$$y_0 = l_0 \cos \lambda_0,$$

and

$$z = z_0 = l_0 \sin \chi_0.$$

Substitution gives an equation relating the shear strain to the instantaneous length and initial parameters:

$$\left(\frac{l}{l_0}\right) = 1 + 2\gamma \cos \lambda_0 \sin \chi_0 + \gamma^2 \sin^2 \chi_0. \tag{9.1}$$

The shear stress on the slip plane in the slip direction is:

$$\tau = \frac{W}{A} \cos \lambda \sin \chi_0$$

where A is the orthogonal cross-sectional area and W is the load.

But $\qquad\qquad \cos \lambda = l_0/l \, (\cos \lambda_0 + \gamma \sin \chi_0).$

Hence $\qquad\qquad \tau = \frac{W}{A} \frac{l_0}{l} (\cos \lambda_0 + \gamma \sin \chi_0) \sin \chi_0.$ $\qquad$ (9.2)

The shear-hardening curve, defined as the shear stress τ versus shear strain γ of the slip system, is obtained by means of (9.1) and (9.2) from the load versus length data, plus the initial orientations of the slip plane and slip directions. When double slip occurs a more complex analysis is required. Some typical shear-hardening curves are given in Fig. 9.5.

Yielding commences at a critical value of the resolved shear stress, on the slip plane in the slip direction, and is independent of all other stress components. This is Schmid's yield criterion (1924) and should be compared with von Mises' and Tresca's criteria for polycrystalline material (§ 6.5). The slip system and value of the critical resolved shear stress for various pure metals are given in Table 9.1. They fall into two groups. The soft crystals have values of the crss below $1 \, \mathrm{MN/m^2}$ (10^2 psi, 10^{-5} G), comprising all the fcc metals and certain cph metals (Cd, Zn, Mg). The hard crystals have values above $10 \, \mathrm{MN/m^2}$ (10^3 psi, 10^{-4} G) and are composed of the bcc metals, the remaining cph metals and many ionic compounds. The explanation of these differences will be given

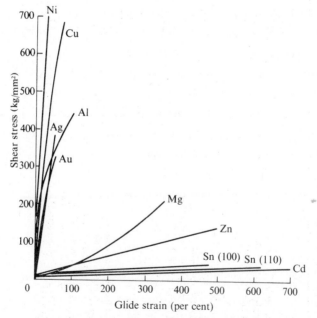

Fig. 9.5. Shear-hardening curves of pure metals at room temperature. (Source as Fig. 9.1.)

Table 9.1. *Slip elements and critical resolved shear stress of pure metals at 20°C*

Crystal structure and slip elements	Metal	Purity (per cent)	Axial ratio c/2a (cph only)	Slip system (bcc and cph only)	Critical resolved shear stress (MN/m²)
Face-centred cubic Slip plane {111} Slip direction $\langle 10\bar{1} \rangle$	Ag	99.999			0.37
		99.99			0.59
	Al	99.995			0.4
	Au	99.99			0.9
	Cu	99.999			0.64
		99.98			0.93
	Ni	99.8			5.8
Body-centred cubic Slip planes {101}, {112}, {123} Slip direction $\langle 111 \rangle$	α–Fe	99.96		{101}, {112} and {123}	28–35
	Mo	99.98		{101}	50
	Nb	99.6		{101}, {112} and {123}	18
	Ta	99.9		{101}, {112} and {123}	50
	W	99.99			~200
Close-packed hexagonal Basal-slip {plane (0001), direction $\langle 11\bar{2}0 \rangle$} Prismatic-slip {plane $\{10\bar{1}0\}$, direction $\langle 11\bar{2}0 \rangle$} Pyramidal-slip {plane $\{10\bar{1}1\}$, direction $\langle 11\bar{2}0 \rangle$}	Cd	99.996	0.943	basal	0.57
		99.996		basal	0.18
	Zn	99.96	0.928	basal	0.93
	Mg	99.95	0.812	basal†	0.44–0.83
				prismatic	39
					not observed
	Zr	99.99	0.795	basal	6.2
				prismatic†	63
	Ti	99.99	0.794	basal	14
		99.9		prismatic†	110
				pyramidal†	92
	Be	99.99+	0.784	basal†	99
					1.3
				prismatic	52

† Preferred slip system.

later in this chapter. Both groups have yield stresses well below those of poly-crystalline material, around $100 \, \text{MN/m}^2$ (10^4 psi), and even further below the theoretical value of $1-10 \, \text{GN/m}^2$ (10^5-10^6 psi) (see § 9.6).

During plastic deformation the yield stress rises, as found with polycrys-talline material. In single crystals strain hardening can be broken down into three stages, which are present to greater or lesser degree in crystals of different lattice structures. Stage I is easy glide with a very low and constant hardening rate (i.e. independent of stress), typically in the range 10^{-5} to 10^{-4} G at room temperature. This is associated with slip on a single slip system. Stage II pro-duces rapid constant hardening, about 4×10^{-3} G, and is due to slip developing on intersecting slip systems. Lastly, in stage III the hardening rate decreases with increasing stress and strain. Some recovery process during this stage op-poses the strain-hardening mechanisms. Stage III is mainly the one existing in polycrystalline materials.

Hexagonal metals exhibit a very long easy glide range (stage I) at all orienta-tions, as shown for various metals in Fig. 9.5 and for cadmium at various temperatures in Fig. 9.6. The latter indicates that the hardening rate increases as the test temperature is reduced. If the crystals are very carefully grown free of

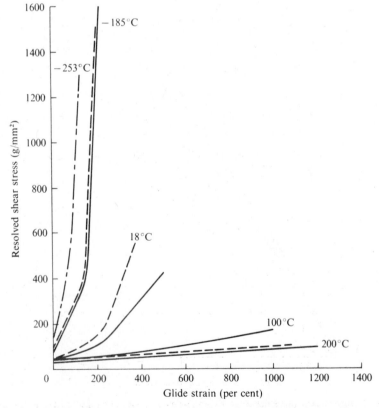

Fig. 9.6. Effect of temperature on the shear-hardening curves (stage I) of cadmium. (Source as Fig. 9.1.) Broken curves obtained at about 100 times faster stressing rate than continuous curves.

all substructure the easy glide stage can be subdivided into two sections with a factor of 4 increase in the hardening rate between the first and second section. This can be seen in the curve for cadmium at 18 °C. The presence of a dislocation substructure (see below) eliminates the very low hardening rate section.

In fcc and some bcc metals all three stages are observed, though stage I exists only in crystals oriented for single slip and then over only small strains. Typical curves for copper at two orientations over a range of temperatures are given in Fig. 9.7. It will be seen that the hardening rate in stage II is independent of temperature.

The appearance of the slip lines is characteristic of each stage of strain hardening, see Fig. 9.8. At first they are long, straight and fairly uniformly spaced, with step heights of 5–10 nm. In stage II they become shorter and become branched, with varied slip heights. In the final stage slip bands with several lines develop with cross slip between the ends of the bands.

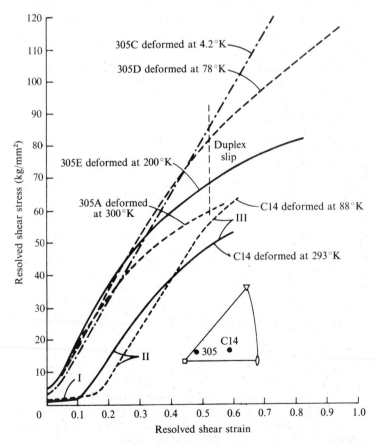

Fig. 9.7. Temperature dependence of the shear-hardening curves of copper crystals at two orientations. (From A. Seeger, in Fisher *et al.*, editors, *Dislocations and Mechanical Properties of Crystals*, Wiley (New York 1957).

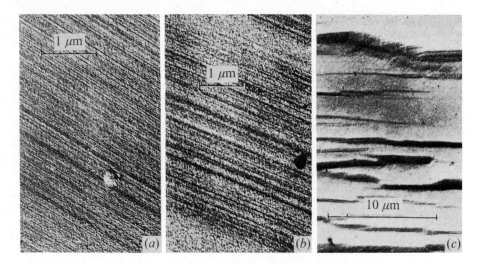

Fig. 9.8. Electron micrographs of slip lines on the surface of a copper crystal. (Source as Fig. 9.7.)

(a) At the end of stage I. Crystal C14 (Fig. 9.7) after 8 per cent shear strain at room temperature.

(b) Stage II. Crystal C14 deformed 20 per cent, repolished and then given an additional 8 per cent shear strain at room temperature.

(c) Stage III. Crystal strained 60 per cent, repolished, and re-strained 10 per cent in shear at room temperature. Slip bands, fragmentation, cross slip and traces of secondary glide can all be seen.

9.4 Slip elements

Slip occurs only on certain preferred crystallographic planes and in certain directions. These are called the slip, or glide, elements. Those for the common metals are given in Table 9.1. The slip planes are usually the close-packed atomic layers (and hence widely separated) and the slip directions are always the close-packed directions. This is intuitively understandable and will be justified in physical terms later.

The glide elements of the three common crystal systems are drawn in their respective structure cells in Fig. 9.9. In fcc the slip plane is (111), or the equivalent octahedral planes {111}, and the slip direction [01$\bar{1}$], or its equivalents ⟨011⟩. These are all close-packed planes and directions. Since there are so many equivalent slip elements, it is clearly difficult to confine slip to only one system to produce easy glide and multiple glide with a high strain-hardening rate is the general rule.

In bcc there are no close-packed planes, but four equivalent close-packed directions ⟨111⟩ in which slip occurs. The slip plane may be one or more of {101}, {112}, or {123}, each of which contains the slip directions. Frequently several planes are operative with a single slip direction, called pencil glide, and the slip plane is difficult to identify. This may explain the dearth of data on the critical resolved shear stress in this class. Again multiple glide, with a high strain hardening rate, is usual.

In cph, the position is more complex. Although called close-packed, there is in fact some variation in the spacing of the atomic layers relative to the atomic

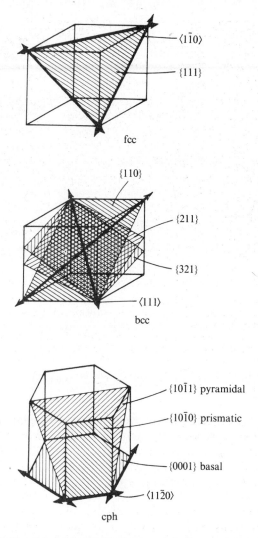

Fig. 9.9. The glide elements in the fcc, bcc, and cph crystal structures.

separation within the layers. The value of $c/2a$ for close-packed rigid spheres is 0.816 where c and a are the unit translations. Elements with greater spacing (Cd, Zn and Mg) slip easily on the close-packed planes (0001) in the close-packed directions $\langle 11\bar{2}0 \rangle$. Elements with less inter-layer spacing (Zr, Ti and Be) have high critical resolved shear stresses for basal slip and the preferred slip is on the prismatic planes $\{10\bar{1}0\}$, but still in the $\langle 11\bar{2}0 \rangle$ directions. Slip on these prismatic planes is also found in the first group when basal slip is prevented by suitable orientation or, if the material is polycrystalline, by grain boundary restraints. Zinc and cadmium have also been observed to slip on $\{11\bar{2}2\}$ planes in the $\langle \bar{1}\bar{1}23 \rangle$ directions, which is unique since slip is not in a close-packed direction.

9.5 Mechanical twinning

Although slip is the most important and universal mode of plastic deformation, there are several others, including twinning, deformation bands, and kinking. Of these, only twinning will be described here.

Twinning produces a change in shape with no change in crystal structure, and in this respect is identical to slip. The change in shape is a simple shear. The deformation does not involve large relative movements of each of the atoms in the twinned volume. The total shear due to twinning is in general much less than that due to slip, and hence twinning is not such an important mode of deformation as is slip. By the small movement of all atoms the same crystal lattice is regenerated in a different orientation to the matrix.

The mechanics of twin formation is shown in Fig. 9.10. The lattice between the planes XX and YY shears into a twin relationship to the rest of the crystal, in other words the twinned part of the crystal is a mirror image in some simple crystallographic plane of the original lattice. The shear movement of a plane in the twin is thus proportional to its distance from the twin plane.

Twinning generally occurs in metals with limited slip systems, for example, close-packed hexagonal, and the new orientation of the twin may be much more favourable for slip. Twinning may also be produced in bcc metals, particularly under high rates of deformation. A photomicrograph of twins produced by deformation in an α-iron crystal is given in Fig. 9.11(*a*). Their width is a few thousand atoms. Stacking faults (see § 2.8) during recrystallisation may cause twins to appear during annealing; this is a feature of many fcc metals. Figure 9.11(*b*) shows annealing twins in bronze. The orientation relationship between twin and matrix is identical with that for deformation twins, though the mechanism of production is entirely different. Annealing of deformation twins may cause them to grow into the surrounding matrix and their width increase markedly.

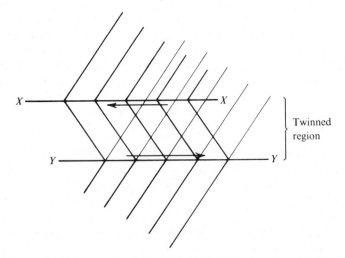

Fig. 9.10. Production of a twin. The atoms between the planes XX and YY have sheared to produce a twin relationship with the parent lattice.

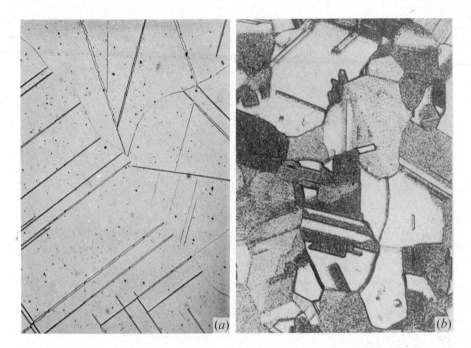

Fig. 9.11. (*a*) Deformation twins (Neumann bands) in α-iron (0.1 % C mild steel) caused by severe impact after cooling in liquid nitrogen (× 220). (By courtesy of J. R. Moon.) (*b*) Annealing twins in a tin bronze (× 220).

Both twin and parent matrices have the same crystal lattice and the same unit cell. A unit cell can be defined by any three non-coplanar rational lattice vectors, that is, vectors whose directions in the crystal are represented by simple, low-valued Miller indices. There must therefore be three non-coplanar vectors which are undistorted by the shearing action which has produced the twin. That is to say they have the same lengths, and the same angles between them, both before and after the twinning operation. The search for these vectors is simplified if it is known whether or not there are any crystallographic planes which remain undistorted. There are in fact two such planes, as can be seen by reference to Fig. 9.12. One is the plane on which the shear takes place. This is

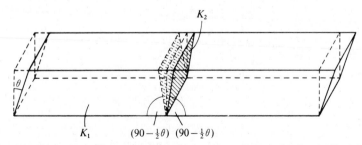

Fig. 9.12. The undistorted planes in a twin. K_1 is the first undistorted plane, and K_2 is the second undistorted plane.

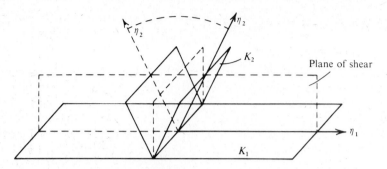

Fig. 9.13. Definition of the twinning directions, η_1 and η_2.

given the notation K_1 and is called the first undistorted plane or the twinning plane. The other plane, K_2, the second undistorted plane, makes an angle with the twinning plane equal to the complement of one half of the angle of shear both before and after twinning. K_1 and K_2 are the only planes which are not distorted by the shearing process, and it is reasonable to suppose that the three undistorted vectors will lie in these planes.

In fact any vector which lies in either of these planes will remain of itself undistorted, but there is an additional requirement that the angles between these vectors be not changed. Referring to Fig. 9.13, two vectors are defined. η_1 lies in K_1, in the direction of the shear; η_2 lies in K_2 and is normal to the line of intersection of K_1 and K_2. η_1 and η_2 now define a new plane, this is called the plane of shear. η_2 makes equal angles, before and after shear, with any vector lying in K_1. If η_2 is one of the three vectors, the other two may lie in K_1, and as two rational vectors will define a rational plane, the twin may be completely defined by K_1 and η_2. Similarly η_1 makes equal angles before and after with any vector lying in K_2, and hence K_2 and η_1 will also completely define the twin.

It is possible to describe three types of twin. A twin is said to be of the first kind when K_1 is a rational lattice plane, that is, a plane of simple, low indices, and η_2 a rational lattice vector. It is of the second kind if K_2 is a rational lattice plane and η_1 is a rational lattice vector. Twins are called compound when K_1 and K_2 are both rational lattice planes and η_1 and η_2 are both rational lattice vectors. Most metals have compound twins, a few have twins of the first kind; twins of the second kind are found only in uranium and some compounds of low symmetry. The twinned lattice can be generated from the parent lattice by a rotation of 180°, in a twin of the first kind, about the normal to K_1, and about η_1 in a twin of the second kind.

In cph metals compound twins are formed; the twinning elements are: $K_1(10\bar{1}2)$ $K_2(\bar{1}012)$ $\eta_1[\bar{1}011]$ $\eta_2[\bar{1}01\bar{1}]$. Zinc has an axial ratio of 1.856, and the angle between K_1 and the basal plane is 47°. Upon twinning a zinc crystal will increase in length parallel to the basal plane, and tension parallel to the basal plane will cause zinc to twin. Magnesium on the other hand has an axial ratio of 1.624 and the angle between K_1 and the basal plane is 43°. Twinning causes a decrease in length parallel to the basal plane, and compression in this direction is needed to cause magnesium to twin. For further details of twinning the reader is referred to a specialist text on the crystallography of deformation.

THEORY OF CRYSTAL DISLOCATIONS

9.6 Theoretical shear strength

Frenkel, in 1926, estimated in a simple manner the critical resolved shear stress in terms of the shear modulus. Figure 9.14(a) shows a section through two close-packed planes of atoms, parallel to the slip direction. If a shear stress is applied as shown in (b), the planes will tend to slide over each other. Suppose the lower plane is held and the upper plane moves homogeneously to the right by a distance x. It is clear that when the upper plane has moved half the interatomic distance it will be in unstable equilibrium between the original position and the next stable position: that is, $\tau = 0$ at $x = b/2$, where b is the interatomic distance in the x direction. Also, τ is again zero at $x = b$ and will have a reversed sign between $x = b/2$ and $x = b$. The curve of τ against x must be something like that shown in Fig. 9.14(c). Frenkel assumed it was a sine wave:

$$\tau = \tau_{max} \sin 2\pi \frac{x}{b}.$$

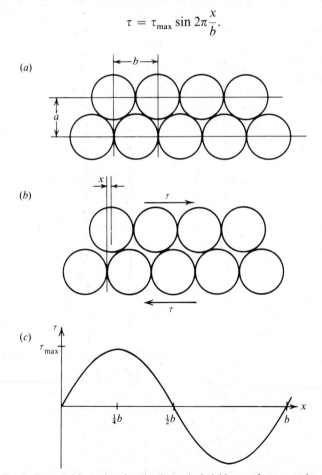

Fig. 9.14. Frenkel's model for estimating the theoretical yield stress for a crystal undergoing homogeneous slip. (a) No applied stress. (b) Applied shear stress τ, producing relative movement x of top plane of atoms with respect to bottom plane. (c) Shear stress τ as a function of displacement x.

The slope at the origin is then

$$\frac{d\tau}{dx} = \frac{2\pi}{b}\tau_{\text{max}}.$$

But this is where the material is behaving elastically, so

$$d\tau = G\frac{dx}{a}$$

where G is the shear modulus and a the interplanar distance.

Hence
$$\tau_{\text{max}} = \frac{G}{2\pi}\frac{b}{a} \simeq \frac{G}{2\pi}. \tag{9.3}$$

This equation gives a theoretical value of the yield stress in terms of the shear modulus, and also shows that slip might be expected to occur more easily on the more widely spaced planes. A more sophisticated treatment (by Mackenzie in 1949) has given a theoretical maximum stress of $G/15\pi$.

Values of G are about $100\,\text{GN/m}^2$ ($10^7\,\text{psi}$), and therefore the theoretical maximum shear stress is about $10\,\text{GN/m}^2$ ($10^6\,\text{psi}$). This is four orders of magnitude larger than the experimental values of about $1\,\text{MN/m}^2$ ($100\,\text{psi}$) found in the soft group of pure metals (§ 9.3), although strengths of this order are found in whiskers. The theory also fails to predict strain hardening. Clearly some drastic modification to the model is required to explain the observed behaviour of single crystals.

Attempts were made to invoke thermal activation of the yield process as a mechanism for producing the observed yield strengths. These failed because the observed temperature dependence of the yield strength is just that of the elastic shear modulus, except at high temperatures when creep processes become active.

It was then realised that whereas Frenkel assumed homogeneous slip, with all atoms in one plane moving simultaneously with respect to those in the neighbouring plane, slip could also take place inhomogeneously by the consecutive movement of rows of atoms on one side of the slip plane with respect to the opposing atoms. This may be likened to the forward movement of a caterpillar. In this way slip would spread gradually in a slip plane at a low stress. The boundary between slipped and unslipped areas of a slip plane is a one-dimensional fault in the crystal lattice, called a dislocation, which will be described in the following section.

The low strength of single crystals was attributed to dislocations in 1934 by several workers (Taylor, Orowan and Polyani). Since the Second World War the quantitative theory of dislocations has developed extremely rapidly and has been able to explain a great number of metallurgical phenomena. Direct experimental confirmation of the existence and behaviour of dislocations came later with the development of several new experimental techniques: etch pits, decoration, X-ray diffraction and, notably, transmission electron and field-ion microscopy. It is now known that the mechanical properties of single crystal and polycrystalline materials are determined by the dislocations which are normally present in large numbers in the crystal lattice and can also be generated under an applied stress (see § 9.10).

9.7 Crystal dislocations

The nature of dislocations can be understood by comparing the atomic arrangements during homogeneous and inhomogeneous slip. Figure 9.15 shows a
section of a crystal taken perpendicular to the slip plane and parallel to the slip
direction : in (*a*), the atoms are in their initial unstressed positions. In (*b*), under a
shear stress τ, slip is shown occurring in a homogeneous mode, as postulated
by Frenkel; at the intermediate stage all atoms above the slip plane have moved
relative to those below by half the interatomic distance. In (*c*), the slip is

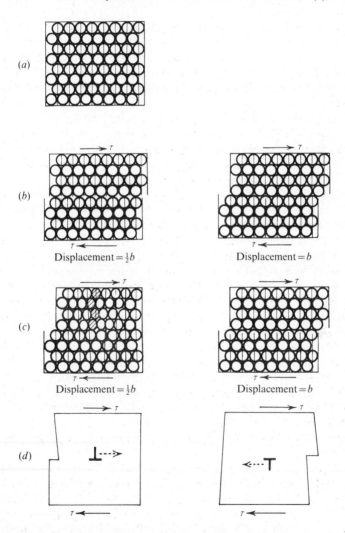

Fig. 9.15. Inhomogeneous slip produced by the glide motion of an edge dislocation. (*a*) Unstressed
crystal. (*b*) Crystal under a shear stress τ, undergoing homogeneous slip. (*c*) Inhomogeneous slip
by means of an edge dislocation. (*d*) Symbol for edge dislocation; note that positive dislocation
moving in one direction produces the same shear as negative dislocation moving in opposite direction.

inhomogeneous and takes place consecutively over the slip plane: at the inter-mediate stage, the atoms on the left have slipped a complete interatomic distance, whilst the others on the right have hardly moved from their initial positions. At the boundary between the slipped and unslipped parts of the slip plane, there is an irregularity in the atomic arrangement, which will run perpendicular to the plane of the diagram: this is called a dislocation, or, to be more specific, an *edge dislocation*. As slip proceeds this dislocation will move towards the right until it emerges at the free surface and the final arrangement is the same as in (*b*), with all the upper atoms advanced one place to the right relative to the lower ones.

The edge dislocation is so named because it is the 'raw' edge of a plane of atoms within the lattice. Thus in Fig. 9.15(*c*), the crystal above the slip plane contains an extra atomic plane (the shaded atoms) which ends on the slip plane to form the dislocation. This is conventionally shown by a symbol ⊥, see (*d*). Alternatively there may be an extra plane below the slip plane, indicated by ⊤. The movement of a ⊥ dislocation from left to right produces the same overall displacement as a ⊤ dislocation moving in the opposite direction.

Another type of dislocation is the screw dislocation, which can also produce inhomogeneous or consecutive slip. It is contrasted with the edge type in Fig. 9.16. Both are boundaries between slipped and unslipped regions of a slip plane, or, looked at the other way, their movement causes a relative shear displacement of the atoms on opposite sides of the glide plane. However, for a screw disloca-tion the boundary is parallel to the slip direction, and its passage causes relative displacement in a direction parallel to itself (not perpendicular, as in an edge type). It is so named because a path taken around it, keeping on the atomic planes, is a helix or screw; that is, *l m n o p* . . . in Fig. 9.16(*a*).

Dislocations will usually be a combination of the pure edge and screw types. Figure 9.16(*c*) shows a slip plane with a central region in which slip has occurred surrounded by an unslipped region. The boundary is a combination of the two types of dislocation. As the dislocation spreads to the boundaries of the crystal, there is slip between the atoms on the two sides of the glide plane. A further generalisation is a dislocation which does not lie in a plane but is a line fault running through the crystal. Thus the line fault *ABCD* in Fig. 9.17 could be formed by an extra plane of atoms *ABCD*, whose perimeter is all edge dislocation (with three separate glide planes). If *CD* now moved to *FE*, the dislocation becomes *ABCFE*, with *ABC* and *EF* edge type, and *CF* screw type, disloca-tions. It is not difficult to conceive a generalised dislocation running through the crystal, such as *XY* in the figure, with edge and screw components on various planes.

The development of thin film transmission electron microscopy by Hirsch *et al.* and Bollmann in 1956 has made possible the direct study of dislocations within crystals. Previously their behaviour was largely inferred theoretically, although some confirmation was possible by various experimental techniques using particular materials: X-ray diffraction (Lang and Newkirk in 1958); etch pits (Horn in 1952); decoration by silver in silver halide crystals (Hedges and Mitchell in 1953). Evidence from electron microscopy largely confirmed the theoretical predictions although the situation is rather more complex than had been envisaged. Dislocation behaviour varies greatly from one material to the

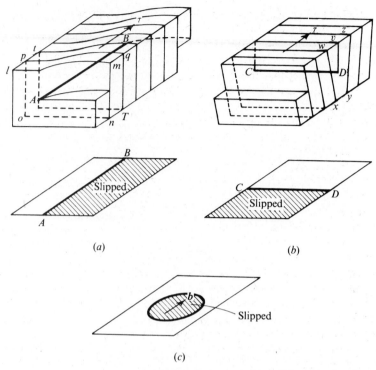

Fig. 9.16. Comparison between the formation of edge and screw dislocations by slip processes. (a) Screw dislocation AB; (b) edge dislocation CD; (c) circular dislocation containing both edge and screw segments.

next, even with the same lattice, due to variation of the stacking fault energy (see § 9.8).

The density of dislocations is expressed as the length of dislocation per unit volume or the number of dislocations intersecting unit area of a random section. For a random distribution the length density will be double the intersection density, although this is within normal margin of error.

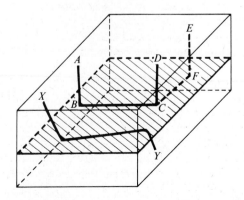

Fig. 9.17. Generalised dislocation, which does not lie in one crystal plane.

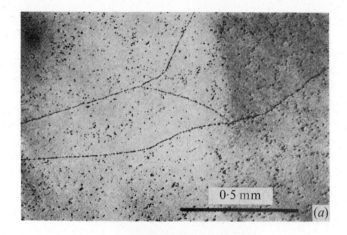

Fig. 9.18. Dislocations in magnesium oxide. (a) As revealed by etch pits in an undeformed crystal. (b) Grown in dislocations anchored by precipitate particles, revealed by electron microscopy. (From J. Washburn, in Thomas and Washburn, editors, *Electron Microscopy and Strength of Crystals*, Interscience (New York 1963).)

Dislocations are present in newly grown crystals and polycrystalline material. The average dislocation density in new ionic and covalent crystals is ~ 10 mm^{-2} and in new metallic crystals 10^3–10^5 mm^{-2}. These may be randomly distributed or some may be in the form of a two-dimensional array or network dividing the grain into sub-grains. Figure 9.18 shows a typical arrangement in MgO revealed by etch pits (*a*) and electron micrography (*b*).

Deforming a material rapidly increases the average density of dislocations, up to 10^{10} mm^{-2} in heavily cold worked metal. The formation of the dislocations depends on the particular metal. In most cubic metals and alloys, see Fig. 9.19, a cell structure develops in which the cell centres are relatively free of dislocations and the cell walls are a tangle of dislocations with a density several times greater than the average figure. The exact scale of this cell structure depends upon the metal, its purity, the temperature and mode of deformation. Typically the cells may be 1 μm across with walls ~ 0.1 μm in thickness. The dislocation density in the walls may be 10^{11} lines/mm^2 or greater, often with no visible dislocations within the cells.

Heating a cold worked metal at first reduces the number of dislocations and then rearranges them into regular two-dimensional networks dividing each grain into sub-grains whose size is typically 1 μm in grains 50 μm in diameter. Figure 9.20 shows a sub-boundary network in a molybdenum–rhenium alloy.

The origin of dislocations, both those found in virgin crystals and those produced in large numbers during deformation, has for long been a matter for speculation. Frank pointed out in 1949 that a screw dislocation intersecting the solid/solution interface provides a mechanism for crystal growth since the angle between the 'cliff face' and the slip step (see Fig. 9.16) allows of stronger adherence of new atoms than does a plane surface. Due to the constant growth rate a radial cliff rapidly becomes a spiral, many of which have been observed, see Fig. 9.21. Various other sources of the original dislocations have now been identified, such as inclusions and thermal stresses. The dislocations formed during deformation were considered at one time to develop at a low stress from the ingrown dislocations, as suggested by Frank and Read in 1950 (see § 9.10). Later the experimental evidence has shown that the additional dislocations originate from grain boundaries, inclusions and surface irregularities. They are then able to multiply during their movement through the lattice by a double cross-slip mechanism first suggested by Koehler. The details of these processes will be described later (§ 9.10).

9.8 Geometric properties of dislocations

A dislocation is specified by the orientation of the line and the magnitude and direction of the associated shear which is called the *Burgers vector **b***. The simplest definition of ***b*** is to consider a dislocation formed by slip over a central area of a slip plane, as in Fig. 9.16(*c*). The Burgers vector ***b*** is then the displacement of the material above the area relative to the material below. It will be noted that a dislocation has a single value of ***b*** all along its length. This approach is not entirely satisfactory as it depends on a particular method of forming the dislocation.

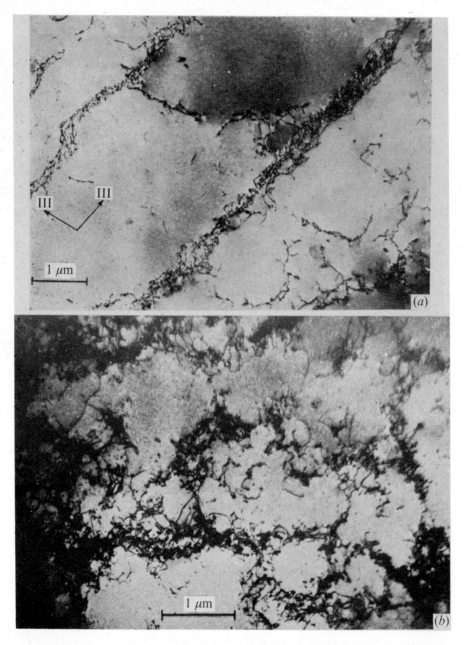

Fig. 9.19. Dislocation cell structure, by electron microscopy. (a) In lightly deformed aluminium. (b) In copper deformed 5 per cent. (From P. R. Swann, in Thomas and Washburn, editors, *Electron Microscopy and Strength of Crystals*, Interscience (New York 1963).)

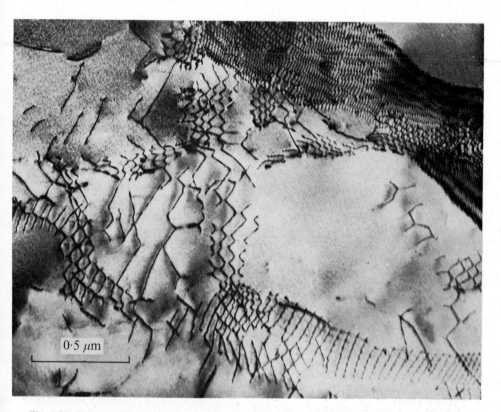

Fig. 9.20. Electronmicrograph of polygonised structure in a molybdenum 34 atomic per cent rhenium alloy, deformed 16 per cent and annealed for one hour at 1050 °C. (From M. J. Witcomb *et al., J. Materials Science,* **3**, 191 (1968).)

A more rigorous definition of the Burgers vector, due to Burgers and Frank, is obtained as follows (see Fig. 9.22): (i) arbitrarily choose an origin O and a direction OA along the dislocation as positive; (ii) make an imaginary circuit around the dislocation (from some point M), in a clockwise direction with respect to OA, by means of a series of steps along the local lattice vectors such that if the crystal were perfect the circuit would close; (iii) due to the presence of the dislocation the path is open and the Burgers vector runs from the starting point M to the finishing point N (some ambiguity exists here, as some authors use the reverse vector).

Since the atoms must move from one mechanically stable position to another as a result of the passing of a dislocation, there are usually only certain discrete values of the Burgers vector, and the corresponding dislocations are called the *characteristic* dislocations of the crystal. Dislocations are called *unit dislocations* if the Burgers vector is equal to the unit lattice vector – in the direction of **b**. A dislocation with a Burgers vector equal to several (integral) lattice vectors is called a *multiple dislocation*. A *partial dislocation* has a Burgers vector which is less than the unit vector, but two or more of whose vectors add to give a unit vector. A planar fault in the crystal lattice is caused by the passage of a single

Fig. 9.21. (*a*) Spiral depressions on the surface of an organic crystal (salol grown from solution in carbon disulphide) (from S. Amelinckx, *J. Chim. Phys.* **50**, 218 (1953)). (*b*) and (*c*) Growth spirals on the (0001) face of a silicon carbide crystal, centred on an emergent screw dislocation (from S. Amelinckx, *J. Chim. Phys.* **49**, 411 (1952)).

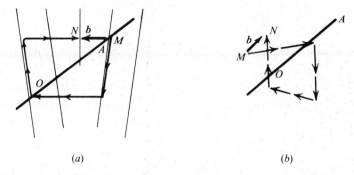

(a)   (b)

Fig. 9.22. Definition of the Burgers vector for (a) an edge and (b) a screw dislocation.

partial dislocation. Dislocations may occasionally occur whose Burgers vectors are not equal to an integral number of unit lattice vectors and are called *imperfect dislocations*.

A dislocation is often specified by giving the direction of its Burgers vector in Miller indices and its magnitude in terms of the unit translations. Thus, in a fcc lattice, a dislocation producing slip from a cube corner to a face centre is written as $(a/2)[110]$, or, in other words, the projections of b on the structure cell axes are $a/2$, $a/2$, and 0.

If a dislocation moves along a plane containing its Burgers vector it produces simple shear displacement and the faces of the slip plane maintain their spacing (conservative motion). This is called *glide* and the plane is a glide, or slip, plane. On the other hand, if an edge dislocation moves over a plane not containing its Burgers vector interstitial atoms or voids will be produced with a consequent change in the slip plane density (see Fig. 9.23). This is called *climb* and it can normally occur only at temperatures high enough for atomic diffusion. Edge dislocations glide or climb but screw dislocations always glide, since their Burgers vector lies along their length. When a screw dislocation gliding on one plane changes to gliding on an intersecting plane it is called *cross slip*.

When one dislocation cuts through another one a step, or *jog*, is formed in each of them, equal to, and in the direction of, the other's Burgers vector, see Fig. 9.24(a). If the Burgers vector of the first dislocation is parallel to the line of

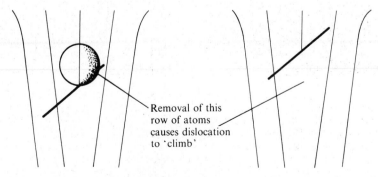

Removal of this
row of atoms
causes dislocation
to 'climb'

Fig. 9.23. Climb of an edge dislocation by vacancy 'condensation'.

the second dislocation, then no jog can be formed in the second dislocation. (Fig. 9.24(*b*)). The jogs represent extra energy and hence a stress is required to drive a dislocation through intersecting, or *forest*, dislocations. The forest stress is usually a major component of the overall stress required to move a dislocation over a slip plane, called the *friction stress*. The jog once formed may give rise to a further component of the friction stress since certain types of jog leave a trail of interstitial atoms or vacancies as they move forward with the dislocation. For

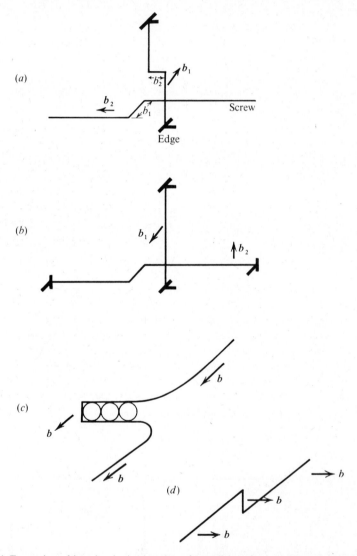

Fig. 9.24. Formation of jogs by the intersection of two dislocations. (*a*) Intersection of edge and screw dislocations produces jogs in each dislocation. (*b*) Intersection of two edge dislocations: in this case b_2 is parallel to the line of dislocation 1, and no jog is therefore formed in 1. (*c*) Jog in screw is edge in character; as screw moves forward edge is left behind, equivalent to the formation of a row of vacancies. (*d*) Jog in edge, also edge in character.

example, a jog in a screw dislocation perpendicular to the glide plane will be a short length of edge dislocation which will have to climb as the screw dislocation glides (Fig. 9.24(*c*)). However, the jog can move conservatively along the screw dislocation. On the other hand a jog in an edge dislocation will remain edge type and travel with the main dislocation. It will also assist climb by providing a point at which vacancies or interstitials can nucleate (Fig. 9.24(*d*)).

Dislocations can form junctions (nodes) within a crystal but can never end except at the surface. Two conditions must be met across a node:

(i) the Burgers vector must maintain its value i.e. $\Sigma \boldsymbol{b} = 0$;

(ii) the line tension forces must be in equilibrium (see § 9.9).

A junction found frequently in the fcc lattice is between three symmetrical dislocations on the (111) plane:

$$\frac{a}{2}[10\bar{1}] + \frac{a}{2}[0\bar{1}1] + \frac{a}{2}[\bar{1}10] = 0. \tag{9.4}$$

This leads to the formation of a hexagonal network.

In fcc and cph lattice a unit screw or edge dislocation in the close-packed planes can divide into two partial dislocations which move as a pair. This is called an *extended* dislocation. Figure 9.25 shows a close-packed plane of atoms A; the next layer of atoms is assumed to be sitting in the B hollows. A dislocation with Burgers vector $\boldsymbol{b}_1$, will displace atom B_1 to B_2, a unit lattice vector. The same displacement is obtained from two partial dislocations $\boldsymbol{b}_2$ and $\boldsymbol{b}_3$ which moves the atoms $B_1 \rightarrow C_1 \rightarrow B_2$. The position C_1 is mechanically stable, but it involves a fault in the stacking arrangement of adjacent close-packed

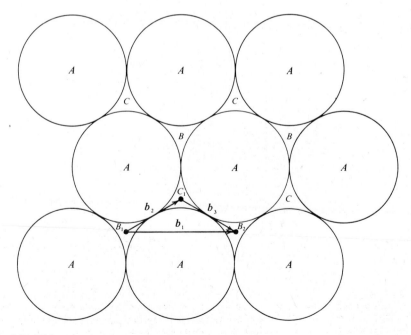

Fig. 9.25. Dislocation in a close-packed plane with Burgers vector $\boldsymbol{b}_1$, $(a/2)[1\bar{1}0]$, splits into two partials with Burgers vectors $\boldsymbol{b}_2$, $(a/6)[1\bar{2}1]$ and $\boldsymbol{b}_3$, $(a/6)[2\bar{1}\bar{1}]$.

planes. The reaction may be written in the Miller indices notation

$$\frac{a}{2}[1\bar{1}0] \rightarrow \frac{a}{6}[1\bar{2}1] + \frac{a}{6}[2\bar{1}\bar{1}].$$ (9.5(a))

Using $a/6$ as the unit of length this may be written simply as

$$[3\bar{3}0] \rightarrow [1\bar{2}1] + [2\bar{1}\bar{1}].$$ (9.5(b))

The two partial dislocations, known as Schockley partials, will repel each other and move apart until the reduction in elastic strain energy is balanced by the increase of stacking fault energy of the atoms in between. Extended dislocations have been observed under the electron microscope, see Fig. 9.26. The separation varies from metal to metal, being about ten atomic spacings in copper and two in aluminium. This is thought to account for many of the differences between these metals since an extended dislocation is only free to move along the fault plane. For an extended screw dislocation to cross slip it must first reunite into a simple unit dislocation; this is shown schematically in Fig. 9.27.

Dislocation reactions can also occur between dislocations on intersecting planes. An important example is that between a pair of partial dislocations on the (111) plane and another pair on the intersecting (11$\bar{1}$) plane. The first members of each pair coalesce into a new partial dislocation which is tied by separate stacking faults to the other partials. The trio, called a Lomer–Cottrell lock, can neither glide nor climb. It is thought to play an important part in the work hardening of fcc metals. In symbols the reactions (in $a/6$ units) are:

in (111) plane: $[30\bar{3}] \rightarrow [2\bar{1}\bar{1}] + [11\bar{2}]$ (a)

in (11$\bar{1}$) plane: $[033] \rightarrow [\bar{1}21] + [112]$ (b)   } (9.6)

at $[1\bar{1}0]$ intersection: $[2\bar{1}\bar{1}] + [\bar{1}21] \rightarrow [110].$ (c)

Fig. 9.26. Extended dislocations and their associated stacking faults in annealed niobium ($\times 3040$). (By courtesy of A. V. Narlikar.)

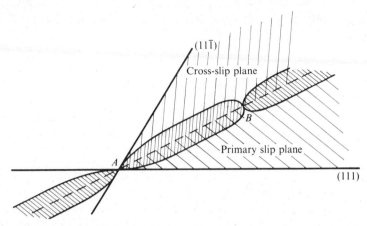

Fig. 9.27. Cross slip of extended dislocation. Between A and B the dislocation has first reunited into a perfect dislocation, before separating again into partials on the cross-slip plane.

9.9 Elastic properties of dislocations

A dislocation produces an elastic strain field in the surrounding crystal from which many of the properties of dislocations can be deduced.

The field of a pure screw dislocation is pure shear and is easily calculated as follows. Consider a thin cylindrical annulus with radius r, thickness dr and unit length, see Fig. 9.28. A screw dislocation of strength b along the axis will shear the ring by an amount b, as shown. The average shear strain is then $b/2\pi r$ and the shear stress

$$\tau_{\theta z} = \tau_{z\theta} = \frac{Gb}{2\pi r}. \tag{9.7}$$

(The stress in Cartesian coordinates is given in the figure.)

Since the field decreases only as $1/r$, the stress is long range. Even at $10^4 b$ the stress is $10^{-5}G$, which is comparable to the yield stress of soft crystals. Strong

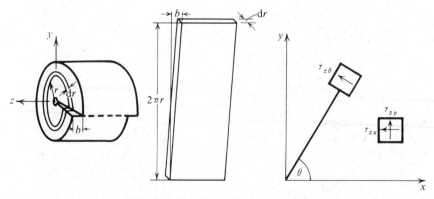

Fig. 9.28. The stress field of a screw dislocation.

$$\gamma = b/2\pi r, \quad \tau_{z\theta} = \frac{Gb}{2\pi} \cdot \frac{1}{r} \quad \text{or} \quad \tau_{zx} = \frac{Gb}{2\pi} \cdot \frac{y}{x^2 + y^2}; \quad \tau_{zy} = \frac{Gb}{2\pi} \cdot \frac{x}{x^2 + y^2}.$$

elastic interaction between dislocations is therefore to be expected. These stresses are named after Taylor who first suggested them as a cause of strain hardening (see § 9.11).

The strain energy of the ring per unit length is

$$\frac{\tau_{z\theta}^2}{2G} \times 2\pi r \, dr = \frac{Gb^2 \, dr}{4\pi r},$$

and the strain energy of the crystal per unit dislocation length is

$$\frac{Gb^2}{4\pi} \int_{R_0}^{R} \frac{dr}{r} = \frac{Gb^2}{4\pi} \ln \frac{R}{R_0} \tag{9.8}$$

where R_0 and R are the lower and upper limits. R_0 is the radius below which Hooke's law is no longer valid and the atomic structure of the material must be considered. The volume within R_0 is called the *core* of the dislocation. The value of R_0 is not critical since the energy is a logarithmic function of it. Hooke's law is generally considered to be valid to about 10 per cent strain, which gives a value for R_0 of about $1.6b$. The upper limit R is the boundary of the crystal or the point at which other dislocations cancel out the stress field.

The stress field of an edge dislocation is more complex to derive. The results are (see Fig. 9.29 for notation):

$$\sigma_{xx} = \frac{-Gb}{2\pi(1-v)} \frac{y(3x^2 + y^2)}{(x^2 + y^2)^2} \qquad \sigma_{yy} = \frac{Gb}{2\pi(1-v)} \frac{y(x^2 - y^2)}{(x^2 + y^2)^2}$$

$$\tag{9.9}$$

$$\sigma_{zz} = v(\sigma_{xx} + \sigma_{yy}) \qquad \tau_{xy} = \tau_{yx} = \frac{Gb}{2\pi(1-v)} \frac{x(x^2 - y^2)}{(x^2 + y^2)^2}$$

and the elastic strain energy per unit length is

$$\frac{Gb^2}{4\pi(1-v)} \ln \frac{R}{R_0}. \tag{9.10}$$

An expression for the elastic strain energy which averages the results for screw and edge dislocations, and which therefore may be applied to the general dislocation, is

$$\frac{Gb}{4\pi(1-v/2)} \ln \frac{R}{R_0}. \tag{9.11}$$

The total energy of the dislocation is obtained by adding to the above expression the energy of the core. This cannot be calculated easily, but it has been estimated that reducing R_0 from $1.6b$ to b will adequately compensate for otherwise neglecting the core energy. Thus the total energy of a dislocation, per unit length, is given by:

$$\frac{Gb^2}{4\pi(1-v/2)} \ln \frac{R}{b} \tag{9.12}$$

and the energy per atom length (b) is:

$$\frac{Gb^3}{4\pi(1-v/2)} \ln \frac{R}{b}. \tag{9.13}$$

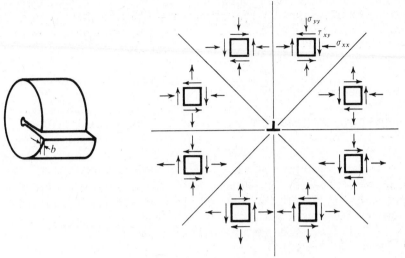

Fig. 9.29. The stress field of an edge dislocation.

$$\sigma_{xx} = \frac{-Gb}{2\pi(1-v)} \cdot \frac{y(3x^2 + y^2)}{(x^2 + y^2)^2}, \quad \sigma_{yy} = \frac{Gb}{2\pi(1-v)} \cdot \frac{y(x^2 - y^2)}{(x^2 + y^2)^2},$$

$$\sigma_{zz} = v(\sigma_{xx} + \sigma_{yy}), \quad \tau_{xy} = \tau_{yx} = \frac{Gb}{2\pi(1-v)} \cdot \frac{x(x^2 - y^2)}{(x^2 + y^2)^2}.$$

Putting in typical values ($G = 3 \times 10^{10} \, \text{N/m}^2$, $b = 0.3$ nm, $v = 0.34$) the elastic field energy per atomic length is, for $R = 10$ mm, $\sim 1.3 \times 10^{-18}$ J (8eV) and for $R = 10^{-3}$ mm, $\sim 6.5 \times 10^{-19}$ J (4eV). The energy is quite large and dislocations would not be present in thermal equilibrium, but, as with grain boundaries, it is a difficult process to eliminate them.

A consequence of the elastic energy is that a dislocation exhibits line tension and tries to shorten itself. The value of the line tension is the energy per unit length, which, as shown, depends to some extent on the environment, that is the range over which the stress field acts. It may be taken as αGb^2, where α is in the range 0.5 to 2.0.

There are several consequences of the result that the energy of a dislocation is proportional to the square of its Burgers vector. Multiple dislocations will always tend to divide into unit dislocations, and their occurrence is rare. Unit dislocations will split into partials, and be separated by a ribbon of stacking fault (Fig. 9.26), provided that the crystal structure will allow the formation of partial dislocations with Burgers vectors at angles less than 45° to the original unit Burgers vector. If the angles are larger than 45° then there is a net increase in energy. This is shown as follows:

$$|\boldsymbol{b}| = |\boldsymbol{b}_1| \cos \theta + |\boldsymbol{b}_2| \cos \theta$$

where $|\boldsymbol{b}|$ is the magnitude of the Burgers vector $\boldsymbol{b}$ of the unit dislocation, and $\boldsymbol{b}_1$ and $\boldsymbol{b}_2$ refer to the partial dislocations, each (by symmetry) making an angle θ with $\boldsymbol{b}$.

Therefore

$$|\boldsymbol{b}_1| = |\boldsymbol{b}_2| = \frac{|\boldsymbol{b}|}{2 \cos \theta}.$$

For there to be a reduction in energy

$$|b_1|^2 + |b_2|^2 = \frac{2|b|^2}{4\cos^2\theta} < |b|^2,$$

that is $\cos^2\theta > \frac{1}{2}$, or $\theta < 45°$. As mentioned in § 9.8, pairs of partials separated by a ribbon of stacking fault are called extended dislocations.

The final consequence of the energy being dependent upon the square of the Burgers vector is that unit dislocations will normally have the smallest possible Burgers vector, which is the distance of closest approach of like atoms. This explains why the slip direction, which is the direction of the Burgers vector of the dislocations producing the slip, is always the atomically close-packed direction.

The stress fields of two dislocations, or of a dislocation and another source of stress such as an impurity, will interact since the combined strain energies will be different from the sum of the separate energies:

$$\frac{(\sigma_1 + \sigma_2)^2}{2G} \neq \frac{\sigma_1^2}{2G} + \frac{\sigma_2^2}{2G}.$$

There is thus a force between the field sources, which can be determined by differentiation of the function for the combined strain energies. A simpler but less rigorous approach will be adopted here.

Consider an element of a crystal containing a slip plane of unit area over which a dislocation with Burgers vector b passes, Fig. 9.30. Let τ be the shear stress in the direction of b and F be the force per unit dislocation length acting perpendicular to the dislocation. The force arises from the work done on the crystal as it is deformed. Hence

$$\int_A^B Fl\,dx = \tau b,$$

i.e. per unit length $F = \tau b$ (a)

and per atomic length $f = \tau b^2.$ (b)   (9.14)

This is called the Mott and Nabarro formula.

It follows that two parallel screw dislocations, whose stress fields are given by (9.7), exert a force on each other along their line of centres equal to

$$F_r = \frac{Gb^2}{2\pi r} \quad \text{per unit length,} \tag{9.15}$$

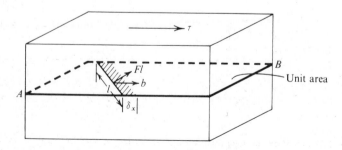

Fig. 9.30. The force on a dislocation in a shear stress field.

where r is the separation. The force is repulsive between like dislocations. A typical value at $r = 10^{-3}$ mm is 1.6×10^3 Gb^2 or 0.6 mN/m.

For two parallel edge dislocations, positioned as in Fig. 9.31, there is a force along the glide plane arising from the shear component of the stress fields, (9.9), given by

$$F_x = \frac{Gb^2}{2\pi(1 - v)} \frac{x(x^2 - y^2)}{(x^2 + y^2)^2},$$ (9.16)

and a force perpendicular to the glide plane, that is, in the climb direction, due to the direct stress component (σ_{xx})

$$F_y = \frac{Gb^2}{2\pi(1 - v)} \frac{y(3x^2 + y^2)}{(x^2 + y^2)^2}.$$ (9.17)

The glide force F_x will be repulsive between like dislocations on the same glide plane and on different glide planes as long as $x > y$. But if $x < y$ there is a change of sign and like dislocations are attracted into alignment directly above each other. Typical plots of F_x versus position on the same and adjacent glide planes are given in Fig. 9.31.

When a dislocation splits into partials, the partial dislocations tend to separate due to this repulsive force, creating a ribbon of stacking fault between them. Associated with the fault is an energy per unit area γ, and as the partials move apart they lower their elastic energy, but the energy due to the stacking fault is increased. An equilibrium width w is reached when the elastic repulsion is just balanced by the attraction in trying to minimise the fault energy. This is given approximately by:

$$\frac{Gb^2}{2\pi w} \simeq \gamma$$

or

$$w \simeq \frac{Gb^2}{2\pi\gamma}.$$ (9.18)

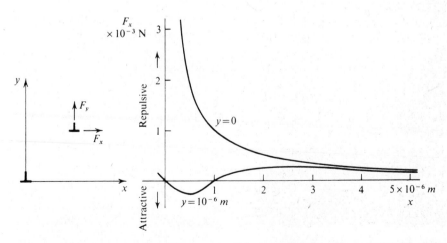

Fig. 9.31. The forces per unit length between parallel edge dislocations. F_x acts in the glide plane, F_y acts normal to this plane and will try to cause the dislocations to climb. Curves for F_x drawn assuming $Gb^2/2\pi(1 - v) = 10^{-9}$ N.

(The above assumes both partials are screws; this is, in fact, impossible and the exact formula for the width of an extended dislocation depends upon the angle between the Burgers vector of each partial and the line of the dislocation.) It can be seen that crystals with a low stacking fault energy will have wide extended dislocations. The width of extended dislocations has a profound effect on many material properties.

The repulsion between similar dislocations on a glide plane leads to a characteristic arrangement of dislocations driven against an obstacle such as a grain boundary. A typical pile-up is shown in Fig. 9.32. The spacing is small at the head of the array and widens towards the tail. Large stresses concentrated at its head play an important role in initiating yield across grain boundaries (see § 11.5) or fracture (§ 10.15). The pile-up can be likened to a queue of people all pushing on the person ahead.

A dislocation approaching such a 'pile-up' on the same, or a closely parallel, plane will experience a repulsive force. An extra stress must be applied to overcome this 'back-stress' and keep the dislocation moving. It is this back-stress from a pile-up which is the principal cause of work hardening. It may cause an approaching dislocation to 'cross slip' on to another slip plane and so avoid the pile-up. However, if the dislocation is extended, it will find it difficult to transfer to another plane. Pile-ups therefore tend to be more serious in crystals with low stacking fault energy, leading to higher rates of work hardening, and also greater

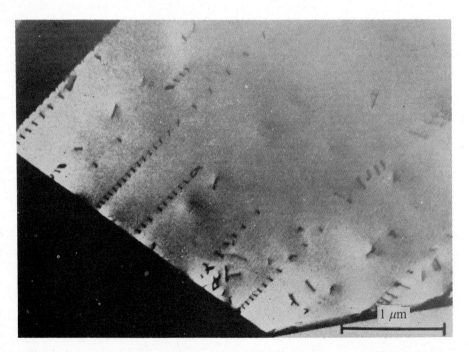

Fig. 9.32. Dislocation pile-ups at a twin boundary in a lightly deformed copper 4.5 per cent aluminium alloy. (From P. R. Swann, in Thomas and Washburn, editors, *Electron Microscopy and Strength of Crystals*, Interscience (New York 1963))

susceptibility to stress-corrosion cracking (dislocation pile-ups form a point of enhanced corrosive attack).

The reason for the climb force F_y is simply explained: under a compressive stress the extra plane of atoms forming an edge dislocation tends to remove itself and thereby reduce the strain energy of the surrounding stress field. Climb is a thermally activated process, with an activation energy identical to that of self-diffusion. This is because climb, like diffusion, is vacancy controlled; a row of vacancies settling along the edge of the extra half-plane will cause it to rise by one atom spacing.

The stress around dislocation pile-ups may be relieved by climb, and this is one of the principal mechanisms in the process of 'recovery' which occurs in the early stages of annealing. Climb processes are also important in creep, preventing the formation of pile-ups at elevated temperatures and thus removing the principal source of work-hardening stress.

The attractive force between dislocations on adjacent parallel glide planes gives rise to an important phenomenon called *polygonisation* (described by Cahn in 1949). In the simplest case, shown schematically in Fig. 9.33, a crystal is

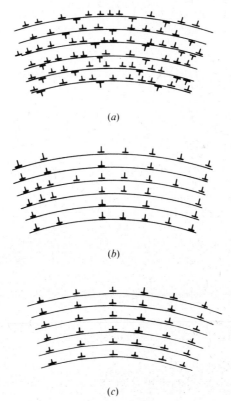

(a)

(b)

(c)

Fig. 9.33. Dislocation rearrangement during polygonisation. (a) Dislocations randomly distributed in the glide planes of a bent crystal. (b) Dislocations have rearranged themselves by glide, dislocations of opposite sign have been annihilated. (c) Dislocations climb to develop walls perpendicular to the glide planes.

Fig. 9.34. Polygonised walls of dislocations in a silicon carbide crystal, revealed by etch pits. (From S. Amelinckx and G. Strumane, in *Silicon Carbide* edited by J. R. O'Connor and J. Smittens. © 1960, Pergamon Press. Reprinted with permission.)

bent into an arc of a circle, producing an irregular array of edge dislocations. When the material is heated so that climb can occur, the edge dislocations rearrange themselves one above the other, perpendicular to the slip plane. The curved lattice becomes a polygon, with about 1° lattice tilt across each boundary, depending on the dislocation density. This is shown by means of etch pits in Fig. 9.34. The effect was first observed by X-rays: the Laue spots become elongated in cold worked metal but break up into discrete spots when the metal is annealed. More complex two-dimensional networks are formed during recovery of polycrystalline material, involving several sorts of dislocation produced by generalised deformation. One example of such a low angle boundary network was shown in Fig. 9.20. The presence of sub-grains, about 1 μm diameter, has a marked effect on the mechanical properties.

Dislocations and solute atoms will also interact on account of their stress fields. Where the solute atom causes direct stresses, interaction is only with edge-type dislocations. An oversize substitutional atom or an interstitial atom is attracted towards the underside of an edge dislocation where the lattice is in tension. Sometimes an atom also produces shear distortion and it then reacts with screw dislocations. An important example is found in body-centred iron, where the interstitial atoms carbon and nitrogen react strongly with both screw and edge dislocations and are responsible for the sharp yield point in steel (see § 11.5).

9.10 Sources and multiplication of dislocations

It was noted earlier (§ 9.7) that dislocations present in virgin crystals are not sufficient in number to provide the large plastic strains involved in yielding nor would they produce large amounts of slip on a few closely spaced planes, as is found in slip bands.

Frank and Read in 1950 suggested how the original dislocations could act as sources capable of generating very much larger numbers of dislocations. Two variants were proposed, shown in Fig. 9.35. In (*a*), now known as the Frank–Read source, a portion *AB* of a dislocation lies in a possible slip plane, with both its ends pinned due to the dislocation leaving the slip plane. When a shear stress is applied, the dislocation first bows out like a bubble, and then closes back on itself; finally it forms a closed loop free of the pinned ends and at the same time the dislocation between *A* and *B* is regenerated. The dislocation loop spreads to the boundaries of the slip plane (or up to any obstacles) whilst the pinned dislocation generates further loops. In this way slip lines with step heights of several thousand atomic spacings can be formed. The stress to operate the source is that required to overcome the line tension *T* of the dislocation as it is bowed out. The critical condition is when it has reached the semicircular shape. Resolving forces for equilibrium (the equation is the same as that for a thin cylindrical shell under internal pressure):

$$\tau b \Lambda = 2T$$

$$\tau = \frac{2T}{b\Lambda} \simeq \frac{Gb}{\Lambda} \tag{9.19}$$

where Λ is the length between the pinned ends of the dislocation. At $\Lambda = 10\,\mu m$ τ equals 3×10^{-5} G, about the yield stress of the soft group of metal crystals (§ 9.3). This calculation assumes there is no friction stress opposing dislocation motion. An alternative scheme, shown in Fig. 9.35(*b*), has one end of the

Fig. 9.35. Multiplication of grown-in dislocations by the Frank–Read mechanisms, viewed normal to the slip plane.

dislocation pinned and the other end at a free surface. Under a shear stress the free end moves round the pinned end many times to produce large slip on the plane.

The Frank–Read mechanism has been observed only infrequently and it seems that the as-grown dislocations are not the normal source of dislocations during yielding. Instead, dislocations are nucleated at grain boundaries and at surface or internal stress concentrators, such as an inclusion. The screw dislocations (or screw component) then multiply by a double cross-slip mechanism, which was proposed by Koehler in 1952 and subsequently observed in both non-metals (e.g. in LiF by Johnston and Gilman in 1960) and metals (in Fe–3% Si by Low and Turkalo in 1962). The mechanism involves the Frank–Read type of generation but, since it occurs in new, moving dislocations rather than the initial ones, it is not called a Frank–Read source.

The possible variations of double cross-slip multiplication are shown in Fig. 9.36. A segment of a screw dislocation AB which is gliding in plane 1 cross slips into an intersecting plane and then cross-slips again into plane 2 parallel to the first plane. Two jogs P and Q are formed. If the jogs are very small they will be able to move with the dislocation by forming vacancies or interstitials (Fig. 9.36(a)). If the jogs are fairly large this process is too slow and they effectively pin the dislocation at these points. The dislocation segment then proceeds to multiply in the Frank–Read manner (Fig. 9.36(b)). Further double cross slip

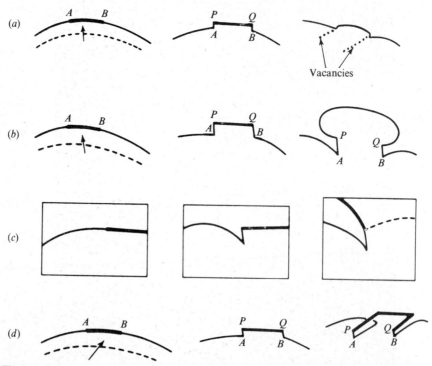

Fig. 9.36. Dislocation multiplication mechanisms based on cross slip. (a) Cross-slip step is only a few interatomic distances, vacancy trails are created. (b) Cross-slip step is large, and dislocation segments operate quite independently of one another. (c) As in (b) but one end of the segment is at a crystal surface. (d) Cross-slip step is insufficiently large for segments on the different slip planes to move independently. Sessile loops are left behind.

will produce a collection of close-spaced slip lines, that is, a slip band. In the second variant, the segment which cross slips is at the end of a dislocation on the surface of the crystal. Only one jog is formed and multiplication is by the second type of Frank–Read mechanism (Fig. 9.36(c)).

If the jogs produced by double cross slip are of intermediate height, a third effect is observed, shown in Fig. 9.36(d). The jog is too large to move by point production but too small for the dislocations on planes 1 and 2 to free themselves from each other's field and act as sources. Instead, as the screw dislocations advance, they remain connected to the jogs by edge dislocations, of opposite sign on the two planes. These dipole dislocations are observed in quantity in strain hardening materials. Figure 9.37 shows all three effects occurring in Fe–3 % Si.

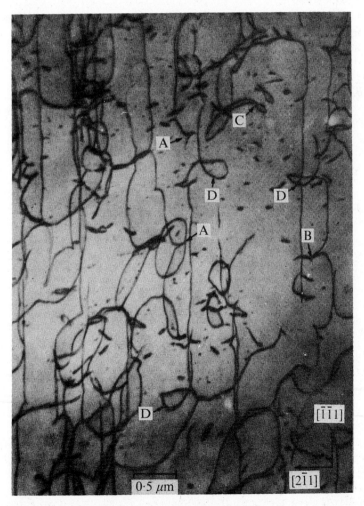

Fig. 9.37. Double cross slip and trails of debris (as shown in Fig. 9.36(d)) in an Fe–3 %Si alloy. The foil is parallel to the (011) slip plane. Note: dipole trails at *A*; pinching off of trails at *B*, *C*; single ended sources from large jogs at *D*. (Reprinted with permission from J. R. Low and A. M. Turkalo, *Acta Met.* **10**, 215 (1962), Pergamon Press.)

9.11 Elastic limit of pure metals

It is surprising that, although dislocation theory originated from the need to explain the low value of the critical resolved shear stress in pure metal, it is still not possible to state unequivocally what is the determining factor for yield. Much greater progress has been made in defining the factors affecting the crss in various alloys (see Chapter 11). This is due to the complexity of the interaction between dislocations and the sensitivity of the crss of pure crystals to the dislocation arrangement and to any impurity atoms present.

In the initial theories it was assumed that dislocations were present in virgin crystals, that the stress required to move a dislocation in an otherwise perfect lattice (now called the Peierls–Nabarro stress) was negligible, and that yield was determined by the long range elastic interaction between an array of dislocations, the Taylor stresses. When plastic flow had removed the initial supply of dislocations, further dislocation pairs would form.

Since then, various attempts have been made to calculate the Peierls–Nabarro stress. This has not been easy as it is critically dependent on the width of the dislocation core, which is determined by several unknown factors, for example the relationship between shear stress and strain for the atoms in the core. It would appear that the original assumption that this stress is not the yield stress is correct in most metals but it may be important in ionic and covalent crystals. It has also been realised that the formation of dislocation pairs would require the local stress to reach the theoretical value for homogeneous shear and the Frank–Read mechanism of dislocation multiplication from the as-grown dislocations was suggested. The operating stress may control the yield. In addition to the Taylor stresses, various other obstacles to dislocation motion have been discovered, either interaction with other dislocations or with point defects. These developments will be considered briefly in the following paragraphs. Finally a radically different approach has been put forward which considers the dynamic behaviour of new dislocations, the as-grown dislocations playing no part in the yield. This is based on extensive experimental study of LiF by Johnston and Gilman but is applicable to other ionic crystals, covalent crystals, and some metals (§ 9.13).

Across a dislocation the relative positions of the atoms on opposite sides of the slip plane change by the Burgers vector, as shown in Fig. 9.38. The dislocation width is defined as the span over which the displacement changes from $-b/4$ to $+b/4$. Widths of one or two atomic spacings are regarded as narrow, while over five are wide. The factor tending to increase the width is the long range elastic strain energy and opposing it is the misalignment energy of the atoms in the core. The Peierls stress decreases rapidly with increasing width. It is therefore clear why slip is on the close-packed planes; since the plane spacing is higher, the alignment energy is lower and the dislocation width is higher. This also explains the ease of plastic flow in metals compared with non-metals: the metallic bond is non-directional and the alignment forces are low compared with ionic or covalent bonds.

Using a simple shear stress displacement law, Peierls obtained a dislocation width of $b/1 - v$, typically 1.5 atomic spacings. The corresponding friction stress is 3.6×10^{-4} G. However a different assumption for the stress-displacement

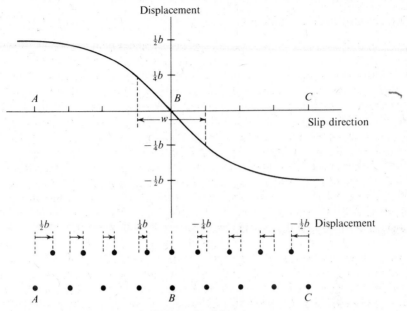

Fig. 9.38. Width, *w*, of a dislocation core.

law can give a slightly wider dislocation and a very much lower Peierls stress: at only three atoms width it is reduced by a factor of 10^4.

The various static effects which may be responsible solely or together for the elastic limit of pure crystals can be summarised as follows:

(*a*) Peierls–Nabarro stress.

(*b*) Taylor long-range elastic stress.

(*c*) Frank–Read source operating stress.

(*d*) Forest stress (required to cut other dislocations intersecting the glide plane).

(*e*) Junction stress (required to separate two attractive dislocations on intersecting planes which have formed a junction).

The yield stress is controlled by the largest of these. In non-metallic crystals, and in bcc metals at low temperatures, the Peierls stress is high and is believed to be the controlling factor. In other pure metals the Peierls stress is low and one or more of the other factors must determine the yield stress. As these depend on the density and distribution of other dislocations, the yield stress cannot be predicted for pure metal crystals, and the wide variation of experimental results is understood. The effects of microstructure, impurity and alloy content, and previous cold work are so great (see Chapter 11) that the strength of pure metal single crystals is, however, only of academic interest.

9.12 Dislocations in non-metals

Dislocations are expected to occur in any crystalline solid, and indeed have been found in most. The principles enunciated for metallic crystals are adhered to, but additional complications often arise. These are, to some extent, due to the fact that there is usually more than one atom associated with each crystal lattice point, and these atoms are often of different species. The Burgers vector then becomes more than one interatomic distance in length, and dislocation energies ($\propto b^2$) rise very rapidly. These high dislocation energies limit the number of naturally occurring dislocations in such materials, and the more complicated stress fields render the fewer dislocations more difficult to move. These facts directly account for the frequently found limited ductility of non-metallic crystals. Because the subject of dislocations and plastic flow in non-metallic crystal structures is so complicated, it can be given only a brief treatment here, and just a few examples will be given to illustrate the main points of difference from dislocation in metals.

In ionic solids a complication arises because similarly charged atoms must not become nearest neighbours, even momentarily, during the deformation process. Sodium chloride crystallises in the fcc lattice structure (shown in Fig. 3.2(a)) and its Burgers vector is $(a/2)\langle 1\bar{1}0\rangle$. (The shortest interatomic distance is $(a/2)\langle 001\rangle$, but this is from a sodium to a chlorine atom.) The glide plane for dislocations should be, as in fcc metals, $\{111\}$, but it is found to be $\{110\}$, as shown in Fig. 9.39(a). A similar discrepancy is found in the caesium chloride lattice structure which is simple cubic (Fig. 3.2(b)). Here the slip system should be $\langle 001\rangle$ $\{001\}$, but is in fact $\langle 001\rangle$ $\{110\}$ (Fig. 9.39(b)). The reason for choice of slip system in these structures is not known; it has been suggested that the Peierls–Nabarro force is much lower on the chosen system than on the crystallographically expected one. A consequence is that in the NaCl structure there are only two, and in the CsCl structure only three, independent slip systems, and crystals cannot undergo arbitrary changes of shape. This accounts for the brittleness of many ionic solids, as mentioned in § 7.2.

Alumina is hexagonal (Fig. 7.12) and the slip system is the expected $\langle 11\bar{2}0\rangle$ $\{0001\}$, but the dislocation is extended normal to the slip plane, as the aluminium and oxygen ions are on different planes. During the passage of a dislocation the ions move along a zig-zag path, and different ions are moving in different directions at a given time, by a process known as 'synchroshear' (described by Kronberg in 1957): see Fig. 9.40. The Peierls–Nabarro force for Al_2O_3 is high, and alumina is one of the strongest solids. Similar shear processes are believed to occur in other complicated ionic crystals and in some intermetallic compounds.

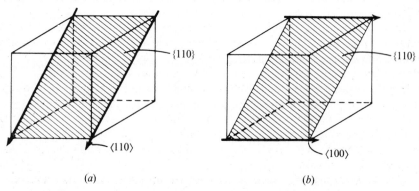

Fig. 9.39. The glide elements in ionic crystals: (a) NaCl, (b) CsCl.

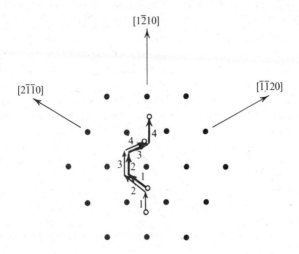

●   Layer of oxygen ions

o→   Motion of next layer of oxygen ions

o➡   Motion of interstitial layer of aluminium ions

Fig. 9.40. 'Synchroshear' in Al_2O_3. The paths followed by the interstitial aluminium atoms, and by the next layer of oxygen atoms, during slip, are not identical, though the final transition is the same. A dislocation in Al_2O_3 is thus extended normal to the slip plane. (After Kronberg.)

Covalent materials have strongly directed bonds, low values of Poisson's ratio v, and hence narrow dislocations with large Peierls–Nabarro forces. In addition dislocations with some edge component will have unsatisfied (dangling) bonds: for example germanium and silicon crystallise in the diamond cubic (fcc) structure, and as expected, the slip system is $\langle 1\bar{1}0 \rangle \{111\}$. Pure edge dislocations do not occur; only pure screw and the '60 degree' dislocation, with the Burgers vector at 60° to the dislocation line. The bond-distortion energy of other dislocations is too high. The dangling bonds of a 60° dislocation are shown in Fig. 9.41. These dangling bonds affect the electrical properties, as described in Chapter 14.

Dislocations in germanium and silicon are reluctant to move at ordinary temperatures and the materials are brittle. Ductility may be introduced by raising the temperature, irradiating with light, applying an electric field, or by doping with donor impurities. All of these will provide more electrons in the conduction band, and thus render the material more metallic in character. This has been advanced as a possible explanation for the increase in ductility. The motion of dislocations in covalent materials is found to be a thermally activated process. Activation energies are a few electron volts ($\sim 10^{-18}$ J) but do not appear to be related to the band gap energy, as would be required by the above explanation.

Covalent compounds such as indium antimonide, with the zinc blende structure (fcc), have two types of 60° dislocations, one lying on a plane of indium

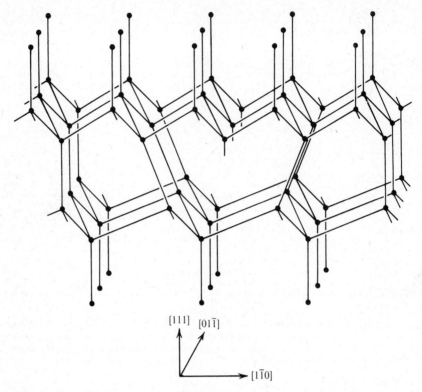

Fig. 9.41. Structure of a 60° dislocation in the diamond cubic structure. Its Burgers vector lies at 60° to the line of the dislocation. A pure screw dislocation can be formed in this structure, but a pure edge dislocation cannot.

atoms and one lying on a plane of antimony atoms (see Fig. 9.42). The dynamics of the two types of dislocation are different, leading to interesting effects.

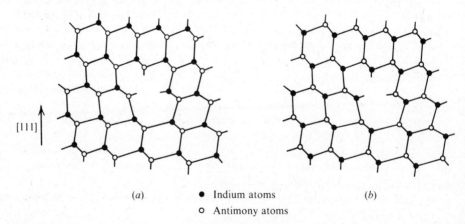

Fig. 9.42. 60° dislocation in InSb (zinc-blende structure). (a) Dislocation ending on a plane of indium atoms. (b) Dislocation ending on a plane of antimony atoms.

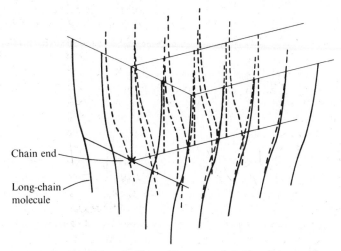

Fig. 9.43. An edge dislocation formed by a row of chain ends in a polymer single crystal.

Dislocations have been observed in polymer single crystals, and are associated with chain ends within the crystal (Fig. 9.43). They play no part in the plasticity properties of polymers, which, as discussed in Chapter 8, are determined by viscous effects in amorphous or glassy material, and will not receive further mention here.

9.13 Dynamical theory of yielding

The yielding of lithium fluoride and its dislocation dynamics have been extensively studied by Johnston and Gilman from 1956 onwards. This material has a simple lattice structure (NaCl type, see Fig. 2.14), a very low as-grown dislocation density (5×10^2 dis/mm^2) and a reagent is available which develops dislocation etch pits in such a way that the movement of dislocations can be followed.

It has been found that dislocations are easily formed around surface defects and their supply is not a limiting factor in yielding. The as-grown dislocations must be effectively pinned as they play no part in yielding. Appreciable movement of fresh dislocations occurs when the stress reaches some value (~ 4MN/m^2) (this is confusingly called the yield stress in their early papers). The dislocation velocity increases rapidly with stress, approximately as the twenty-fifth power, but at high stresses it levels off at near the speed of shear waves. This is shown in Fig. 9.44. At the lower velocities, around 0.1 mm/s, which are typical of those occurring in normal tensile testing, edge dislocations move fifty times faster than the screw components. At some stress (~ 5 MN/m^2) screw dislocations begin to cross slip and multiply by the mechanism proposed by Koehler (see § 9.10, Fig. 9.36). A moving screw dislocation always trails behind it numerous line defects which are narrowly spaced dipoles of positive and negative edge dislocations formed at intermediate height jogs, see Fig. 9.45. Each line now breaks up into a series of short lengths which are prismatic (rectangular) loops capable of acting as Frank–Read sources when the applied stress is high enough. These dislocations multiply in turn by the Koehler mechanism and a glide band forms.

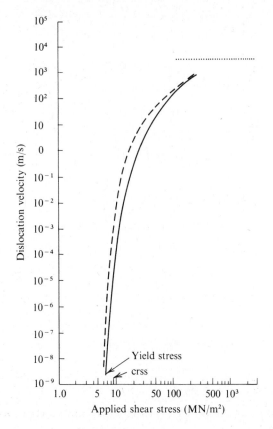

Fig. 9.44. Dislocation velocity as a function of stress, for edge and screw components, in lithium fluoride. − − − edge components, —— screw components, · · · · velocity of $(110)[\bar{1}10]$ shear waves = 3.6×10^3 m/s. (From W. G. Johnston and J. J. Gilman, *J. Appl. Phys.* **30**, 129 (American Institute of Physics, 1959).)

Although the line defects are formed by advancing screw dislocations at all stresses, and hence mark the path of a dislocation, these will only generate fresh dislocations at higher stresses. There will be a range of jog heights and dipole separation and the rate of forming new dislocation loops increases rapidly with stress, see Fig. 9.46. Similar behaviour has been observed in Ge, Si, Al_2O_3 and in some bcc metals.

In measuring the critical resolved shear stress of a crystal, a testing machine imposes a given straining rate which is taken up partly by elastic deflection of the crystal (and machine) and partly by plastic straining of the crystal. Initially only elastic straining occurs and the stress rises, then some dislocations begin to move and significant yielding occurs when the dislocations begin to multiply rapidly. All the straining imposed by the machine can now be taken up by the dislocation multiplication and extension is at almost constant stress. The crss according to this view is a dynamic phenomenon, contrasting strongly with the static effects listed above. The shape of the stress–strain curve is dependent on the stress–velocity dependence and on the multiplication rate. In some cases a pronounced yield drop is produced.

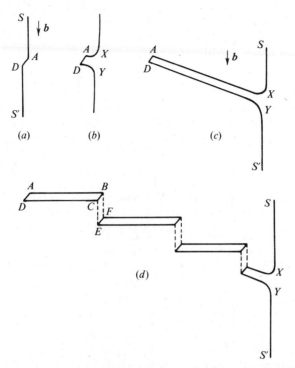

Fig. 9.45. Trail formation by a jog on a screw dislocation. (*a*)–(*c*) shows the formation of a double trail *AX* and *DY*, due to a sessile jog *AD* on a moving screw dislocation *SS'*. (*d*) shows that if the dislocation is not pure screw, the trail will break up by glide of *AB* and *DC*, and cross glide and annihilation of *CE* and *BF*. (From W. G. Johnston and J. J. Gilman, *J. Appl. Phys.* **31**, 632 (American Institute of Physics, 1960).)

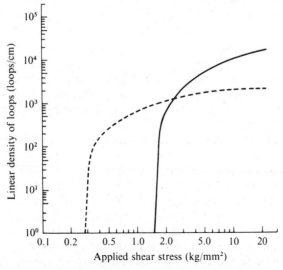

Fig. 9.46. The effect of stress on dislocation multiplication. —— hard crystal, – – – soft crystal. (From W. G. Johnston and J. J. Gilman, *J. Appl. Phys.* **30**, 129 (American Institute of Physics, 1959).)

The theory is developed as follows. When a dislocation moves completely across a single crystal, the relative displacement of the two parts of the crystal separated by the glide plane of the dislocation is b. If the length of the crystal normal to the glide plane is l, then the shear strain produced is b/l. If the dislocation does not move completely across the crystal, but sweeps out an area A', then the shear strain is less than b/l, and may be written as $A'b/Al$ where A is the full area of the glide plane. Clearly $Al = V$, the volume of the crystal, and $A' = \lambda x$, where λ is the length of the dislocation, and x is the distance moved by it. The contribution of this one dislocation to the shear strain rate is given by

$$\delta \dot{E} = \frac{b\lambda v}{V}$$

where v is the velocity at which the dislocation moves. If there exist several types of dislocation, the total strain rate in a particular direction will be:

$$\dot{E} = 0.5 \sum b_i \lambda_i v_i \tag{9.20}$$

where λ_i is the total mobile length per unit volume of the ith type of dislocation with Burgers vector $\boldsymbol{b}_i$ moving at velocity v_i. The factor 0.5 averages the various orientations of slip planes. Normally there will be only two types of dislocation, screw and edge (or 60° for diamond cubic structure) with the same Burgers vector, and (9.20) can be written:

$$\dot{E} = 0.5b[\lambda_s v_s + \lambda_e v_e]. \tag{9.21}$$

If one type of dislocation moves much more rapidly than the other, then the strain rate is controlled by the slow moving dislocations. This statement at first appears to contradict (9.21), but, supposing the edges are 50 times faster than screws, as for LiF, the edge segments move rapidly to the surfaces of the crystal and are lost, trailing at their ends screw segments whose length increases until the only mobile dislocations are screws running the width of the glide planes (see Fig. 9.47).

The strain rate is thus given by

$$\dot{E} = 0.5b\lambda v, \tag{9.22}$$

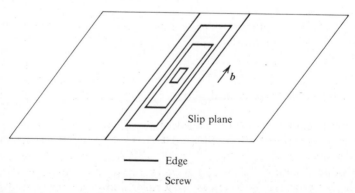

Edge

Screw

Fig. 9.47. Successive stages in the expansion of an originally square dislocation loop, initially composed of equal lengths of edge and screw segments. Edge components have higher velocities, and are rapidly lost at the edges of the crystal, leaving only screw components to participate in the deformation process.

where λ is now the length per unit volume of mobile dislocation and v is the velocity of the slowest dislocations. Experimentally v is found to be a function of stress and temperature:

$$v = \left(\frac{\sigma}{\sigma_0}\right)^n \qquad (9.23)$$

where σ_0 contains the temperature dependence, often for non-metallics in the form of an activation energy, and n is an exponent which is small for crystals with a high Peierls–Nabarro force, and large for materials with a low Peierls–Nabarro force. The value of n is 1–2 for covalent solids such as silicon and germanium, 35–50 for ionic solids such as LiF and MgO, 2–35 for bcc metals, and ~ 200 for fcc metals. The length of mobile dislocation increases from its initial value λ_0, as a function of the strain and, possibly, the stress.

When a strain rate is imposed upon a crystal the deformation is initially elastic, but as the stress rises some dislocations begin to move. As they move they create more dislocations, λ increases, and as the stress is still rising, v increases until dislocation movement accounts entirely for the strain rate. The elastic contribution to the strain rate can now be forgotten, and (9.22) holds. Further deformation causes an increase in λ, and hence v can decrease, thus the stress necessary to maintain the strain rate decreases. The stress–strain curve has gone through a maximum (upper yield point).

After a certain amount of deformation, the active length of dislocation ceases to rise, due to immobilisation of dislocations in pile-ups, and the flow stress may then be expected to be constant. However the stress in (9.23) is the actual stress experienced by the dislocations, and is equal to the applied stress σ_a minus the work-hardening back-stress, $q\varepsilon$, where $q = d\sigma/d\varepsilon$. Thus in order to keep v constant, σ must rise and the stress–strain curve passes through a minimum (lower yield point).

Equation (9.22) now becomes:

$$\dot{\varepsilon} = 0.5b\lambda(\varepsilon)\left(\frac{\sigma_a - q\varepsilon}{\sigma_0}\right)^n \qquad (9.24)$$

whence

$$\sigma_a = q\varepsilon + \sigma_0\left(\frac{2\dot{\varepsilon}}{b\lambda(\varepsilon)}\right)^{1/n}. \qquad (9.25)$$

$\lambda(\varepsilon)$ depends upon the initial density of mobile dislocations λ_0, and upon the dislocation multiplication rate α

$$\lambda(\varepsilon) = \lambda_0 + \alpha\varepsilon^m$$

and in addition both α and m may be functions of the stress. Thus general solutions for all materials are impossible. For LiF, however, α appears to be constant and $m = 1$.

Johnston calculated in 1962 the effect of different values of λ_0 and of n on the stress–strain curves in LiF (Figs. 9.48 and 9.49). It can be seen that the requirements for a sharp yield drop are a low initial density of mobile dislocations and a low value of n. Both of these conditions are satisfied in many non-metallics, where a high elastic energy limits the natural dislocation density, and in bcc metals where dislocations are readily pinned by interstitial impurities, and high Peierls–Nabarro forces give low values of n. The application of this theory to the sharp yield point in steels is described in § 11.5.

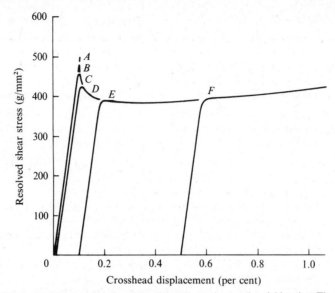

Fig. 9.48. Effect of initial density of mobile dislocations, λ_0, on the yield point. The curves were computed with identical values of all parameters except λ_0. Dislocations/cm^2: A 10^2, B 10^3, C 10^4, D 10^5, E 10^6, F 5×10^6. (From W. G. Johnston, *J. Appl. Phys.* **33**, 2716 (American Institute of Physics, 1962).)

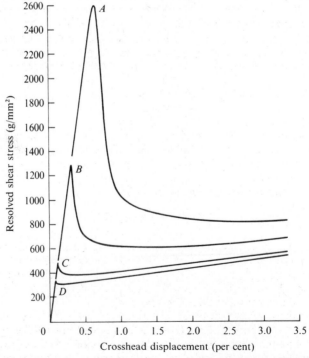

Fig. 9.49. Effect of dislocation velocity exponent on stress–strain curves, calculated with identical values of all other parameters. $\lambda_0 = 10^3/\text{cm}^3$. A: $m = 3$; B: $m = 5$; C: $m = 16.15$ (LiF); D: $m = 50$. (From W. G. Johnston, *J. Appl. Phys.* **33**, 2716 (American Institute of Physics, 1962).)

QUESTIONS

1. The axis of a single crystal of zinc makes an angle of 44° with the normal to the basal plane, and of 46° to the slip direction $\langle 11\bar{2}0 \rangle$. The critical resolved shear stress is 0.5 MN/m². At what tensile stress will deformation by slip begin? What is the shear strain after the crystal has elongated by 25 per cent?

2. The axis of a copper single crystal makes the following angles with three slip systems:

   47° to the normal of $(11\bar{1})$, 43° to $[101]$,
   68° to the normal of $(1\bar{1}1)$, 22° to $[110]$,
   35° to the normal of (111), 55° to $[10\bar{1}]$.

   The crystal yields at an applied tensile stress of 1 MN/m². Calculate the resolved shear stress on each of the above three slip systems, state which system will be operative, and thus determine the critical resolved shear stress for the crystal.

3. Along what direction must a fcc crystal be stressed to produce initially (a) double, (b) triple, and (c) quadruple glide?

4. Calculate for the example given in Question 2 above, the tensile stress and elongation at which double glide will commence.

5. Explain, with reference to appropriate stereograms, the difference between primary, critical, conjugate and cross-slip systems.

6. Slip is found to occur on basal (0001) or prismatic $\{10\bar{1}0\}$ planes in cph metals, according as the axial (c/a) ratio is large or small. Explain this behaviour, and calculate at what axial ratio you would expect slip to be equally likely on both systems.

7. A magnesium crystal is subjected to compression at room temperature such that the compression axis is parallel to $[0001]$. If the stress to nucleate twinning on $(10\bar{1}2)$ is ten times greater than the critical resolved shear stress for slip on (0001), does the initial deformation of the crystal occur by slip or by twinning?

8. Hexagonal crystals with high axial (c/a) ratios usually twin when tension is applied to the basal plane, whereas if the axial ratio is low, tension normal to the basal plane is required to cause twinning. At what axial ratio do you expect this change in behaviour to occur?

9. Prove, by the use of Burgers circuits or otherwise, that
   (a) a dislocation cannot end within a crystal,
   (b) the Burgers vector of a dislocation is invariant,
   (c) where several dislocations meet at a node, the sum of their Burgers vectors is zero.

10. Prove that if a closed loop of dislocation moves so as to change its area projected on a plane normal to the Burgers vector, mass transport must have taken place.

11. Networks of dislocations may be formed in most types of crystals. The reaction which is responsible for these networks in fcc crystals is

$$\frac{a}{2}[10\bar{1}] + \frac{a}{2}[0\bar{1}1] + \frac{a}{2}[\bar{1}10] = 0.$$

Deduce the appropriate network-forming reactions for bcc and cph crystals.

12. Slip in bcc crystals occurs in the $\langle 111 \rangle$ direction on $\{110\}$, $\{122\}$, and $\{123\}$ planes. What dislocation reactions lead to the formation of extended dislocations on each of these systems? To what extent would you expect the formation of extended dislocations to influence the choice of slip system in a bcc crystal?

13. One of the most difficult problems in dislocation theory is the calculation of the contribution to the energy of the dislocation core. One approach is to regard the core as a hollow tube, with a surface energy γ per unit area. Deduce expressions for the equilibrium radius r_0 of the core, and the total energy per unit length, of a screw dislocation. What are the values of r_0 and energy per unit length, if $b = 0.3$ nm, $\gamma = 1$ J/m², and $G = 3 \times 10^{10}$ N/m²?

14. Demonstrate how tensile or compressive stresses acting on planes normal to the slip plane, can be relieved by non-conservative (climb) motion of edge dislocations. Calculate the force per unit length on an edge dislocation, when subjected to a tensile stress σ acting along its Burgers vector.

15. Glide dislocations in fcc crystals form extended dislocations by the following reaction:

$$\frac{a}{2}[10\bar{1}] \rightarrow \frac{a}{6}[11\bar{2}] + \frac{a}{6}[2\bar{1}\bar{1}].$$

Calculate the width of the resultant stacking fault in aluminium and copper.

	Al	Cu
Lattice parameter	0.36 nm	0.40 nm
Shear modulus	2.5×10^{10} N/m^2	3.6×10^{10} N/m^2
Stacking fault energy	0.02 J/m^2	0.004 J/m^2
Poisson's ratio	0.31	0.33

16. Show how a vertical row of edge dislocations produces a low-angle grain boundary. Relate the angle of the boundary (misorientation across the boundary) to the dislocation spacing. How does the energy of such a boundary change with angle?

17. Obtain an expression for the energy of a circular loop, radius r, of dislocation of Burgers vector b. What is the energy required to form such a loop in the presence of a shear stress τ acting in the plane of the loop and parallel to b? Calculate the critical stress needed to cause the loop to expand.

18. One of the factors which determines the elastic limit of crystals is the presence of 'forest' dislocations intersecting the active slip plane. Jogs are created whenever a slip dislocation cuts through a forest dislocation. The work done in moving a dislocation through the 'forest' must equal the energy required to create the jogs. Calculate the yield strength of well-annealed aluminium and copper crystals, using the data of Question 15, and assuming a 'forest' density of 10^2 lines/m^2.

19. (a) Two parallel screw dislocations of the same sign, lie on intersecting slip planes ($1\bar{1}1$) and ($\bar{1}11$) of a fcc crystal. Their Burgers vectors are parallel to the line of intersection of the slip planes. The dislocations are extended. Calculate the minimum length of one dislocation that can be activated as a Frank–Read source due to the stress field of the other dislocation.

   (b) If the two dislocations are of opposite sign, and lie on parallel ($1\bar{1}1$) slip planes a distance of 10^3 Burgers vectors apart, what stress is necessary to move one past the other? (The shear modulus is 3×10^{10} N/m^2.)

20. A 10 mm cube of aluminium is sheared at the rate of 1 mm/s^1. The density of mobile dislocations is 10^8 lines/m^2. What is the average dislocation velocity in the metal? (The lattice parameter of aluminium is 0.36 nm.)

FURTHER READING

E. Schmid and W. Boas: *Plasticity of Crystals*. Hughes (1950).

A. H. Cottrell: *Dislocations and Plastic Flow in Crystals*. Oxford University Press (1953).

W. T. Read: *Dislocations in Crystals*. McGraw-Hill (1953).

R. Maddin and N. K. Chen: Geometrical aspects of plastic deformation of single crystals. *Progress in Metal Physics* **5** (1954).

J. C. Fisher *et al.* (eds.): International conference at Lake Placid: *Dislocations and Mechanical Properties of Crystals*. Wiley (1957).

W. G. Johnston and J. J. Gilman: Dislocation velocities, dislocation densities and plastic flow in LiF. *J. Appl. Phys.* **30**, 129 (1959).

W. G. Johnston and J. J. Gilman: Dislocation multiplication in LiF crystals. *J. Appl. Phys.* **31**, 632 (1960).

H. G. van Bueren: *Imperfections in Crystals*, 2nd edition. North Holland (1961).

N.P.L. Symposium (1963): *The Relation between the Structure and Mechanical Properties of Metals*. 2 vols. London, H.M.S.O. (1963).

G. Thomas and J. Washburn (eds.): First Berkeley International Materials Conference: *Electron Microscopy and Strength of Crystals*. Interscience (1963).

J. Friedel: *Dislocations*. Pergamon (1963).

S. Amelinckx: *The Direct Observation of Dislocations*. Academic Press (1964).

J. and J. R. Weertman: *Elementary Dislocation Theory*. Collier–Macmillan (1964).

F. R. N. Nabarro, Z. S. Basinski and D. Holt: The plasticity of pure single crystals. *Advances in Physics*, **13**, 595 (1964).

A. H. Cottrell: *Theory of Crystal Dislocations*. Blackie (1964).

D. Hull: *Introduction to Dislocations*. Pergamon (1965).

J. Christian: *Theory of Transformations in Metals and Alloys*. Pergamon (1965).

P. H. Hirsch and J. S. Lally: The deformation of magnesium single crystals. *Phil. Mag.* **12**, 595 (1965).

Z. S. Basinski and F. Weinberg (eds.): Ottawa International Conference: Deformation of crystalline solids. *Canadian J. Phys.* **45**, parts 2–13 (1966).

F. R. N. Nabarro: *Theory of Crystal Dislocations*. Oxford University Press (1967).

J. P. Hirth and J. Lothe: *Theory of Dislocations*. McGraw-Hill (1968).

10. Physics of fracture

10.1 Introduction

In considering the mechanical properties of metallic materials emphasis has been placed on plasticity as the most important physical phenomenon. Other classes of materials are largely brittle and fracture is the controlling factor in their strength. This chapter introduces the physics of fracture; fatigue and creep fracture will not be included. Although experimental and theoretical knowledge of fracture has increased rapidly in the last thirty years, much remains to be determined in this field.

The subject of fracture mechanics has received particular attention in the last few decades on account of the brittle fracture of many mild steel ships, the loss of two Comet aircraft (1954) by sudden catastrophic failure of the high strength aluminium alloy pressure cabin (initiated by a fatigue crack) and the ever present desire to produce tough materials with still higher strength to weight ratios. Mild steel is normally ductile and only fractures after considerable plastic flow and energy absorption. However, it was forcibly shown to nautical engineers during the Second World War that in some circumstances it can fail catastrophically by brittle fracture without significant plastic flow. Brittle fracture is particularly dangerous in a welded structure which forms a single continuous medium through which a crack can propagate without interruption. In view of its importance to a wide range of engineering, the second part of this chapter is devoted to brittle fracture in mild steel.

The terminology of fractures is most confusing and it is important to start by defining terms, which are unfortunately not universally agreed. There are three methods of classifying fractures. The first, which will be called here the *kind* of fracture, relates to its mechanics, that is, how the loads are applied to the structure and how they are redistributed during the fracturing process. The kinds of fracture are: ductile, brittle, creep, fatigue and adiabatic shear. Ductile fracture means the fracture proceeds slowly as the external loads supply energy to extend the crack, a process which will involve considerable plastic flow of the material adjacent to the crack. The crack is in stable equilibrium. Brittle fracture means the crack spreads rapidly with the energy for propagation being obtained from the elastic energy of the surrounding material, rather than from the external loads. There is little plastic deformation during crack propagation and the crack is unstable. In creep fracture the crack is produced by applying the loads for a very long time and in fatigue fracture it is formed by the repeated fluctuations of the load. Adiabatic shear fracture is produced by very rapid loading such that the heat generated by plastic straining has no time to flow throughout the crystal. Only the ductile and brittle kinds of fracture will be considered here in detail.

In the second classification, which will be called here the *mode* of fracture, the fractures are distinguished by their microscopic characteristics: fibrous (often called ductile), cleavage (often called brittle), intercrystalline and shear.

380

In fibrous fractures the surface is rough and fibrous due to the crack growing by the joining up of numerous microcracks with heavy plastic flow and tearing. Because of the high energy involved this mode of fracture is usually of the ductile kind, and the two terms are often used synonymously. But this is incorrect and will cause confusion; for example, high strength aluminium sheet has been known to give a brittle kind of fracture although the fracture surface is fibrous. In cleavage fractures the crack follows definite crystal planes of low indices, leaving relatively bright faceted surfaces with little or no plastic deformation. Because of the low energy involved, cleavage fractures are mostly of the brittle kind and for this reason tend to be called brittle, but here also the loose use of terms is to be deplored. Intercrystalline fractures are those where the crack follows along the grain boundaries – the previous two modes are transcrystalline. They are usually brittle when due to the segregation of impurities or a second phase at the grain boundary. Creep fractures also follow the grain boundaries. Shear fracture is found only in single crystals, see § 10.4.

Finally, in the third classification, which has been called the engineer's, fractures are called ductile or brittle according to the amount of plastic flow present; for example, the reduction in thickness of a plate at the fracture surface or the increase in length across the fracture may be measured.

In this book, the terms ductile and brittle fractures will always refer to the kind of fracture, that is, whether the crack grows slowly with external energy supply or fast without it. In describing fracture modes the terms fibrous and cleavage will be used, not the oft-used terms ductile and brittle. For the third classification, the term fracture ductility will be used – high, low or nil, as appropriate.

It is also timely to recall the definitions of ductile and brittle materials given in Chapter 5; a ductile material necks before fracture in a standard tensile test at room temperature, whereas a brittle material fractures without necking. In a ductile material, the fracture kind will normally be ductile and the mode fibrous. However, by altering the temperature, state of stress or rate of loading, a brittle kind of fracture may be induced, with probably a cleavage mode.

10.2 Theoretical fracture stress

The theoretical strength can be calculated in terms of the surface energy and elastic modulus as follows. Suppose the interatomic forces per unit area vary with atomic spacing x in the general way shown in Fig. 10.1 (see Chapter 3). The curve is approximated by a sine wave with wavelength λ:

$$\sigma = \sigma_\theta \sin 2\pi \frac{x}{\lambda} \tag{10.1}$$

where σ_θ is the theoretical strength. The work done per unit area in pulling the atoms apart is then:

$$\int_0^{\lambda/2} \sigma \, dx = \int_0^{\lambda/2} \sigma_\theta \sin 2\pi \frac{x}{\lambda} \, dx = \frac{\lambda \sigma_\theta}{\pi}.$$

This work appears as surface energy $2S$ of the fracture, that is,

$$\sigma_\theta = \frac{2\pi S}{\lambda}. \tag{10.2}$$

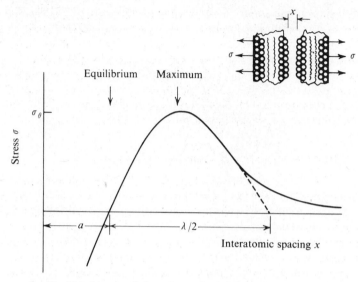

Fig. 10.1. Interatomic force (per unit area) versus atomic spacing.

For the initial extension Hooke's law applies:

$$\delta\sigma = E\frac{\delta x}{a}$$

where a is the unstressed interatomic distance. Differentiating (10.1) gives:

$$\left(\frac{d\sigma}{dx}\right)_{x=0} = \sigma_\theta\frac{2\pi}{\lambda}\left(\cos 2\pi\frac{x}{\lambda}\right)_{x=0} = \sigma_\theta\frac{2\pi}{\lambda}.$$

Hence
$$\sigma_\theta = \frac{E}{2\pi}\frac{\lambda}{a}. \tag{10.3}$$

Eliminating λ from (10.2) and (10.3) gives the theoretical strength

$$\sigma_\theta = \left(\frac{ES}{a}\right)^{\frac{1}{2}}. \tag{10.4}$$

(N.B. Some workers integrate (10.1) from 0 to $\lambda/4$, making the theoretical strength $(2ES/a)^{\frac{1}{2}}$.)

The surface energy of solids can be estimated from measurements taken in the liquid state and are typically around 1 J/m² (10³ erg/cm²). For $E = 100$ GN/m² (15 × 10⁶ psi) and $a = 0.3$ nm the theoretical strength is

$$\sigma_\theta \approx 20\,\text{GN/m}^2\,(3 \times 10^6\,\text{psi}).$$

The theoretical strength can also be expressed solely as a function of the elastic modulus. The surface energy can be shown to be roughly equal to $aE/100$ (this is equivalent to putting $\lambda \approx a$ in (10.1)). The theoretical strength is then given by $E/10$, or in other words, materials fail theoretically at ~ 10 per cent strain.

In practice most brittle materials fracture at around $10^{-3}E$ or lower; metals yield at a similar stress and fracture after large plastic deformation at about $10^{-2}E$. However very thin fibres of some materials such as glass and hard drawn steel have strengths approaching the theoretical value $E/10$. Whiskers of most materials, that is, almost perfect single crystals, which can be grown under special conditions, also have nearly theoretical strength. The problem to be considered here is why the experimental fracture strength of most materials is one or two orders of magnitude lower than the theoretical value.

10.3 Brittle fracture of glass and ceramics: Griffith theory

Scientific study of fracture stems from Griffith's classic investigation in the 1920s of the strength of glass. At room temperature glass is elastic up to the point at which it fails by brittle fracture, that is, a Hookean solid. Griffith found that the strength of a potash glass fibre approached the theoretical value when measured within a few seconds of drawing, but it decreased over a period of a few hours to an equilibrium value which was dependent on the fibre diameter, as shown in Fig. 10.2 (curve 1). The curve can be expressed approximately by

$$\sigma = A + B/d \tag{10.5}$$

where d is the fibre diameter, and A and B are constants.

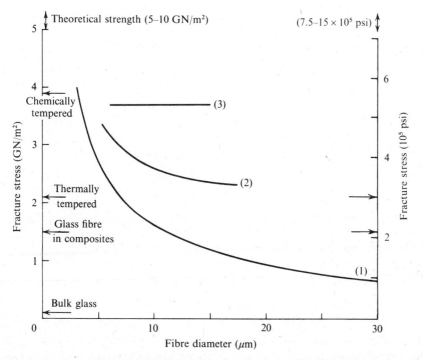

Fig. 10.2. Experimental fracture stress of glass fibres versus diameter. (1) Griffith's equilibrium values reached some hours after drawing. (2) Otto's curve obtained four hours after production. (3) Thomas's results independent of time when stored *in vacuo*.

Griffith suggested that the low strength of glass in the equilibrium state is due to the presence of cracks, which are assumed to form spontaneously after drawing, and that fracture occurs when one of them is able to extend rapidly by the conversion of the elastic strain energy of the adjacent material into the surface energy of fracture. This energy interchange is the basic concept of Griffith's theory which has survived many controversies on the actual strength of glass and the origin of the cracks. In recent years it has also been applied in a modified version to brittle fracture in metals (§ 10.14).

Using Inglis's solution of 1913 for the stress and strain distribution around an elliptical crack of length $2c$ in an infinite plate under tension σ (see Fig. 10.3), Griffith derived the decrease of strain energy in the surrounding material due to the presence of the crack (Δ per unit plate thickness)

$$\Delta = \frac{\pi c^2 \sigma^2}{E} \tag{10.6}$$

where E is Young's modulus. The minor axis of the ellipse has been reduced to its zero limit in obtaining this expression. This is double a simple estimate, due to Zener in 1948, in which the energy release is considered to be confined to a cylinder of radius c around the crack, with the stress being fully relieved to zero.

The surface energy of the crack is $4cS$. For the crack to spread into a fracture the differential of these two terms will be equal:

$$\frac{\pi \sigma_c^2 2c\, \delta c}{E} = 4S\, \delta c,$$

giving the Griffith fracture stress

$$\sigma_c = \left(\frac{2ES}{\pi c}\right)^{\frac{1}{2}}. \tag{10.7}$$

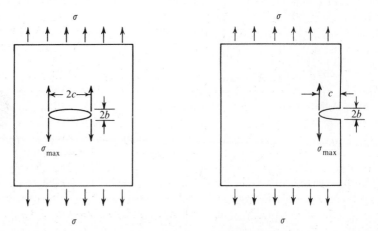

Fig. 10.3. Maximum stress at extremities of the major axis of an elliptical hole or surface crack.
$$\sigma_{\max} = \sigma(1 + 2c/b) = \sigma(1 + 2(c/\rho)^{\frac{1}{2}})$$
(ρ = radius of curvature at end of major axis).

Alternatively the calculation can be made in terms of the stress. The maximum tensile strength at the tip of an elliptical crack was found by Inglis to be:

$$\sigma_{max} = \sigma\left(1 + \frac{2c}{b}\right) \tag{10.8a}$$

where c and b are the major and minor axes respectively. Or, in terms of the radius at the end of major axis:

$$\sigma_{max} = \sigma\left(1 + 2\left(\frac{c}{\rho}\right)^{\frac{1}{2}}\right) \tag{10.8b}$$

$$\approx 2\left(\frac{c}{\rho}\right)^{\frac{1}{2}}\sigma.$$

If it is assumed (Orowan in 1948) that the minimum value of ρ is the interatomic spacing a, and that the theoretical strength is given by (10.4), the fracture stress in the presence of a crack is:

$$\sigma_c = \left(\frac{ES}{4c}\right)^{\frac{1}{2}}. \tag{10.9}$$

This differs only by a small numerical factor from Griffith's expression, (10.7). Although Griffith initially visualised internal flaws it was later realised that these were more probably on the surface. The preceding theory still applies except that c now becomes the crack depth.

According to the theory, the fracture strength is inversely proportional to the square root of the inherent crack length. Figure 10.4 plots numerical values for two values of the surface energy, 0.2 and 1.0 J/m², typical of glasses; a representative elastic modulus has been taken at 70 GN/m². The influence of the form of the crack on the stress concentration is shown in Fig. 10.5, which is based on (10.8). A circular hole causes an elastic concentration factor of three, in contrast with the very high values (dependent on the crack length) resulting from a natural crack which is assumed to have a tip radius approximately equal to the atomic spacing.

As already noted, Griffith considered the initial strength of his glass fibres, measured within minutes of drawing, was the theoretical strength and the later equilibrium values were due to the presence of cracks which formed spontaneously by some unspecified molecular rearrangement. The observed variation of the equilibrium strength with diameter (Fig. 10.2) is attributed to the scatter in crack length about its mean value, given that the probability of formation of a crack with a specified length decreases exponentially with length. Statistical theory then predicts that, if the strength is determined by the longest crack present, it will decrease with increasing specimen size, as will the scatter about the mean. Both predictions were confirmed by experiment. Extrapolating the curve of equilibrium fracture strength to zero diameter, when no cracks will be present, gives a value of 10 GN/m² (1.6 × 10⁶ psi) which agrees with the theoretical value, taking $E = 60$ GN/m² and $S = 0.2$ J/m² (extrapolated from values for liquid glass).

Many workers have found the strength of glass fibres to vary with the diameter and it became accepted for a time that this was the fundamental effect in glass fibres, overlooking the initial high values found by Griffith. Various erroneous

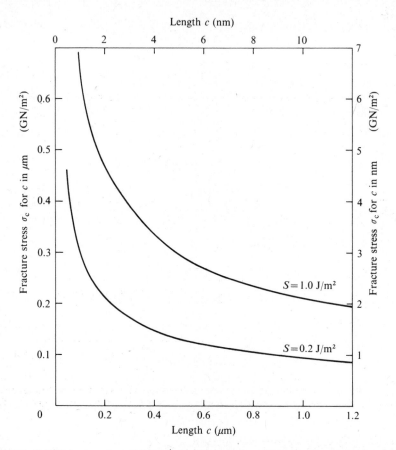

Fig. 10.4. Griffith fracture stress $(2ES/\pi c)^{\frac{1}{2}}$ versus (external) crack depth c. S = the surface energy of the fracture and E = Young's modulus (taken as 70 GN/m²). (Note: 1 GN/m² = 1.45 × 10⁵ psi, 1 J/m² = 0.068 ft lb/ft².)

theories were advanced to explain the phenomena. Recently (1960) Thomas has shown that, given very careful handling, the strength of glass fibres is high (around 3.5 GN/m², 0.5 × 10⁶ psi) and independent of size and time when stored *in vacuo*, see Fig. 10.2 (curve 3). In air the strength drops slightly: 14 per cent after 128 days at zero humidity and 30 per cent after 128 days in 100 per cent humidity, much less than the reductions found by Griffith in only a few hours. The low conventional strength is due to surface cracks formed by accidental damage. When Proctor removed the surface with hydrofluoric acid the strength of an 8 mm diameter bar was raised from about 70 MN/m² (10⁴ psi) to 2.8 GN/m² (4 × 10⁵ psi), close to the value for a virgin fibre. At the same time etch pits appeared on the surface suggesting that initially sharp cracks were enlarged and rounded by the etchant with consequent reduction of the stress concentration, as shown in Fig. 10.6.

It is not surprising that no direct evidence of surface cracks has been found since, even when the strength has dropped to 0.1 GN/m² (typical of untempered commercial glass), the required crack length is only 0.3 μm, equal to 1000

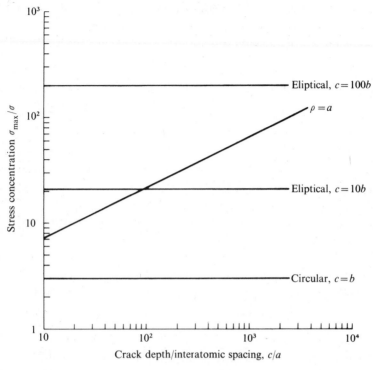

Fig. 10.5. Elastic stress concentration factor due to various shaped cracks (see Fig. 10.3. for notation). c = major axis (crack semi-length), b = minor axis, ρ = crack tip radius, a = interatomic spacing.

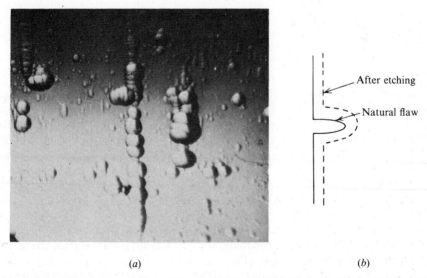

Fig. 10.6. Glass surface after etching with hydrofluoric acid. (*a*) Etch pits ($\times 110$). They may form on incipient surface cracks as shown schematically in (*b*) (by courtesy of B. A. Proctor, *Applied Materials Research*, **3**, 24 (1964)).

atomic spacings, and the opening probably only a few atomic spacings. However, surface cracks are consistent with the evidence that a compressive surface layer raises the strength markedly. (The various thermal and chemical techniques of toughening glass, raising the median strength from 0.1 to $4\,\text{GN/m}^2$, were described in § 7.9.)

It is currently thought that the high strength of virgin fibres is still not the theoretical strength, but is down by a factor of ten. Various theories are now being advanced to explain this; they range from the pre-existence of flaws inherent in the glass structure, to the yielding of the glass prior to fracture.

The fracture of crystalline ceramics has received only limited investigation. As stated in Chapter 7, polycrystalline ceramics are brittle and fail with a brittle kind of fracture; some typical values of the tensile strength and modulus of rupture for various ceramics were given in Table 7.2. The effect of grain size and porosity on the modulus of rupture of alumina has been investigated by Spriggs and Vasilos in 1963 (see § 7.6 and Fig. 7.15). The experimental points fitted the relations:

$$\left.\begin{array}{lll} \text{(porosity } P) & S \propto e^{-\text{const } P} & (a) \\ \text{(grain size } d) & S \propto d^{-\frac{1}{2}} & (b) \end{array}\right\} \tag{10.10}$$

where S is the modulus of rupture or maximum fibre stress at fracture in bending (cf. § 5.6). The dependence on grain size was less marked than is found for metals undergoing brittle fracture, where the strength is proportional to $d^{-\frac{1}{2}}$ (see § 10.4). However, the evidence on ceramics is slim.

A consequence of the Griffith cracks initiating fracture is that the strength of a brittle solid shows wide scatter about the mean value and depends on the size of the specimen. Statistical theory is required to consider the probability of cracks of various lengths and some of the problems of using ceramics for engineering structures are discussed in § 11.10. The strength in compression is higher by a factor of three or four (cf. yield stress of materials which is independent of the sign of the stress).

10.4 Fracture of metals

Single crystals of metals often show no separate fracture stage. Separation occurs after extensive slip on a few parallel slip bands, one of which finally parts, leaving a rough worked surface. Close-packed hexagonal magnesium is an example, with slip and separation occurring on a basal plane. Alternatively, multiple slip and twinning leads to the formation of a neck or wedge which finally separates.

In cleavage fracture, which some metallic crystals undergo, particularly at low temperatures, certain crystal planes of low indices separate to give a fairly bright smooth surface. There is normally little plastic deformation associated with this mode of fracture. Body-centred cubic metals (e.g. Fe, Ni, W) are most prone to it, usually on the $\{001\}$ planes, but not including the alkali metals (Li, Na, K, ...). Iron is an example of particular technological importance. In the cph metals, zinc cleaves easily at $196\,^\circ\text{C}$ and with difficulty at $20\,^\circ\text{C}$; cleavage is on the basal plane and occurs even after plastic bending of the crystal. Magnesium will also cleave with difficulty on a number of planes. The fcc metals

Table 10.1. *Cleavage planes in crystals*

Crystal lattice	Material	Cleavage plane	Critical normal stress (MN/m^2)	Temperature (°C)
bcc	α-Fe	(001)	270	−185
	Mo	(001)		< −100
	Cr	(001)		
	W	(001)		
cph	Zn	(0001)	1.8–2.0	−185
	Mg	$\begin{cases} (0001)(10\bar{1}1) \\ (10\bar{1}2)(10\bar{1}0) \end{cases}$		
Rhombohedral body centred	Bi	$\begin{cases} (111) \\ (11\bar{1}) \end{cases}$	3.2 6.8	−80, 20 20
	Sb	(111)	6.5	20
Cubic	NaCl	$\begin{cases} (001) \\ (011) \end{cases}$	2.2, —	20, −190 −190
	LiF	(001)		

fail by cleavage only in fatigue. Table 10.1 gives the rather limited experimental data.

Polycrystalline metals may also have no distinct fracture stage in a tensile test but neck down to a point (Chapter 5). Zinc, lead and silver do this in an ordinary tensile test and many other metals will do so under hydrostatic compression. Less ductile metals form a crack in the neck; it starts in the centre perpendicular to the stress axis, grows with further plastic deformation and then changes over to 45° producing the well known cup and cone (see Fig. 5.3(a). Puttick has shown (see Fig. 5.3(b)) that in the first stage of fracture numerous cavities form at microscopic inclusions which then grow and coalesce into a macroscopic crack. The mechanics of the last stage of shearing is not understood.

Metals that can fracture by cleavage as single crystals will also fail in a similar way in the polycrystalline state at low temperature. This fracture mode is very dangerous in engineering structures as it leads to sudden and catastrophic brittle fracture. The phenomenon is not confined to metals with any particular crystallographic lattices but in practice it is most important in bcc metals (e.g. Fe, W, Mo, Cr, V). In some circumstances cleavage fracture occurs at temperatures well above that at which it occurs under uniform uniaxial stress, as in a tensile test. For example, mild steel will only give a cleavage failure below about −150 °C in a normal tensile test, but in a notched bar tensile test this may occur at room temperature (see below).

The effect of grain size on cleavage fracture is shown in Fig. 10.7. Above a critical diameter the cleavage stress is given by

$$\sigma_c = \sigma_0 + Kd^{-\frac{1}{2}} \qquad (10.11a)$$

where d is the grain diameter and σ_0 and K are constants. This law was first found in 1953 by Petch for ferrite at −196 °C (see Fig. 10.11) and has since been shown to have wider application. In this range it is thought that the crack nucleation is the limiting factor. Below the critical grain diameter, the law

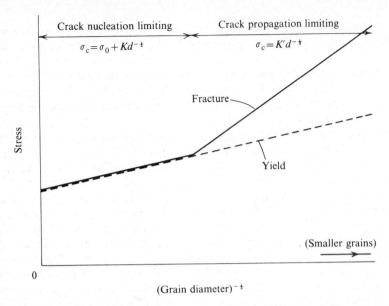

Fig. 10.7. The effect of grain size on the cleavage fracture stress of polycrystalline metals. Below a critical grain size some yielding precedes fracture. (For actual results on mild steel see Fig. 10.11.)

changes to

$$\sigma_c = K'd^{-\frac{1}{2}} \qquad (10.11b)$$

and crack propagation becomes the critical factor. At the same time some plastic strain occurs prior to fracture, as shown in Fig. 10.7 by the yield stress curve dropping below the fracture stress.

In view of its great technological importance the remainder of this chapter will be devoted to the problem of brittle fracture of mild steel.

BRITTLE FRACTURE OF MILD STEEL

10.5 Introduction

Brittle fracture of mild steel structures was not unknown before 1939: over the previous sixty years some thirty major failures of storage tanks, bridges and ships are recorded, although the details are limited. Steel manufactured by the Bessemer process (§ 12.2) was known to be particularly prone to this kind of failure and was not accepted for certain types of structure; however, the new practice of using oxygen instead of air for blowing the converter has moderated this defect. During the Second World War brittle fracture took on a new and serious significance: over one fifth of the 5000 American merchant vessels built during the war had developed cracks by 1946, many of them catastrophic. These ships were welded and the crack travelled at high speed (1300 m/s) across many plates, sometimes splitting the ship in two. A classic example was the tanker SS *Schenectady*, which broke in two amidships whilst in still water in the fitting out dock. The loads were static and low (60–90 MN/m², ~10⁴ psi), and the ambient temperature below 5 °C. A historical review of brittle fractures in service has been given by Biggs (1960).

When brittle fractures in ships were first reported, the welding was suspected as the likely cause since it was a comparatively new process in ship construction. The residual stresses were measured and shown to be high, but it was concluded that if they were the basic cause of brittle fracture no welded structure could survive. Nor was it clear why the fracture should run across the main plates rather than along the welds. One important consequence of using welding is that the structure becomes continuous and a crack can run without a break. In riveted ships many cracks are found to have been arrested by the rivet holes, or at worst to have spread across only a single plate. Nevertheless, more than a dozen riveted merchant ships are reported to have broken in half since 1900.

Material taken from fractured ships was always found to be ductile and within specification when tested in a standard tensile test at room temperature. When it was realised that the fracture mode was cleavage, tests were made with notched bars and at reduced temperatures, around 0 °C, at which the fractures were taking place. Brittle fractures were then obtained in the laboratory, although at a high average stress. Prior to this time, Izod and Charpy tests (see § 5.8) were standard practice only on heat treated steels – the energy absorbed being measured at room temperature. Although mild steel was known to give brittle cleavage fractures in tensile tests at low temperatures (below about − 150 °C), it was not realised that in notched bar tests on thick plates these could occur up to room temperature or higher (40 °C). Various notched bar tests have now been devised to determine the transition temperature range below which brittle fracture occurs; one of them, the Tipper test, is described below.

Although these tests produce brittle fracture of mild steel at near ambient temperatures, the stress required is high, around 0.5 GN/m^2, whereas service failures occur at stresses below the lower yield stress (~ 0.2 GN/m^2). This may be because the initiation of a crack from a machined notch requires a high stress whilst the propagation stage occurs at a low stress. The Robertson test sets out to investigate only the propagation phase, the crack being initiated by means of an impulsive blow. The source of the crack in service remains obscure. A possible solution has been found by Greene and Weck, who have shown that the presence, prior to welding, of a small defect (e.g. a saw cut) near the prepared edge of a heavily restrained plate has a dramatic effect on the strength after the weld has been made. Fuller details of these various tests will be given later.

Apart from the high rate of crack propagation and lack of plastic deformation adjacent to the crack, brittle fractures show a characteristic pattern in the form of chevron markings, as shown in Fig. 10.8. These are ridges and corresponding

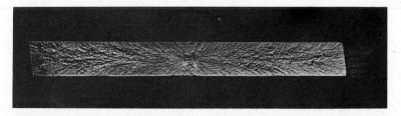

Fig. 10.8. Brittle fracture surface, showing initiation point at internal weld defect and typical chevron pattern. (By courtesy of R. Weck and The Welding Institute.)

valleys formed on the fracture surfaces as a result of the joining up of numerous microcracks which develop ahead of the propagating main fracture. The chevrons always point towards, and are a useful method of identification of, the origin of the fracture. Another characteristic appearance of brittle fracture produced in laboratory notch tests is a small (thumbnail) region of fibrous fracture at the root of the notch. Service failures do not usually show this region of plastic flow and the possible explanation is given in § 10.10.

10.6 Transition temperature range

In the investigation of brittle fracture a variety of notched bar tests have been devised, alongside the original Izod and Charpy tests; the specimens are loaded in bending or tension. Various parameters are measured, other than the direct one of crack stability. The principal ones are: fracture mode (or appearance); fracture ductility (measured as the reduction in thickness at the root of the notch); and energy absorbed. When these parameters are plotted against temperature, see Fig. 10.9, it is found that they all change their values rapidly over a band of temperature, known as the transition temperature range. Within the broad range it is possible to identify narrower bands in which individual parameters are changing rapidly. The width of the overall band and of the transition ranges of the individual parameters varies from one type of test to another and also with the material under test. In addition, the scatter of results is large in the transition range and the width tends to increase with the number of tests made.

It is often convenient to specify a single transition temperature rather than a transition range, although it must be understood that there is usually not a

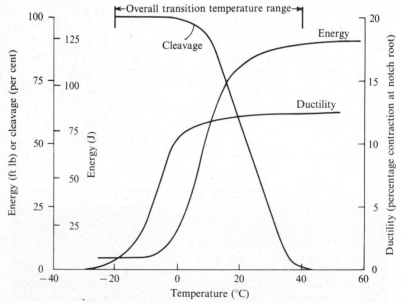

Fig. 10.9. Ductile–brittle transition temperature range of a typical mild steel. The overall band includes various narrower bands associated with particular parameters.

sharp discontinuity in the property considered. This may be chosen at the middle of the temperature band or at the temperature at which a parameter has a particular value, for example 20 J ($\sim$ 15 ft lb) of energy absorbed, or 1 per cent lateral contraction at the notch root. Of the large and confusing number of parameters and criteria which have been used, it appears that there are basically only two physically distinct transitions:

(*a*) Fracture appearance transition, defined as the change from fibrous to cleavage mode of fracture propagation. Criteria associated with this transition are: fracture appearance (say 50 per cent cleavage); high energy absorption (say, at the top 'knee' of the energy transition curve).

(*b*) Ductility transition, defined as the change from fracture initiation with prior plastic deformation to initiation without it. Criteria for this transition are: 100 per cent cleavage (that is, the elimination of the fibrous thumbnail at the notch root); low level of energy absorption (say, in the bottom knee of the energy transition curve).

In considering transition temperatures from different tests using different criteria, it is important that only those associated with the same transition (fracture appearance or ductility) should be compared if any logical deductions are to be made. The temperature difference between the fracture appearance and ductility transitions is dependent on the test details and in some cases they may nearly coincide. However the fracture appearance transition is never at a lower temperature than the ductility transition.

The variety of transition temperature ranges and even greater number of transition temperatures within them has led to much conflicting data and discussion as to which is the best test, test specimen and criterion. By 'best' is meant that which will most accurately indicate liability to brittle fracture in service at any operating temperature. Following extensive tests on material from fractured structures the Tipper test has been claimed to do this and of the many tests in the field it will be singled out for description (§ 10.8), together with the Robertson test, which is of rather a different character (§ 10.9).

The problem of specifying a test is made more contentious since the various transition temperature ranges tend to lie in the region 0 °C to 40 °C where most structures have to operate and it has not been possible, without economic penalty, to lower the transition temperatures of steels to temperatures at which the differences in detail would be unimportant.

10.7 Unnotched tensile test

Data from unnotched tests on ferrite and mild steels over a temperature range from ambient down to -250 °C have been obtained, in 1959, by Hahn and his co-workers. Typical results for a coarse-grained mild steel with 0.22 per cent carbon are plotted in Fig. 10.10. In addition to the usual parameters, the number of ferrite grains containing microcracks was noted, either at 10 per cent strain or after fracture (whichever occurred first).

Six temperature ranges can be identified in terms of the properties as follows:
Region *A* (*RT* to -10 °C). The material is typically ductile (see reduction of area curve) with a cup and cone fracture occurring in the neck.
Region *B* (-10 °C to -130 °C). The material still necks but the fracture now

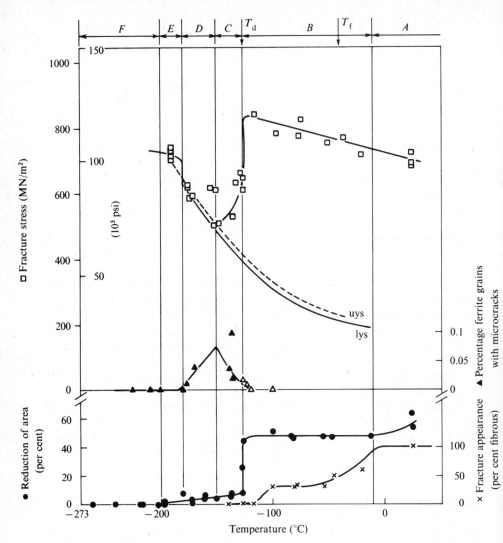

Fig. 10.10. Unnotched tensile test data for 0.22 per cent carbon steel (grain size 100 μm) against temperature. (After G. T. Hahn, B. L. Averbach, W. S. Owen and M. Cohen, in *Conference on Fracture* (Swampscott, 1959).)

starts as fibrous and finishes as cleavage. This is shown by the per cent fibrous curve. The 50 per cent fibrous corresponds to the fracture appearance transition (T_f).

Region C ($-130\,°C$ to $-150\,°C$). At the boundary with B the fracture stress and ductility (as shown by the reduction of area) drop rapidly. The fracture stress remains above the lower yield stress, however. The B–C boundary temperature corresponds with the ductility transition (T_d). Across the region, the number of cracked ferrite grains increases rapidly; in tests on coarse grained ferrite a value of 2 per cent was reached. The microcracks are mostly about the length of the grain and occur throughout the specimen.

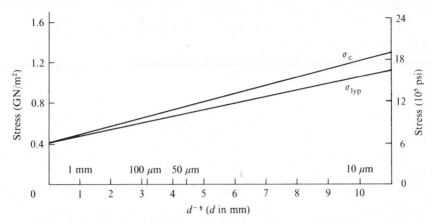

Fig. 10.11. Cleavage strength and lower yield stress of ferrite versus grain size at $-196\,°C$. Lines are based on results for mild steel and two irons. (After N. J. Petch, *J.I.S.I.* **173**, 25 (1953).)

Region D ($-150\,°C$ to $-180\,°C$). The fracture stress coincides with the lower yield stress, but fracture is always preceded by some yielding. The number of microcracks, which occur in the yielded grains only, decreases across the region.

Region E ($-180\,°C$ to $-200\,°C$). Fracture now occurs without prior yielding at around the extrapolated values of the upper yield stress. No microcracks are found, the first one formed presumably spreading into fracture.

Region F (below $-200\,°C$). Fracture occurs at stresses well below the yield stresses and is closely associated with twinning.

At a given temperature the cleavage fracture stress is dependent on the grain size, in accordance with (10.11), and independent of the composition. Fig. 10.11, which is for cleavage at $-196\,°C$, shows that the fracture stress is approximately doubled as the grain size is reduced from a very coarse grain size ($250\,\mu m$) to a very fine grain ($25\,\mu m$).

10.8 Notch tensile test (Tipper test)

The Tipper notch tensile test is one of several tests that have been devised to measure the brittle fracture characteristics of mild steel. It is claimed – Baker and Tipper in 1956 – to give an accurate indication of the temperature below which a material is liable to brittle fracture in service.

The Tipper test piece is shown in Fig. 10.12; it has two notches cut in opposite sides of a rectangular bar. The procedure is to cool it below the test temperature before fitting into a tensile machine, and to apply the load slowly to that at fracture the specimen has reached the desired temperature, as measured by a thermocouple clipped to it.

The appearance of the fracture surfaces is noted, that is, the ratio of fibrous to cleavage fracture areas; also measured are the ductility as given by the reduction in plate thickness mid-way between the notches, and the fracture load. The test is repeated over a range of temperatures.

Typical results are plotted in Fig. 10.13(a). The transition temperature range, which is obtained either from the fracture appearance or the ductility, both

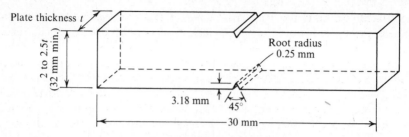

Fig. 10.12. Tipper notched-bar tensile test piece.

criteria agreeing closely, extends over about 5 °C in good cases and over 20 °C in bad ones. A high width/thickness ratio, preferably 2 or $2\frac{1}{2}$, was found to give narrow transition ranges, and a thick plate of a killed steel may be machined down to this ratio to bring the fracture load within the capability of normal

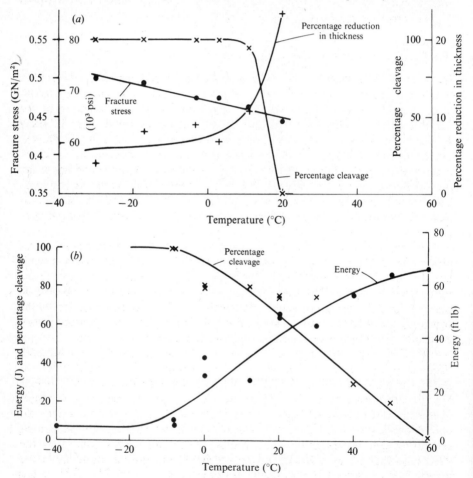

Fig. 10.13. Tipper (*a*) and Izod (*b*) test results on 14.7 mm mild steel plate. (After J. F. Baker and C. F. Tipper, *Proc. Instn Mech. Engrs* (*Lond.*) **170**, 65 (1956).)

testing machines. An important feature of the test is that the notch dimensions have been shown to be not critical – in sharp contrast with the Charpy and Izod tests.

Following numerous tests on casualty material Tipper concluded that there was an excellent correlation between the transition temperature for 50 per cent cleavage and service behaviour. Further, it was shown that in composite test pieces of different materials welded together a brittle fracture running in material below its transition temperature was arrested on entering material above its transition temperature.

Unfortunately other workers have not adopted this test and many favour a transition temperature at which the energy absorbed in the Charpy test is 15 ft lb ($\sim$ 20 J). Tipper carried out Izod and Charpy V-notch tests on casualty material at the same time as doing her own tests and the comparison is shown in Fig. 10.13(*b*). They gave no sharp transition temperature based either on the energy absorbed or the fracture appearance, and no correlation either with the Tipper transition temperature or, more vital, with service behaviour.

10.9 Crack propagation stress (Robertson test)

In the Tipper test the fracture stress (maximum load/original area between notches) is equal to, or greater than, the tensile strength of the material, as obtained in a conventional tensile test at room temperature. This is shown by the curve for the fracture stress in Fig. 10.13(*a*). Yet the nominal stress in many service failures is known to have been less than the lower yield stress. The explanation may lie in the two stages of any fracture: crack nucleation and crack propagation. In the Tipper test both stages are necessary, but in actual structures the crack may have formed previously and the condition for crack propagation would then be the important one. The Robertson test of 1951 sets out to measure the stress for crack propagation (or arrest) as a function of the temperature.

The test specimen is shown in Fig. 10.14. The test piece proper measures about 305 × 76 mm (12 × 3 in) by the full plate thickness, with one side semi-circular and drilled with a 10 mm diameter hole. From this hole a saw cut is made 0.5 mm thick and 10 mm deep towards the specimen centre. To save on plate material, extension pieces of mild steel are welded to the long edges of the test piece. Prior to the actual test a load is applied to yield the extension pieces, ensuring a uniform stress distribution in the actual test. Liquid nitrogen (-196 °C) is fed to the hole and a small gas flame applied to the other end. When a steady state is reached with one end at about -70 °C and the other at 60 °C, the temperatures along the specimen are noted: the gradient is about 4–8 °C per cm (10–20 °C per inch) at the centre. A heavy mass is now placed against the hot end, the transverse stress applied and a bolt gun fired at the cold end. A crack runs from the saw cut into the warmer material and stops; the remaining section then yields. The arrest position is determined either from the yield markings or by opening up the specimen and noting the focus of the arrested crack. Arrest temperatures are determined within 1 °C. The test is repeated with different applied stresses.

A typical curve of stress versus arrest temperature is shown in Fig. 10.15.

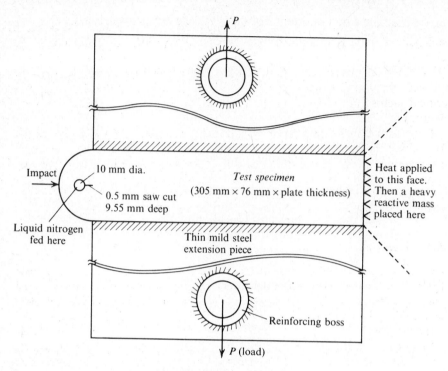

Fig. 10.14. Robertson test piece.

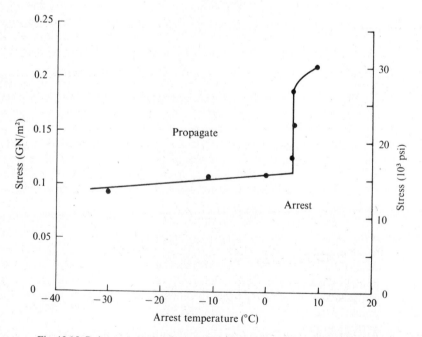

Fig. 10.15. Robertson test results on 15.8 mm (0.625 in) thick mild steel plate.

At low temperatures a small stress, which is about 100 MN/m² (15 000 psi), is sufficient to propagate a crack. However, above a transition temperature the propagation stress rises sharply. This temperature lies between −20 °C and 20 °C and does not correspond with other transition temperatures. Experiment showed that the test piece geometry and the temperature gradient were not critical factors in determining the propagation stress or the transition temperature.

Robertson concluded that it was possible to guarantee to avoid brittle fractures if the designer chose a working stress which was less than the crack propagation stress below the transition temperature. Since this involves approximately halving the allowable working stress this solution to the brittle fracture problem does not appear to be economically possible, unless the propagation stress can be increased significantly by metallurgical means.

10.10 Greene effect

Two characteristics of brittle failures in service which have not easily been reproduced in the laboratory are the absence of plastic deformation at the origin of the crack and the low applied stress. Greene in 1949 and Weck in 1952 have achieved them by welding plates with a fine saw cut in or adjacent to the prepared edge.

In Greene's tests, see Fig. 10.16, two $\frac{3}{4}$ inch thick plates, 30 × 18 inches, (∼2 × 80 × 45 cm) were butt-welded together after making a jeweller's saw cut in one of the prepared edges, transverse to the weld and $\frac{3}{16}$ or $\frac{7}{16}$ inch (∼5 or

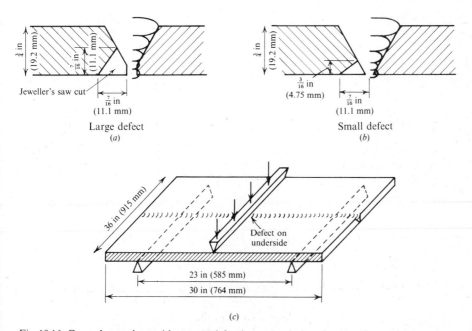

Fig. 10.16. Greene's test plates with saw-cut defect in prepared edge of weld. (a) Large defect (before welding); (b) small defect (before welding); (c) bend test (after welding).

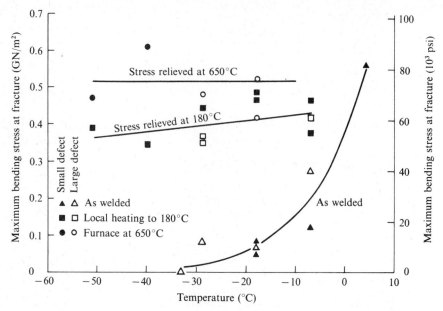

Fig. 10.17. Greene's test results, showing effect of stress relieving on fracture stress. (After T. W. Greene, *Welding Res. Suppl.* **28**, 193s (1949).)

11 mm) through the plate. Some 26 pairs of plates were taken from a single large plate with a 15 ft lb (20 J) Charpy V-notch transition at 12 °C, and welded using a common procedure. Some were then tested directly, and the others given a stress relieving treatment, either by local heating of the weld to about 180 °C or heating overall in a furnace to 650 °C. Testing consisted of three-point loading the welded plates as a beam with the weld along the span and the saw cut at mid-section on the tension side. Tests were made between −50 °C and +5 °C. Fig. 10.17 shows the maximum fibre stress at fracture against the temperature. It can be seen that the fracture stresses were exceedingly low, even zero, in the as-welded specimens but they recovered after heat treatment.

In Weck's experiments two $\frac{3}{4}$ in (2 cm) thick plates were welded along their outer edges into a 16 × 8 in (40 × 20 cm) opening in a solid frame cut from 4 in (10 cm) plate, see Fig. 10.18. Mid-way along, and 2 in (5 cm) back from, the prepared edge of the test weld, one of the plates had a $\frac{1}{2}$ in (12.7 mm) diameter hole from which two jeweller's saw cuts were made $\frac{1}{8}$ in (3.2 mm) deep parallel to the weld. A reinforcing rib was then welded across the frame and to one of the plates. After making the test weld, the temperature of the whole jig was reduced by placing it in a cold chamber. Three materials were tested; coarse- and fine-grained mild steel and a low alloy steel. The coarse-grained steel cracked spontaneously at −8 °C in the two specimens tested whilst the others did not crack at −65 °C, even when heavily hammered. Izod tests showed no sharp transition in the coarse-grained steel and the energy absorbed was 10 ft lb (14 J) at −8 °C; in the other two steels sharp transitions occurred at about −45 °C and −75 °C respectively. Weck concluded there was no correlation between liability to spontaneous fracture and the Izod transition temperature.

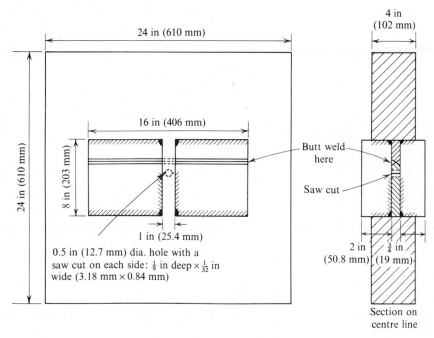

Fig. 10.18. Weck restraining frame.

A possible explanation of these experiments is that the saw cut strain hardens the material which strain ages during the welding. This raises the yield stress at the root of the cut and allows a cleavage mode fracture to be nucleated there. The theory is given in more detail in § 10.13.

10.11 Plane strain fracture toughness K_{Ic}

In 1970, following ten years of research, a new method of toughness testing was agreed and tentative test specifications issued (ASTM E399–70T and BSI DD3:1971). A new material property, or an old one in a new guise, is measured called the plane strain fracture toughness, symbol K_{Ic}. K_{Ic} measures the resistance to crack extension under plane strain conditions at the crack front, these being the most favourable for crack propagation. In many practical situations plane strain conditions do not exist and stress relief by plastic flow occurs, rather than crack extension. K_{Ic} is thus a conservative measure of crack toughness. A large research programme, sponsored by NASA and ASTM, has been primarily directed to the fracture mechanics of high strength, low notch toughness materials, such as maraging steels (see § 12.9), used for rocket fuel tanks. The yield stress, close to the tensile strength, is typically $2.0\,\mathrm{GN/m^2}$ (3×10^5 psi) and the K_{Ic} value around $2.5 \times 10^3\,\mathrm{N\,mm^{-2}\,mm^{\frac{1}{2}}}$ (7×10^4 psi in$^{\frac{1}{2}}$). This is in marked contrast with the properties of the low and medium strength steels under discussion here, for which typical values for a pressure vessel steel are a yield stress of $0.5\,\mathrm{GN/m^2}$ (7×10^4 psi) and a K_{Ic} of $6.5 \times 10^3\,\mathrm{N\,mm^{-2}\,mm^{\frac{1}{2}}}$ (1.9×10^5 psi in$^{\frac{1}{2}}$). The ratio K_{Ic} to σ_y, on which the theory and test method are based, is higher by over a factor of 10 in the latter, and the size of the test piece required to achieve plane strain conditions is proportional to the square of the ratio! Thus, although this test method is not immediately applicable to the brittle fracture of mild steel, it has such significance to other higher strength materials as to demand a brief summary here, especially as the concepts are likely to be extended soon to deal with plasticity effects in more ductile materials.

The proportions of the proposed test specimen for loading in three point bending are shown in Fig. 10.19; an alternative compact tension specimen will not be discussed. The absolute size is discussed below, but a minimum thickness of 6.4 mm (0.25 in) is set. If B is the thickness, the width (depth) W equals $2B$ and the length is $4.2W$ (giving a loading span $4W$). At mid-length a notch is machined in one edge extending across half the width (i.e. $a \simeq B$); straight-through and chevron designs are permitted, with a root radius of 0.08 mm or less. Knife edges at the entry to the notch are machined integrally with specimen or detachable ones can be screwed on. These are for locating a displacement gauge to detect the opening of the notch during loading. The specimen is now subjected to fatigue loading until a crack forms at the notch root and travels at least 1.3 mm or 5 per cent of the overall length of the notch and fatigue crack, whichever is the greater. Certain limits on the maximum loading cycle are set to avoid damage to the material ahead of the crack. The advance of the crack can be observed on the faces of the specimen and its exact length checked after fracture. The specimen

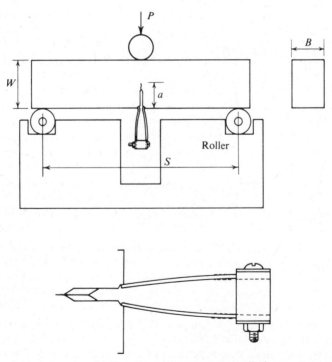

Fig. 10.19. Specimen proportions for K_{Ic} testing; enlargement shows displacement gauge on integral mounting.

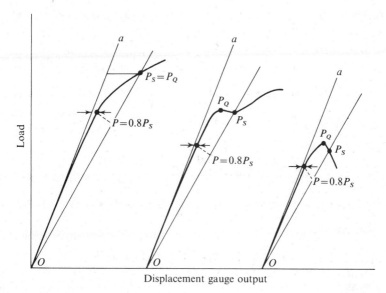

Fig. 10.20. Principal types of load displacement records. Note that slope OP_S is exaggerated for clarity.

is now ready for three point bending during which the outputs of the load transducer and the displacement gauge are recorded automatically, for example, on an X–Y plotter.

Typical load displacement curves are shown in Fig. 10.20. Ideally, for a linear elastic solid (i.e. Hookean), the curve is linear up to the point at which the crack starts to extend and fracture occurs. Real materials exhibit some plasticity, and the load (P_Q) at which the crack has extended 2 per cent, with limited plastic flow, is determined as follows. A secant is drawn through the origin with 5 per cent less slope than the tangent to the curve at the origin. The intersection with the curve is P_S. If every point on the curve prior to P_S is at a lower load (i.e. there is no maximum load prior to P_S), P_S is taken as P_Q and used in calculating K_{Ic}. Where there is a maximum prior to P_S, this is taken as P_Q. To check on the plastic flow prior to this point, it is decreed that the deviation of the actual curve from the ideal curve (tangent to the curve at the origin) at $0.8P_S$ must not exceed 25 per cent of that at P_S; if it does, the test is invalid.

The value of K_{Ic} is now obtained from

$$K_{Ic} = 0.0348 \frac{P_Q S}{BW^{\frac{3}{2}}} \left[2.9 \left(\frac{a}{W} \right)^{\frac{1}{2}} - 4.6 \left(\frac{a}{W} \right)^{\frac{3}{2}} + 21.8 \left(\frac{a}{W} \right)^{\frac{5}{2}} \right.$$
$$\left. - 37.6 \left(\frac{a}{W} \right)^{\frac{7}{2}} + 38.7 \left(\frac{a}{W} \right)^{\frac{9}{2}} \right],$$

where a is the crack length measured from the notched edge of the specimen and B, W and S are respectively the thickness, width (depth) and span. If the load P_Q is in newtons and all lengths in mm, K_{Ic} is obtained in N mm^{-2} mm$^{\frac{1}{2}}$ (For pound-inch units the numerical factor is unity and K_{Ic} is in psi in$^{\frac{1}{2}}$). The value of the function (a/W) is 2.66 at the nominal value $a = 0.5W$.

It is now necessary to check that the specimen size has been large enough to give a valid K_{Ic} result; until this is proven the parameter obtained above is strictly only a qualified value (given the symbol K_Q). Undersized specimens do not achieve plane strain at the crack front and erroneously high values of fracture toughness are recorded. It has been shown empirically that the specimen thickness (B) and crack length (a) shall be at least $2.5(K_{Ic}/\sigma_y)^2$, where σ_y is the 0.2 per cent offset yield stress. If K_{Ic} is found to be higher than estimated when selecting a specimen size, it is possible that this conditions has not been met and the test must be repeated with a larger specimen.

The ratio K_{Ic}/σ_y, which may be called the relative toughness, is probably more important as a material property than K_{Ic} itself. It, or more specifically its square, is a measure of the size of plastic zone which forms ahead of the crack prior to cracking, and it thus indicates the size of specimen and crack length for which this zone is negligible and plane strain conditions can be considered to exist, with maximum liability to crack extension. If the ratio is below 5 mm$^{\frac{1}{2}}$ (1 in$^{\frac{1}{2}}$) a material must be regarded as having low relative toughness. Some typical values of K_{Ic}/σ_y and $2.5(K_{Ic}/\sigma_y)^2$ for various materials are given in Table 10.2. The very wide range in values will be noted. Thus the maraging steel can be tested with the minimum sized specimen ($B = 6.4$ mm) and the very tough pressure vessel steel would require a specimen 490 mm (20 in) thick! This result has been obtained by extrapolation from results on 300 mm (12 in) thick specimens at temperatures below room temperature. The sensitivity of

Table 10.2. K_{Ic} and K_{Ic}/σ_y values of some materials at room temperature

Material	$\sigma_y(0.2\%)$ 10^3 N/mm^2 (10^3 psi)	K_{Ic} 10^3 N mm^{-2} mm$^{\frac{1}{2}}$ (10^3 psi in$^{\frac{1}{2}}$)	K_{Ic}/σ_y mm$^{\frac{1}{2}}$ (in$^{\frac{1}{2}}$)	$2.5(K_{Ic}/\sigma_y)^2$ mm (in)
Maraging steel	1.8 (260)	2.4 (68)	1.3 (0.26)	4.3 (0.17)
Aluminium alloy (7075–T6)	0.54 (79)	0.94 (27)	1.7 (0.34)	7.4 (0.29)
Titanium alloy (6% Al–4% V)				
(a) normal heat treatment	1.06 (154)	2.3 (65)[†]	2.1 (0.42)	11 (0.44)
(b) high yield	1.10 (160)	1.2 (35)	1.1 (0.22)	3.0 (0.12)
High strength alloy steel (0.45% C–0.5% Ni–1.1% Cr)				
(a) normal heat treatment	1.46 (212)	3.1 (90)[†]	2.1 (0.42)	11 (0.45)
(b) high yield	1.71 (249)	1.6 (45)	0.91 (0.18)	2.1 (0.082)
Medium carbon steel (AISI 1045, normalised)	0.258 (37.5)	1.7 (50)	6.6 (1.3)	110 (4.4)
Pressure vessel steel (ASTM A533B, quenched and tempered)	0.47 (68)	6.6 (190)[‡]	14 (2.8)	490 (20)

[†] Result not valid due to undersize specimen.
[‡] Result by extrapolation.

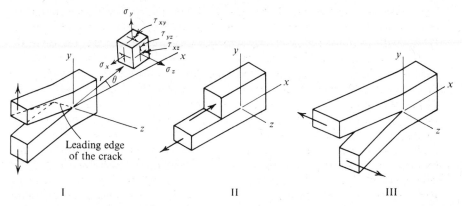

Fig. 10.21. Basic modes of crack loading: I opening mode; II edge sliding mode; III tearing mode.

the titanium alloy and high strength alloy steel to heat treatment is also indicated: a slightly higher yield stress is accompanied by a sharp drop in relative toughness.

This test has a theoretical basis in elastic fracture mechanics which shows that the parameter K_{1c} is a fundamental physical property of a material, in a similar way that the yield stress is. The value obtained is not specific to one particular test configuration, as are most of the other tests of fracture (e.g. Izod, Charpy, Tipper), which may be labelled empirical. Three modes of loading a crack are illustrated in Fig. 10.21: I opening mode, II edge sliding and III tearing. A generalised crack will be subjected to all three types of load simultaneously. Only the opening mode needs to be considered here since this is the most critical for crack growth. The important concept is that the stress field distribution just ahead of the crack (crack tip stress field) is determined entirely by the crack surfaces and is not dependent on remote boundaries and loading forces. These only contribute to the magnitude of the stresses (i.e. stress intensity). For example the tensile stress at a point ahead of the crack is given by

$$\sigma_x = \frac{K_1}{2\pi r^{\frac{1}{2}}} \cos\frac{\theta}{2} \left[1 - \sin\frac{\theta}{2}\sin\frac{3\theta}{2} \right]$$

where r and θ are the polar coordinates of the point. Similar equations exist for σ_y, σ_z and the shear stresses, all of the form stress $= K_1 f(r, \theta)$. The constant K_1, is called the *stress intensity factor* for the opening mode (I). It is a function of the overall specimen, that is specimen geometry and crack length, and the loads. It can be obtained analytically, for example, the expression given above for K_{1c} gives the formulae for this particular test specimen. Other expressions have been derived for many other shapes and loading conditions. It can also be obtained experimentally. The *fracture criterion* is that the crack will extend when the stress field reaches some particular state for a material. Since the distribution is constant, this means that the stress intensity K_1 reaches a critical value for a particular material, called K_{1c}. (The basis of the symbolism should now be apparent!) The temperature and rate of increase of the stress field must be appropriate. The elegance of this approach is that it is entirely independent of the physical phenomena leading to crack extension. For example, crack growth may be dependent on the maximum stress at some point ahead of the crack front, or the volume of material above some stress level. Since any such phenomenon occurs within the crack tip stress field of known distribution, all these criteria must be related to the stress level as determined by the stress intensity factor. Further information is given in the ASTM Special Technical Publications, nos. 381, 410 and 463.

10.12 Metallurgical factors

Increasing the carbon content, while keeping other factors as constant as possible, raises the transition temperature ranges of steel. Figure 10.22 shows typical notched bar energy values versus temperature for carbon steels over the range 0.01 to 0.67 per cent carbon. It will be seen that both the width of the transition range and the average-energy transition temperatures increase with carbon content. Silicon, except in small quantities, also raises the transition temperatures, similar to carbon. Manganese has the reverse effect and a high manganese to carbon ratio is desirable.

A small ferrite grain size lowers the transition temperature. In Fig. 10.23 the 20 ft lb (27 J) Charpy transition temperature for two steels is shown decreasing linearly with the ferrite grain size. Grain size is dependent on the chemical composition, the final rolling or normalising temperature and the plate size. Large plates are particularly susceptible to brittle fracture, partly on account of their large grain size.

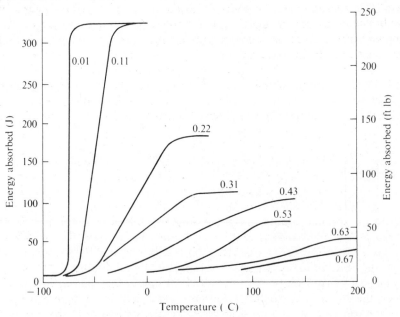

Fig. 10.22. Effect of carbon content of plain carbon steels on V-notch Charpy transition curves. Figures against each curve indicate weight per cent of carbon. (After J. A. Rinebolt and W. J. Harris Jnr., *Trans. A.S.M.* **43**, 1175 (1951).)

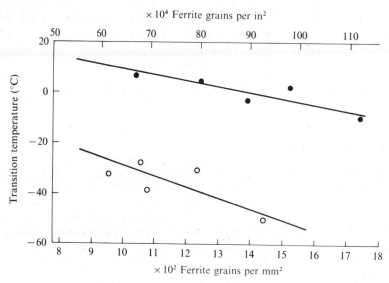

Fig. 10.23. Effect of ferrite grain size on transition temperature. ● 0.25% carbon, 0.45% Mn; ○ 0.21% carbon, 0.75% Mn. (After R. H. Frazier *et al.*, *Ship Structure Committee Report*, SSC-53 (1952).)

Cold work, especially when followed by strain ageing, is a potent factor in raising transition temperatures. Figure 10.24 shows the effect on the energy–temperature curves of straining up to 10 per cent and ageing for one month. This effect is the reason for the periodic annealing of lifting chains required by law. The brittle behaviour of steel parts made by cold working has been reported often and care must be taken that any manufacturing process does not give rise to a potentially dangerous product.

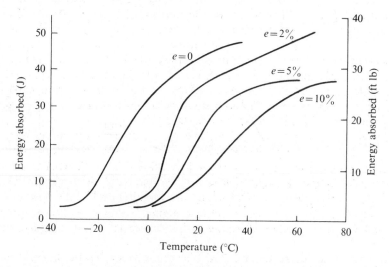

Fig. 10.24. Effect of prestrain (*e*) and ageing for one month on keyhole Charpy test. (After M. Gensamer *et al.*, *Ship Structure Committee Report*, SSC-9 (1947).)

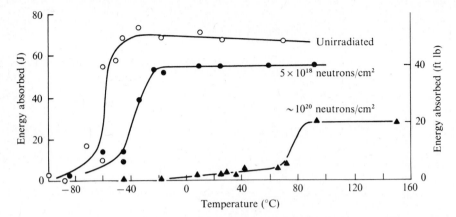

Fig. 10.25. Effect of neutron irradiation (>1 MeV) on fracture toughness of high strength carbon–silicon steel. Subsize Izod specimens irradiated at 95 °C (max.). (After R. G. Berggren and J. C. Wilson, *Oak Ridge National Lab. Report* CF-56-11-1 (1957).)

Neutron irradiation also raises the transition temperature. Typical results are shown in Fig. 10.25; the dose is measured in terms of the number of neutrons per unit volume times their mean velocity times the time. This fact and the use of thick plates has made the design of the reactor pressure vessel one of the critical areas in the engineering of nuclear power.

10.13 Macroscopic theories

In the theory of fracture started by Ludwik in 1923 and subsequently developed by Davidenkov in 1936 and Orowan in 1945, plastic flow and fracture were held to be independent phenomena with yielding obeying a critical shear stress criterion and fracture dependent on the maximum principal stress. The behaviour of a ductile metal subjected to a simple tensile test is then determined by the relative values of the two functions of the strain: $Y(e)$, representing the yield stress, and $F(e)$, representing the fracture stress, as indicated in Fig. 10.26(*a*). At zero strain the yield stress is below the fracture stress and plastic flow occurs when the applied stress equals $Y(0)$. As the metal is strain hardened the value of Y increases but F is assumed to be constant or decreasing. Fracture occurs when the $Y(e)$ and $F(e)$-curves intersect, that is, at a strain e_1.

If the state of stress is changed, the relative positions of the curves for yield and for fracture are altered. Superimposing a compressive stress R perpendicular to the axis will lower the axial stress at yield to $Y(e) - R$, according to the maximum shear stress criterion (§ 6.5); on the other hand, the stress for fracture is unchanged (maximum principal stress criterion). Figure 10.26(*b*) now applies: the intersection of the curves occurs at a greater strain e_2, corresponding to increased ductility, as found in practice with hydrostatic pressure or in a drawing die.

Conversely, tensile stresses perpendicular to the axis will raise the yield curve, giving reduced ductility. A circumferential notch on a tensile test piece produces such a stress distribution, due to the restraint on radial contraction of the material

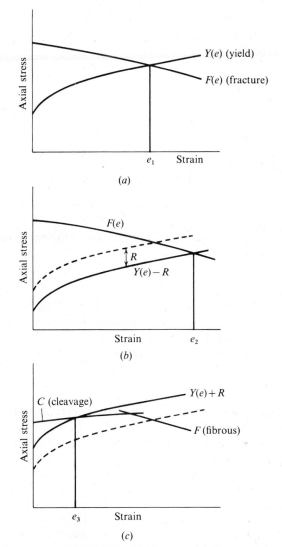

Fig. 10.26. Ludwik–Davidenkov theory of fracture.

(a) Y and F curves represent yield and fracture stresses in simple tension. Fracture occurs at strain e_1.

(b) With a superimposed compressive radial stress R the yield curve is shifted by R to $Y - R$ (maximum shear stress criterion). Fracture now occurs at e_2 ($> e_1$).

(c) Davidenkov introduced cleavage and fibrous fracture curves. Superimposing a tensile radial stress R causes the yield curve to move to $Y + R$ and a cleavage fracture at e_3.

in the notch section by the unyielded material in the full section. This was the original theory of notch brittleness put forward by Ludwik.

It was realised that, whilst this theory explained the reduction of ductility, no account was given of the change of fracture mode, and in 1936 Davidenkov proposed an additional curve $C(e)$ to represent fracture by cleavage, see Fig. 10.26(c); the $F(e)$ curve now represents the axial stress for fibrous fracture.

Raising the yield curve by means of a circumferential notch will now reduce the ductility and also produce a cleavage mode of fracture. That is, when the $(Y(e) + R)$-curve intersects the cleavage curve (C) before meeting the fibrous fracture curve (F).

The tendency to cleavage fracture at low temperature is also explained, at least qualitatively, since the yield stress of materials which are susceptible to cleavage fracture has been found to increase sharply with reducing temperature. The $Y(e)$-curve for room temperature is shifted upwards to $Y'(e)$ for low temperatures and intersects the cleavage curve before the fibrous fracture curve.

In the years from 1945 to 1948 Orowan developed the macroscopic approach to fracture theory in several respects. Firstly, he showed that the fibrous fracture at a given strain does not occur at a definite value of the maximum principal stress; the invariant F-curve is therefore incorrect.

More important for the study of brittle fracture, he and his co-workers showed that there is a limit to the raising of the $Y(e)$-curve by the use of sharper and deeper notches. Prior to this work, it had been believed that a notch had only to be made deep and sharp enough to induce cleavage fracture in all materials known to be capable of this mode of fracture (for example, at low temperatures). By considering the reverse problem of a rigid indenter being pushed into the surface of an ideal plastic material they were able to use earlier results of continuum plasticity theory to demonstrate that the maximum principal stress can only reach three times the tensile yield stress. This ratio is called the plastic constraint factor. This problem has already been discussed in Chapter 6 (§ 6.9). Thus only those materials for which the cleavage curve C is not more than three times the yield curve $Y(e)$ (at the test temperature) can be made to undergo cleavage fracture by introducing a sharp notch.

Other factors than the temperature and plastic restraint so far considered may raise the yield curve. It is known, for example, that the yield point of mild steel is increased at high rates of straining; thus once a brittle fracture has started to run at speed the velocity effect may become the dominant factor rather than the triaxial stress system. Another possibility is that the value of the upper yield stress is dependent on the volume of material which is being stressed. Finally strain ageing (§ 5.5) is known to raise the yield stress to as much as five times the normal value. This could be the explanation of the Greene effect (§ 10.10) where the saw cut deforms the material locally and this becomes aged during the welding.

Important historically are numerous experiments, around 1930, by Kuntz who attempted to determine the fracture stress curve, which he called 'technical cohesive stress'. Tensile specimens were strained by various amounts before having a standard 60° V-notch cut in them and then pulled to fracture. The fracture stress was multiplied by a factor to take account of the ratio of the constraint factor in an ideal notch to that in the standard notch. The technical cohesive stress was found to increase and then decrease with strain, later interpreted by Davidenkov as a separate rising C-curve and a falling F-curve. Unfortunately, the analysis assumed that an ideal notch would give infinite restraint and the conclusions must be disregarded.

It is now intended to consider what factors determine the cleavage fracture stress.

10.14 Irwin–Orowan theory

Irwin (in 1947) and Orowan (in 1948) suggested that the brittle fracture of metals by cleavage obeys a Griffith-type criterion. That is, internal flaws are present that will grow if the release of elastic strain energy of the structure is greater than the energy of the new fracture surfaces. In contrast with glass, which is elastic to fracture and has a true surface energy due to surface tension, metals deform plastically in a thin layer (0.5 mm) adjacent to a cleavage fracture. Irwin and Orowan suggested that the energy of this plastic layer be included in the effective surface energy, when in fact it becomes the dominant term.

The Griffith condition for the fracture stress σ_c now becomes (cf. (10.7)):

$$\sigma_c = \left(\frac{2E\gamma}{\pi c}\right)^{\frac{1}{2}}$$

where $2c$ is the internal crack length, E is Young's modulus and γ is the effective surface energy.

Several experimental techniques have been used to measure the surface plastic work. It has been estimated from the distortion of X-ray diffraction spots as the surface is progressively etched away. A value of 2×10^3 J/m^2 was obtained for ship's plate in the transition range (cf. ~ 1 J/m^2 true surface energy). Alternatively, it has been deduced from measurements of the temperature rise, since the majority (85–95 per cent) of the work of plastic flow is converted into heat; values in the range 8–16×10^3 J/m^2 were obtained. The major part of this energy is expended in crossing grain boundaries where the metal tears between one cleavage plane and the next. Within a grain the effective surface energy is between 1 and 10 J/m^2, which is of the same order as the estimated true surface energy. Surface energy remains the great unknown parameter which is fundamental to brittle fracture theory.

The crack length corresponding to an effective surface energy of 10^4 J/m^2 and a fracture stress of 0.4 GN/m^2 (6×10^4 psi) is almost 10 mm. It is unlikely that flaws of this length are normally present in steel, although grain boundaries weakened by segregation would provide them in special circumstances. If in the initial stages of cracking only the true surface energy is involved, say 1 J/m^2, the critical crack length reduces to 1 μm. Even this size of flaw is unlikely in pure metals. Current theories have therefore postulated that the cracks are formed from dislocations which have piled-up at an obstacle, such as a grain boundary.

10.15 Dislocation theory of fracture

Current theories of cleavage fracture postulate the formation of microcracks by the coalescing of slip dislocations, instead of the existence of inherent flaws previously assumed. Various mechanisms for the conversion of slip dislocations into crack dislocations have been proposed, but none is universally accepted. If the resulting crack reaches the critical length without plastic flow reducing the stress concentration, it will grow rapidly in the Griffith manner. Fracture has thus to be preceded by yielding – in agreement with the experimental evidence (see Fig. 10.10). An outline of one dislocation theory of brittle fracture will be given here.

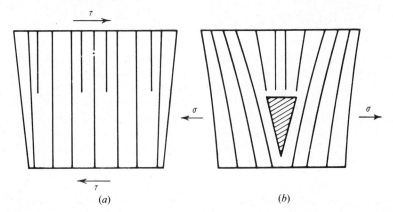

Fig. 10.27. Zener method of crack nucleation by coalescing of edge dislocations on a glide plane (1948). (a) Array of edge dislocations held up at barrier; (b) coalesced to form dislocation crack.

In 1948 Zener first proposed that a crack might form by the coalescing of a number of edge dislocations in the same glide plane. The mechanism is illustrated in Fig. 10.27: dislocations piled-up at an obstacle on the glide plane coalesce to form a dislocation crack in a plane perpendicular to the glide plane. It will be clear that shear stress on the glide plane is necessary to force the dislocations to coalesce but tensile stress on the crack plane is required to extend the microcrack into a fracture. Stroh calculated in 1957 the shear stress τ_m to cause n piled-up dislocations to coalesce into a microcrack:

$$\tau_m = \tau_f + \frac{12\gamma}{nb} \tag{10.12}$$

where τ_f is the friction stress, γ is the effective surface energy and b is the Burgers vector. He also showed that the tensile stress σ_p necessary to propagate the fracture is given by

$$\sigma_p \geqslant \frac{2\gamma}{nb}. \tag{10.13}$$

(This is the equivalent to the Griffith criterion in dislocation terms.) It follows that, even with a uniaxial stress distribution, where the tensile stress is double the maximum shear stress, the stress to cause the dislocations to coalesce is more than sufficient to cause the dislocation microcrack to grow. This disagrees with the experimental evidence that a hydrostatic tensile stress component (making $\sigma > 2\tau$) has an important effect in inducing brittle fracture.

Another method of forming dislocation cracks which avoids the preceding objection has been proposed by Petch, and independently by Cottrell (both in 1958). The microcracks are formed by the coalescing of edge dislocations on two intersecting planes, see Fig. 10.28. In the bcc lattice, a slip dislocation with Burgers vector $(a/2)[\bar{1}\bar{1}1]$ gliding in the (101) plane meets a dislocation with Burgers vector $(a/2)[111]$ gliding in a $(10\bar{1})$ plane and forms a dislocation $a[001]$ on the (100) plane, which is equivalent to a dislocation crack in the per-

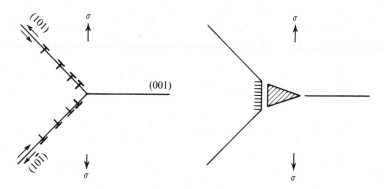

Fig. 10.28. Cottrell method of crack nucleation by the coalescing of edge dislocations on two intersecting glide planes (1958).

pendicular plane (001). The reaction is

$$\frac{a}{2}[\bar{1}\bar{1}1] + \frac{a}{2}[111] \rightarrow a[001].$$

There is a release of elastic energy by this reaction so no applied stress is required to coalesce the dislocations and nucleate the microcrack. The critical factor is whether it grows. There is no energy released by the corresponding reaction in the fcc system.

According to this theory, brittle fracture in mild steel occurs whenever the number of dislocations generated in a grain at yield is sufficient to form a dislocation crack, by the method just described, which is long enough to propagate at the prevailing stress level (that is, the yield stress) by the Griffith mechanism. Yielding is postulated to occur when the stress is sufficient to propagate yield across the grain boundaries. (This is described in more detail in § 11.5.) Below the yield stress, mobile dislocations in a few prematurely yielded grains move to the grain boundaries where they pile up. At the yield stress, the stress generated ahead of the dislocation pile-up is able to generate new dislocations in the adjacent grain. The yield stress is given by (see (11.12)):

$$\tau_y = \tau_f + \tau_s x^{\frac{1}{2}} d^{-\frac{1}{2}} \tag{10.14}$$

where τ_f is the friction stress, τ_s the stress to create mobile dislocations at a point a distance x across a grain boundary, and $2d$ is the grain size.

The number of dislocations produced in a grain with a freely active source (that is, requiring negligible stress to activate it) is (see (11.10)):

$$n = \frac{\pi d(\tau - \tau_f)}{Gb}. \tag{10.15}$$

This assumes a pile-up forms at the grain boundary until the back stress counterbalances the applied stress acting on the source. No relief of the pile-up stresses by generating other dislocations beyond the boundary is assumed. If this is the number of dislocations required to penetrate the first boundary (that is, out of the prematurely yielded grain), a similar number will have to be generated in the next grain to penetrate the next boundary. Yielding thus consists of a series of dislocation avalanches in each grain, n dislocations per grain, across the metal.

According to (10.13) a dislocation crack formed from these dislocations would have the critical size for Griffith type propagation if

$$\tau = \frac{\sigma}{2} \geqslant \frac{\gamma}{nb}.$$

Calling the critical value for propagation τ_p

$$\tau_p = \frac{\gamma}{b} \frac{Gb}{\pi d(\tau_p - \tau_f)}$$

i.e.

$$\tau_p(\tau_p - \tau_f) = \frac{\gamma G}{\pi d}. \tag{10.16}$$

Brittle fracture occurs if the yield stress, at which the avalanches of dislocations develop, is equal to, or greater than, the crack propagation stress τ_p:

$$\tau_y \geqslant \tau_p.$$

That is, substituting the value of τ_y from (10.14) for τ_p in (10.16):

$$(\tau_f + \tau_s x^{\frac{1}{2}} d^{-\frac{1}{2}}) \tau_s x^{\frac{1}{2}} d^{-\frac{1}{2}} \geqslant \frac{\gamma G}{\pi d},$$

i.e.

$$(\tau_f d^{\frac{1}{2}} + \tau_s x^{\frac{1}{2}}) \tau_s x^{\frac{1}{2}} \geqslant \frac{\gamma G}{\pi} \tag{10.17}$$

is the condition for brittle fracture. If the inequality is the other way the dislocation cracks may be formed but they will not grow rapidly in the Griffith manner, instead plastic flow occurs.

The theory shows a ductile–brittle transition temperature due to the rapid rise of the friction stress τ_f and the source stress τ_s which is known to occur as the temperature is reduced. The transition is also shown to be dependent on the grain size in agreement with experiment. The theory also shows that the condition for brittle fracture is not simply a function of the yield stress, since the parameters appear in different combinations in the equations for yield (10.14) and for brittle fracture (10.17). In addition, much more experimental evidence can be rationalised. This is not possible to include here and reference should be made to the 1958 papers by Cottrell, and Petch and Heslop.

QUESTIONS

1. Distinguish between
   (a) ductile and brittle kinds of fracture;
   (b) fibrous and cleavage modes of fracture;
   (c) high and low fracture ductility.
2. Show that the theoretical fracture stress is given by $\sim (ES/a)^{\frac{1}{2}}$.
3. (a) Give a brief account of Griffith's experimental study of the fracture of glass.
   (b) Outline Griffith's theory of fracture and show that the fracture strength is given by $\sim (ES/c)^{\frac{1}{2}}$.
4. (a) What is the ductile–brittle transition temperature range?
   (b) Distinguish between the fracture appearance and ductility transition.
   (c) Why are the details of the test important and on what criterion would you select a test procedure?
5. Discuss some of the metallurgical problems involved in the design of pressure vessels for nuclear power reactors.

6. Distinguish between crack nucleation and propagation. Describe briefly the Robertson test, its purpose and typical result.
7. Describe how brittle fracture at low applied stress ($<10^8$ N/m², 15×10^3 psi) has been achieved in a laboratory welding test.
8. Give an account of Ludwick's theory of fracture and the modifications by Davidenkov and Orowan.
9. (a) What is the Irwin–Orowan theory of brittle fracture in metals?
    (b) The surface plastic work has been measured at about 10^3 J/m². Estimate the crack length according to theory and comment.
10. Outline the Petch–Cottrell dislocation theory of brittle fracture in mild steel.
11. (a) Outline a method of determining the plane strain fracture toughness K_{1c}, indicating any criteria to be met in proving the result is valid.
    (b) Explain briefly the reason for the symbol K_{1c}.
    (c) For a centre-cracked plate specimen the stress intensity factor is

$$K_1 = 1.77 \frac{Pa^{\frac{1}{2}}}{BW}\left[1 - 0.1\left(\frac{2a}{W}\right) + \left(\frac{2a}{W}\right)^2\right] \text{N mm}^{-2} \text{ mm}^{\frac{1}{2}}$$

where P is the tensile load, $B \times W$ the cross-section, and $2a$ the crack length. For an aluminium alloy with $\sigma_y = 540$ N/mm² and $K_{1c} = 940$ N/mm^{-2} mm$^{\frac{1}{2}}$, determine the average stress for crack extension if $W = 10B = 20a = 100$ mm. What is the minimum value of B and $2a$ for which the above formula is applicable, using this material. Why is this so?

FURTHER READING

Original papers
A. A. Griffith: The phenomena of rupture and flow in solids. *Phil. Trans. Roy. Soc.* A**221**, 163 (1920).
A. A. Griffith: *First Int. Conf. Appl. Mech., Delft*, **55**A (1924).
E. Orowan: *Trans. Inst. Eng. and Shipbuilders, Scotland*, **165** (1945).
T. W. Greene: Evaluation of effect of residual stress. *Welding Res. Suppl.* **28**, 193s (1949).
T. S. Robertson: Brittle fracture of mild steel. *Engineering* (*London*), **172**, 444 (1951).
R. Weck: Experiments on brittle fracture of steel resulting from residual welding stresses. *Welding Res.* **6**, 70r (1952).
T. S. Robertson: Propagation of brittle fracture in steel. *J.I.S.I.* **175**, 361 (1953).
J. F. Baker and C. F. Tipper: The value of the notch tensile test. *I. Mech. E.* (*London*), **170**, 65 (1956).
N. J. Petch: The ductile–brittle transition in the fracture of α-iron, part I. *Phil. Mag.* **3**, 1089 (1958).
N. J. Petch and J. Heslop: The ductile–brittle transition in the fracture of α-iron, part II. *Phil. Mag.* **3**, 1128 (1958).
A. H. Cottrell: Theory of brittle fracture in steel and similar metals. *Trans. A.I.M.E.* **212**, 192 (1958).
H. C. Rogers: The tensile fracture of ductile metal. *Trans. A.I.M.E.* **218**, 498 (1960).
W. F. Thomas: The strength and properties of glass fibres. *Phys. Chem. Glasses*, **1**, 4 (1960).
D. M. Marsh: Flow and fracture in glass. *A.I.M.E. Conf.* (*Maple Valley*) (1962).
J. J. Gilman and D. C. Drucker (eds.): *Fracture of Solids*. Wiley (1963).
R. M. Spriggs and T. Vasilos: Effect of grain size on transverse bend strength of alumina and magnesia. *J. Am. Ceram. Soc.* **46**, 224 (1963).

Books and reviews
E. Orowan: Fracture and strength of solids. *Reports on Prog. in Phys.* **12**, 185 (1948–9).
N. J. Petch: The fracture of metals. *Progress in Metal Physics*, **5**. Pergamon (1954).

E. R. Parker: *Brittle Behaviour of Engineering Structures*. Wiley; Chapman and Hall (1957).

B. L. Averbach (ed.): *Conference on Fracture, Swampscott 1959*. M.I.T. and Wiley (1959).

C. ʹF. Tipper: The brittle fracture of metals at atmospheric and sub-zero temperatures. *Met. Rev.* **2**, no. 7 (1957).

W. D. Biggs: *The Brittle Fracture of Steel*. Macdonald and Evans (1960).

B. A. Proctor: Fracture of glass. *Appl. Mat. Res.* **3**, 28 (1964).

A. S. Tetelman and A. J. McEvily, jr.: *Fracture of Structural Materials*. Wiley (1967).

H. Liebowitz (ed.): *Fracture*. Vols. 1–7, Academic Press: 1 *Microscopic and Macroscopic Fundamentals* (1969); 2 *Mathematical Fundamentals* (1968); 3 *Engineering Fundamentals and Environmental Effects* (1971); 4 *Engineering Fracture Design* (1969); 5 *Fracture Design of Structures* (1969); 6 *Fracture of Metals* (1969); 7 *Fracture of Non-Metals and Composites* (1972).

G. M. Boyd (ed.): *Brittle Fracture in Steel Structures*. Butterworth (1970).

ASTM Special Technical Publications relating to plane strain fracture toughness:

STP 381: *Fracture Toughness Testing and its Applications* (ASTM Symposium, Chicago, June 1964) (1965).

STP 410: W. F. Brown, jr and J. E. Strawley: *Plane strain crack toughness testing of high strength metallic materials*. (1967).

STP 463: W. F. Brown, jr (ed.): *Review of developments in plane strain fracture toughness testing*. (1970).

11. Methods of strengthening materials

DUCTILE METALS

11.1 Introduction

There are only a few methods of converting the weakness (1–10 MN/m^2, 150–1500 psi) of single crystals of metallic elements, discussed above in Chapter 9, into the high strength (100–1000 MN/m^2, 15–150 $\times 10^3$ psi) of materials required for engineering construction. They rely for their success on restricting the movement of dislocations, which have been shown to be responsible for plastic flow and thereby for the allowable working stress of a ductile material (Chapter 5). The detailed dislocation mechanisms by which this is achieved within each method are more numerous and not yet fully elucidated.

The yield stress for the onset of plastic flow and the subsequent strain-hardening rate are both raised by a particular strengthening method but not necessarily in the same ratio, since different dislocation mechanisms are involved. At the same time the fracture behaviour is affected but, in ductile materials, this is normally of secondary importance. The successful strengthening methods are those that give high strength, as measured by the yield stress, tensile strength or hardness, whilst maintaining some ductility and notch toughness.

It is not possible to do justice to all aspects of strengthening in one chapter. Attention will be focussed on the effect of each method on the yield stress and the dislocation mechanisms involved, rather than on the subsequent strain harden-ing. Consideration of the dependence of the yield stress on strain rate and temperature, which can be important in elucidating the validity of theories and to some practical applications, will also have to be curtailed. Further details are available in the original papers and monographs listed at the end of the chapter. Some attention will be paid to the effect of annealing on the various strengthening techniques, that is, the result of raising the temperature of materials for a period.

The methods of strengthening brittle materials, in which fracture is the controlling factor, will be considered in the second half of this chapter.

11.2 Grain boundaries

The first factor contributing to the strength of constructional materials is the presence of grain boundaries, which occur, of course, in all polycrystalline material. Whilst a single crystal has a free surface and can deform as required by a single active slip system, grains in a polycrystal with differing orientations of their lattice and slip systems are forced to conform to the overall strain. As noted in discussing ionic ceramics (§ 7.2) von Mises has shown that at least five independent slip systems are necessary to obtain an unrestricted change. Figure 11.1 shows in polycrystalline α-brass (solid solution of 30 per cent zinc

417

Fig. 11.1. Slip lines in polycrystalline α-brass (Cu–30% Zn) (× 80).

in copper matrix) how the slip systems in adjacent grains are differently orientated. It is also clear that the grains are still coherent, that is, the grain boundaries do not open up. At high temperature and slow strain rate, that is, under creep conditions this is not so and the grain boundaries are a source of weakness.

Figure 11.2 compares the tensile curves for single crystals of various orientations with the tensile curve for polycrystalline material, for fcc copper and cph magnesium. The polycrystalline curve is not the statistical average of the single crystal curves because of the requirement for compatibility of the strain in adjacent grains. In hexagonal metals, slip is easy on the basal planes but high stresses are required to produce prismatic slip (i.e. on the $\{10\bar{1}0\}$ planes) and twinning, and the grain boundaries have a large strengthening influence.

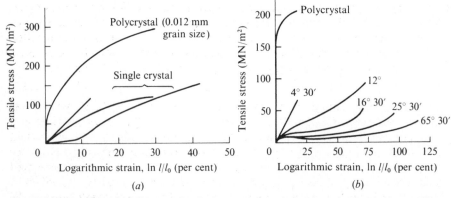

Fig. 11.2. Comparison of tensile curves for single crystals at various orientations, and polycrystalline metals: (a) fcc copper; (b) cph magnesium.

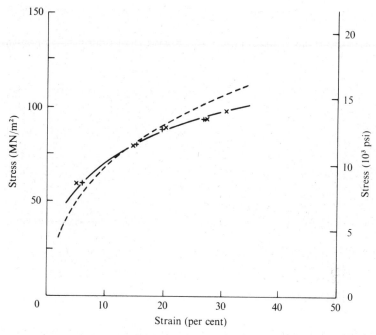

Fig. 11.3. Comparison of theoretical and experimental strain-hardening curves for polycrystalline aluminium. $---$ theory; $-\!\!-\!\!-$ experiment. (After G. I. Taylor, *J.I.M.* **62**, 307 (1938).)

In fcc and bcc metals, with numerous equivalent slip systems, the influence of the grain boundaries is fairly small. Several workers have attempted to derive the polycrystalline tensile curve from the single crystal shear-hardening curve (see Cottrell's book, 1953). The theoretical curve obtained by Taylor in 1938 is compared with the experimental tensile curve in Fig. 11.3.

Strength increases with decreasing grain size. Figure 11.4(*a*) plots the yield curves of Cu–30 % Zn (i.e. α-brass) for various grain sizes and (*b*) gives the yield stress as a function of grain size for several brasses with different zinc contents. In such experiments it is necessary to ensure that other variables are not altering at the same time and masking the effect due to the boundaries. For example, segregation of impurities to the boundaries or the initial dislocation sub-structure may be affected by the treatment used to vary the grain size. The yield stress σ varies with grain size *d* according to the Hall–Petch equation:

$$\sigma = A + Bd^{-\frac{1}{2}}. \tag{11.1}$$

This law was originally found to give the variation of the lower yield stress in low carbon steels (Hall in 1951, Petch in 1953) and in other materials with sharp yield points. It is now known to apply to the initial yield stress and the subsequent flow stress (yield stress at a given plastic strain) of most materials. It will be derived theoretically in § 11.5 from a consideration of the stresses generated ahead of an array of dislocations held up by a grain boundary. The values of *A* and *B*, and their variation with temperature, will depend on the detailed model applicable to each type of material.

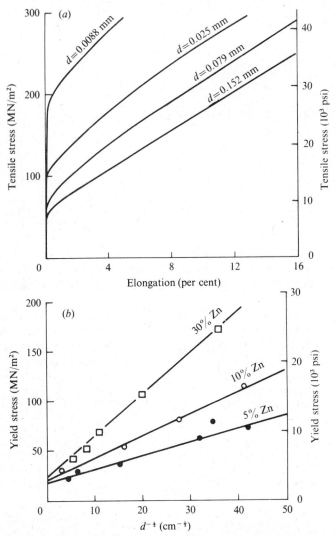

Fig. 11.4. Effect of grain size on the yielding of copper–zinc solid solutions (α-brass). (a) Stress–strain curves at various grain sizes (30% Zn). (b) Yield stress versus (grain size)$^{-\frac{1}{2}}$ at various zinc contents. (After H. Suzuki, *N.P.L. Symposium No. 15*, 518 (1963).)

11.3 Strain hardening

The phenomenon of strain hardening, in which the yield stress rises with increasing plastic deformation (§ 5.2), is exploited as a method of strengthening. Its use is normally limited to pure metals, for which no other method is available, on account of the associated large reduction in ductility. This is illustrated in Fig. 11.5, which compares the yield curves of a ductile metal in three conditions: (a) annealed; (b) after straining beyond the yield stress in a tensile test; and (c) after wire drawing. In the last process the yield stress can be raised above the tensile strength (TS) of the annealed material, due to the compressive radial

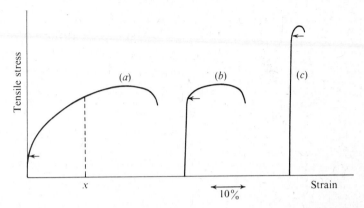

Fig. 11.5. Strengthening by strain hardening (*a*) annealed metal, (*b*) after straining plastically in tension to point *x*, (*c*) after wire drawing. The yield stress is indicated by an arrow in each case.

stresses from the die preventing the onset of necking. The lack of ductility in strain hardened material will be noted. Some strength values for aluminium in the soft, half hard and hard condition (produced by cold rolling) are given in Table 11.1.

During cold working, the grains become elongated in the direction of maximum extension and the crystal lattices rotate into *preferred orientations*. The former effect is illustrated in Fig. 11.6. Any non-metallic inclusions become similarly elongated into fibres and if these lie along the direction of maximum stress their weakening effect is much reduced. This is one reason for the superior mechanical properties of forgings over castings. Material with a preferred orientation of the grain lattices is said to possess a *texture*, and the properties become anisotropic, that is, they vary with direction. Figure 11.7 shows this effect in rolled copper. A texture and anisotropy persist even after recrystallisation (described below). Sheet material with a texture is unsuitable for press work as it does not deform uniformly. On the other hand, textures in transformer laminations greatly enhance the magnetic properties. Preferred orientations have been extensively studied, but will not be treated here.

The theory of strain hardening in pure metal crystals has been briefly outlined in Chapter 9 (§ 9.3 and § 9.9). In polycrystalline metals only stage III of the strain hardening found in single crystals is important since the interaction between the slip systems of adjacent grains inhibits significant amounts of easy glide (stage

Table 11.1. *Properties of strain-hardened aluminium (99.5 %), after cold rolling*

	Yield stress (0.1 % proof) (MN/m^2)	Tensile strength (MN/m^2)	El. (%)	DPN
Annealed	20	90	32	21
Half hard	100	120	8	33
Hard	125	150	5	40

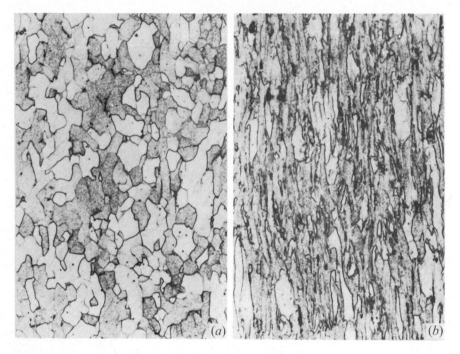

Fig. 11.6. Change of grain shape with cold work, for iron: (a) normalised, equi-axed grains; (b) after 40 per cent strain, elongated grains with indistinct boundaries. ($\times$ 230.)

I) and rapid hardening (stage II). The plastic behaviour of polycrystalline pure metals is thus determined by two opposing factors: rapid hardening produced by slip on multiple systems and strain softening due to cross slip, that is, when screw dislocations change their slip plane from one plane to an intersecting one. This allows them to by-pass obstructions and avoid high energy entanglements with other dislocations. Stacking fault energy, which limits the splitting of dislocations into partials (see § 9.9) and consequently assists cross slip, is therefore an important factor distinguishing the strain-hardening characteristics of different metals.

The density of dislocations increases rapidly with strain, according to the relation

$$\rho = \rho_0 + C\varepsilon^\alpha \qquad (11.2)$$

where ρ_0 is the initial dislocation density and C and α are parameters. Typical values in a metal are: $\rho_0 = 10^3$–10^5 mm^{-2}, $C = 10^{6 \pm 1}$ mm^{-2} and $\alpha = 1.0 \pm 0.5$. After heavy work ($\varepsilon \approx 1$) the number of dislocations rises to around 10^{10} mm^{-2}. These are average densities. The dislocations are actually not uniformly distributed but grouped together in a cellular arrangement, see Fig. 11.8. The cell walls contain a high density of entangled dislocations, several times the average figure, and the cell centres are relatively free of dislocations. The size of the cell decreases with increasing strain to a limiting value around 1 μm. The quantitative theory connecting the yield stress and dislocation structure is complex and at present contentious; it will not be tackled here.

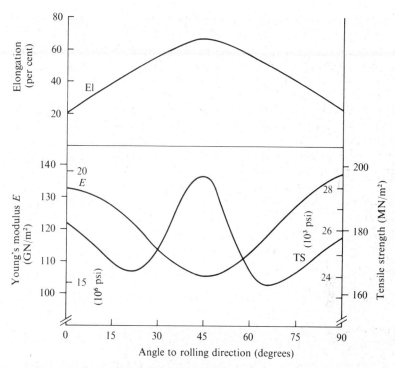

Fig. 11.7. Anisotropy of properties in copper possessing a texture. Cold rolled 96 per cent and annealed at 500 °C for 30 minutes.

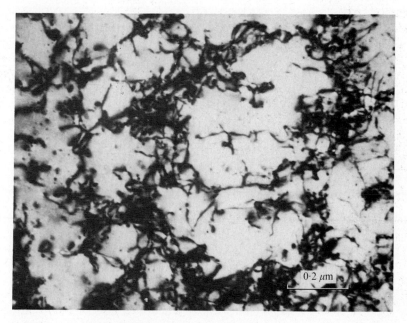

Fig. 11.8. Cellular arrangement of dislocation tangles formed in Mo–34 at.% Re after 16 per cent straining. (By courtesy of M. J. Witcomb.)

When a cold-worked metal is heated, changes occur in the microstructure and properties in two distinct stages, called *recovery* and *recrystallisation*. The driving force is the five per cent of plastic work which is stored internally. Recovery, which occurs over a temperature range for a particular cold worked metal, involves some reduction of the overall dislocation density and the formation of low angle boundaries dividing the grains into sub-grains about 1 μm across with their lattice orientations varying by about 1°. This effect, called polygonisation, was explained by Cahn in 1949 for a bent single crystal in terms of the formation of walls of parallel edge dislocations (see § 9.9 and Fig. 9.33). More complicated networks of dislocations are formed in polycrystals after cold work and annealing; this is shown in Fig. 11.9 for a Mo–Re alloy (cf. Fig. 11.8). Property changes during recovery are usually small, typically about one fifth of the way to restoration of the original properties of the unstrained or fully recrystallised material. Complicating effects may be superimposed, such as strain ageing (§ 5.5).

Above the recovery temperature range, recrystallisation takes place with the formation of new strain-free grains. Unlike recovery, this is a process of nucleation and growth. Recrystallisation is dependent on the degree of cold work and the time–temperature cycle, see Fig. 11.10. Increased cold work decreases

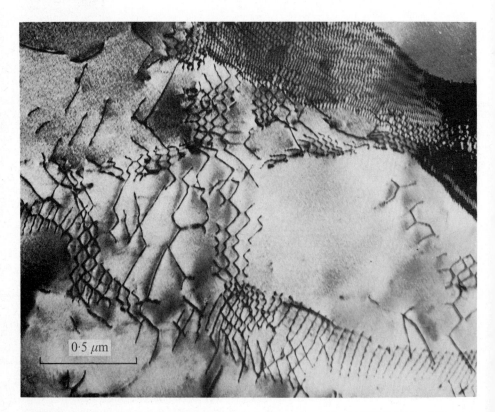

Fig. 11.9. Dislocations in polygonised array formed after recovery. Mo–34 at.% Re strained 16 per cent and annealed for 1 hour at 1050 °C. (By courtesy of M. J. Witcomb.)

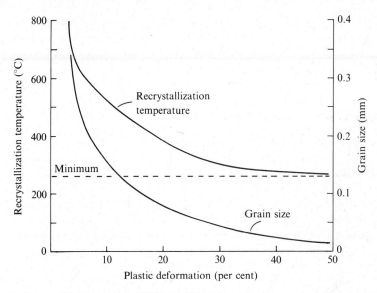

Fig. 11.10. Recrystallisation of brass (70% Cu–30% Zn): temperature and grain size versus prior plastic deformation. Held for 1 hour at any particular temperature.

the temperature required to bring about full recrystallisation in a given time. Any time spent at the recovery temperatures will reduce the stored strain energy and reduce the rate of recrystallisation. The recrystallisation temperature approaches a limiting value with increasing cold work, called the minimum recrystallisation temperature or often, simply, recrystallisation temperature. Its value decreases slowly with increased time at temperature, normally taken as one hour. Values for various metals are listed in Table 11.2; these are about 0.4 to 0.5 the absolute temperature of melting.

The recrystallisation grain size is also a function of the amount of cold work, being large in material which has received only the minimum amount of cold work required to bring about recrystallisation. This is a useful method of growing single crystals but can be a nuisance in structural applications, particularly where non-uniform plastic working is involved. Heavy cold work followed by

Table 11.2. *Minimum recrystallisation temperature (at about 1 hour) for various metals and alloys*

	°C		°C
Aluminium	150	Nickel	620
Beryllium	1290	Platinum	450
Brass (70% Zn–30% Cu)	375	Silver	200
Cadmium	50	Steel (0.2% C)	460
Gold	200	Tantalum	1020
Iron	450	Tin	0
Lead	0	Tungsten	1210
Magnesium	150	Zinc	15

recrystallisation just above the minimum recrystallisation temperature gives desirable fine grains but heating to a higher temperature will cause grain growth. Two other factors which increase the recrystallisation rate are the degree of purity, particularly in very pure metals, and the fineness of the initial grain size prior to cold work.

11.4 Solid solution strengthening

Single-phase materials are hardened by other atoms being taken into substitutional or interstitial solid solution. Rosenhain postulated in 1921 that the strengthening effect was due to the 'roughening' of the slip planes. It is now known that there are several detailed mechanisms by which the solute atoms interact with the dislocations, apart from their indirect effect of changing the density and distribution of the dislocations formed during freezing.

The yield strength, that is, the critical resolved shear stress in single crystals or the proof stress in polycrystals, is raised approximately linearly with the concentration of solute atoms. Figure 11.11 shows the hardening in single crystals of gold–silver and aluminium–copper alloys. The much greater strengthening in the Al–Cu system as compared with Ag–Au will be noted; this corresponds to a larger misfit of atomic diameters (11 and 0.2 per cent respectively) – all four elements are fcc. The strengthening effect for various polycrystalline copper solid solutions is shown in Fig. 11.12. Various workers have tried to rationalise the data for different solute atoms in one solvent by relating the hardening effect to the change of lattice parameter and also to the valency ratio of the solute and solvent atoms. Due to the variety of mechanisms involved, these equations are of limited value.

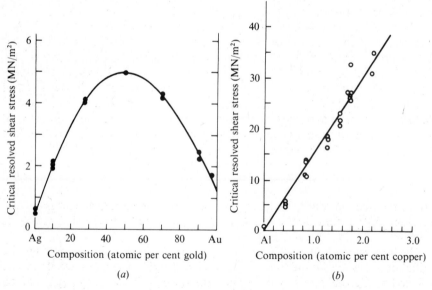

Fig. 11.11. Solid solution strengthening in single crystals of (*a*) silver–gold alloys, (*b*) aluminium–copper alloys.

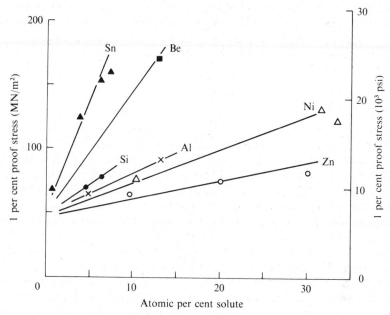

Fig. 11.12. Solid solution strengthening in polycrystalline copper alloys with various elements. (After R. S. French and W. R. Hibbard, *Trans. A.I.M.E.* **188**, 53 (1950).)

Several dislocation mechanisms of solution hardening have been proposed, and it is probable that two or more will be acting simultaneously in an alloy. The resistance to yielding may be increased by pinning dislocations in their initial positions, or by raising the 'friction' stress resisting their movement over a slip plane, or by making it difficult to avoid obstacles by cross slip.

The first theory by Mott and Nabarro, developed between 1940 and 1948, considers the internal stresses created by misfitting solute atoms, which must be overcome before a dislocation can move through the lattice. The theory also covers precipitates of a second phase (see § 11.7). The yield stress is dependent on the magnitude of the internal stress and the mean wavelength over which it fluctuates. The wavelength, which is equal to the separation of the solute atoms Λ, has to be judged in relation to the radius of curvature R to which the internal stress τ_i is able to bend a dislocation. The curvature is given by (see § 9.10 and (9.19)):

$$R = \frac{T}{b\tau_i} = \frac{\alpha G b}{\tau_i} \qquad (11.3)$$

where the line tension $T = \alpha G b^2$ with $\alpha \approx 1$. For increasing coarseness of dispersion of the solute atoms or precipitate, three regimes are distinguished: (a) $\Lambda \ll R$; (b) $\Lambda \approx R$; (c) $\Lambda \gg R$.

The importance of the flexibility of the dislocation was not appreciated in the initial development of the theory. A completely rigid dislocation would not be affected by internal stress fields because there would be as many regions along its length where the internal stress was driving it forward as there would be driving it back. The net friction force only exists because the dislocation is able

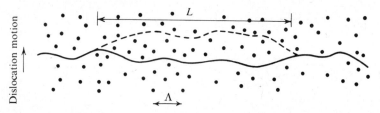

Fig. 11.13. Internal stress field theory of solution hardening (Mott–Nabarro): $\Lambda \ll R$, where Λ is separation of solute atoms and R is radius of curvature to which the stress field can bend a dislocation. A section length $L > \Lambda$ moves forward at one time. • Solute atoms.

to follow to some extent the potential energy valleys and moves forward a section at a time. The length of this loop is an important parameter in determining the yield stress, which according to the theory is a maximum when the dispersion equals the dislocation curvature.

For $\Lambda \ll R$, which is typical of solid solutions, the dislocation cannot bend around each individual stress centre but follows a line which minimises the sum of its inherent elastic energy and its potential energy due to interaction with the stress field, see Fig. 11.13. Forward movement occurs when a section of the dislocation several times larger than Λ jumps into the next equilibrium position. The details of the argument for determining the section length will not be reproduced here. The yield stress is calculated as

$$\tau = 2.5 G \varepsilon^{\frac{4}{3}} c \qquad (11.4)$$

where c is the atomic concentration and ε the misfit : $r_b = r_a(1 + \varepsilon)$. (The earlier version of this theory gave a smaller yield stress : $\tau = G\varepsilon^2 c$.) The linear dependence on the concentration is in agreement with many experimental results, such as those given in Fig. 11.11, and the $\frac{4}{3}$ exponent of the misfit is confirmed by some experiments in which various solute atoms (Mn, In, Sn) were added to copper single crystals in concentrations of 10^{-2} to 10^{-3}, see Table 11.3. However, the hardening is only $7 \, \text{GN/m}^2$ while the theory predicts a value of $100 \, \text{GN/m}^2$. The discrepancy was less, only a factor of two, with the Al–Cu crystals (Fig. 11.11(b)). The discrepancy may be due to clustering of the solute atoms, which effectively increases Λ and makes (11.5) applicable, or to the influence of thermal energy assisting the dislocation over barriers. Experiments show that there is roughly a factor of three between the yield stress at absolute zero and room temperature.

Table 11.3. *Solid solution hardening of copper crystals with various solute atoms*

	Mn	In	Sn
ε	0.1	0.26	0.28
$\dfrac{d\tau}{dc}(\text{GN/m}^2)$	0.343	1.11	1.42
$\varepsilon^{-4/3}\dfrac{d\tau}{dc}(\text{GN/m}^2)$	7.4	7.3	6.7

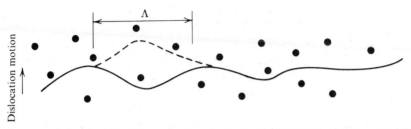

Fig. 11.14. Internal stress field theory of precipitation hardening (Mott–Nabarro): $\Lambda \approx R$ (cf. Fig. 11.13). Section length $L \approx \Lambda$ moves forward at one time. ● Stress centre.

For $\Lambda \approx R$, which is considered to represent a precipitation-hardened alloy at maximum hardness (discussed in § 11.7) a dislocation is able to bend around individual stress centres and a section length $L \approx \Lambda$ moves forward independently, as shown in Fig. 11.14. The yield stress then becomes

$$\tau = 2G\varepsilon c. \tag{11.5}$$

Finally when $\Lambda \gg R$, which represents a dispersion-hardened material (discussed in § 11.8), a dislocation moves forward by passing between the obstacles leaving a closed dislocation loop behind, as shown in Fig. 11.15. This mechanism was suggested by Orowen in 1947. The yield is then given by

$$\tau \approx \frac{2Gb}{\Lambda}. \tag{11.6}$$

Details of the dislocation mechanisms of precipitation hardening and dispersion hardening will be considered later.

In the Cottrell theory of solution hardening, dislocations are pinned by the solute atoms forming a group or 'atmosphere' around them, thereby reducing the elastic stress fields. As mentioned in § 9.9, the stresses around a positive edge dislocation are reduced by an oversize substitutional atom, or any interstitial atom, located below it, see Fig. 11.16. If the solute atom distorts the lattice unequally in different directions, that is, produces a shear strain, it will also react with the shear component of a dislocation field, as exists around a screw dislocation. This type of interaction is not to be expected from substitutional solute

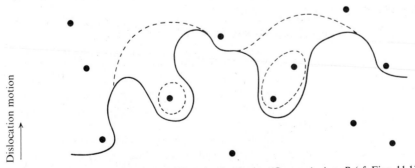

Fig. 11.15. Internal stress field theory of dispersion hardening (Orowan): $\Lambda \gg R$ (cf. Figs. 11.13, 11.14). Dislocation passes between obstacles leaving loop behind.

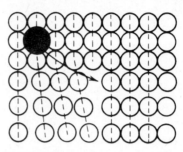

Fig. 11.16. Oversize substitutional atom is attracted to position below a positive edge dislocation, reducing the stress field energy. Dislocation becomes pinned by solute atom.

atoms in symmetrical lattices (e.g. fcc and bcc), but an interaction can occur through the hydrostatic field exerting a torque on a screw dislocation and converting it into an edge type. When the solute atoms have grouped around a dislocation it is locked or pinned, since the released strain energy must be supplied to separate the dislocation from its atmosphere. However, once the interaction is overcome by an applied stress the dislocation is able to move at a lower stress, tending to produce a sharp yield point. The magnitude of the effect will depend on several factors, which are particularly favourable for carbon and nitrogen in solution in iron. The theory of sharp yield points will be discussed in the next section.

Several other hardening mechanisms have been suggested. A chemical interaction (Suzuki) is present because solute atoms lower the stacking fault energy in a lattice and are attracted to extended dislocations. Lowering of the stacking fault energy causes the dislocation to become further extended and inhibits cross slip. The importance of this effect to strain hardening is being increasingly appreciated. There is an electrical interaction (Cottrell and Nabarro) between solute ions and the electric charge associated with the stress field of a dislocation. Finally, there is a geometric interaction (Fisher): instead of a random distribution, solute atoms may form local clusters or, conversely, arrange themselves regularly in the host lattice – short range order. Any local cluster or order is destroyed by a dislocation passing through and the increased energy has to be supplied.

11.5 Sharp yield point

The sharp yield point in mild steel and the related phenomenon of strain ageing have already been described in macroscopic terms (§ 5.5). Similar effects, though less pronounced, are found in several other metals of normal purity, notably bcc Mo, Wo, Ta, and in some alloys, such as Al–Mg, Cu–Zn, Cu–Sn. They occur in polycrystalline metal and, to a lesser extent, in single crystals. Figure 11.17 shows a sharp yield point and strain-ageing effect in a single crystal of iron containing 0.005 per cent carbon, and Fig. 11.18 shows a sharp yield in polycrystalline tantalum.

The theory of yielding and strain ageing in iron and low carbon steel, begun by Cottrell and Bilby in 1949, was one of the first successes of dislocation concepts. In the light of further experimental evidence the theory has been revised and extended (by Hahn in 1962, by Cottrell in 1963). It will be given

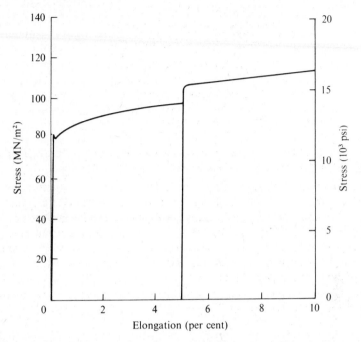

Fig. 11.17. Sharp yield point and strain ageing in an iron crystal with 0.005 per cent carbon, strained 5 per cent and aged for 3 hours at 200 °C. (After H. Schwartzbart and J. R. Low, *Trans. A.I.M.E.* **185**, 637 (1949).)

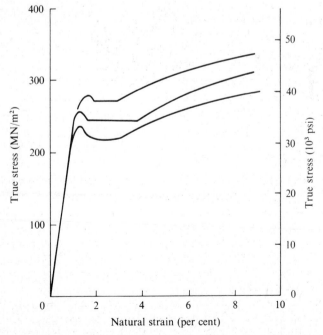

Fig. 11.18. Sharp yield point in polycrystalline tantalum with various sub-structures (293 °K). (After D. Hull, I. D. McIvor and W. S. Owen, *N.P.L. Symposium No. 15*, 596 (1963).)

here at some length because of its scientific interest and technological importance in mild steel; it can be omitted by the reader requiring a discussion of the strengthening methods rather than details of the dislocation mechanisms and theory. Earlier, the sharp yield point was attributed to some unspecified hard film which was thought by Dalby in 1913 to form at the grain boundaries.

The starting point of the Cottrell and Bilby theory is the strong pinning of the initial dislocations by clustering of carbon and nitrogen atoms, as briefly described in §11.4. In interstitial solid solution in iron, these atoms have a particularly high interaction energy with both screw and edge dislocations, causing them to 'condense' into a single row of atoms alongside each dislocation. The potential energy valley is then very narrow and a high stress is required to move the dislocation away from its atmosphere. This was postulated to occur at the upper yield stress; the lower yield stress was then the stress required to propagate the dislocations across grain boundaries.

The interaction energy is obtained by treating the solute atom as an elastic inclusion. If (R, α) are the polar coordinates of the inclusion from the centre of an edge dislocation ($\pi > \alpha > 0$ on the upper side of a positive edge dislocation), it is given by:

$$U = \frac{A \sin \alpha}{R} \tag{11.7}$$

where $A = 4Gb\varepsilon r^3$, with r the radius of the solvent atom and $(1 + \varepsilon)r$ that of the solute atom. The energy is positive for an oversize atom in the compressed region above a dislocation and negative when in the extended region below. The equation breaks down at the centre of the dislocation where Hooke's law is no longer valid, and a minimum value of R is taken: $R_0 \simeq 0.2$ nm.

For carbon in iron, with $G \approx 80$ GN/m^2, $2r = b = 0.25$ nm and $\varepsilon = 0.33$ (based on volumetric changes of martensite), Cottrell and Bilby theory gives

$$A = 5 \times 10^{-29} \text{ Nm}^2$$

and $\qquad U_{min} = -A/R_0 \simeq -2.5 \times 10^{-19}$ J $(1.6$ eV$)$.

Other methods of arriving at the maximum binding energy suggest this result is too high by a factor of 3, due to the use of linear elasticity theory. A similar interaction energy is predicted between carbon and nitrogen atoms and screw dislocations in iron, on account of the shear distortion produced by the particular positions in the unit cell occupied by the solute atoms.

The concentration of solute atoms at a position with binding energy U is normally given by a Maxwellian distribution:

$$c = c_0 e^{U/kT}$$

where c_0 is the average concentration and k is Boltzmann's constant. However, for carbon and nitrogen in iron $U \gg kT(kT = 0.025$ eV at room temperature) and the formula no longer applies. (Clearly the concentration cannot rise above unity.) Instead of forming a diffuse atmosphere, the carbon and nitrogen atoms condense in a single line alongside the dislocation at the position of maximum binding. Maximum pinning occurs when there is one interstitial atom per atomic length of dislocation. At greater concentrations the atmosphere is not so localised, the potential energy valley less steep and the pinning stress less. For a dislocation density ρ mm^{-2}, the number of optimum sites per unit volume is

$\rho/b\,\mathrm{mm}^{-3}$, equivalent to an average atomic concentration of ρb^2. For $\rho = 10^8\,\mathrm{mm}^{-2}$ and $b = 0.25\,\mathrm{nm}$, the concentration is 6×10^{-6}, which compares reasonably with the observed values of the carbon content for maximum upper yield stress, which is in the range 10^{-3} to 10^{-5}.

The stress to move a dislocation away from a condensed line of solute atoms is obtained by differentiating (11.7). At a distance x from its initial position, see Fig. 11.19, the dislocation has an interaction energy:

$$U(x) = -\frac{AR_0}{x^2 + R_0^2}.$$

Hence the attractive force per atomic length of dislocation is:

$$f(x) = \frac{dU(x)}{dx} = \frac{2AR_0 x}{(x^2 + R_0^2)^2}.$$

This is a maximum at $x = R_0/\sqrt{3}$: $f_{max} = \dfrac{0.65A}{R_0^2}.$

At yield, the applied shear stress overcomes the interaction force. Using (9.14) for the applied force on atomic length of dislocation:

$$\tau_0 b^2 = \frac{0.65A}{R_0^2},$$

and the tensile yield stress is: $\sigma_0 = 2\tau_0 = \dfrac{1.3A}{b^2 R_0^2}.$ (11.8)

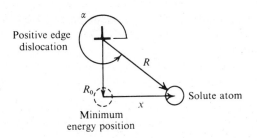

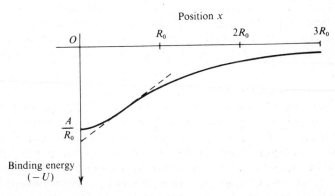

Fig. 11.19. Interaction energy of solute atoms and edge dislocation: $U = A \sin \alpha/R$. Slope indicated by dashed line represents stress σ_0 to separate dislocation and solute atom.

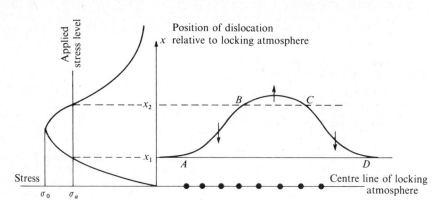

Fig. 11.20. Unlocking of pinned dislocation by applied stress and thermal energy. Direction of stress on each segment of dislocation is shown by the arrows.

Taking $A = 1.3Gb\varepsilon r^3 \simeq 1.5 \times 10^{-29}\,\mathrm{N\,m^2}$ (allowing for the overestimate of the binding energy), gives

$$\sigma_0 \simeq 10\,\mathrm{GN/m^2}.$$

This is the theoretical upper yield stress at zero temperature according to Cottrell and Bilby.

At finite temperatures, thermal energy markedly assists the applied stress to unlock a dislocation from its atmosphere. Due to the applied stress σ_a, the dislocation moves off-centre from the locking atmosphere to position x, where the restoring stress equals the applied stress (see Fig. 11.20). A loop $ABCD$ is thrown forward by thermal activation and reaches position x_2 at which the restoring stress is again equal to the applied stress (but it is now decreasing with x). Above a critical length of loop BC the rest of the dislocation is drawn over the energy barrier and away from its locking atmosphere. Cottrell and Bilby calculated that the stress for unpinning varied rapidly with temperature in agreement with the observed dependence of the yield stress (but see below). There is a tenfold decrease between $0\,^{\circ}\mathrm{K}$ and room temperature, that is, the theoretical yield stress reduces to about $1\,\mathrm{GN/m^2}$. They also developed a kinetic theory of strain ageing in agreement with experiment, based on the return of the yield point in overstrained metal when the solute atoms diffuse to the freed dislocations.

Experimental values for the upper yield stress are very dependent on avoiding premature yielding due to local stress concentrations such as are produced by loading grips, changes of cross-section or inclusions. Hutchinson in 1957 and 1963 obtained values up to $500\,\mathrm{MN/m^2}$ (73×10^3 psi) with 0.03 per cent carbon steel and $2\,\mu\mathrm{m}$ grain size. It was claimed that the close agreement between the theoretical and experimental values confirmed the theory that the upper yield stress coincided with the unpinning of locked dislocations.

Certain evidence, however, does not support this theory. For example, some slight scratching can be tolerated without destroying the yield point, and, significantly, the upper yield stress is a function of the grain size, as shown by the experimental results on 0.12 per cent carbon steel given in Fig. 11.21. The preceding theory does not include any grain size effect. It has also been observed

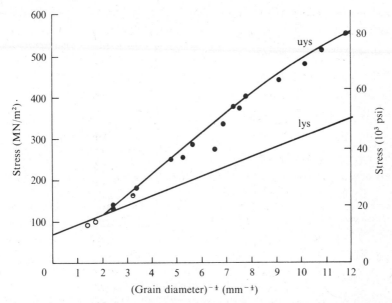

Fig. 11.21. Dependence of upper and lower yield stress of mild steel (0.12 % C) on grain size at 18 °C. NB At large grain size the yield stresses merge. (Points for lys are omitted as they showed little scatter from the line drawn.) (After N. J. Petch, *Acta Met.* **12**, 59 (1964).)

that some isolated grains have yielded prior to the upper yield stress, which is associated with the propagation of slip across the grain boundaries rather than the first unpinning of dislocations.

Petch proposed in 1964 the theory that the upper yield stress is associated with the penetration of grain boundaries, as indeed is the lower yield stress. The difference between the two lies in the number of grains which are yielding. At the uys only a limited number of grains contain mobile dislocations (created at local stress concentrators, such as inclusions) and these are therefore moving fast due to the imposed overall strain rate on the specimen. The friction stress opposing dislocation motion is known to increase with dislocation velocity (proportional to the logarithm of the strain rate), and consequently the applied stress required to penetrate grain boundaries is raised by this amount. As the number of grains with active dislocation increases, the velocity drops and with it the friction stress and the applied stress required for crossing grain boundaries. Finally a Lüders band is formed and the number of active grains remains constant as it advances through the specimen at the lower yield stress. It is shown below, (11.12), that the stress for a slip band to penetrate grain boundaries is given by

$$\sigma = \sigma_f + kd^{-\frac{1}{2}}$$

where σ_f is the friction stress, d is the grain diameter and k is a constant which depends on the ease of generating dislocations. According to Petch, σ_f consists of two terms

$$\sigma_f = \sigma_0 + \Delta\sigma_0 \log_{10}\frac{1}{Nd^3}. \qquad (11.9a)$$

σ_0 is the friction stress with uniform straining in all grains and the second term takes account of the velocity effect (logarithmic law) due to only N grains being active per unit volume (Nd^3 is thus the active volume per unit specimen volume and $1/Nd^3$ the increase in the dislocation velocity over homogeneous straining). $\Delta\sigma_0$ is the increase in friction stress due to a tenfold increase in strain rate. The upper yield stress is then

$$\sigma_u = \sigma_0 + kd^{-\frac{1}{2}} + \Delta\sigma_0 \log_{10}\frac{1}{N_0 d^3} \qquad (11.9b)$$

where N_0 is the initial number of active grains per unit volume containing stress concentrators. The first two terms represent the lys and the last term represents the difference between the uys and lys due to the velocity effect. Experimental confirmation of the theory has been obtained by Petch, but it will not be given here. The value of N_0 lies in the range 10–100 grains/mm³. For a typical grain size of 50 μm, or 10^4 grains/mm³, this means that one in 10^2 to 10^3 grains are yielding before the uys is reached, when break out into adjacent grains occurs.

Turning now to the lower yield stress, Cottrell in 1953 first suggested that this is the stress required to extend slip bands from one grain to the next. Over the intervening years some of the details have been modified. For example it was initially thought that the lys was the stress to spread a small plastic zone consisting of a few grains across the grain boundaries; it is now considered that the lys corresponds to the forcing of the grain boundaries ahead of a developed Lüders band. Again, it was originally thought that the propagation involved unpinning of dislocations from their atmospheres; it is now found that at ambient and low temperatures new dislocations are being generated in the next grain adjacent to the boundary. Basic to all these ideas is the dislocation pile-up at a grain boundary and the way it amplifies the applied stress up to the point at which dislocations become mobile in the next grain.

The characteristic spacing of a number of dislocations on one slip plane being forced by an applied stress up against an obstacle, such as a grain boundary, is shown in Fig. 11.22. Each dislocation in the array is being pushed forward

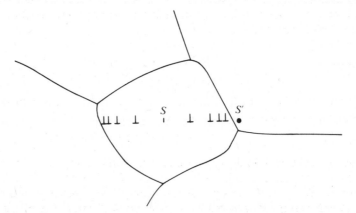

Fig. 11.22. Dislocation pile-up at grain boundary. The dislocations are coming from the Frank–Read source S. At the lower yield stress the shear stress ahead of the pile-up is able to operate a further source S' in the adjacent grain and yield propagates across all the grains.

with a force τb per unit length by the applied stress τ; being like dislocations they are repelling each other and the leading dislocation is also repelled by the obstacle's stress field. The situation may be likened to an unruly queue of people trying to push its way on to an overfull bus! From energy considerations (virtual work principle) the total forward stress field on the leading dislocation is $n\tau$. (This assumes that only the leading dislocation interacts with the obstacle. Consider the work done as the array of dislocations, *en bloc*, moves forward a small distance.) The number of dislocations in the pile-up will increase until the back-stress produced at the source S just cancels the applied stress, assuming negligible stress is required to operate the source. For a large number in the pile-up this has been calculated in 1951 by Eshelby, Frank and Nabarro:

$$n = \frac{\pi L \tau}{Gb} \qquad (11.10a)$$

where L is the distance from the source to the obstacle. (This formula is for screw dislocations, multiply by $1 - v$ for edge type.) It was also shown that the separation of the first two dislocations is

$$y \simeq \frac{Gb}{\pi n \tau}. \qquad (11.10b)$$

The stress generated a distance x ahead of the leading dislocation is approximately (for $L > x > y$)

$$\tau_p = \left[1 + \left(\frac{L}{x} \right)^{\frac{1}{2}} \right] \tau. \qquad (11.11a)$$

If there is a friction stress opposing movement of dislocations over the slip plane (e.g. Peierls–Nabarro stress, forest dislocations, etc.) the stress ahead of the pile-up is reduced to

$$\tau_p = \tau + (\tau - \tau_f) \left(\frac{L}{x} \right)^{\frac{1}{2}}. \qquad (11.11b)$$

Rearranging and putting L equal to the grain diameter d (it may be argued that $L = d/2$ or even $d/4$ but the numerical factor is unimportant to the argument since x is never specified):

$$\tau = \frac{\tau_f + \tau_p (x/d)^{\frac{1}{2}}}{1 + (x/d)^{\frac{1}{2}}},$$

i.e.
$$\tau = \tau_f + x^{\frac{1}{2}} \tau_p d^{-\frac{1}{2}}. \qquad (11.12)$$

The experimental law for the variation of the lower yield stress with grain size had been shown by Petch to be:

$$\tau_{lys} = A + Bd^{-\frac{1}{2}} \qquad (11.13)$$

and it was initially held from comparison of the two equations that A represented the friction stress and B equalled $x^{\frac{1}{2}} \tau_p$, where τ_p was the stress to unlock dislocations from their atmospheres of carbon atoms. The known variation of the lys with temperature was attributed to variation of τ_p due to thermal activation (discussed above).

However, further experiments in which the lys was measured as a function of the grain size over a range of temperatures (below ambient) showed that it is

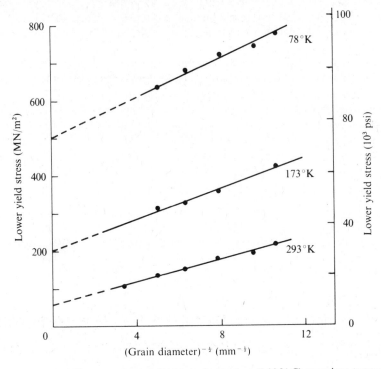

Fig. 11.23. Lower yield stress versus grain size in Armco iron (0.03 % C) at various temperatures. (After M. M. Hutchison, *Phil. Mag.* **8**, 121 (1963).)

the term A which is altering with temperature, not the constant B. This is clearly shown by the results on Armco iron reproduced in Fig. 11.23; the intercepts represent A and the slopes B. Other evidence also suggested that τ_p could not always be equated with the unpinning stress. For example the fact that Petch's law held for both the lower yield stress and the flow stress (yield stress at a nominal strain, say 5 per cent) is difficult to explain since clearly the flow stress does not involve unpinning.

It is now accepted (see Cottrell's 1963 paper) that the lower yield stress in polycrystalline iron is the stress to cause slip at the front of a Lüders band to propagate across the grain boundaries. At ambient and low temperatures the stress at the head of a dislocation pile-up in one grain forms new dislocations close to the boundary of the adjacent grain (i.e. $\tau_p = \tau_d$ in (11.12) where τ_d is the stress to generate a dislocation). At higher temperatures, or with only weakly pinned dislocations, the dislocation pile-up frees the dislocations in the adjacent grain from their atmospheres of carbon atoms (i.e. $\tau_p = \tau_u$ in (11.12) where τ_u is the unpinning stress at the relevant temperature). The former process is independent of temperature whilst the latter is thermally activated. Thus at elevated temperatures the unpinning process can occur at a lower applied stress than the dislocation generation mechanism and it takes over. This is shown by Fig. 11.24 in which the parameter $B (\propto \tau_p)$ is plotted against temperature for high purity iron (0.001 per cent C + N) in three conditions of dislocation locking. At low temperatures B is constant whilst the generation mechanism is operative ($\tau_p = \tau_d$). At high temperatures B decreases with temperature, due to increasing

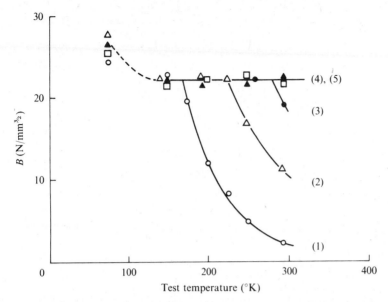

Fig. 11.24. Dependence on temperature of parameter B in Petch's equation for the lower yield stress for high purity iron (0.001 % C + N). Various degrees of dislocation pinning have been produced by different heat treatments: (1) ○ Quenched, (2) △ Q + 1 h at 140 °C, (3) ● Q + 2 h at 140 °C, (4) □ Q + 12 h at 140 °C, (5) ▲ Furnace cooled. (After R. M. Fisher, Ph.D. thesis, Cambridge University (1962).)

assistance by thermal energy in unlocking dislocations ($\tau_p = \tau_u$). The changeover occurs at a higher temperature for more strongly pinned dislocations, and with furnace cooling is above the temperature range of the experiment.

The parameter A, or the intercept term of the lys–grain size relationship, represents the Peierls–Nabarro stress, which in bcc crystals is temperature sensitive. Whilst $\tau_p = \tau_d$, only this term is responsible for the variation of the lys with temperature. When $\tau_p = \tau_u$, both A and B vary with temperature.

A dynamical theory of yield point phenomena in iron has been developed in 1962 by Hahn, which applies the work of Johnson and Gilman on the yield behaviour of LiF to bcc metals. The dynamical theory of yielding has already been described in §9.13 and only a brief recapitulation with special reference to iron and mild steel will be given here. The theory was developed in the first instance for single crystals but its concepts can be extended to cover polycrystalline material. There are similarities to Petch's theory described above and it is to be expected that a unified picture combining features of both will shortly emerge.

The three elements of the dynamical theory of discontinuous yielding are:

1. Low initial number of mobile dislocations. In bcc metals the in-grown dislocations (10^4–10^6 dis/mm^2) are immobilised by locking, after Cottrell, to give ~ 1 dis/mm^2 initially.

2. A dislocation velocity which increases with stress (cf. Fig. 9.40):

$$v = \left(\frac{\sigma}{\sigma_0}\right)^n \tag{11.14}$$

where σ_0 is the stress for unit velocity. It is taken here as 180 MN/m² for v equal to 1 mm/s.

Values of n measured directly in Fe–3% Si crystals and derived from measurements of the strain sensitivity of the upper yield stress of mild steel are 30 ± 5 at room temperature, increasing at lower temperatures.

As the number of dislocations increases, a linear strain-hardening law is assumed:

$$\Delta\sigma = q\varepsilon_p,$$

and hence

$$v = \left(\frac{\sigma - q\varepsilon_p}{\sigma_0}\right)^n.$$

The value of q is about 3–4 GN/m².

3. Rapid dislocation multiplication, involving double cross slip of screw dislocations and trailing lines of edge dislocation dipoles, which break up into prismatic loops and subsequently generate further dislocations by the Frank–Read type mechanism (see Fig. 9.45 and §9.13). The number of active dislocations has been roughly approximated by

$$\rho = f(\rho_0 + C\varepsilon_p^a) \tag{11.15a}$$

where f represents the proportion of dislocations which are mobile (typically 0.1), ε_p is the plastic strain, ρ_0 is the initial dislocation density (dis/mm²), and C and a are constants. Numerical values have been measured (in polycrystalline material) as follows:

Decarbonised iron $\qquad \rho = 0.1(\rho_0 + 1.6 \times 10^7\varepsilon_p^{0.8})$, $\qquad$ (11.15b)

0.2 per cent carbon steel $\qquad \rho = 0.1(\rho_0 + 4.7 \times 10^5\varepsilon_p^{1.5})$. $\qquad$ (11.15c)

The first of these will be used in the following to define the multiplication rate.

Neglecting the elastic strain component, the imposed strain rate equals the number of mobile dislocations moving with velocity v and contributing strain b (the Burgers vector), see (9.22):

$$\dot\varepsilon = 0.5bv\rho. \tag{11.16}$$

Combining (11.14), (11.15a) and (11.16) gives

$$\sigma = q\varepsilon_p + \sigma_0\left[\frac{\dot\varepsilon}{0.5bf(\rho_0 + C\varepsilon_p^a)}\right]^{1/n} \tag{11.17a}$$

Stress–strain curves for this model are shown in Fig. 11.25 taking values for the constants representative of iron: $n = 35$, $b = 0.248$ nm, $f = 0.1$, $\dot\varepsilon = 0.02$/min, $\sigma_0 = 186$ MN/m², $q = 3.44$ GN/m² and ρ_0 covering the range from 1 to 10^6 dis/mm². It will be seen that for small ρ_0, corresponding to pinned initial dislocations, there is clearly defined upper and lower yield stress, but at high ρ_0 (e.g. no dislocation pinning, or after some plastic straining) there is little yield drop. Decreasing n, the other factors being held constant, increases the size of the yield drop (see §9.13).

The preceding curves were calculated on the basis of uniform material and homogeneous straining. For real materials with local variations in ρ_0 and σ, the yield point commences at some points and then spreads into the surrounding material. Thus non-uniformity will modify the yield curves, reducing the upper yield stress and producing a Lüders strain at constant stress.

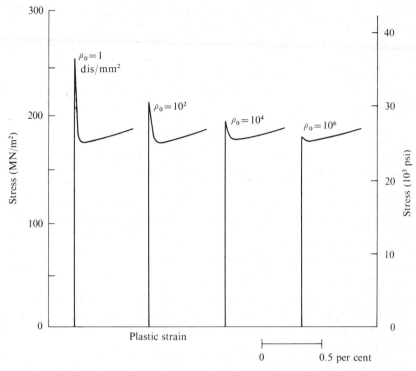

Fig. 11.25. Yield drop according to the dynamical theory of yielding: (11.17a) with $\dot\varepsilon = 0.02/\text{min}$, $n = 35$, $q = 3.44\,\text{GN/m}^2$, $\sigma_0 = 186\,\text{MN/m}^2$, $b = 0.248\,\text{nm}$, and multiplication rate according to (11.15b). (After G. T. Hahn, *Acta Met.* **10**, 727 (1962).)

Of the various aspects of the yield point phenomena which the dynamical theory claims to rationalise, only the delay time (§ 5.5) will be mentioned here. Rearranging (11.17a) and putting σ constant gives

$$t = \int \frac{\sigma_0^n \, d\varepsilon_p}{0.5bf(\rho_0 + C\varepsilon_p^a)(\sigma - q\varepsilon_p)^n}. \qquad (11.17b)$$

Examples of the strain–time response to an applied stress of $240\,\text{MN/m}^2$ are given in Fig. 11.26. For ρ_0 small the initial strain rate is small but it increases rapidly due to multiplication. For a typical strain gauge with sensitivity of 10^{-4} to 10^{-6} the initial strain is not detected and there appears to be a finite delay time before measurable yield develops (see (b)). At high values of ρ_0 the strain rate is high initially and no delay time is detected.

11.6 Equilibrium two-phase alloys

Most materials consist of more than one phase and the following sections will discuss the strengthening mechanisms in various two-phase metallic systems. A generalised discussion could embrace a very wide range of materials under this heading. For example, fibre-reinforced metals or ceramic–metal mixtures are equilibrium two-phase structures and similar principles apply as are

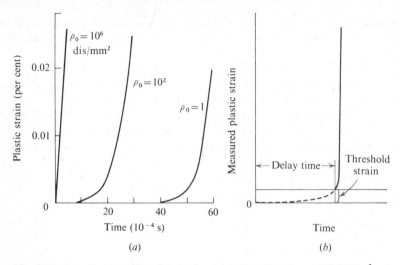

Fig. 11.26. Delay time characteristics of dynamic model: (11.17b) with $\sigma = 240$ MN/m^2, $\sigma_0 = 184$ MN/m^2, $n = 30$, $b = 0.248$ nm, and multiplication rate according to (11.15b). (Reprinted with permission from G. T. Hahn, *Acta Met.* **10**, 727, Pergamon Press, 1962.)

considered here. However they will be treated separately (the second half of this chapter). Other materials that might be considered from the point of view of two-phase aggregates are: traditional ceramics with a glass phase surrounding quartz grains; new ceramics with porosity providing the second phase; polymers with crystalline and amorphous regions.

In all binary systems the properties are a function of the properties of the individual phases and, even more important, how the phases are arranged or distributed. For structural purposes the best mix consists of a brittle phase and a ductile phase. The strong brittle phase should not be continuous but be dispersed in the weaker ductile phase, either as small dispersed particles or as fibres. In the former the hard particles are strengthening the ductile matrix, which is the basic load carrier; this occurs in conventional metallic systems. In the latter the hard strong fibres are the load carrying phase, whilst the ductile phase protects their surfaces from damage, which would nucleate fracture, and also limits the spread of any fracture in the brittle phase by absorbing energy as the crack propagates. This will be discussed further in § 11.12.

The equilibrium microstructures of metallic binary alloys, which are obtained on slow cooling from a single-phase liquid or solid state, have been extensively described in Chapter 4. This section will give examples of the resultant property changes due to the second phase. There are two ways in which the properties of nominally binary alloys can be greatly improved. The addition of small quantities of other elements will often affect the kinetics of phase changes, and hence the distribution of the phases, without significantly altering their composition or individual properties. Important examples are found in cast irons where the distribution of the graphite is determined by the presence of small quantities of other elements such as silicon or nickel. Cast irons will be considered in the following chapter. The other very important technique is heat treatment, for example, precipitation hardening and quench hardening (see

§ 11.7 and § 11.9). In brief, these processes are methods of forming a fine dispersion of one phase in a matrix of the other, which gives optimum mechanical properties.

A typical equilibrium structure of a two-phase alloy consists of primary grains of one phase mixed with some eutectic, or eutectoid, grains composed of two phases, frequently in alternate layers. Each phase may be a pure element, a solid solution or an intermediate compound. The properties of the first two have just been discussed. Intermediate compounds have in general complex crystal structures with no close-packed planes for easy slip. They are therefore strong but brittle. When both phases in the alloy are soft and ductile there is relatively little change of properties with composition, although some strengthening from the presence of the eutectic grains may occur, since the lamellae are equivalent to a very fine grain material. Figure 11.27 shows the hardness versus composition of Pb–Sn alloys (soft solders); the relevant equilibrium diagram was given in Fig. 4.14. The peak hardness occurs at the eutectic composition, around 60% Sn–40% Pb.

A more pronounced effect is found when one phase is weak and ductile while the second phase is strong and brittle. There is an increase in strength and hardness as the amount of the eutectic (or eutectoid) grains increases relative to the weak primary grains. Beyond the eutectic composition the alloy will tend to be brittle owing to the second phase providing the matrix. Plain carbon steels can be considered in this class, although at low carbon content the picture is complicated by the sharp yield point phenomenon. The primary grains are relatively soft and ductile iron (ferrite) and the second phase is hard and brittle

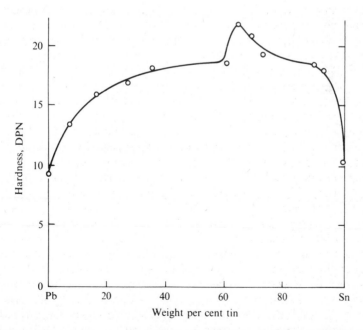

Fig. 11.27. Hardness versus composition of a binary alloy with two ductile phases: Pb–Sn, as cast. (For equilibrium diagram see Fig. 4.14.) (After M. A. Meyer.)

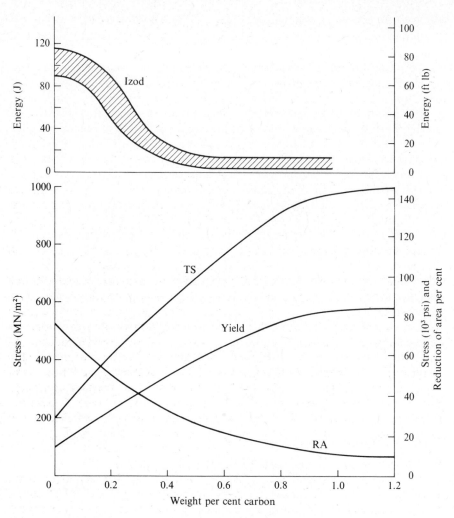

Fig. 11.28. Mechanical properties of binary alloy with ductile phase (ferrite) and a brittle phase (cementite): plain carbon steel (Fe–Fe$_3$C) in the normalised condition. (For equilibrium diagram see Fig. 4.25.)

intermediate compound Fe$_3$C (cementite). The variation of mechanical properties with composition is shown in Fig. 11.28; the iron–carbon equilibrium diagram was given in Fig. 4.25.

11.7 Precipitation hardening

Precipitation hardening is a method of strengthening one phase by means of another phase finely dispersed in it. The arrangement of the phases is obtained by precipitation from a supersaturated solid solution. Age hardening was the original term for describing the process before the mechanism was understood.

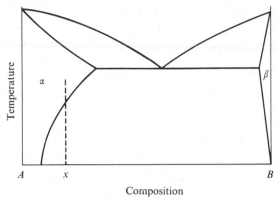

Fig. 11.29. Equilibrium diagram for precipitation hardening alloy.

It is still used when the precipitation is brought about at, or slightly above, room temperature.

For an alloy to be capable of being precipitation hardened, the equilibrium diagram must be of the form shown in Fig. 11.29. A suitable alloy composition (x) exists as a single-phase solid solution α at an elevated temperature and a second phase β is formed at room temperature. On slow cooling, this would be precipitated as fairly large agglomerates, mainly at the grain boundaries. In order to precipitate a fine dispersion, the alloy is quenched from the high temperature single-phase condition, temporarily suppressing the formation of the second phase. This is called the solution treated (or annealed) condition, and the material is relatively soft and ductile. Formation of the second phase as a fine dispersion throughout the α-matrix takes place either spontaneously over a period at room temperature or by heating to a temperature somewhat below the single-phase region. This is called ageing or precipitation treatment. It increases the strength and hardness at some expense to the ductility.

Aluminium–4 % copper was the first alloy to be treated in this way (by Wilm in 1911) and is still widely used. The equilibrium diagram (see Fig. 4.23) shows the precipitating phase to be an intermediate compound $CuAl_2$ (θ-phase), but there are actually several stages in the precipitation and other phases are formed first, see below. The strengthening effect of ageing for various times is shown for polycrystalline material in Fig. 11.30 and for single crystals in Fig. 11.31. The maximum hardness of the polycrystals and the maximum crss (critical resolved shear stress) of single crystals are double the values for solution-treated material. Further ageing causes a reduction of the hardness and crss, the process being called overageing. It corresponds with coarsening of the precipitate and a change in the dislocation mechanism of hardening (see § 11.8). Fortunately the material does not overage at room temperature. Many other aluminium alloys have been developed over the years for greater strength, machinability, ease of casting or corrosion resistance. Copper often plays a leading role in their heat treatment and the additional elements influence the kinetics of the precipitation process.

Hardening can be increased further by plastically deforming the material in the solution-treated state and then precipitating. This technique is also

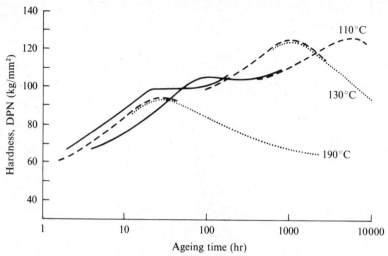

Fig. 11.30. Precipitation hardening of polycrystalline Al–4% Cu: hardness and precipitate structure versus time at various temperatures. (Solution treated: 540 °C for two days.) ——GP-1; ---
θ'' (GP-2); · · · · θ'. (Data from J. M. Silcock *et al., J.I.H.* **82**, 239 (1953–4).)

used to enhance the dimensional stability, probably by removing the residual stresses created on quenching. Figure 11.32 shows the hardening effect in beryllium–2.2% copper, an alloy used for galvanometer suspensions where nearly perfect elasticity is required. The microyield stress (2×10^{-6} plastic strain) was raised from 400 MN/m^2 to 700 MN/m^2 by intermediate working.

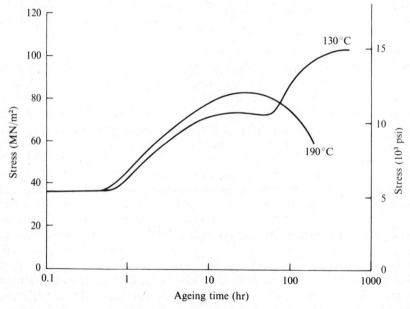

Fig. 11.31. Precipitation hardening of single crystal Al–4% Cu: critical resolved shear stress versus time. (After D. Dew-Hughes and W. D. Robertson, *Acta Met.* **8**, 156 (1960).)

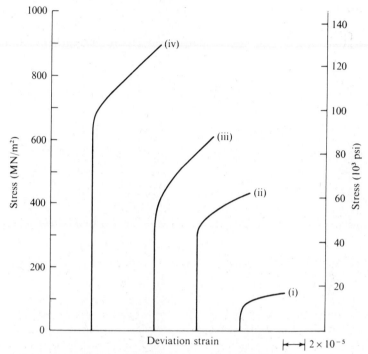

Fig. 11.32. Microyielding of 2.2 per cent beryllium–copper in various conditions:
(i) Solution treated ST (quenched from 825 °C): PS = 150 MN/m². MYS = 70 MN/m².
(ii) ST and cold drawn CD (15 per cent reduction of area): PS = 585 MN/m², MYS = 320 MN/m².
(iii) Precipitation hardened (ST and 3 hours at 300 °C): PS = 860 MN/m², MYS = 390 MN/m².
(iv) ST, CD and heat treated (3 hours at 300 °C); PS = 1070 MN/m², MYS = 670 MN/m².
Curves show deviation from Hooke's law. PS = 10^{-3} off-set, MYS = 2×10^{-6} off-set. (After C. S. Smith and R. W. Van Wagner, *A.S.T.M.* **41**, 825 (1941).)

Precipitation hardening can be applied to a large number of alloy systems besides aluminium alloys. One important example is the 'Nimonic' series of nickel–chromium alloys, strengthened by the precipitation of titanium and aluminium compounds. Another is the newly developed high strength iron–nickel alloys, known as maraging steels; these will be discussed in § 12.9.

The explanation of ageing as due to sub-microscopic (optical) precipitates obstructing slip was proposed in 1919 by Merica, Waltenberg and Scott. Subsequently the details of the structural changes taking place during precipitation of many alloys have been exhaustively established, initially by deduction from electrical resistivity and X-ray measurements, and latterly by direct evidence from thin film electron microscopy. Several metastable precipitates are formed in sequence before the final, equilibrium, phase develops. In the aluminium–copper system they are known as: Guinier–Preston zones, θ'' (or GP-2 zones), θ' and θ-phases. The phases present at each stage of the hardening of Al–4 % Cu are marked in Fig. 11.30, and thin-film transmission electronmicrographs are given in Fig. 11.33. It was once held that GP zones and θ'' were only 'local concentrations' of solute atoms and not distinct phases, but it is now argued that there is no difference in kind between the successive precipitates. GP zones consist of discs of copper atoms, one atom thick (0.36 nm) by about 25 atoms (8 nm) in diameter: see Fig. 11.33(*a*). θ''-precipitate, which gives the maximum hardness, is composed of plates, 1–10 nm thick by 10–150 nm diameter, with a typical spacing of 10–15 nm. In the micrograph, Fig. 11.33(*b*), the plates appear as needles as their planes are perpendicular to the photograph; the dark regions around some plates are due to strain fields. On further ageing,

Fig. 11.33. Intermediate precipitates on ageing supersaturated solid solution of Al–4% Cu; solution treated by heating to 540 °C and water quenching.

(*a*) Guinier–Preston zones (white streaks) after ageing for 16 hours at 130 °C (× 500 000). (By courtesy of R. B. Nicholson.)

(*b*) θ'' phase after ageing for 5 hours at 160 °C. Coherency of matrix and precipitate produces strain fields (dark areas). This condition required for maximum hardness (× 160 000). (By courtesy of R. B. Nicholson.)

(*c*) θ' particles after ageing 200 hours at 200 °C (× 40 000). (By courtesy of J. D. Boyd.)

θ'' converts to θ'-phase in larger precipitates and the strain field disappears – see Fig. 11.33(c). At still higher temperature the equilibrium θ-phase is formed as discs, up to 5 μm in diameter; their size increases and number decreases with time and temperature, see Fig. 11.34.

Precipitates do not form homogeneously but favour dislocation sites, such as sub-grain boundaries and miscellaneous loops and helices formed on quenching. Grain boundaries also strongly influence the developed structure, sometimes producing a thin film of precipitate and sometimes a precipitation-free zone, depending on the particular alloy. The quench rate is a factor through its effect on the dislocation structure and through the number of vacancies retained to room temperature. (The equilibrium number of vacancies decreases with decreasing temperature.) The vacancies play an important role in assisting the diffusion of solute atoms during precipitation. It will be appreciated that although the basic idea of precipitation hardening is simple there is plenty of scope for experimental investigation of the optimum procedure in a particular alloy.

Several dislocation theories of precipitation hardening have been advanced and it is probable that several mechanisms are simultaneously contributing to the total resistance to dislocation movement. The interactions between a dislocation and a precipitate can be divided into short range and long range, the former occurring as the dislocation passes through the precipitate and the latter when the dislocation is about the same distance away from a precipitate as the precipitate spacing. The long range interaction was considered in the first theory, by Mott and Nabarro in 1948, which has already been presented in connection with solid solution hardening (§ 11.4). In the early stages of precipitation (GP and θ'' in Al–Cu) the long range stress fields are large due to the coherency of the precipitate lattice with that of the matrix. Mott and Nabarro predict a yield stress for spherical precipitates at the maximum hardness given by (see (11.5)):

$$\tau = 2G\varepsilon c \qquad (11.18)$$

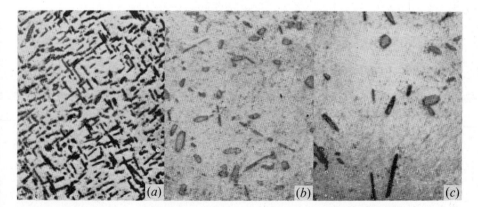

Fig. 11.34. Equilibrium θ phase precipitated from supersaturated solid solution of Al–4% Cu; solution treated by heating to 540 °C and quenching, aged at 450 °C for (a) 1 minute; (b) 24 hours; (c) 48 hours ($\times$ 1050). (From D. Dew-Hughes and W. D. Robertson, *Acta Met.* **8**, 147 (1960).)

where ε is the misfit and c the atomic concentration ($\approx$ volume fraction of precipitate). The long range stress fields of disc or lenticular shaped precipitates have also been considered. Typical values (for Al–4% Cu θ'' zones) are $\varepsilon = 0.1$, $c = 0.015$ and $G = 25\ \text{GN/m}^2$; the crss to overcome the coherency stress field according to (11.18) is then $75\ \text{MN/m}^2$. This is the same order of magnitude as experimental values of the crss at room temperature, $100\ \text{MN/m}^2$.

The more important factor in precipitation hardening is the short range interaction as a dislocation is forced through a precipitate, as first considered by Kelly and Fine in 1957. There are two principal terms contributing to the cutting resistance. First, additional energy γ_s is required for the increase of precipitate to matrix interface, as shown for spherical particles in Fig. 11.35; this term will also include the energy of dislocations produced at the interface due to misfits of the Burgers vector in the precipitate and matrix. Secondly, where the precipitate has an ordered structure and the Burgers vector of the matrix is not equal to the repeat distance of the ordered precipitate, additional energy γ_p is required on the slip plane of the precipitate. This is similar to the Fisher strengthening term due to ordering in solid solutions (§ 11.4). For spherical particles it has been shown from energy balance considerations that the upper limit of the crss is given by

$$\tau = \frac{\gamma f^{\frac{1}{2}}}{b} \tag{11.19}$$

where γ is the total increase of energy due to the precipitate being sheared one atomic distance ($\gamma = \gamma_s + \gamma_p$) and f is the volume fraction of precipitate. Difficulties arise in evaluating γ. The interface energy γ_s has been estimated from the heat of reversion (i.e. as the precipitate is taken back into solution at elevated

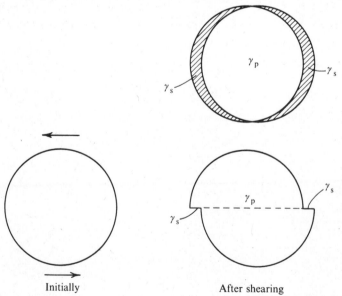

Initially   After shearing

Fig. 11.35. Increase of energy due to shearing of precipitate: γ_p associated with slip plane, γ_s with particle–matrix interface.

temperature). It was assumed that

$$\gamma_s = \frac{\Delta E}{3b^2}. \tag{11.20}$$

For aluminium–silver alloys values of γ_s were in the range 0.1–0.2 J/m^2. The corresponding value of γ_p due to disordering the Ag$_3$Al precipitate was estimated as ~ 0.015 J/m^2. Taking representative values: $\gamma = 0.15$ J/m^2, $f = 0.02$, $b = 0.3$ nm, the stress to shear a precipitate according to (11.19) is 70 MN/m^2, which is the same order as the experimental value for the crss at room temperature.

The final factor which can contribute to the strength will come from the solid solution hardening from the residual atoms in solid solution. The contribution to the crss of Al–4% Cu crystals decreases from 30 MN/m^2 when solution treated to 5 MN/m^2 when GP-2 zones (θ'') start to form (10 days at 130 °C). This may be compared with the measured crss (Fig. 11.31) which was 100 MN/m^2 after precipitation.

Extensive experimental evidence has been obtained to investigate the theory of precipitation hardening (see the 1963 review by Kelly and Nicholson), but quantitative proof has not proved easy or conclusive due to the variety of possible interactions, the difficulties of estimating the parameters such as volume concentration and surface energies (γ_s, and γ_p), the non-spherical shape of the precipitates, and the influence of thermal energy. The latter has not been considered here but it may be noted that it lowers the crss at finite temperatures by a factor which depends on the particular mechanism operating. In view of the complexities no attempt will be made here to summarise experimental evidence put forward by protagonists of the various theories. It seems that the different mechanisms are responsible for the successive stages of the hardness–time curve in the way shown in Fig. 11.36. Initially there is some solid solution hardening, which decreases as the precipitates form. GP-1 zones cause the coherency stress to rise rapidly whilst the cutting stress remains secondary, due to the small

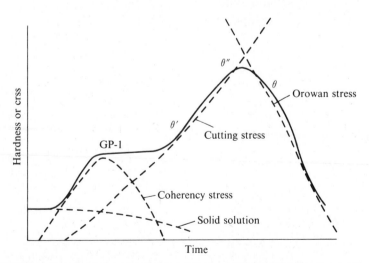

Fig. 11.36. Interplay of various precipitation-hardening mechanisms leading to successive stages in the hardness versus time curve.

precipitate size. The first plateau on the hardening curve occurs as the coherent stress passes its peak and the cutting stress begins to rise. The cutting stress continues to rise as the precipitate changes to large sizes of θ'' and θ', with the coherency stress now negligible. Finally the hardness decreases due to the intervention of the Orowan mechanism, by which dislocations pass between the precipitates, now larger and more widely spaced. This mechanism is described in the following section on dispersion hardening.

11.8 Dispersion hardening

In dispersion hardening the increased strength is again due to the presence of dispersed particles in a matrix but the dispersion is coarser than is considered in precipitation hardening and the characteristics of the slip lines and strain hardening are different. A dispersion-hardened single crystal behaves like polycrystalline material, that is, yielding is gradual, with slip occurring on several slip systems, independent of crystal orientation, and there is no critical resolved shear stress according to Schmid's law. It is generally accepted that the particles are non-deforming unlike those in precipitation hardening.

Dispersion-hardened materials are produced in several ways. For academic study, a precipitation hardened alloy can be overaged. For commercial materials, powdered metallurgy and internal oxidation methods are used. Examples of the former are SAP (sintered aluminium powder), in which aluminium oxide particles are sintered in a matrix of aluminium, and TD nickel, in which thoria is dispersed in nickel. In the internal oxidation technique, a solid solution is formed and the solute atoms are subsequently oxidised by heating in an oxygen atmosphere. Examples are silicon or aluminium in copper and aluminium or magnesium in silver, much used for electrical contacts.

The differences in strain-hardening and slip-line appearance between precipitation and dispersion hardening are illustrated in Figs. 11.37 and 11.38. Figure 11.37 shows the stress–strain curves for single crystals of Al–3.7% Cu treated to give various precipitates. With GP-1 and GP-2 (θ'') zones, slip occurs

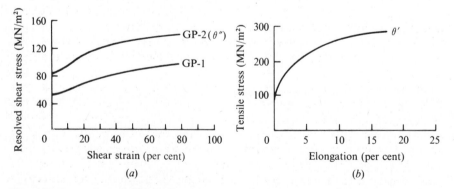

Fig. 11.37. Comparison of yield curves of (*a*) precipitation-hardened and (*b*) dispersion-hardened single crystals of Al–3.7% Cu at room temperature. In (*a*) GP-1 and GP-2 (θ'') zones allow single slip, leading to a shear-hardening curve. In (*b*) θ' precipitates cause multiple slip leading to a tensile stress–strain curve as for polycrystals. (After R. J. Price and A. Kelly, *Acta Met.* **12**, 159 (1964).)

Fig. 11.38. Slip lines in Al–4 % Cu with various precipitates: solution treated by heating to 540 °C and quenching (× 80). (From D. Dew-Hughes and W. D. Robertson, *Acta Met.* **8**, 156 (1960).)
(a) supersaturated solid solution, 8 per cent strain: well defined straight lines.
(b) GP-1 zones (50 h at 130 °C), 20 per cent strain: finer straight lines.
(c) θ'' (GP-2) (100 h at 130 °C), 20 per cent strain: less distinct wavy lines.
(d) θ'' and θ' (500 h at 130 °C), 20 per cent strain: indistinct lines.

on a single primary system and the tensile test data can be analysed into a shear-hardening curve. With θ' and θ precipitates, deformation is immediately by multiple slip, that is, the crystal deforms like a polycrystalline material, and the data are plotted as tensile stress versus elongation. The corresponding changes in the appearance of the slip line are shown in Fig. 11.38. As the matrix loses copper by precipitation, the coarse slip clusters typical of solid solutions give way to more numerous and finer lines, characteristic of pure metals; the waviness of the lines is probably due to dislocations avoiding Guinier–Preston zones. With θ'' and θ' precipitates the lines are very indistinct, due to the onset of multiple slip.

The established theory of dispersion hardening is due to Orowan who, in 1948, suggested that when the spacing of the particles is large in relation to the flexibility of a dislocation, the dislocations are able to pass between them, as shown schematically in Fig. 11.39. It will be seen that a dislocation loop is left around each particle as the dislocation line advances. The theory preceded the invention of the Frank–Read source, which is very similar in its operation. The yield stress, which is that required to bend the dislocation into a semicircle with a diameter equal to the gap between the particles Λ, is given by (9.19):

$$\tau = \frac{2T}{b\Lambda} = \frac{2\alpha Gb}{\Lambda} \qquad (11.21)$$

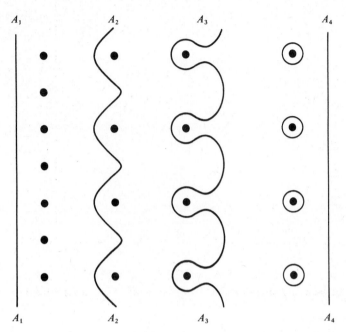

Fig. 11.39. Orowan mechanism for a dislocation to pass through an array of dispersed particles.

where $T = \alpha G b^2$ is the dislocation line tension. The numerical value of α lies between 0.5 for $\Lambda = 10^2 b$ and 2.0 for $\Lambda = 10^3 b$.

In experiments designed to check alternative theories of dispersion hardening, the critical resolved shear stress was measured on single crystals of aluminium–copper alloys containing particles of θ-phase. The results are plotted in Fig. 11.40, where the abscissa is the reciprocal of the particle spacing. For 4 per cent copper, the points mainly lie on a straight line through the origin, in accordance with Orowan's theory. For 3 per cent copper and 5 per cent copper there is a great amount of scatter, particularly with the latter. This was correlated with the deviation of the particles from an ideal spherical shape which was assumed in calculating the spacing Λ. Taking $G = 2.5\,\text{GN/m}^2$, $b = 0.28\,\text{nm}$ and $\alpha = 2$, the theoretical slope is 27.5 N/m which agrees excellently with the observed value of 28.6 N/m for the best straight line through the 4 per cent copper points (excluding the group of points for wide particle spacing which corresponded to plate-shaped precipitates).

Another theory involving cross slip of the dislocation to by-pass the particles has been advanced by Hirsch, in 1957, and developed by Ashby, in 1964. This will account for the tangle of dislocation which has been observed with the electron microscope to form around the particles, for example, as shown in Fig. 11.41. A detailed examination of the mechanism proposed indicates that the yield is still limited by the bowing of the dislocation, as proposed by Orowan, before cross slip can occur.

A contrary theory which requires the particles to fracture has been argued in 1960 by Ansell and Lenel. The particles are sufficiently widely spaced to allow

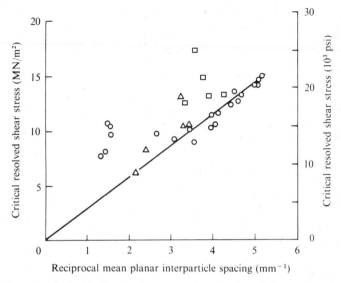

Fig. 11.40. Dependence of critical resolved shear stress on spacing of dispersed particles (θ phase in Al–Cu). $\triangle$ 3% Cu, $\bigcirc$ 4% Cu, $\square$ 5% Cu. (Source as Fig. 11.34.)

the Frank–Read sources to operate and produce a pile-up of dislocations on each particle, which eventually fractures. The yield stress is given by

$$\tau = \left(\frac{\mu b \sigma_f}{2\Lambda}\right)^{\frac{1}{2}} \tag{11.22}$$

where σ_f is the fracture stress of the dispersed particle.

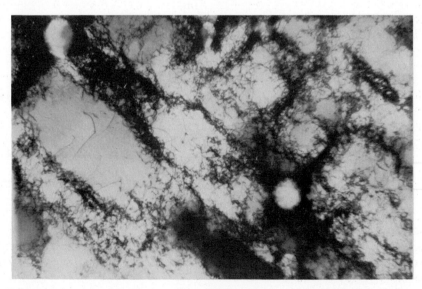

Fig. 11.41. Dislocation tangles formed around SiO_2 particles (white spheres) in a copper matrix after deformation ($\times$ 16 700). (By courtesy of J. Humphreys.)

It might be thought that experimental evidence on the dependence of the crss on Λ would decide unequivocally between the mechanism according to Orowan and that according to Ansell and Lenel. However, in practice it is difficult to determine Λ for a given experiment and the same evidence has been used to support both theories! For spherical, uniformly dispersed particles the weight of evidence supports Orowan's theory, but for flake-like dispersions the evidence inclines towards Ansell and Lenel.

Another possibility, which has not been examined, is that a pile-up of dislocations, after Ansell, leads to cross slip, after Hirsch. Equation (11.22) would then apply with σ_f replaced by the stress for cross slip.

The observed high rate of strain hardening is attributable to the back-stress exerted by the dislocation loops which are left behind both by the Orowan mechanism and by the cross-slipping process. It is difficult to see how a fractured particle can lead to a high strain-hardening rate.

11.9 Quench hardening

The oldest method of strengthening by heat treatment is quench hardening, which is carried out on steels, It involves a special type of phase change during the quench which does not require the diffusion of atoms to nucleate and grow new grains. Instead, the original lattice is sheared into the new phase. This is called a *martensitic reaction* and the method is also called strengthening by martensitic transformation. Martensitic reactions occur in several metals and alloy systems, see Table 11.4, and the transformation product is called martensite irrespective of its particular composition. The discussion here will be confined to the quench hardening of steel, as this is the most important example. After quenching there is normally a second stage of heat treatment, called tempering, in which ductility and notch toughness are restored at the expense of strength and hardness; this will be discussed in Chapter 12.

It will be recalled (§ 4.8) that at high temperatures iron has a fcc crystal structure which dissolves up to 1.7 per cent carbon in interstitial solid solution, this being called austenite. At room temperature iron is bcc (ferrite) and can only take into solution about 0.025 per cent carbon. On cooling austenite slowly to allow equilibrium states to develop, the carbon atoms are able to diffuse and

Table 11.4. *Occurrence of martensitic transformations*

Structural change on cooling	Alloy system
fcc $\rightarrow$ cph	Co, Fe–Mn
fcc $\rightarrow$ bcc	Fe–Ni
fcc $\rightarrow$ bct	Fe–C; Fe–Ni–C; Fe–Cr–C; Fe–Mn–C
fcc $\rightarrow$ fct	In–Tl, Mn–Cu
bcc $\rightarrow$ cph	Li, Zr, Ti, Ti–Mo, Ti–Mn
bcc $\rightarrow$ fct (?)	Cu–Zn, Cu–Sn
bcc $\rightarrow$ distorted cph	Cu–Al
bcc $\rightarrow$ orthorombic	Au–Cd
tetragonal $\rightarrow$ orthorombic	U–Cr

form the intermediate compound Fe_3C (cementite). On quenching austenite above a critical cooling rate (see Chapter 12) the equilibrium changes are suppressed and the fcc lattice changes rapidly by a shearing mechanism into bct (body-centred tetragonal); this is bcc with one axis overlength, the distortion being due to the carbon atoms remaining in interstitial solution. This phase is called *martensite*.

The axial ratio of the structure cell and the strength vary with carbon content as shown in Figs. 11.42 and 11.43 respectively. With no carbon present, the martensite is bcc but it is rather harder than ferrite because it consists of many fine needle-shaped grains in a highly stressed condition, resulting from the shearing mode of formation. With carbon present, a supersaturated solution is formed, progressively distorting the crystal lattice and increasing the hardness up to a maximum of 900 DPN at 0.7 per cent carbon. The corresponding yield stress (0.6 per cent off-set) is 1.6 GM/m² (2.3×10^5 psi), although due to the brittleness this figure can only be realised in very carefully loaded laboratory test pieces.

The phase change from austenite to martensite is a complex shearing of the lattice which requires each atom to move only a short distance relative to its nearest neighbours. No diffusion occurs. Nuclei of martensite are formed and grow with the speed of sound into the austenite until an obstruction is reached. Low carbon steels appear to form needles and the higher carbon steels, plates. New nuclei form and expand into the remaining austenite, which is rapidly divided up into smaller homogeneous volumes. Martensite is thus a collection of small plate-shaped or needle-shaped crystallites and a typical microstructure is shown in Fig. 11.44. High internal stresses are developed due to the mode of transformation.

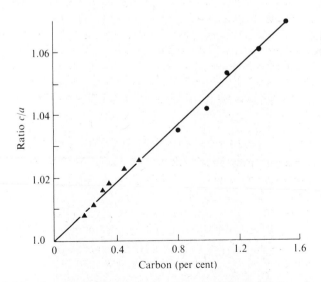

Fig. 11.42. Axial ratio of body-centred tetragonal martensite as a function of carbon content. (● for martensite powder. Below 0.6 per cent carbon results obtained with alloy steel ▲ to avoid tempering during quench.) (After G. V. Kurdjumov, *J.I.S.I.* **195**, 26 (1960).)

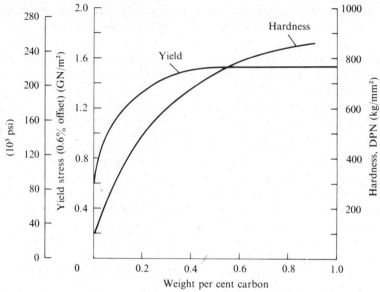

Fig. 11.43. Yield stress and hardness of martensite as a function of carbon content.

The detailed atomic rearrangement which takes place in the interface as austenite transforms to martensite is exceedingly complex. An exposition will not be given here. Suffice it to note that the rearrangement has to satisfy three requirements: the change of macroscopic shape; the change in crystal structure;

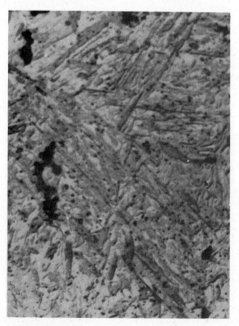

Fig. 11.44. Martensite micrograph, 0.4 per cent steel, heated to 850 °C and water quenched (× 1000).

and an interface plane (called the habit plane) which, because it is common to both phases, remains undistorted (invariant plane strain). These cannot be met by a simple shear parallel to the interface plane and two components are required. The first gives the required crystal lattice change and the second, which is fine scale twinning, or slip bands, within the martensite plates produced by the first shear, gives an approximately invariant plane strain. The development of the fine structure in martensite is shown schematically in Fig. 11.45.

The tetragonal crystal structure of martensite is explained as follows. The carbon atoms in austenite are in the interstitial octahedral positions with six nearest neighbour iron atoms, as shown in Fig. 11.46(*a*); the presence of the carbon atoms dilates the fcc lattice but does not distort it. In martensite, which is bcc in the absence of carbon, the interstitial positions are irregular octahedral, that is, there are again six neighbours but two of them are slightly closer than the others, see Fig. 11.46(*b*). Whereas in the fcc lattice there is one interstitial position per lattice point, in bcc there are three, which divide into three sets, each set with the shortened axes parallel to one of the cube axes. Owing to the shearing nature of the transformation, irrespective of the detail, all the carbon atoms will arrive from the fcc lattice into one of the sets in the bcc lattice. The

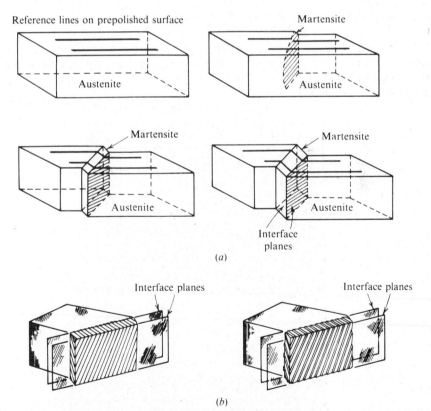

Fig. 11.45. Two components of displacement in martensitic transformation of austenite. (*a*) Homogeneous shear displacement. (*b*) Fine shears within the transforming regions: left, fine-scale slip; right, fine-scale twinning. (From M. Cohen, *Trans. A.I.M.E.* **224**, 638 (1962).)

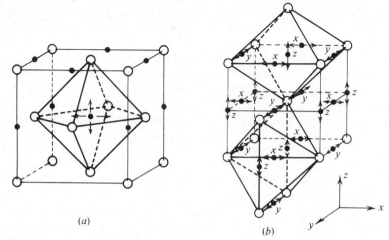

Fig. 11.46. Interstitial sites for carbon atoms in iron lattice: (*a*) In fcc (austenite) the octahedron of nearest iron atoms is symmetrical. (*b*) In bcc (martensite without carbon) each octahedron has a fore-shortened axis (as indicated) parallel to one of the three cell axes. During transformation the carbon atoms move into one of the three sets causing bc tetragonal structure. (Source as Fig. 11.45.)

distortion produced by an oversize interstitial atom will be greatest in the direction of the shortened axis (36 per cent expansion against 4 per cent contraction at right angles). Hence the lattice becomes tetragonal, not an expanded bcc lattice as would develop if the sites were randomly occupied.

The mechanism of hardening in steel has been a subject of great interest and speculation for hundreds of years, but it is only recently that the detailed atomic movements have been elucidated and dislocation theory developed to the point at which a satisfactory theory could be formulated. Several possible causes for the exceptional hardening effect of carbon in bcc iron have been advanced: (i) solid solution hardening, (ii) lattice strains caused by the shearing transformation, (iii) fine particle size of the martensite, (iv) precipitation hardening by carbon atoms diffusing to form small precipitates of equilibrium phase, (v) strengthening of the atomic bonding. Winchell and Cohen in 1962 carried out a series of elegant experiments, and decided that the hardening is a composite effect, but they singled out solid solution hardening as particularly effective because the non-symmetrical distortion of the lattice interacts strongly with dislocations, whose length is limited by the fine structure. Similar conclusions were reached by Kurdjumov in 1960.

BRITTLE MATERIALS

11.10 Introduction

In contrast to metals, which contain glissile dislocations and deform plastically, non-metallic materials fracture without appreciable flow at tensile stresses which are low compared with their theoretical strengths. Their compressive strengths are an order of magnitude higher and brittle materials have long been used in compression. The low tensile strength is due to the presence of flaws which grow into brittle fractures by the Griffith mechanism (Chapter 10). To strengthen such materials, which are desired for their thermal and chemical stability, it is necessary to reduce the size of the flaws and increase the energy absorbed during crack propagation (cf. (10.7)). Improved manufacturing techniques have recently increased the strength of new ceramics through reduced porosity and finer grain size (§ 7.6). However, removal of the flaws is never likely to be sufficient in itself for a really high tensile strength material, because the low notch toughness would mean a high risk of catastrophic failure initiated by some local defect or irregularity. In the last decade a great deal of interest has been taken in trying to combine high strength and high fracture toughness. The problem is common to ceramics, which are inherently brittle, and to very high strength metals, in which glissile dislocations have been anchored by one means or another.

Various techniques have been used for many years for utilising the high compressive strength of brittle materials. Structures can be designed so that all applied stresses are compressive: two well known examples are concrete dams and stonework arch bridges. In reinforced concrete, a composite material has been developed such that all tensile stresses are carried by steel rods and all compressive stresses by concrete; this achieves high overall load carrying ability at minimum cost. These techniques fall within the realms of structural engineering rather than materials science and will not be treated here. Glass has been converted into a useful structural material by developing residual stresses (that is, stresses that remain after all external loads have been removed) which put the surface layer into compression. This can be done by quenching the glass from an elevated temperature or by various chemical means; these have been described in § 7.9.

Turning to polycrystalline ceramics, it is possible, but not probable, that alloying will produce five independent slip systems which are required for plastic flow, as described in § 7.2, thus increasing the energy of fracture. A more promising solution, already largely achieved, is the development of composite materials in which very high strength fibres of ceramic or hard drawn metal are held in a matrix. The matrix can be ductile, so that crack propagation is prevented by plastic flow, or brittle, when crack propagation is prevented by weak interfaces between the fibres and the matrix. This occurs in the well known glass reinforced plastics (GRP) better called resin bonded glass fibres. Composite materials will be discussed below.

A technique for introducing some ductility, which pre-dates fibre composites, is the formation of cermets consisting of equi-axed particles of ceramic held in a metallic matrix. They were developed originally as cutting tool materials, but have recently been considered for high temperature structures. The presence of the metal phase provides just sufficient plasticity for greatly increased tool life but cermets remain brittle and low in notch toughness. They have been discussed in § 7.5.

Faced with materials of intermediate or high strength but no ductility the designer must radically alter his design methods, which have been developed over the years for ductile metals. It is no longer sufficient to calculate average stresses on sections, hoping that plastic flow will relieve local stress concentrations. A detailed analysis of the stress distribution is required to determine the maximum local stress which can initiate fracture. Allowance must also be made for the wide scatter of strength values about the mean, and statistical methods applied to achieve economic designs with an acceptable failure rate. Surface finish and manufacturing techniques must be controlled much more precisely than with metals, and differential thermal expansion between components must be allowed for. In spite of such precautions, the use of brittle materials in critical structural applications is hazardous. An example of the difficulties has been the development of silicon nitride for gas turbine blades. Its superior strength and corrosion resistance at high temperature over metallic blades make this a very desirable design change, but in spite of much effort it has not been possible to overcome the low resistance to impact, and blade life has been so unsatisfactory that the change has never been made. However, less onerous applications of silicon nitride are now being found.

11.11 Fibre composite materials

Composite materials are the latest development in the search for lightweight, high strength, materials with high fracture toughness. This section gives an introduction to composite materials; the following section treats the scientific principles. The limited amount of experimental work in support of the theory will not be given here (see the review by Cratchley, 1965).

In these materials very strong but brittle fibres, which may be continuous or short lengths, are embedded in a metal or plastics matrix. The fibres can be hard drawn metal wires, metal or ceramic whiskers (with few mobile dislocations), glass fibres, or ceramics filaments, such as boron or carbon. The fibres are the load carrying members and the matrix is there to transfer the load into them, to protect their surface and to raise the energy for crack propagation, thus preventing a brittle kind of fracture (cf. § 10.1). This is in complete contrast to the various strengthening methods for metals, where the metal matrix is strengthened, for example by the presence of precipitates of a second phase. In composites the matrix itself is not strengthened or hardened by the presence of the fibres – though this statement needs some qualification when the thickness of the matrix between the fibres is very small and some strengthening occurs by plastic restraint. The terms 'fibre reinforced composites' or 'fibre reinforced plastics' are therefore rather misleading, though widely used.

The high fracture toughness of composites is due to one of two mechanisms, depending on the matrix. With a metal matrix the fracture energy is high on traversing the ductile metallic phase. With a plastics matrix, which is itself brittle, the adhesion of the interface between the fibre and matrix must be weak so that it opens up in front of an advancing crack, thus blunting it. This will be described in more detail later.

With a metal matrix and ceramic fibres the high strength and negligible creep can be maintained to high temperatures, around 1000 °C, and a possible application is gas turbine blading. With a resin matrix higher strength to weight ratios are possible but the maximum temperature is limited to about 200 °C, although with new high polymers under development (see § 8.11) this may be raised to 400 °C in a few years. Their application to aircraft structures will lead to a significant improvement in performance.

Resin bonded glass fibre materials have been used on an increasing scale since their introduction in World War II for radomes (covers transparent to radar), greatly helped by the discovery (*c.* 1950) of thermoset polyesters which polymerise without the need for pressure or heat and with no reaction by-product (§ 8.2). The practical strength of glass fibre is up to $1.7 \, GN/m^2$ (2.5×10^5 psi) and the applications, with some notable exceptions (e.g. rocket motor casings), have involved only moderate tensile strength, up to about $1.0 \, GN/m^2$ (1.5×10^5 psi) for 70 per cent by volume of unidirectional fibres. Glass fibre plastics have been chosen more on account of their flexible and easy manufacturing techniques and their durability without protective coatings than for their strength characteristics.

The scale of research has increased sharply with the discovery of ceramic whiskers (e.g. SiN, Al_2O_3 and SiC) and fibres (e.g. C and B), which combine very high strength, up to $20 \, GN/m^2$ (3×10^6 psi), very high elastic modulus, up to $700 \, GN/m^2$ (100×10^6 psi) (cf. glass $60 \, GN/m^2$, (8.7×10^6 psi)) and low specific gravity, in the range 2–4. Epoxy or other resins are used for maximum strength and adhesion. At the same time the methods of achieving high fracture toughness are better understood. The importance of having a high elastic modulus in very high strength materials will be appreciated from Fig. 11.47, which shows the effect of 1.6 per cent strain in the initially straight wing spars of an aircraft. This is the reason for the superiority of carbon fibres over glass – there is almost an order of magnitude increase in Young's modulus – and accounts for the tremendous effort now being made to produce carbon fibre at an economic price.

The types and properties of various fibres are given in Table 11.5. They cover whiskers, fibres of glass, ceramic or polymer, and hard drawn metallic wires. The whiskers have the highest strength, above $10 \, GN/m^2$ ($>10^6$ psi), about a factor of three below their theoretical strength $E/10$. The remainder have strengths between 1 and $10 \, GN/m^2$ (1.5–15×10^5 psi). Care must be taken that the strength is not reduced by surface damage and this is one of the purposes of the matrix. Similar strengths have been achieved in large specimens but their notch toughness is low and a slight scratch would trigger a brittle fracture with a release of energy capable of pulverising the whole material.

For maximum strength of a composite the fibres must be arranged parallel to the direction of the stress. Where there is biaxial stress distribution a laminate

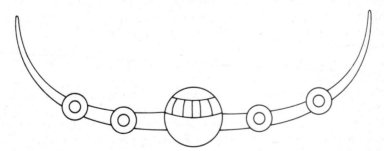

Fig. 11.47. High strength, high strain, material can lead to design problems. The wing spar boom, initially straight, is shown at 1.6 per cent strain. (From J. E. Gordon, *Proc. Roy. Soc. A.* **282**, 16 (1964).)

Table 11.5. *Properties of various fibres, composites and comparative metals.*
(NB The important parameter of fracture toughness is not given here)

	Sp. gr.	E (GN/M^2)	TS (GN/m^2)	E/sp. gr. (GN/m^2)	TS/sp. gr. (GN/m^2)
Group 1: ceramic whiskers					
Graphite	2.2	700	20	320	9.1
Silicon nitride	3.2	400	7	125	2.2
Silicon carbide	3.2	500	7	155	2.2
Alumina	4.0	420	14	105	3.5
Group 2: glass, ceramic or polymer fibres					
Carbon (RAE type 1)	2.0	460	1.7	230	0.9
(RAE type 2)	2.0	260	2.9	130	1.5
Boron	2.5	420	2.5	170	1.0
Asbestos	2.5 ·	190	6	76	2.4
Mica	2.7	230	3	85	1.1
Nylon-6,6	1.1	5	0.8	4.5	0.7
E glass: as drawn	2.5	60	3	24	1.2
in use	2.5	60	1.7	24	0.7
Group 3: hard drawn metal wires					
Piano (0.9 % C)	7.8	210	4	27	0.5
Stainless steel	7.9	200	2.4	25	0.3
Molybdenum	10.3	365	2.1	35	0.2
Tungsten	19.3	345	2.9	18	0.1
Group 4: composites and other high strength materials					
Steel alloy – current	7.8	200	1.3	25	0.17
Al alloy	2.8	70	0.6	25	0.21
Ti alloy	4.5	115	1.0	26	0.22
Beryllium	1.8	300	0.5	170	0.28
Steel alloy – future	8	200	5	25	0.65
70 % glass fibre epoxy, type 1	2	350	1.3–5*	175	0.6–2.5*
type 2	2	200	2.3–5*	100	1.2–2.5*

* Current and projected values for uniaxial lay up.

can be used with the fibres in adjacent layers arranged orthogonally. Normally
the two layers are combined as the warp and weft in a woven cloth, either using a
continuous yarn or the less expensive woven rovings. The fibre content in
each direction is then half that for unidirectional fibres. Cheaper still but giving
a lower fibre content are mats composed of randomly arranged short fibres.
Where the fibres are arranged orthogonally the material is orthotropic, with
three orthogonal planes of symmetry, and stress analysis has to be done
accordingly, following methods already developed with plywood. A promising
development, which overcomes the limitation of low fibre content, is sheet
reinforcement with alternate layers of brittle and plastic material. For three-
dimensional strength, randomly arranged chopped fibres are used, but clearly
the fibre content is very low and the reinforcement only marginal.

It is confidently expected that composites with uniaxial strength/weight
ratios at least a factor of 2, possibly 4, times that of conventional high strength
light alloys will develop within the next decade; the final line of Table 11.5
makes the point. However, many difficulties have still to be overcome before
commercial exploitation is assured, particularly as regards costs. This applies

to the manufacture of the fibres, especially whiskers, and to accurately aligning them in a matrix without damage. Fully automatic methods have to be developed if costly and variable hand laying is to be eliminated. Joining methods will have to be developed, but do not appear to present insuperable difficulties. The greatest uncertainty in properties appears to be the fatigue behaviour. Repeated plastic flow of the matrix or sliding at the fibre/matrix interface could cause failure. Finally many design problems associated with radically new manufacturing methods and with the very high strains involved have yet to be investigated.

11.12 Theory of fibre composites

Composites with continuous brittle fibres, arranged uniaxially, will be considered first. When the composite is loaded the strains in the fibres and matrix will be equal and the modulus of the composite in the elastic range is given by

$$E_c = V_f E_f + V_m E_m \tag{11.23}$$

where V_f and V_m are the volume fractions of the fibre and matrix respectively ($V_f + V_m = 1$). This equation has been found experimentally; theoretically it is the lower limit for the composite modulus, applicable when Poisson's ratios of the components are the same. In the elastic range the stress on the fibres is E_f/E_m that in the matrix.

   If the matrix is a metal it will begin to yield when the strain reaches

$$\varepsilon_y = \frac{\sigma_{ym}}{E_m}$$

where σ_{ym} is the matrix yield stress. The mean composite yield stress is then

$$\sigma_{yc} = E_c \varepsilon_y = \left(1 + \frac{V_f E_f}{V_m E_m}\right) V_m \sigma_{ym}. \tag{11.24}$$

Above this stress, the composite modulus decreases to the value

$$E'_c = V_f E_f + V_m \frac{d\sigma_m}{d\varepsilon_m} \qquad (a)$$
$$\simeq V_f E_f \qquad\qquad\quad (b) \Bigg\} \tag{11.25}$$

where $d\sigma_m/d\varepsilon_m$ is the strain hardening rate of the matrix. This decreases with increasing strain to around $E_m/100$ and can usually be neglected. As mentioned above, a high value of E_c and E'_c are important in high strength materials, and these equations indicate the requirement for high E_f and V_f.

   As regards the tensile strength of the composite, two regimes exist depending on whether the fibre volume fraction is above or below some value V_{min}. Above V_{min} the fracture of a fibre at the fibre fraction stress σ_{tf} leads ideally to the immediate failure of the composite at that section and the strength is

$$\sigma_{tc} = V_f \sigma_{tf} + (1 - V_f)\sigma'_m \tag{11.26}$$

where σ'_m is the matrix flow stress at the fibre fracture strain.

   For V_f below V_{min} there is sufficient strength in the matrix to carry the load as the fibres break. At V_{min} the transferred load just increases the stress on the matrix from σ'_m up to the matrix tensile strength σ_{tm}, and

$$(1 - V_{min})(\sigma_{tm} - \sigma'_m) = \sigma_{tf} V_{min}$$

i.e.

$$V_{min} = \frac{\sigma_{tm} - \sigma'_m}{\sigma_{tf} + \sigma_{tm} - \sigma'_m}. \qquad (11.27)$$

The composite tensile strength is then

$$\sigma_{tc} = V_m \sigma_{tm} = (1 - V_f)\sigma_{tm}. \qquad (11.28)$$

The variation of composite strength with fibre content is shown in Fig. 11.48: at low fibre content the tensile strength drops with increasing fibre content in accordance with (11.28); above V_{min}, the strength rises according to (11.26). The composite is only stronger than the matrix when the fibre content V_{crit} is exceeded:

$$\sigma_{tm} = V_{crit}\sigma_{tf} + (1 - V_{crit})\sigma'_m$$

i.e.

$$V_{crit} = \frac{\sigma_{tm} - \sigma'_m}{\sigma_{tf} - \sigma'_m} \simeq \frac{\sigma_{tm}}{\sigma_{tf}}. \qquad (11.29)$$

It may appear surprising that the addition of high strength fibres does not immediately increases the composite strength, but the explanation lies in the different strains at which the two components fail. Some figures for V_{crit} are given in Table 11.6 for various matrix materials at several fibre strength levels. It will be seen that for typical strengths for continuous fibres, around 2GN/m²

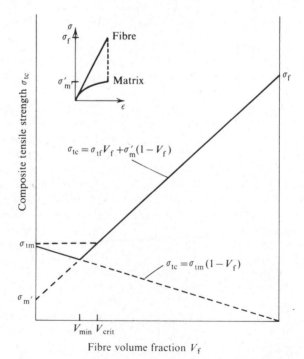

Fig. 11.48. Theoretical variation of composite strength with volume fraction for continuous uni-directional brittle wires. (After A. Kelly and G. J. Davies, *Met. Rev.* **10** (37), 1 (1965).)

Table 11.6. *Critical volume fractions (V_{crit} per cent) of continuous fibres required to reinforce a matrix (from (11.29)). (After A. Kelly and G. J. Davies)*

Matrix		Fibre fracture stress σ_{tf}			
Material	Flow stress[†] σ_m' Tensile stress σ_{tm}	690 MN/m² (10^5 psi)	1.70 GN/m² (2.5×10^5 psi)	3.45 GN/m² (5×10^5 psi)	6.90 GN/m² (10^6 psi)
Aluminium	$\sigma_m' = 27.5$ MN/m² (4000 psi) $\sigma_{tm} = 82.5$ MN/m² (12 000 psi)	8.3	3.2	1.6	0.8
Copper	$\sigma_m' = 41.4$ MN/m² (6000 psi) $\sigma_{tm} = 206$ MN/m² (30 000 psi)	25.5	9.8	4.8	2.4
Nickel	$\sigma_m' = 62.0$ MN/m² (9000 psi) $\sigma_{tm} = 310$ MN/m² (45 000 psi)	39.6	15.0	7.3	3.6
18:8 Stainless steel	$\sigma_m' = 172$ MN/m² (25 000 psi) $\sigma_{tm} = 448$ MN/m² (65 000 psi)	53.4	17.8	8.4	4.1

[†] σ_m' is matrix flow stress at which fibre fractures (assumed here to be independent of fibre strength).

(3×10^5 psi), a considerable volume of reinforcement is necessary in matrices of nickel or stainless steels before any improvement occurs. Although the theoretical upper limit of packing for parallel cylinders is 91 per cent, 80 per cent is a practical maximum value above which the composite strength decreases due to damaging of the fibres at contacts.

Consider now uniaxial discontinuous fibres. These may be either staple fibre made by deliberately chopping fibre produced in continuous form to reduce the cost of processing into cloth, or whiskers whose length is limited by the manufacturing technique. The theory will also have significance to normally continuous fibres since at high stress they often break into progressively shorter lengths due to the presence of flaws scattered along the fibre. The strain is no longer uniform across the composite section due to the irregularity at the fibre ends and shear strains and stresses are produced in the matrix on planes parallel to the fibre axis. The stress in a fibre is always zero at its end and builds up along its length as a result of shear stress acting on the fibre/matrix interface. The rate of increase in fibre load P is

$$\frac{dP}{dz} = 2\pi r_f \tau_z \tag{11.30}$$

where r_f is the fibre radius and τ_z the shear stress on the interface at a distance z from the end. Approximate solutions to the elastic stress field in the matrix have been derived but they underestimate the maximum shear stress which occurs at the end of the fibre. It has been shown experimentally that this is very dependent on the shape of the end and for a square end is approximately equal to the maximum longitudinal stress in the fibre ($E_f \varepsilon_c$). Since the fibre has been selected for high strength, the matrix fails long before the fibre fractures, either by plastic yielding in a metal matrix or by shear failure of the fibre/matrix interface bond in a resin matrix.

If the interface shear stress has a constant value τ, (11.30) integrates to

$$P = 2\pi r_f \tau z$$

i.e.

$$\sigma_f = \frac{2\tau z}{r_f}. \tag{11.31}$$

The fibre stress increases linearly from the fibre end up to the value $E_f e_c$. The corresponding strain distribution (for a metal matrix) is shown in Fig. 11.49: in the matrix close to the fibre end the strain exceeds the mean composite strain due to plastic flow; in the fibre the strain increases linearly from the end until it reaches the mean composite strain.

The length over which the load is transferred into the fibre is called the *transfer length*. As the stress on the composite is increased the transfer length and the maximum fibre stress increases, as shown in Fig. 11.50, until a limit is reached either because the transfer regions meet at the middle of the fibre, and no further transfer of stress can take place, or the fibre fractures. For the latter objective to be reached, the fibre length must be greater than a minimum value called the *critical fibre length* l_c, given by (11.31):

$$\sigma_{tf} = \frac{\tau l_c}{r_f},$$

that is the critical aspect ratio is

$$\frac{l_c}{d} = \frac{\sigma_{tf}}{2\tau} \tag{11.32}$$

where d is the fibre diameter. $l_c/2$ is then the critical transfer length.

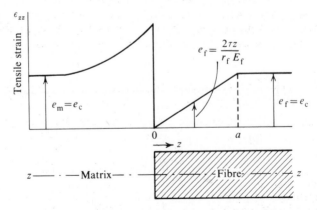

Fig. 11.49. Theoretical strain distribution near a fibre end. Strains shown are in the direction of the fibre, on the fibre axis. e = average tensile strain in matrix.

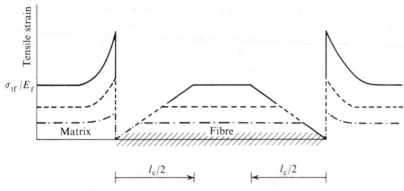

Fig. 11.50. Strain distribution in a fibre of finite length as a function of the matrix plastic strain.

The value of τ for an ideal plastic material is given by the yield stress τ_{ym} and the metal flows around the end of the fibre. For a work-hardening metal it has been found that the strain hardening of the matrix at the fibre ends is very much greater than the average value and the appropriate value of τ is the ultimate shear strength, i.e. $\sigma_{tm}/2$. For a brittle material, such as epoxy or polyester plastic, the bond between the fibre and matrix fails and the matrix slides off the fibre end forming a cavity. The bond breaks first at the tip and advances along the fibre – the fracture process being similar to that observed between a rigid glue spot and a rubber sheet when the latter is stretched. The value of τ is then the friction stress. It has been suggested that the normal stress at the interface is a result of the contraction of the plastic on polymerising and the friction stress has been related to the coefficient of friction, the plastic's yield stress and the thickness of the matrix between fibres, but no experimental support for these ideas has been found. Values of the friction stress in epoxy or polyester resins are known to vary widely from less than 5 MN/m^2 (725 psi) (especially if an incompatible protective size or moisture is present), to around 35 MN/m^2 (5×10^3 psi) for glass fibres and 100 MN/m^2 for carbon fibres. Too high a bond strength is undesirable as it prevents delamination ahead of a crack and resultant notch toughness, see below. Some typical values of the critical aspect ratio and fibre length for various fibre/matrix combinations are given in Table 11.7 based on (11.32). It will be seen that the critical aspect ratios

Table 11.7. *Typical values of critical aspect ratios and fibre lengths from (11.32) and estimated values of* τ

Fibre	Tungsten wire	Glass fibre	Carbon fibre	Ceramic whisker
Matrix	Copper	Epoxy	Epoxy	Metal
Fibre diameter, d (μm)	2000	7.5	7.0	2
Fibre strength, σ_{tf} (GN/m^2)	3	2	2	15
Flow or bond strength, τ (MN/m^2)	80	10–35	70	50
Critical aspect ratio: $l_c/d = \sigma_{tf}/2\tau$	19	100–30	70	150
Critical fibre length, l_c (mm)	38	0.75–0.2	0.5	0.32

are in the range 20–150 and useful aspect ratios should be ten times these figures. This is not always easy to obtain in growing whiskers.

The average fibre stress in a fibre of length $l\,(>l_c)$ just prior to fracture is

$$\bar{\sigma} = \frac{1}{l}\int_0^l \sigma_f\,dz$$

$$= \frac{1}{l}\left[\sigma_{tf}(l - l_c) + \frac{\sigma_{tf}}{2}l_c\right],$$

i.e. $\bar{\sigma} = \sigma_{tf}(1 - l_c/2l).$ (11.33)

The equation shows that for maximum utilisation of fibre strength, the length must be long in relation to the critical transfer length, i.e. $l_c/l \ll 1$; from (11.32) l_c can be kept short by making τ high and d small (but see below on notch toughness).

The important relationship for composite strength is obtained by substituting $\bar{\sigma}$ for σ_{tf} in (11.26):

$$\sigma_{tc} = \sigma_{tf}V_f\left(1 - \frac{l_c}{2l}\right) + \sigma'_m(1 - V_f).$$ (11.34)

This applies for $V_f > V_{min}$ so that failure of one fibre leads immediately to failure of the composite. The ratio of the strength of a composite with discontinuous fibres to that with continuous fibres is always less than unity, as might be expected, and decreases with increasing fibre volume fraction. In the limit of $V_f = 1$ the ratio is

$$\frac{(\sigma_{tc})_{dis}}{(\sigma_{tc})_{cont}} = 1 - \frac{l_c}{2l}.$$ (11.35)

This is plotted in Fig. 11.51 from which it can be seen for $l > 4l_c$ the decrease in strength is marginal. For realistic values of V_f (i.e. $V_f < 0.8$) the ratio lies nearer unity, that is, in the region shown hatched.

The value of V_{min} for discontinuous fibres depends, as for continuous fibres, on whether or not the matrix can carry the load transferred from the fractured fibres.

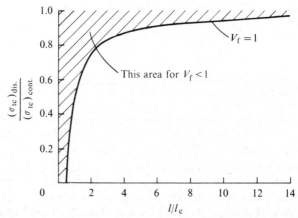

Fig. 11.51. Comparison of the strength of composites with discontinuous and continuous fibres.

Now, however, not all the fibres fail together since fibres with their ends within $\pm l_c/2$ of the critical cross-section are carrying less than the fibre fracture stress and cannot be fractured because the available length for transfer of stress is less than the critical value. The volume fraction of fully stressed fibres is given by $(1 - l_c/l)V_f$ and hence V_{min} is (by (11.27)):

$$(1 - l_c/l)V_{min}\sigma_{tf} = (\sigma_{tm} - \sigma'_m)(1 - V_{min})$$

i.e.
$$V_{min} = \frac{\sigma_{tm} - \sigma'_m}{\sigma_{tf}(1 - l_c/l) + \sigma_{tm} - \sigma'_m}. \tag{11.36}$$

Below V_{min} the composite tensile strength is the sum of the matrix strength plus the contribution from the unbroken fibres with an average strength of $\sigma_{tf}/2$:

$$\sigma_{tc} = (1 - V_f)\sigma_{tm} + \frac{l_c}{l}V_f\frac{\sigma_{tf}}{2}. \tag{11.37}$$

Owing to the second term it is possible for the strength of composites with discontinuous fibres to increase continuously from $V_f = 0$; that is if

$$(1 - V_f)\sigma_{tm} + \frac{l_c}{l}V_f\frac{\sigma_{tf}}{2} > \sigma_{tm},$$

i.e.
$$\sigma_{tf} > \frac{2l}{l_c}\sigma_{tm}.$$

For longer fibres (relative to l_c) or greater σ_{tm} (relative to σ_{tf}) this inequality will not apply and a finite V_f is required to achieve a strength equal to the pure matrix, as for continuous fibres. Using (11.34):

$$\sigma_{tf}V_{crit}\left(1 - \frac{l_c}{2l}\right) + \sigma'_m(1 - V_{crit}) = \sigma_{tm},$$

i.e.
$$V_{crit} = \frac{\sigma_{tm} - \sigma'_m}{\sigma_{tf}(1 - l_c/2l) - \sigma'_m}. \tag{11.38}$$

The critical fibre length l_c and the matrix to fibre shear stress τ can be evaluated experimentally by means of (11.34). Differentiation with respect to V_f gives:

$$\frac{d\sigma_{tc}}{dV_f} = \sigma_{tf}\left(1 - \frac{l_c}{2l}\right) - \sigma'_m$$

$$= (\sigma_{tf} - \sigma'_m) - \frac{\sigma_{tf}l_c}{2}\frac{1}{l}. \tag{11.39}$$

From experimental values of composite strength σ_{tc} versus fibre content V_f at several fibre lengths the value of $d\sigma_{tc}/dV_f$ is obtained as a function of $1/l$. Figure 11.52 gives some results for 2 mm tungsten waves in a copper matrix at two temperatures. It will be seen that at high aspect ratios (low d/l) the points lie on a straight line in accord with the theory. The slope of the lines is $-(\sigma_{tf}l_c/2d)$ and σ_{tf} is obtained from the intercept, since $\sigma'_m \ll \sigma_{tf}$; hence l_c. Using (11.32), the matrix/fibre shear stress τ is given by $\sigma_{tf}^2/4s$ where s is the slope. Numerical values of l_c and τ obtained from the data of Fig. 11.52 are given in Table 11.8. (It may be noted that measurements of the longest fibre which can be pulled out of the matrix give values of τ about double.) Figure 11.53 compares the values of τ

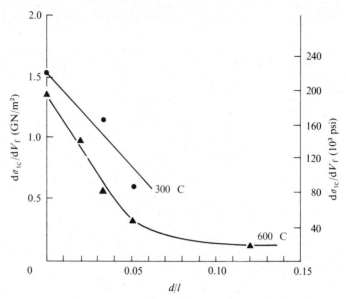

Fig. 11.52. Evaluation of the critical transfer length for 2 mm tungsten wires in copper. (After A. Kelly and W. R. Tyson, *Proc. 2nd Int. Mat. Symposium, California* (1964).)

with the tensile strength of the copper matrix and confirms the statement made earlier that in a work hardening metal the interface shear stress is about half the tensile strength.

Besides incorporating very high strength fibres, composite materials have the great attraction that they can also be designed for high notch toughness. In this they have the advantage over most high strength materials. The notch toughness can arise in various ways. With a metal matrix it is due to the plasticity of the matrix which reduces the stresses at the front of a crack and prevents the propagation of a brittle fracture from one fibre to the next. With a brittle matrix, either the crack is deflected by delamination of the composite or the

Table 11.8. *Values of l_c and τ_c derived from the experimental data of Fig. 11.52*

			300 °C	600 °C
neg. slope $s = \dfrac{\sigma_{tf} l_c}{2d}$	(GN/m²)		15.2	21.7
intercept σ_{tf}	(GN/m²)		1.5	1.35
$\dfrac{4s}{\sigma_{tf}} = l_c$	(mm)		40	64
$\dfrac{\sigma_{tf}^2}{4s} = \tau$	(MN/m²)		37	21

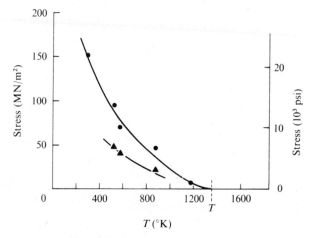

Fig. 11.53. The copper matrix/tungsten wire shear strength τ compared with the tensile strength at various temperatures. (Source as Fig. 11.52.)

energy of fracture is made high by designing for a proportion of the fibres to pull out of the matrix instead of fracturing.

Delamination occurs as follows. Ahead of a crack there is a stress component σ_x acting on planes transverse to the crack propagating stress, see Fig. 11.54. This component, a maximum about a tip radius ahead of the advancing crack, can cause a weak interface to open up, which then prevents further crack propagation, as shown in Fig. 11.55. This effect has been observed in resin bonded glass fibres.

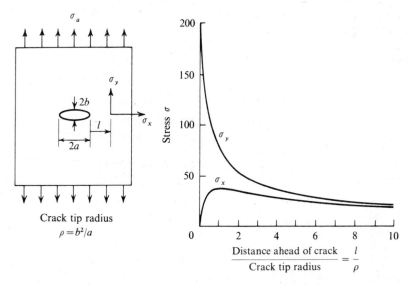

Fig. 11.54. Tensile stresses ahead of a crack tip for a remotely applied unit uniaxial stress ($\sigma_a = 1$) and $a/b = 100$. (After J. Cook and J. E. Gordon, *Proc. Roy. Soc.* **A282**, 508 (1964).)

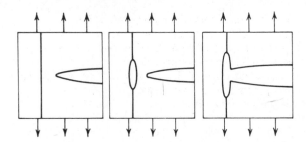

Fig. 11.55. Delamination and blunting of a crack. (Source as Fig. 11.54.)

The final method of achieving high notch toughness is to make the fibre length equal to the critical fibre length. Then all the fibres pull out the matrix instead of breaking and it is a simple calculation to show that, assuming the shear stress at the interface is maintained, the work done per unit area is

$$2\gamma = \tfrac{1}{12}\sigma_f l_c V_f. \tag{11.40}$$

The strength is naturally reduced to half the value obtained with very long fibres ($\gg l_c$). For maximum toughness l_c should be made as long as possible, provided that the fibre lengths are sufficient to ensure $l = l_c$.

QUESTIONS

1. Contrast the problems of strengthening ductile and brittle materials.
2. Discuss briefly the methods of strengthening metals. (The dislocation mechanisms are not required.)
3. Discuss the influence of grain boundaries on the strength of metals, contrasting fcc and cph materials. Quote the Hall–Petch equation.
4. (a) What changes occur in the microstructure and properties as a result of cold rolling?
   (b) What is its principal disadvantage as a method of strengthening?
   (c) What changes occur in heating and describe the critical strain techniques of growing single crystals?
5. Outline the Mott and Nabarro theory of solid solution hardening. Mention briefly alternative theories.
6. (a) Describe the heat treatment and successive changes of microstructure in a precipitation-hardened alloy.
   (b) Discuss the various dislocation mechanisms contributing to the strength and indicate their importance in the successive stages of the hardness versus ageing time curve.
   (c) Briefly what is the effect on the precipitate distribution and the properties of (i) additional elements, (ii) cold work in the solution treated state?
7. (a) Contrast normal yielding and sharp yielding, including a note on the Lüders band and strain ageing.
   (b) Outline Cottrell's dislocation locking theory of the yield point phenomenon in low carbon iron.
   (c) What evidence suggests that the uys is not associated with dislocation unlocking stress, and describe briefly Petch's theory.
8. Develop the Cottrell–Petch theory of the lower yield stress in mild steel and show how the experimental evidence on the variation of the lys with temperature indicates that in normal circumstances the friction stress rather than the unlocking stress is varying and that new dislocations are formed ahead of the Lüders band.

9. Outline Hahn's dynamic theory of the sharp yield point.
10. (a) Distinguish between precipitation hardening and dispersed particle hardening.
    (b) What is Orowan's theory of the dislocation mechanism involved in the latter?
    (c) What rival theories have been proposed?
11. (a) Describe the microstructural changes during the quench hardening of steel, including some details of the martensitic transformation.
    (b) Why do the carbon atoms distort the bcc iron lattice along only one axis to form a body-centred tetragonal structure cell?
    (c) Discuss briefly the possible mechanism of hardening.
12. (a) Distinguish between a fibre reinforced metal and a dispersed particle hardened metal. Compare the roles of the fibres, particles and matrix in both cases.
    (b) Distinguish three classes of 'fibre' used in composite materials and consider briefly their relative merits.
    (c) Why is a fibre with a high elastic modulus desirable?
13. (a) Sketch a typical curve of strength versus volume fraction of fibres in a composite material.
    (b) Derive an expression for the minimum volume fraction of fibres V_{crit} necessary to achieve strengthening with (i) continuous fibres (ii) short fibres.
14. (a) Define the critical transfer length l_c for a fibre in a composite and discuss the parameter τ for different types of matrix material.
    (b) How have values of l_c been measured experimentally?
15. Explain three possible mechanisms responsible for the high notch toughness of composite materials.

FURTHER READING

Original papers

E. Orowan: Fracture and strength of solids. *Repts. Progr. in Physics*, **12**, 185 (1948–9).

E. O. Hall: The deformation and ageing of mild steel. *Proc. Phys. Soc. Lond.* **B64**, 747 (1951).

N. J. Petch: The cleavage strength of polycrystals. *J.I.S.I.* **174**, 25 (1953).

A. Kelly and M. E. Fine: The strength of an alloy containing zones. *Acta Met.* **5**, 365 (1957).

D. Dew-Hughes and W. D. Robertson: Dispersed particle hardening of aluminium–copper alloy single crystals. *Acta Met.* **8**, 147 (1960).

D. Dew-Hughes and W. D. Robertson: The mechanism of hardening in aged aluminium–copper alloys. *Acta Met.* **8**, 156 (1960).

C. S. Ansell and F. V. Lenel: Criteria for yielding of dispersion strengthened alloys. *Acta Met.* **8**, 612 (1960).

G. T. Hahn: A model for yielding with special reference to the yield point phenomena of iron and related bcc metals. *Acta Met.* **10**, 727 (1962).

P. G. Winchell and M. Cohen: The strength of martensite. *Trans. A.S.M.* **55**, 347 (1962).

M. M. Hutchison: The temperature dependence of the yield stress of polycrystalline iron. *Phil. Mag.* **8**, 121 (1963).

A. H. Cottrell: Discontinuous yielding. *NPL Symposium* No. 15, 456. H.M.S.O., London (1963).

N. J. Petch: The upper yield stress of polycrystalline iron. *Acta Met.* **12**, 59 (1964).

M. Ashby: The hardening of metals by non-deforming particles. *Z. Metall.* **55**, 5 (1964).

P. M. Hazzledine and P. B. Hirsch: A critical examination of the long range stress theory of work hardening. *Phil. Mag.* **15**, 121 (1967).

Books and reviews

A. H. Cottrell: *Dislocations and Plastic Flow in Crystals.* Clarendon Press (1953).

Relation of Properties to Microstructure: A.S.M. Symposium. Novelty, Ohio (1954).

G. V. Kurdjmov: Phenomena occurring in the quenching and tempering of steel. *J.I.S.I.* **195**, 26 (1960).

Strengthening mechanisms in solids: A.S.M. Seminar (1960).

D. McLean: *Mechanical properties of metals*, chapters 5 and 6. Wiley (1962).

M. Cohen: The strengthening of steel. *Trans A.I.M.E.* **244**, 638 (1962).

A. Kelly and R. B. Nicholson: Precipitation hardening. *Progr. in Mat. Sci.* **10**, 151 (1963).

The relation between the structure and mechanical properties of metals, vols. 1 and 2. NPL Symposium No. 15. H.M.S.O., London (1963).

R. M. Ogorkiewicz: Glass fibre plastics. *Chart. Mech. Engr* (April 1964).

J. E. Dorn and J. D. Mote: On the plastic behaviour of polycrystalline aggregates. *Mat. Sci. Res.* **1** (1965).

A. Kelly and G. J. Davies: The principles of fibre reinforced metals. *Metall. Rev.* **10**, no. 37 (1965).

D. Cratchley: Experimental aspects of the fibre reinforcement of metals. *Metall. Rev.* **10**, no. 37 (1965).

A. Kelly: *Strong Solids.* Clarendon Press (1966).

J. E. Gordon: *The New Science of Strong Materials.* Pelican Original A920 (1968).

Designing with brittle materials. *Conf. Mech. Engr. (Lond.)* (December 1968).

12. Iron–carbon system: cast irons and steels

12.1 Introduction

Although specific ceramics and polymers have been discussed in their respective chapters, the structures and properties of metallic materials have been treated so far in general terms. This chapter considers in some detail specific alloys based on the iron–carbon system, that is, cast irons, plain carbon steels, and alloy steels. These materials form the bulk of engineering constructional materials, some 25 million tons of steel alone being produced each year in Britain, 100 million tons in the USA, and 400 million tons throughout the world. They occupy their leading position on account of three factors: the abundance of iron in the earth's crust, amounting to 5 per cent; the relative ease with which the ore can be reduced by carbon to the metallic state; and the versatility of iron–carbon alloys, which can be produced with a great range of desirable mechanical properties. Corrosion is their chief drawback.

Relevant data have already been presented. The iron–carbon, or iron–iron carbide, equilibrium diagram was introduced in Chapter 4, and is repeated for ease of reference in Fig. 12.1. The names of the phases and their crystal structures were summarised in Table 4.1. The sharp yield point and strain ageing phenomena in low carbon steels were treated in § 5.5 and § 11.5, and the mechanism of quench hardening, when austenite is transformed to martensite, in § 11.9.

12.2 Manufacture and impurity elements

The manufacture of iron and steel is a subject rather outside the scope of this book on the theory of materials. However, the objective has been to understand materials and correctly apply them to engineering ends, and it is not inappropriate to use a chapter on the technology of cast iron and steel to illustrate the general observation that the manufacturing methods influence the composition and properties of the product. Pure elements or binary alloys without any impurities are usually found only in laboratories (except for semiconductor devices).

Iron is found in nature combined into oxides, carbonates and sulphides, mixed up with a large amount of earthy matter, called gangue. The iron has to be separated from the gangue, reduced to the metallic state, and finally alloyed with the correct amount of carbon. This involves several complex processes at very high temperatures (1450 °C) and the final product will always contain several impurities alloyed with the iron and some non-metallic inclusions.

The main stages in manufacture are as follows. In a blast furnace, the ore is reduced with coke, and separated from the gangue with limestone, to form pig iron, which contains up to 10 per cent carbon and other elements. Pig iron is then converted either to cast iron or to steel. Cast iron is made in a cupola where the pig iron is slightly refined with coke, limestone and scrap iron until

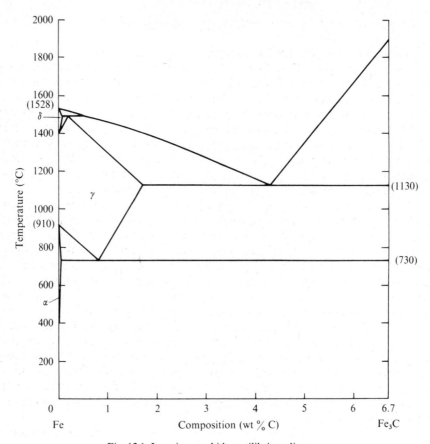

Fig. 12.1. Iron–iron carbide equilibrium diagram.

it contains from 1.7 per cent to about 5 per cent carbon and some manganese, silicon, sulphur and phosphorus.

Steel is made in several ways, all involving the removal of carbon and other impurities by oxidation, followed by deoxidation and the addition of carbon to the required level. In the Bessemer process, air or oxygen is blown right through molten pig iron and scrap steel contained in a small conical furnace (15 tonne capacity). By contrast, in the open hearth process (Siemens–Martin), producer gas and air are heated by the outgoing gases and then blown over the surface of the molten charge contained in a large shallow crucible furnace (100 tonne capacity). In the former method, the heat comes from the oxidation of the impurities, whilst in the latter additional heat comes from combustion of the producer gas.

During the removal of carbon and other impurities by oxidation, the molten iron absorbs oxygen, both as dissolved oxygen and as iron oxide. During solidification, the equilibrium condition changes and the oxygen reacts to form carbon monoxide, which comes out of solution and, if not able to escape, forms blowholes. The final stage is therefore to deoxidise by adding silicon, manganese or aluminium and, where necessary, to add the required quantity of carbon as the alloying element.

The degree to which the deoxidisation is taken has a marked effect on the structure of the ingot, which shows in the final product after rolling and forming operations. In rimming steel, gas is allowed to evolve and this causes the ingot to form a surface layer free of impurities, which segregate to the core. There is little contraction during cooling due to the entrapped gas, and no pipe develops at the centre of the ingot. Subsequent rolling will close any cavities formed. In killed steels, the deoxidisation is complete, no gas evolves and there is no segregation. On the other hand, a pipe is formed and has to be discarded, the steels contain higher impurities, as a result of deoxidisation, and the surface quality is not so good. A normal compromise is the use of semi-killed steel.

The furnaces are lined with bricks which are either acid (silica) or basic (dolomite). At the temperature involved, it is not possible to find a stable inert material at an economic price and the linings react with the charge to some extent, limiting the compositions of the ores and the reactions which are allowable. Thus high silicon ores would react with basic linings and manganese cannot be added to acid furnaces. The type of lining partly determines the final product, which is consequently classed as acid or basic steel. Basic steels were once considered inferior, due no doubt to the cheaper grades being of this type. However, given good practice either type should give a high quality product.

In view of the complexity of the production processes it is not surprising that steel contains, besides iron and carbon, some residual impurity elements, notably manganese (up to 1 per cent), silicon (up to 0.3 per cent), sulphur and phosphorus (together up to 0.05 per cent). The first two are beneficial to properties, whilst that the latter two are generally undesirable, except that sulphur is sometimes added in free machining steels.

Manganese, which is added as a deoxidiser, forms carbides and a portion is present combined in with the cementite, increasing the proportion of pearlite for a given carbon content. It also combines with sulphur to form a high melting point compound which appears as concentrated inclusions in the steel. In the absence of an excess of manganese, sulphur would react to iron sulphide, an insoluble low melting compound which forms a film around the pearlite grains and makes the steel brittle at elevated temperatures (hot-short).

Silicon is present in the ores and is added as a deoxidiser. It forms oxide, which mainly floats into the slag, although some remains in the melt as inclusions and slightly degrades the properties of the steel. Silicon also goes into solution in the ferrite, marginally increasing the strength and ductility. Lastly, phosphorus forms a solution in the ferrite and in quantities above about 0.05 per cent (depending on the carbon content) causes low notch toughness (Izod value).

The technology of iron and steel production has advanced rapidly in the last decade and is continuing. The principal developments have been the increased productivity from blast furnaces, with decreased coke consumption. This has been due to raising the operating temperatures and pressures, and to using oxygen, which is now available in tonnage quantities. At the same time several rival methods of ore reduction, known loosely as direct reduction, have been developed and are used in places where smaller production units than blast furnaces are required. The use of continuous casting in place of ingot moulds is leading to increased productivity and quality. In steel making, the use of tonnage oxygen in Bessemer converters or, more recently (1952), in top blown converters of the LD type (Linz–Donawitz) has improved the quality of the product and reduced the cost. In the LD converter, oxygen is fed direct to the steel–slag interface, where the important reactions take place, avoiding turbulence which leads to entrapment of slag in the steel. Finally, vacuum melting, or vacuum treatment of the liquid steel, has become common practice for high quality forging steels, on account of the lower hydrogen content (responsible for hairline cracking) and the reduced inclusion content. The elimination of the final deoxidisation stage and the reduction in ingot faults, which often do not show up until a late stage in the fabrication process, have more than paid for the cost of vacuum treatment.

12.3 Cast irons

Cast irons cover a wide range of structures and properties, with the common factor that they are iron–carbon alloys with above 1.7 per cent carbon, usually in the range 2–5 per cent. This gives them significantly lower melting and casting temperatures than steels. Their differences arise from the state of the carbon, which may be combined with iron to form cementite, as indicated by the $Fe–Fe_3C$ equilibrium diagram, or may be present as graphite in various forms. Cementite is only a metastable state and decomposes to graphite and ferrite when the carbon content is high, the cooling rate slow, or certain other elements are present, especially silicon and nickel. The types of cast iron include white, grey, malleable and nodular. The invention of nodular cast iron since the Second World War has given a range of alloys which combine the cheapness and ease of casting normally associated with cast irons with the strength and ductility expected of mild steels.

White cast iron has the carbon combined into the intermediate compound cementite and the microstructures, containing ferrite and cementite phases, can be deduced from the $Fe–Fe_3C$ equilibrium diagram (Fig. 12.1). For a typical composition around 3 per cent carbon, the cementite is formed on solidification in the eutectic grains, and subsequently by precipitation in the solid state from the austenite, initially during cooling from the eutectic to eutectoid temperature, and finally during the eutectoid transformation to pearlite. The ferrite is formed only in this last stage. A typical photomicrograph is given in Fig. 12.2. White cast iron is hard, brittle, and wear resistant on account of the massive cementite. It is not used directly in complete castings due to its brittleness and poor machinability, but only as an intermediate step in the production of malleable cast iron. A surface layer of white cast can be formed on top of grey iron where surface hardness and wear resistance is required, as in railway wagon wheels. This is obtained by more rapid cooling of the iron in contact with metallic chill plates placed in the sand mould.

Grey cast iron, which forms the majority of castings, has its carbon mainly in the form of graphite flakes which develop on casting, usually due to the presence of silicon. The flakes are joined together in clusters, as shown in Fig. 12.3, and are embedded in a steel-like matrix. Typical microstructures are shown in Fig. 12.4. The degree of dissociation of the cementite to graphite can be controlled.

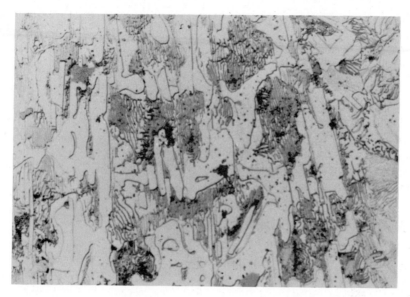

Fig. 12.2. White cast iron microstructure (Fe–3 % C). Etched in 4 per cent picral (× 330). Carbon is present as cementite. White areas are eutectic cementite and dark areas are pearlite.

In the more common pearlitic grey cast iron, the amount of silicon is sufficient at normal cooling rates to dissociate all cementite except that formed during the eutectoid reaction to pearlite (Fig. 12.4(*a*)). In ferritic cast iron the silicon content is higher, or the cooling rate slower, and all the carbon is present as graphite flakes in a matrix of ferrite grains (Fig. 12.4(*b*)). Intermediate structures can also be obtained.

Phosphorus is also present in many cast irons, up to 1.5 per cent. This forms an intermediate phase Fe_3P which solidifies as a eutectic with ferrite (or with ferrite and cementite) at about 960 °C, well below the cementite–austenite eutectic point at 1150 °C. Thus only this small quantity of phosphorus lowers the freezing point significantly, that is, improves the fluidity and reduces shrinkage. The phosphide eutectic can be distinguished from pearlite in the microstructure on account of its coarser lamellae, when using the standard picric acid and alcohol etchant (4 per cent picral).

A grey iron casting may be heat treated for three purposes: (i) to remove residual stresses caused by non-uniform cooling of the casting (also done by ageing in the atmosphere for a year); (ii) to convert from the pearlitic to the ferritic condition, thereby reducing hardness and improving machinability (subsequently the process may be reversed); (iii) to harden by quenching and tempering. Stress relief is done at 620–650 °C, annealing at 700–760 °C, and quenching at 850–880 °C.

The mechanical properties are determined by the size and distribution of the graphite flake clusters rather than by the matrix. The graphite introduces planes of weakness and stress concentrations, which lead to low strength and no ductility in tension but considerable strength and plastic flow in compression. It also improves machinability, reduces friction and improves damping.

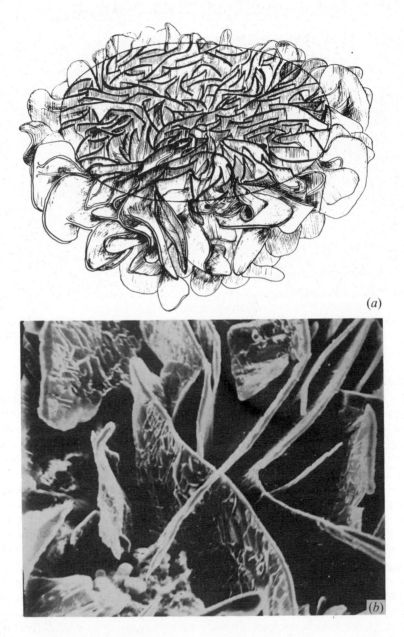

Fig. 12.3. Graphite flake cluster in grey cast iron. (*a*) Schematic, showing intercept with plane surface (from H. Morrogh, *J.I.S.I.* **206**, 1 (1968)). (*b*) Graphite skeleton in coarse graphite iron after etching metal away (scanning electronmicrograph ×350). (By courtesy of the British Cast Iron Research Association.)

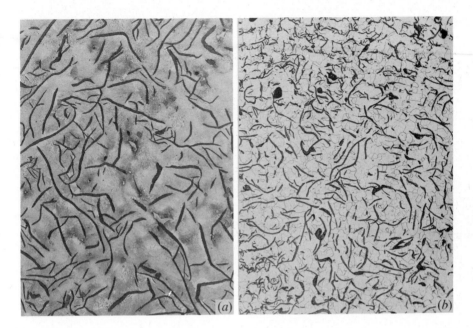

Fig. 12.4. Grey cast iron microstructures (× 80). Graphite flakes, formed during casting, in a matrix of (*a*) pearlite, and (*b*) ferrite. Etched in (*a*) 4 per cent picral, (*b*) 5 per cent nital. (By courtesy of G. N. J. Gilbert and the British Cast Iron Research Association.)

Typical values for the mechanical properties of grey cast irons are given in Table 12.1. It is important to note that the properties in a thin section are liable to be markedly different from those in the thicker parts, owing to the sensitivity of the structure to the cooling rate. The addition of nickel reduces this effect. By 'inoculating' the melt with artificial nuclei of calcium silicide, the graphite structure is refined and superior castings are obtained, known as Meehanite iron.

In malleable cast iron, graphitisation is brought about by heat treatment in the solid state, rather than during solidification as in grey iron, and equi-axed particles of graphite, known as temper carbon, are formed in a matrix of ferrite. The graphite modules may be compact spheres or loose and irregular; a typical micrograph is given in Fig. 12.5. The process is to heat a white cast iron (with all the carbon combined into cementite) to around 850 °C for up to a week in a neutral atmosphere. The improved mechanical properties, especially ductility, due to the spherical particles justify the extra time and cost. By stopping the process before complete graphitisation has occurred, or by subsequent heat treatment, the matrix surrounding the temper carbon can be made pearlitic. This increases the hardness and wear resistance at the expense of ductility. Another variant, common in Europe, is called *whiteheart* malleable cast iron; this is formed on heating in an oxidising atmosphere so that the carbon is largely removed from the surface layer which merges into a substrate of pearlite with graphite nodules. The castability of the initial white iron tends to be greater than for blackheart but the final ductility is less. The various names come from

Table 12.1. *Typical mechanical properties of cast irons*

Material	Representative composition C. Si (per cent)	Microstructure	Yield stress (0.1 per cent set) (MN/m²)	Tensile strength (MN/m²)	Ductility: elongation (per cent)	Hardness: Brinell	Compressive strength (MN/m²)	Modulus of rupture (MN/m²)	Remarks
White cast iron: low carbon	2.75 C, 1.0 Si	Free cementite and pearlite		250–300	0	400–550		500–700	Hard, brittle, nonmachining but wear resistant. Used for malleable irons and local 'chill' surfaces on grey iron.
high carbon	3.25 C, 0.25 Si			300–450	0	450–600		450–550	
Grey cast iron (ordinary grades): as cast, pearlitic	3.25 C, 2.0 Si	Graphite flake clusters in pearlite/ferrite matrix	100–200†	150–250	0.5	180–240	600–800	350–450	Graphitisation promoted by high C, high Si and low cooling rate. Low price; easily cast and machined; low strength and toughness; high damping capacity and compressive strength. Basic grade for iron castings.
annealed, ferrite			85–140†	125–200	0.5–1.0	100–150			
Grey cast iron (high strength): as cast, pearlitic	2.75 C, 2.25 Si	Fine graphite flake clusters in pearlite matrix	200–275†	300–400	0.5	210–320	750–1000	450–650	More expensive to cast and machine than ordinary grades. Less sensitive to section size.
Malleable cast iron: 2.5 C, 0.8 Si blackheart		Graphite nodules (temper carbon) in ferrite matrix. Whiteheart is decarbonised near surface	260–300	350–400	10–20	110–140			Anneal white cast iron for 2–7 days at 850–950°C in neutral atmosphere (for blackheart) or oxidising (for whiteheart). Good strength, high ductility and toughness. Widely used in vehicle, agricultural and general engineering.
whiteheart			280–320	400–450	5–10	120–220			
Spheroidal graphite cast iron: as cast, pearlitic, normalized, pearlitic	3.5 C, 2.0 Si Mg treatment	Graphite nodules in pearlite/ferrite matrix	300–400	600–750	3	240–290	1000–1250	900–1000	Graphite nodules formed on casting. Combines ease of casting of cast iron with the mechanical properties of mild steel. Widely used in engineering e.g. crankshafts, turbine casings, gears, brake drums, machine components.
annealed, ferritic			400–550	750–925	5	260–320	750–900	900–950	
			200–300	400–450	10–25	130–170			

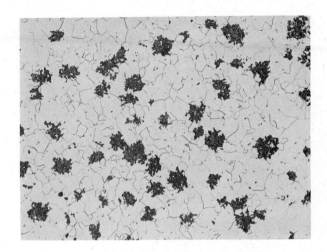

Fig. 12.5. Blackheart malleable cast iron microstructure, 5 per cent nital (× 80). Graphite nodules (temper carbon) in ferrite matrix, formed by heat treatment of white cast iron. (By courtesy of G. N. J. Gilbert and the British Cast Iron Research Association.)

the appearance of the fractured material. Typical values of the mechanical properties are given in Table 12.1.

Nodular cast iron, developed since 1946, also has spherical particles of graphite in a ferrite or pearlite matrix, but these are formed during solidification. The addition, just prior to casting, of magnesium and/or cerium, alloyed with a carrier such as nickel, alters the mechanism of graphite formation to give spherical particles in place of the flake clusters formed in grey iron. As cast, the matrix is largely pearlitic, but this can be altered by heat treatment. On heating to 850–900 °C, the graphitic carbon dissolves in the austenite until equilibrium conditions are attained. On air cooling the austenite changes to pearlite, but on very slow cooling, or holding below the critical temperature range at 680–700 °C for about ten hours the carbon returns to the graphite nodules and the matrix is ferrite. Typical microstructures are shown in Fig. 12.6. In thin sections, an oil quench will transform the austenite to martensite, which can be tempered to give the required combination of strength and ductility as in heat-treated steels (§ 12.7). Nodular cast iron, also known as SG (spheroidal graphite) or ductile cast iron, has almost steel-like strength and ductility, yet retains the advantages of castability associated with conventional cast irons. The exact properties, see Table 12.1, depend on the composition and the form of the matrix: ferritic nodular iron has high ductility with tensile strength of about 400 MN/m^2 (55–70 × 10^3 psi) and the pearlitic iron has rather lower ductility and higher strength, up to 1.0 GN/m^2 (150 × 10^3 psi).

Finally, the austenitic cast irons must be mentioned in which the addition of nickel, chromium, copper or manganese extends the temperature range in which austenite is stable down to room temperature and below. These irons have superior corrosion and heat resistance, and other special properties, such as low expansion or non-magnetism.

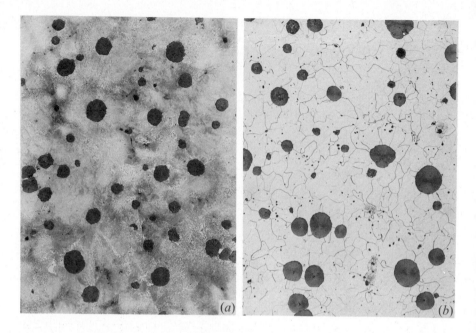

Fig. 12.6. Nodular cast iron microstructures (× 80) (*a*) Tearlitic: graphic modules in a tearlite matrix, formed during casting. (*b*) Ferritic: graphite nodules in ferrite matrix, formed by heat treatment of (*a*). Etched in (*a*) 4 per cent picral, (*b*) 5 per cent nital. (By courtesy of G. N. J. Gilbert and the British Cast Iron Research Association.)

12.4 Plain carbon steels

As regards mechanical properties, plain carbon steels conveniently divide into four main groups. The irons and dead soft mild steels have carbon contents below 0.15 per cent (plus some Mn, Si, S and P); these are used in constructions where strength is relatively unimportant but easy fabrication by cold forming and welding are advantageous. The second group are the mild steels with 0.15 to 0.3 per cent carbon which are used for structural and machine applications where strength up to 500 MN/m² (7.5 × 10⁴ psi) and fracture toughness are important but they retain good plasticity, machinability, weldability and a low price. Structural shapes such as I-beams, channels and angles fall in this group. Both groups are used in the normalised or hot rolled condition, or sometimes cold rolled. The microstructure consists of ferrite with increasing amounts of pearlite as the carbon content is raised (see Fig. 4.27).

Although the strength and hardness of steel in the normalised condition continue to rise with higher carbon content, the fracture toughness drops rapidly. Medium carbon steels, within the range 0.3 to 0.6 per cent carbon, are therefore hardened and tempered to obtain optimum mechanical properties, although their use in the normalised condition is not entirely ruled out. They are used for crankshafts, axles, railway wheels and gears amongst other components where there is a combined need for strength (up to about 800 MN/m², 1.2 × 10⁵ psi), toughness and wear resistance. The plain carbon steels can only be successfully heat treated in small sections with rapid quenching, and for larger or more intricate components alloy steels with greater hardenability are required.

The final group consists of the tool steels with about 0.6 to 1.4 per cent carbon where the resistance to abrasive wear and the ability to maintain a cutting edge are paramount rather than the conventional mechanical properties for structural and machine applications. These steels are used in the hardened and tempered condition. Although the hardness in the as-quenched state is fairly constant over the carbon range, the wear resistance increases at the expense of toughness (see § 7.5 for the history of tool materials). Steels from this group are also used for helical and leaf springs, and for high strength cold drawn wire.

Soft iron and high carbon steels are also used on account of their magnetic properties, see Chapter 15. The rest of this section will review the first and second groups of steels used in the normalised or hot rolled condition. The following sections will consider the heat treatment of steels.

The highest purity iron is made for research purposes by zone refining, that is, traversing a melted zone repeatedly in one direction along a specimen. The carbon content can be reduced to 20 ppm and other elements, mainly oxygen and nitrogen, to less than 10 ppm in total. Typical room temperature stress–strain curves are given in Fig. 12.7, for wire specimens after 99 per cent reduction of area and annealing at various temperatures for one hour, as indicated.

The purest iron available commercially, known as ingot iron or Armco iron (after the American Rolling Mill Company), contains about 0.02 per cent carbon and under 0.1 per cent impurities. Wrought iron, now little made, contains a similar amount of carbon but around 0.4 per cent of impurities, distributed as fine threads of slag. This characteristic structure is the result of the unique method of manufacture which involves forming bars, cutting them up into a stack and then rolling them together. Ingot iron microstructure consists entirely of ferrite grains but owing to the rapid solution strengthening effect of carbon it has a relatively high yield stress (200 MN/m^2) and a TS of 300 MN/m^2

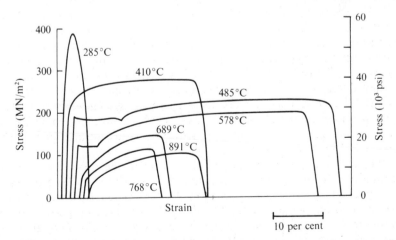

Fig. 12.7. Zone-refined polycrystalline iron (20 ppm C). Stress–strain curves for wire specimens which have been cold drawn to 99.1 per cent reduction of area and annealed 1 h at the temperatures indicated. (After E. Bull Simonsen and J. M. Dossin, *J.I.S.I.* **202**, 380 (1965).)

Table 12.2. *Typical mechanical properties of plain carbon steels (actual values show wide variation with detailed composition and history)*

Material	Carbon content (per cent)	Condition	Lower yield stress (MN/m²)	Tensile strength (MN/m²)	Ductility: elongation; reduction of area	Hardness: DPN	Izod notch toughness (J)	Remarks, applications
Zone refined iron	0.002	A (700 °C)	50	100	60% El			Annealed 1 h at 700 °C after 99% cold drawing. 40 grains/mm². Research work.
'Armco' ingot iron	0.02–0.04	A (925 °C)	200	300	45% El 75% RA	90	135–165	Sheet and wire, often galvanised or stove enamelled. Cores for DC magnetic circuits. Electrical conductors.
Mild steel: dead soft	0.05–0.15	N (900 °C)	250	400	40% El 70% RA	120	80–120	Sheets for deep drawing, drawing and press work. Wire for fencing and nails.
		CR (50%)	none	550	2–4% El	200		CR = Cold rolled
Mild steel	0.15–0.3	N (850 °C)	300	450	35% El 50% RA	140	40–80	Boiler plates and tubes. Structural sections (e.g. I-beams, channels, angles). Bolts, rivets, concrete reinforcing rods. Surface hardening (carburising). General engineering.

Type	%C	Condition			El % & RA %		Impact	Uses
Medium carbon steel	0.3–0.5	N (800 °C)	400	600	25% El 35% RA	200	7–27	Machinery components such as crankshafts, connecting rods, camshafts, axles. Higher carbon for gears, high strength forgings. Benefits of heat treatment obtained only in small sections or near surface. (Figures for 15 mm dia.)
		Q (800 °C)	800	950	0% El 0% RA	700	0	
		QT (800/500 °C)	600	800	15% El 45% RA	375	27–55	
		QT (800/600 °C)			25% El 55% RA	275	40–70	
Tool steel: low carbon	0.6–0.9	N (750 °C)	500	850	15% El 30% RA	250	~4	Springs; tools for woodworking; mower blades; shovel teeth, ploughs. Railway wheels and rails. Benefits of HT only obtained in small specimens or near surface (Figures for 15 mm dia.)
		Q (750 °C)			0% El 0% RA	850	0	
		QT (750/500 °C)	1100	1200	4% El 6% RA	450	~7	
		PCD (90%)		2000	4% El			High strength wire. PCD = patented and cold drawn.
Tool steel: medium carbon	0.9–1.2	N (850 °C)	550	950	10% El 20% RA	300	~2.7	Chisels, punches, forging dies, saws, milling cutters, drills, picks. Use quenched and stress relieved (150–250 °C). The latter process reduces tendency to spontaneous cracking without reducing hardness and wear resistance.
		QSR (750/200 °C)		1300	0% El 0% RA	900	0	
		QT (750/500 °C)	1250		2% El 4% RA	480	~5.5	
Tool steel: high carbon	1.2–1.5	N (900 °C)	550	800	4% El 4% RA	340	~2.0	Turning and planing tools, reamers, gauges, razor blades. Higher carbon content increases wear resistance at the expense of toughness, with no hardness change. SR = stress relieve.
		QSR (750/200 °C)			0% El 0% RA	900	0	

$1 \text{ MN/m}^2 = 145 \text{ psi}$ $1 \text{ J} = 0.735 \text{ ft lb}$

(45 000 psi). The ductility is large, with 40–50 per cent elongation and 70–80 per cent reduction of area. It is extensively used for stamping and deep drawing where the relatively low strength is not a disadvantage. Prior rolling eliminates objectionable Lüders bands or stretcher-strain marks. It galvanises and takes stove enamel well. Ingot iron sheet is also used in many electrical applications, such as transformers, due to its high permeability and low remanence; the addition of 4 per cent silicon raises the resistance to eddy currents (see Chapter 15).

Steels in the range 0.05 to 0.3 per cent carbon are called mild steels and they combine good mechanical properties in the form of strength, ductility and notch toughness, with ease of fabrication by machining, hot or cold forming, and welding. Furthermore, their price is less than any comparable material. They form 'home base' in the selection of constructional metals from which one moves only with good reason. The microstructure now contains cementite in the form of pearlite grains (see Fig. 4.25). The higher the carbon content the higher the proportion of the hard pearlite grains, and the greater the strength and less the ductility and notch toughness of the steel. Dead soft mild steel, with 0.05 to 0.15 per cent carbon, has a tensile strength of 250–350 MN/m^2 (35–50 × 10^3 psi), slightly above that of ingot iron. Its relatively low strength is compensated by the ease of fabrication, for example, by cold forming and welding. Ordinary mild steel has between 0.15 and 0.3 per cent carbon and combines good strength and fracture toughness with the ability to be fabricated, machined and welded, all at an economic price. It is standard material for structural and machine components, including structural sections (e.g. I-beams, channels), plates, tubes and many other general engineering parts. Typical parameters are: lower yield stress 300 MN/m^2 (45 000 psi); TS 400–550 MN/m^2 (60–80 000 psi); 35 per cent elongation and 50 per cent reduction of area.

Medium carbon steel with 0.3 to 0.5 per cent carbon has increased strength but its fracture toughness falls sharply; it finds limited use where the lower fracture toughness is compensated by the higher wear resistance, for example in railway wheels and tracks. It can be heat treated to produce surface hardness. The mechanical properties of plain carbon steels are summarised in Table 12.2.

Mild steel is usually supplied 'as rolled', that is, it has been hot worked just above the recrystallisation temperature and then allowed to cool in still air. This gives a microstructure which is similar to that in the normalised condition, that is, cooled in still air from just above the critical temperature range. The ferrite and pearlite are relatively strain free. Black steel is still coated with oxide scale formed during rolling. Bright steel has been pickled to remove the scale and may have been lightly cold rolled to improve the surface finish and dimensional tolerance. This process can lead to undesirable strain ageing and raising of the transition temperature for brittle fracture (see Chapter 10).

Grain size is an important parameter of the microstructure of steels, affecting the properties. There are two grain sizes to be considered, closely related to one another: the austenite grains existing above the critical temperature range and the ferrite grains (with some pearlite) which are formed from them on cooling to room temperature. The austenitic grain size is also important in the formation of martensite by quenching (§ 12.6). When carbon steel is heated through the critical temperature range, austenite is nucleated at the boundaries of the pearlite

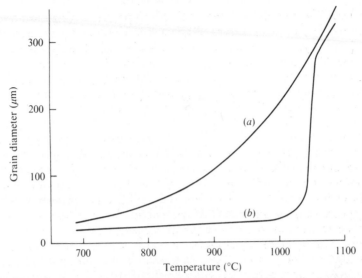

Fig. 12.8. Austenitic grain size formed on heating above the critical temperature range: (*a*) inherently coarse grained, (*b*) inherently fine grained.

and grows initially into the pearlite and then into the ferrite, until only austenite grains are present. Within limits, there is an increase in the number of grains; that is, grain refinement occurs. Further heating causes the larger austenite grains to grow at the expense of the smaller ones. In some steels, known as (*inherently*) *fine grained*, the growth is discontinuous, little change occurring initially as the temperature is raised above the upper critical temperature but later occurring rapidly. *Inherently coarse grained* steels grow more steadily with temperature, see Fig. 12.8. On cooling there is no change of grain size in the austenitic region, but on passing through the critical range ferrite is nucleated at the austenitic grain boundaries, and grains of ferrite and, later, pearlite grow. The size is dependent on the original austenite grain size and the cooling rate. If the austenite grains are large, ferrite also nucleates within the grains and forms plates on specific crystal planes, called *Widmanstätten structure*. Large ferrite grains and particularly any Widmanstätten structure lead to low notch toughness. Large grain size can be refined by heating to just above the critical temperature range. This does not require any prior working, of course, and must be distinguished from recrystallisation following cold work, which is the only method of grain refinement available in most materials; the minimum recrystallisation temperature of iron is about 400 °C. Various simple techniques are available for determining the austenite grain size other than direct observation under a hot stage microscope, but they will not be detailed here.

In the production of high strength steel wire, plain carbon steel with about 0.8 per cent carbon is converted to very fine pearlite by isothermal transformation from austenite (see § 12.5), called patenting, and then heavily cold drawn. Reduction of area may be as high as 80 or 90 per cent. A low temperature (250 °C) treatment is sometimes given to raise the 0.1 per cent proof stress, which in cold worked material is low in relation to the TS. Typical strengths range from

1.5 to 3.0 GN/m^2 (2 to 4 $\times$ 10^5 psi), as high as any other type of steel. The ductility is low (4 per cent elongation) and the notch toughness not considered, but these are unimportant in many applications, mainly due to the small cross-sections involved.

HEAT TREATMENT OF STEEL

12.5 Isothermal transformation of austenite

Plain carbon steels with carbon content above 0.4 per cent are responsive to heat treatment in useful sized sections and are normally used in the quenched and tempered condition, that is, after continuous rapid cooling from the austenitic state and reheating to some temperature below the critical temperature range. The phases are rearranged as compared with the annealed condition, giving a better combination of mechanical properties. In particular, the notch toughness is raised. By the addition of alloying elements, steels with below 0.4 per cent carbon can be heat treated in significant sections (see § 12.8). Before discussing the microstructures and properties obtained with various cooling rates, the transformation of austenite at constant temperature will be considered. The complexities of continuous cooling are more easily understood after studying isothermal transformation.

Isothermal transformation of austenite was first investigated by Davenport and Bain in 1930. Small thin specimens of steel are required, say 6 mm diameter by 1 mm thick, so that their temperature can be changed very rapidly. A specimen is first brought to the austenitic state by heating above its critical temperature range and holding it there long enough to ensure the carbides are completely taken into solution. It is then transferred rapidly into a molten lead or salt bath at a temperature T_s below the equilibrium austenite region. After a time t it is moved quickly into a cold water bath. The specimen is then prepared for microscopic examination, from which it is possible to deduce the phase changes that occurred at the intermediate temperature T_s. The experiment is repeated with various values of t and T_s.

The room temperature microstructures of eutectoid steel after transformation for various times at a temperature slightly below the eutectoid temperature are as shown in Fig. 12.9. Initially there is a single light etching phase. After a time, a dark etching constituent appears which is identified as pearlite, see (a); this forms at points on the original austenitic grain boundaries. The amount of pearlite increases with transformation time, see (b) and (c), until none of the light etching phase is present. It appears that the original austenite is retained during the initial rapid cooling to the intermediate temperature and that a measurable time is required for the nucleation and subsequent growth of pearlite from austenite. The light etching component of the room temperature microstructure is martensite to which any remaining austenite is transformed during the final quenching to room temperature (see below). The pearlite is unchanged during this quench as ferrite and cementite are equilibrium phases over the temperature range.

Progress of the transformation at different temperatures is plotted in a diagram with log(time) as the abscissa and temperature as the ordinate. This is called an

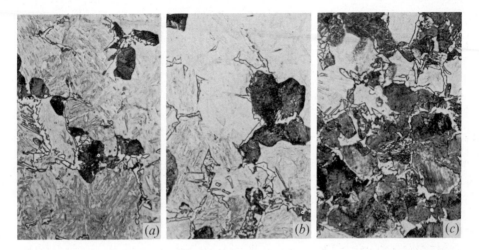

Fig. 12.9. Isothermal transformation of austenite to pearlite at 600 °C. (*a*) 1 s, 10 per cent transformed to pearlite; (*b*) 2 s, 50 per cent transformed; (*c*) 4 s, 75 per cent transformed. Carbon steel was austenised at 840 °C and quenched to 600 °C; after holding for the times indicated it was quenched to room temperature. (Etched in nital × 500.)

isothermal transformation diagram or TTT diagram (temperature–time–transformation). The original term, 'S-diagram', was based on incomplete information on its shape. A typical diagram for eutectoid steel is given in Fig. 12.10. The left-hand curve *ABC* marks the time at each temperature required for transformation of austenite to begin; the right-hand curve *A'B'C'* is the time for complete transformation.

Transformation just below the eutectoid temperature (720 °C) is very slow to start and takes very long to complete: the end product is coarse pearlite, that is, thick lamellae of ferrite and cementite (see for example Fig. 4.27(*d*)). Lowering the temperature at first shortens the transformation times and produces finer pearlite, which finally can only be resolved in an electron microscope. Below about 550 °C the transformation times lengthen again and the transformation product changes progressively to *bainite* (Fig. 12.11). This process is used commercially with some steels and is known as *austempering*. Like pearlite, bainite is a mixture of ferritic and carbide phases, but the mechanism of formation is different.

During pearlite formation, local concentration of carbon atoms leads to the precipitation of carbide plates, which grow edgewise into the austenite matrix by diffusion of the carbon atoms to them. The surrounding austenite becomes depleted of carbon and can transform directly to ferrite. Alternate lamellae of carbide and ferrite are formed in this way to give the typical pearlite structure. The rate controlling step is the diffusion of the carbon atoms. Details of bainite formation have not yet been fully elucidated, but it is thought that plates of ferrite are formed first and carbide particles are precipitated subsequently. The rate controlling step is now the diffusion of iron atoms from the austenite to the ferrite regions. An electron microscope is required to resolve the phases, see Figs. 12.11(*a*) and (*b*) which show bainite growing at 450 °C; the untransformed background in (*a*) is martensite formed in the final quench. This is called

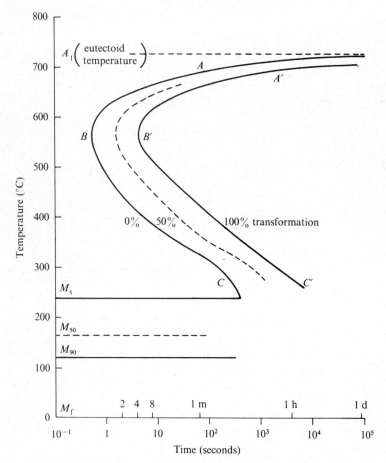

Fig. 12.10. Isothermal transformation diagram for eutectoid steel.

upper bainite. The particles of cementite are clearly visible in a matrix of ferrite. At lower temperatures, around 250 °C, the details of the transformation mechanisms alter again and the product is now called *lower bainite*, see Figs. 12.11(c) and (d). The structure is much finer, although the carbide particles can just be resolved in the ferrite matrix. However, the carbide is now ε-carbide, which has a close-packed hexagonal structure and composition 8.4 per cent carbon, not cementite with an orthorhombic structure and 6.7 per cent carbon (see § 4.8). Unlike pearlite, which grows as equi-axed grains within the austenite, bainite forms as plates or needles, signs of which remain after full transformation and lead to the descriptions 'feathery' for upper bainite and 'acicular' for lower bainite. Although the structure of the transformation product becomes finer with reduced temperature, it is found that where two products are being formed simultaneously, say fine pearlite and upper bainite, the product which is associated with the lower temperature, here upper bainite, is the coarser.

The changeover in the transformation product from pearlite to bainite as the temperature drops below the knee of the TTT diagram requires some

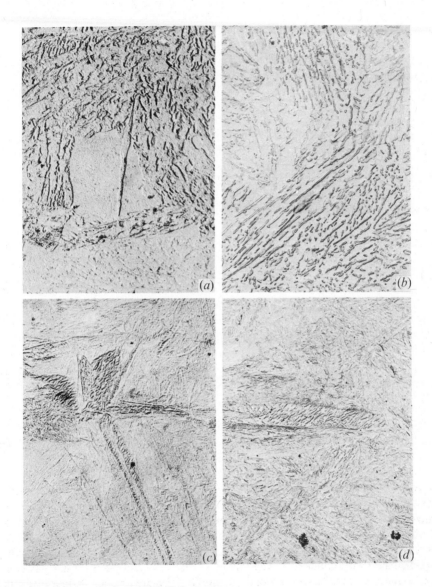

Fig. 12.11. Isothermal transformation of austenite to bainite (× 3750). (By courtesy of J. R. Blank and the British Steel Corporation.)
(*a*) Upper bainite 50 per cent transformed at 450 °C.
(*b*) Upper bainite 100 per cent transformed at 450 °C.
(*c*) Lower bainite 10 per cent transformed at 250 °C.
(*d*) Lower bainite 100 per cent transformed at 250 °C.

explanation. The knee is actually a composite of two knees, one of which represents transformation to pearlite and the other to bainite. In alloy steels the two curves are often well separated from each other (see Fig. 12.23). The formation of any new phase is driven by a reduction in the free energy (cf. § 2.9). However, the nucleation of a new phase requires energy for the interface and some supercooling is necessary below the equilibrium change temperature. As the supercooling is increased, the size of stable nuclei decreases, since the release of free energy is per unit volume and need for interface energy is per unit surface area, and they form more frequently; that is, the transformation time decreases. However, the atoms have to diffuse into their new positions and as the temperature is lowered the rate of diffusion rapidly decreases, causing the transformation times to lengthen. The net result of these various factors is a knee-shaped curve on the transformation diagram. In plain carbon steels, it so happens that the pearlite and bainite reaction curves overlap each other. In alloy steels the presence of alloying elements influences the diffusion of carbon atoms, usually slowing down the rate, and the pearlite and bainite knees become separated.

The transformation diagram also includes two horizontal lines, marked M_s and M_{90}, which indicate the start and 90 per cent completion of martensite formation. The rapid shearing of austenite into martensite has already been described (§ 11.9). This can only occur when the formation of pearlite or bainite at higher temperatures has been avoided by rapid quenching. Transformation to martensite then starts at M_s, here about 250 °C, and the proportion increases as the temperature is lowered, independently of the time factor. The last traces of austenite are very persistent, especially with high carbon content or alloy steels, and the value of M_f at which transformation is fully completed is very dependent on the sensitivity of the experimental method. Unless a finite amount of retained austenite is specified it can be assumed that M_f has been taken at about 3 per cent retained austenite. 99 per cent transformation will be 150–200 °C lower; for example, with eutectoid steel M_{90} is about 100 °C, M_{97} about 0 °C and M_{99} about -193 °C. The retained austenite plays an important role in tempering (§ 12.7), and also affects the magnetic properties (Chapter 15).

A new technique for following the course of transformation into martensite is clearly necessary in place of that used for pearlite and bainite. One method is to use a dilatometer, since there is about 4 per cent volumetric expansion into martensite from austenite. Another is to reheat the specimen to 350 °C for 15 seconds prior to the final quench, as this tempers the martensite already present and distinguishes it from the martensite formed subsequently during a final quench to room temperature. Micrographs obtained by this technique are given in Fig. 12.12, which shows the progressive transformation to martensite as the temperature is reduced.

The M_s and M_f lines should strictly not be shown on an isothermal transformation diagram since transformation is occurring athermally. However under special circumstances it is possible to observe the time dependency of the nucleation of martensite which occurs by thermal activation. Each nucleus then grows very rapidly into the surrounding austenite until stopped by some obstacle.

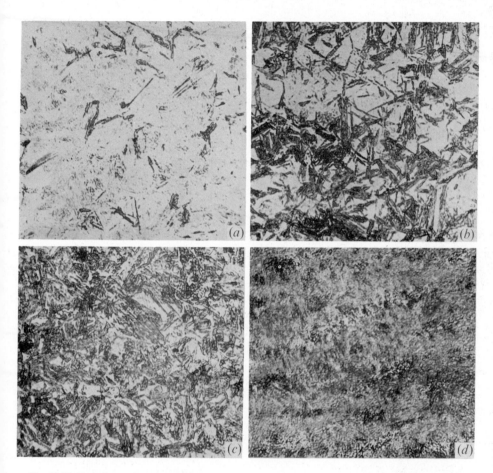

Fig. 12.12. Transformation of austenite to martensite (× 790). 0.4 % C, 1.5 % Ni steel (BS En24) was austenitised at 835 °C and quenched to: (a) 320 °C; (b) 310 °C; (c) 290 °C; (d) 230 °C. The steel was then heated to 350 °C for 20 seconds before quenching to room temperature to temper (and darken) the martensite. (By courtesy of International Nickel Ltd.)

The values of M_s and M_f are mainly controlled by the carbon content, although alloy elements forming substitutional solid solution (in the austenite), notably manganese and chromium, also affect them. The variation with carbon content is shown in Fig. 12.13.

For a hypoeutectoid steel, the transformation diagram is as shown in Fig. 12.14. Transformation to ferrite, at the austenite grain boundaries, now occurs if the specimen is held at a temperature within the critical temperature range – up to the limit shown by the equilibrium diagram for proeutectoid ferrite. Below the eutectoid temperature ferrite is formed first and at some time, shown by the carbide line, cementite is also precipitated, in the form of pearlite grains. The two processes may then overlap. As the temperature of transformation is reduced, the carbide line approaches the start of transformation line, showing that the amount of proeutectoid ferrite is becoming less and the composition of the pearlite is becoming richer in ferrite, moving away from the equilibrium

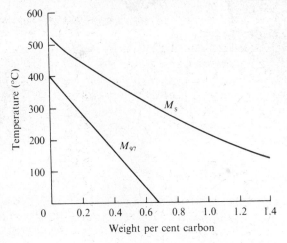

Fig. 12.13. M_s and M_{97} versus carbon content.

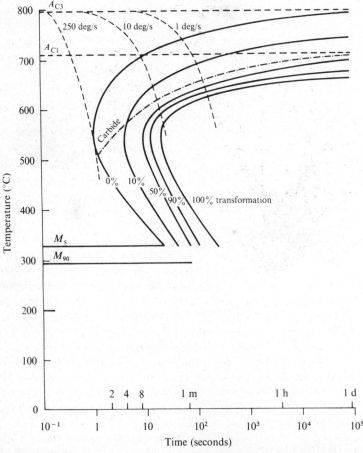

Fig. 12.14. Hypoeutectoid steel isothermal transformation diagram: 0.34% C 0.75% Ni (BS En12 1% Ni steel). Without the nickel present the curves would be moved one decade to the left making measurements at short times difficult. (The continuous cooling lines refer to Fig. 12.15.)

composition. The data in Fig. 12.14 are for an alloy steel with 0.34 per cent carbon and 0.75 per cent nickel. The effect of the nickel is to slow down the reaction rates without altering the phases that are formed – although the nickel atoms will be in substitutional solid solution at a very low concentration. Without the nickel, a 0.34 per cent plain carbon steel would have reaction times less than half those shown and the experimental technique described above becomes inadequate. However the principles of the transformation are unaffected. The use of alloying elements for reducing the transformation rates is discussed further in § 12.8.

12.6 Continuous cooling transformation of austenite

In the conventional heat treatment of steel, austenite is transformed into martensite during continuous rapid cooling from above the critical temperature range. For hypereutectoid steels (>0.9 per cent carbon) it is sufficient to heat just above the eutectoid temperature, since this will remove all the soft ferrite phase and the lower temperature reduces the tendency to crack on quenching. It is possible to derive an approximate constant cooling rate diagram from an isothermal diagram but it is not very accurate, nor quantitatively significant, since the cooling rate is not constant in practice: the temperature varies linearly with log(time). The inaccuracy comes from the methods of constructing the diagram. One method assumes the transformation product after continuous cooling for, say, four seconds is the same as that which would be formed isothermally after four seconds at some intermediate temperature. This is not very accurate however the intermediate temperature is chosen, but by taking the intermediate temperature at 75 per cent of the actual temperature drop, a constant cooling rate diagram is obtained which shows, at least qualitatively, the main characteristics of continuous cooling transformation. The 75 per cent, rather than 50 per cent, makes some allowance for the very much shorter nucleation times at lower temperatures, at least down to 550 °C. This method has been used on the isothermal transformation data of Fig. 12.14 to produce Fig. 12.15.

The lines giving the start and finish of transformation are no longer knee shaped but end abruptly at BB'. No transformation occurs at temperatures between BB' and M_s because, with constant cooling rates, it is not possible to enter the 'underbelly' of the isothermal diagram where bainite is formed. A little bainite may be formed in the region of BB' due to the overlapping of the bainite and pearlite formation. In some alloy steels, where the pearlite and bainite knees are well separated and pearlite formation occurs only after long nucleation times, it is possible to form only bainite on continuous cooling.

The microstructures of a medium carbon steel at various cooling rates are as follows, see Fig. 12.16. At slow cooling rates, around 1 °C per second, transformation is to some ferrite and mostly coarse pearlite, starting at the intersection of the cooling line with AB and completing at the intersection with $A'B'$. At faster rates, finer pearlite is formed, see (b). At rates around 50 °C per second, the cooling line intersects BB': transformation begins at the grain boundaries of the austenite to fine pearlite, possibly with some bainite, and ceases before completion on the line BB', at around 500 °C. The untransformed austenite now

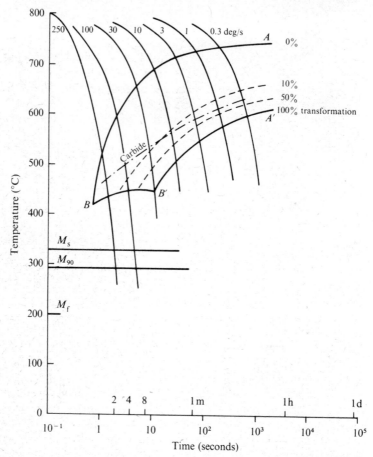

Fig. 12.15. Approximate continuous cooling transformation diagram. 0.34% C, 0.75% Ni. Derived from isothermal transformation data in Fig. 12.14.

remains until the temperature drops to the M_s line around 350 °C, when it begins to convert to martensite athermally. The reaction goes to nominal completion at M_f. The final microstructure consists partly of pearlite and partly of martensite, as shown in (c), and is called *split transformation*.

At high cooling rates, above about 300 °C per second, the cooling curve misses AB altogether and transformation is entirely to martensite between the M_s and M_f temperatures, as already described (see (d)). The minimum cooling rate to give a fully martensite structure is called the *critical cooling rate* and is an important parameter of a steel as the best heat treatment practice requires that the quenching rate shall exceed it. (This is discussed further in § 12.8.) The critical cooling rate for plain carbon steels varies with carbon contents as plotted in Fig. 12.17. The high values at low carbon contents are the principal reason why mild steels cannot usefully be heat treated in other than very thin sections, except for surface hardening.

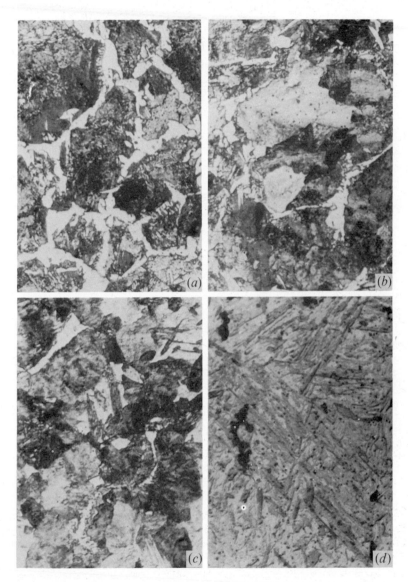

Fig. 12.16. Microstructures of medium carbon steel at various cooling rates. Etched in 5 per cent nital (× 500).

(*a*) Slow cooling: ferrite (white) and coarse pearlite.
(*b*) Medium cooling: a little ferrite (white) and fine pearlite.
(*c*) Fast cooling: very fine pearlite (dark) and martensite: split transformation.
(*d*) Very fast cooling: martensite.

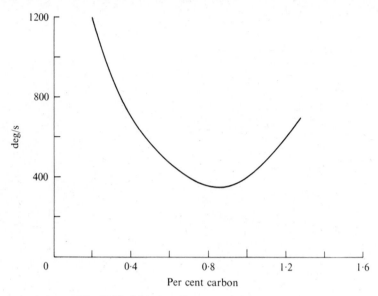

Fig. 12.17. Critical cooling rates of plain carbon steels.

The critical cooling rate and the position of the transformation curves on the TTT diagram are dependent, amongst other variables, on the austenite grain size (see § 12.4). Large austenite grains do not transform as rapidly as fine grains, since nuclei tend to form at grain boundaries, but the martensite plate or needle size are larger and the final properties are not so good. Overheating to produce large austenite grains is thus not a desirable method of reducing the critical cooling rate.

12.7 Tempering

Martensite has great hardness, depending on the carbon content (see Fig. 11.43), but no ductility or notch toughness. It is also liable to crack spontaneously due to the high internal stresses created by the differential cooling between the centre and surface of a part during quenching and by the volumetric and shear strains on transformation. To restore some ductility and notch toughness, at the expense of the hardness, it is necessary to heat to a temperature below the critical temperature range, and cool in the furnace or air to room temperature. The operation is called *tempering* (or *drawing*). Atoms are able to diffuse at the tempering temperature and the equilibrium phases start to form again, although their arrangement is quite different from the normalised condition. The reaction rate is initially high, decreasing as the reciprocal of time, and the time at temperature is not a critical factor after a few minutes, but a period of about an hour is usual to obtain a uniform temperature through the specimen.

Typical changes in the mechanical properties are shown in Fig. 12.18, called a tempering chart. This is usually provided by the supplier of a steel to aid in the selection of the correct tempering temperature for a required combination of mechanical properties. Tempering is normally done between 450 °C and 650 °C,

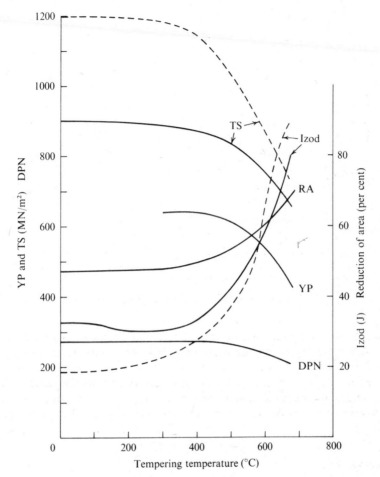

Fig. 12.18. Tempering chart of 1 per cent nickel steel. 286 mm diameter bar. —— Oil quenched from 850 °C, – – – water quenched from 850 °C.

the range in which the notch toughness and ductility rise rapidly whilst the hardness and strength drop slowly. Heating to only 200 °C is sufficient to relieve internal stresses and remove the tendency to cracking. (N.B. This effect is seldom indicated on the tempering chart.) Another method of reducing some of the tendency of steels to crack and distort on quenching directly to martensite is to quench in two stages, first a rapid quench to slightly above the M_s temperature, and later, after a period to allow quenching stresses to be relieved, a mild quenching to room temperature. The martensite is formed relatively uniformly throughout the specimen in this way. The process relies on the long period required to form bainite at a temperature slightly above M_s (see Fig. 12.10). It is called *martempering* and should be distinguished from austempering (see above) and maraging, described later (§ 12.9).

Between 250 °C and 350 °C there is a marked depression in the notch toughness, which is not reflected in the strength and ductility curves. This is the *brittle tempering range* and must be avoided. In alloy steels with nickel, chromium or manganese, there is also a drop of notch toughness after tempering between 550 °C and 600 °C, followed by slow cooling. This effect, called *temper brittleness*, can be avoided by water quenching from the tempering temperature or adding 0.5 per cent molybdenum to the steel. These phenomena must not be confused with *blue brittleness* of mild steel which is a marked loss of ductility at 300 °C, which makes this temperature unsuitable for working. The theories of these various effects will not be included here.

Three stages of tempering are distinguished. In the first stage, between 100 and 150 °C, high axial ratio, high carbon martensite changes to low axial ratio (1.012), low carbon (0.3 per cent) martensite and carbide. The carbide, probably ε-carbide (as in lower bainite), precipitates at the twin boundaries within the martensite plates (see § 11.9). The specific volume decreases by an amount depending on the carbon content: 0.1 per cent at 0.4 per cent carbon rising linearly to 0.6 per cent at 1.0 per cent carbon (cf. about 4.3 per cent expansion from austenite to martensite). Changes in low carbon martensite are less well marked and are not so well understood: there will be stress relaxation and, according to Nutting, precipitation of Widmanstätten carbide. In high carbon steels, hardening due to the precipitation of carbide exceeds the softening due to the reduction in carbon content of the martensite and there is a net hardening in this stage. Some first stage tempering occurs during the quenching, particularly in medium carbon steels with high M_s temperatures. This leads to inhomogeneity since the first-formed martensite is tempered while the later material is not.

In the second stage of tempering, around 250 °C, any retained austenite converts to lower bainite, that is, ε-carbide and ferrite. This involves a local expansion of 3 per cent by volume and causes very high internal stresses. The notch toughness drops during this stage.

In the final and most important stage of tempering, which develops rapidly between 300 °C and 400 °C, the ε-carbide and the carbon still in solution are converted into small plates of cementite throughout a matrix of ferrite, which eventually recrystallises. The structure is called *sorbite* and is that normally required in a heat-treated steel. There is a specific volume reduction in this stage similar to that in the first stage. It will be observed that sorbite is composed of the same phases, ferrite and cementite, as are found in normalised steel but their arrangement is quite different. As will be shown below (§ 12.8) this microstructure of finely dispersed cementite in a ferrite matrix has the superior mechanical properties; for example, at the same hardness level sorbite has greater ductility and notch toughness than a pearlitic microstructure.

At still higher temperatures, approaching 700 °C, and after some hours, the cementite grows into fewer and larger spherical particles, a structure known as *spheroidite*. This degree of tempering is beyond that normally required.

Photomicrographs of tempered martensite at various stages are given in Fig. 12.19. With an electron microscope the structure of sorbite can be resolved. With an optical microscope, although there is a slight change of appearance with standardised preparation, the structure only becomes resolved when spheroidite is formed.

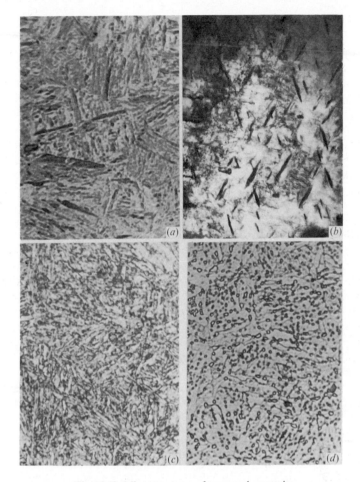

Fig. 12.19. Microstructures of tempered martesite.
(a) 0.35 % C, water quenched from 850 °C: martensite (× 670).
(b) 0.2 % C, quenched and tempered 1 hour at 250 °C: Widmanstätten cementite in martensite
matrix (× 40 000). (By courtesy of E. Tekin.)
(c) 0.35 % C, water quenched 850 °C and tempered 2 hours at 600 °C: sorbite (× 670).
(d) 0.35 % C, water quenched 850 °C and annealed 2 days at 660 °C: spheroidite (× 670).

12.8 Mass effect and hardenability

Optimum mechanical properties in a heat-treated steel are obtained after
tempering a fully martensitic structure without any pearlite or bainite present
due to split transformation during quenching. Although it is desirable to exceed
the critical cooling rate at all points in the object being quenched, this is not
always achieved and 90 per cent martensite at the centre of the specimen is often
considered adequate. The cooling rate obtained will depend on the quenching
medium and agitation, which determine the heat transfer coefficient, and
the size of the specimen, or more specifically, the cross-section. It may be
necessary to restrict the severity of the quench on account of distortion or

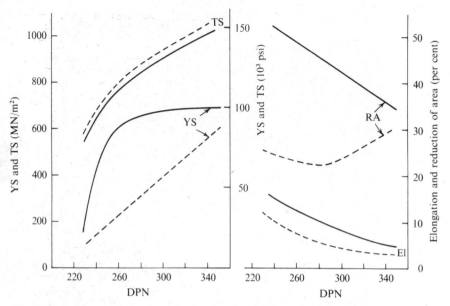

Fig. 12.20. Tensile properties of tempered martensite (sorbite) compared with pearlite of the same hardness obtained by continuous cooling at different rates. (YS is stress for 10^{-4} set). ——— tempered martensite, – – – pearlite.

cracking, particularly for irregular pieces. Thus, for a material of given composition and with a certain quenching procedure, there is a maximum cross-section which can be fully quenched. Larger cross-sections will be slack quenched at their centres and will suffer a progressive drop in the properties obtainable, even after tempering. This limitation is called the *mass effect*, in spite of the fact that it is really a section effect.

The superior properties of steels after full quenching and tempering are shown in Figs. 12.20 and 12.21. Figure 12.20 compares the tensile properties of martensite tempered at various temperatures, with those of pearlite obtained by continuous cooling at different rates. At any hardness value the yield stress (for 10^{-4} set) and the ductility as shown by the elongation and reduction of area to fracture, are greater for the tempered martensite. Figure 12.21 shows the notch toughness versus the tensile strength for various bar sizes and tempering temperatures. At any strength value, the Izod value of the smaller bars is above that of the larger ones. The former will have been fully quenched even at the centre, while the latter are slack quenched, that is, cooled below the critical cooling rate. A similar result is shown in the tempering chart of Fig. 12.18, which compares properties after water and oil quenching of bars. It might be thought that the effects of slack quenching could be eliminated by tempering at a lower temperature, but the preceding discussion of the structural changes occurring in the various stages of quenching and tempering will have shown that the structures obtained are quite different by the two routes.

The maximum diameter of round bar of a particular steel which can be heat treated to quoted values of the mechanical properties is called the *limiting ruling section* in British specifications. Ideally, a material has a single value of the limit-

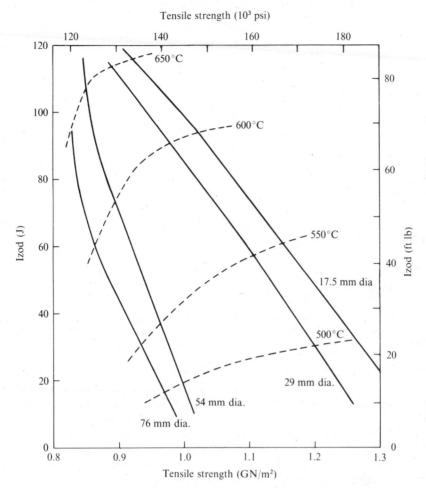

Fig. 12.21. Variation of mechanical properties due to 'mass effect'. Larger diameter bars, with split transformation structure on the axis, have lower notch toughness after tempering to a given strength level. ($1\frac{1}{2}\%$ Mn–Mo–C steel.)

ing ruling section, through which a fully martensitic structure is achieved with a known quenching procedure. As mentioned above, in practice some slack quenching is allowed in larger sections and the consequent lower properties tolerated. Specifications therefore quote several strength and toughness levels and the limiting ruling sections for each.

Figure 12.22 compares the mass effect in a plain carbon steel and in an alloy steel: hardness is plotted against radial position in quenched bars of increasing size. Owing to the high critical cooling rates of plain carbon steels (see also Fig. 12.17) only very small cross-sections can be successfully heat treated, unless only surface hardness is required; in this example, the limiting ruling section is about 19 mm (0.75 in). By adding alloying elements such as nickel and chromium, the critical cooling rate can be reduced and the limiting ruling section correspondingly increased, here to about 100 mm (4 in).

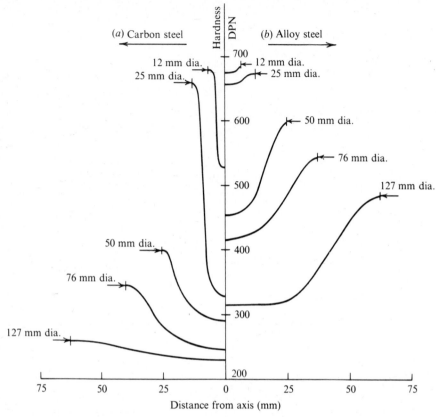

Fig. 12.22. Hardness distribution across quenched bars of various diameters: (*a*) 0.45 % C steel, (*b*) 0.40 % C, 1 % Cr–Va steel.

Alloy steels have been developed which are superior to plain carbon steels in various ways, depending on the application. Classes of alloy steel exist, amongst others, for cutting tools, for magnets, for high or low temperature service, and for use in corrosive environments. It is not the intention to discuss alloy steels except to touch briefly on one class developed for their superior room temperature strength, ductility and notch toughness. These are known as machine, or constructional, steels (see, for example, BS970 En series of steels).

The primary function of the alloying element in these steels is to reduce the critical cooling rate and permit the development of tempered martensite in large cross-sections. All alloying elements, with the exception of cobalt, have a greater affinity for carbon than does iron. The carbon atoms tend to concentrate around the alloy atoms, forming embryonic carbides, and are reluctant to diffuse away. This reduction in the diffusion rate of carbon atoms slows down the transformation to pearlite, but has little effect on bainite formation. Figure 12.23 shows the TTT diagram of a typical alloy steel (3 per cent nickel, 1 per cent chromium, 0.3 per cent carbon) with the pearlite curves shifted far along the time axis. The martensite hardness is still determined by the carbon content, independently of the alloy elements, and Fig. 11.43 holds. The phase changes on

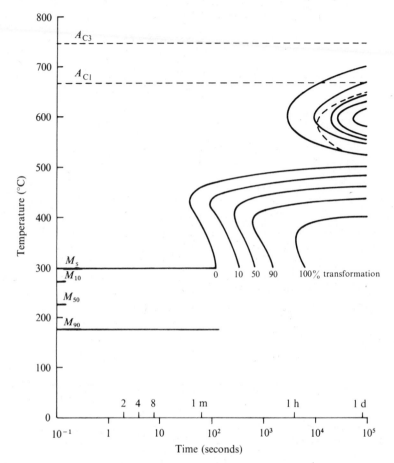

Fig. 12.23. Isothermal transformation diagram of a typical alloy steel (3% Ni 1% Cr 0.3% C to En23). *NB* The reaction to pearlite is retarded more than that to bainite, cf. Fig. 12.14.

tempering occur at a slower rate and at higher temperatures than for plain carbon steels. The phases formed are in general the same, although their compositions are slightly changed.

In selecting machine steels it is usual to employ the concept of hardenability, particularly in America. *Hardenability* of a steel is a measure of the ease of quenching to a fully martensitic structure. It is thus another approach to mass effect, critical cooling rate and limiting ruling sections. Hardenability is not concerned with the hardness of the martensite formed, only with the cooling rate to achieve full transformation. A low carbon alloy steel can have greater hardenability than a high carbon plain steel, although the hardness of their respective martensites will be in the reverse order. Like some other terms, it is not well chosen.

Two tests of hardenability are used. In the Grossmann method, the micro-structure is observed in several quenched bars of various diameters. The position of the 50 per cent martensite zone, which is more easily seen than the 100 per cent

Table 12.3. *Quench severity factor H*

	Agitation		Coolant		
Component	Coolant	Air	Oil	Water	Brine
Nil	Nil	0.02	0.3	1.0	2.2
Moderate	Nil		0.4	2.0	
Violent	Nil		0.8	4.0	7.5
Nil	Spray		1.5	10.0	

martensite region, is determined in each bar and the *critical diameter* (for 50 per cent martensite) for the quenching procedure used is the bar size in which the zone is on the axis. The *ideal* critical diameter (for 50 per cent martensite on axis) is then the critical diameter for an ideal quench with an infinite heat transfer rate. This is obtained by means of Fig. 12.24 and Table 12.3, which gives the value of a quench severity parameter H for various quenching procedures. The correlation between ideal diameters for 50 per cent, 95 per cent and 99.5 per cent martensite, based on assumed transformation characteristics, is given in Fig. 12.25. From these graphs it is possible to determine the structure at the centre of any bar size with any quenching procedure, at least approximately.

The commonest measure of hardenability is the Jominy end-quench test. This requires much less time and material than the Grossmann method but the results are not so directly related to practice. A one inch (25 mm) diameter bar, four inches (100 mm) long, is austenised just above the critical temperature

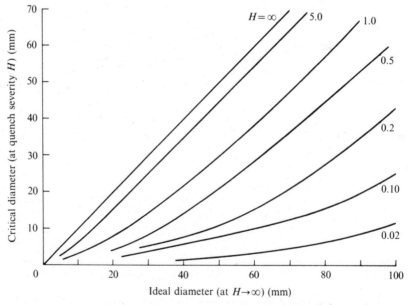

Fig. 12.24. Relationship between critical diameter (50 per cent martensite) at quench severity H (see Table 12.3) and the ideal diameter at $H \to \infty$. (After M. A. Grossmann, M. Asimow and S. F. Urban.)

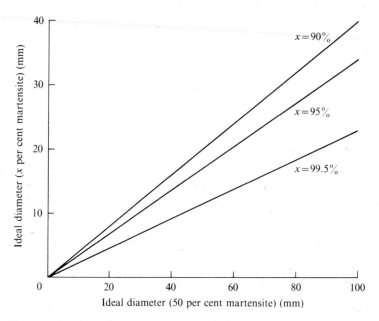

Fig. 12.25. Relation between ideal diameters ($H \to \infty$) for various martensite contents on axis. (Typical transformation characteristics are assumed.)

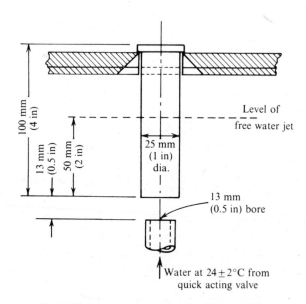

Fig. 12.26. Jominy end-quench test of hardenability.

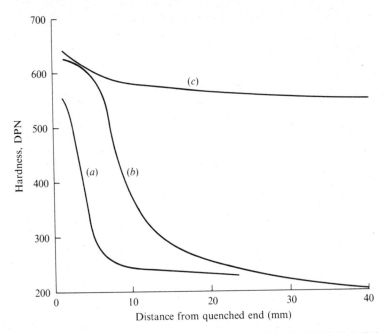

Fig. 12.27. End-quench curves (Jominy) for: (*a*) plain carbon steel (0.4% C 0.7% Mn); (*b*) low alloy steel (1% Ni 0.35% C, cf. Fig. 12.14) BS En12; (*c*) alloy steel (3% Ni 1% Cr 0.3% C, cf. Fig. 12.23) BS En23.

range and then quenched by a water jet impinging on one end, see Fig. 12.26. After cooling, a flat, 0.015 inch (0.4 mm) deep, is ground along the side and the hardness measured every $\frac{1}{16}$th inch (1.5 mm). Typical results are given in Fig. 12.27. The 0.4 per cent carbon plain carbon steel has very low hardenability with the hardness dropping off rapidly at about 4 mm from the quenched end – in agreement with the data on quenched bars plotted in Fig. 12.22. The low alloy steel (1 per cent nickel, 0.35 per cent carbon) has slightly better hardenability with the median hardness occurring at about 10 mm. The third curve shown is for 3 per cent nickel, 1 per cent chromium, 0.3 per cent carbon, which has high hardenability. The isothermal transformation diagrams of the latter two steels were given in Fig. 12.14 and Fig. 12.23 respectively and they confirm that the lengthened reaction times and increased hardenability go with increasing alloy content.

Correlation of end-quench test results with the quenching behaviour of round bars can only be approximate because of the non-linear cooling rates involved. Inhomogeneity of the material across the section of a bar is also an important factor. By equating the mean cooling rate from 700 °C to 500 °C the equivalent positions along a Jominy specimen and in oil-quenched bars have been established, see Fig. 12.28.

If it is desired to establish the Jominy position for certain split transformation microstructures (e.g. 50 per cent martensite), these can be observed directly, or an empirical relationship can be used between the percentage loss of hardness and the microstructure as shown in Fig. 12.29.

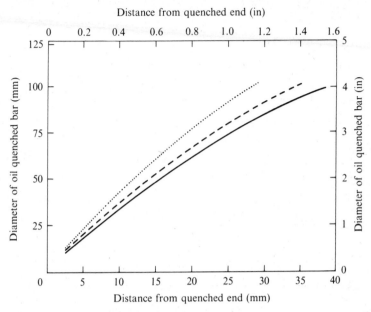

Fig. 12.28. Correlation of oil-quenched bars and position in Jominy end-quench specimen (for equal average cooling rate 700–500 °C). —— axial position; – – – $r/b = 0.5$; · · · · $r/b = 0.8$.

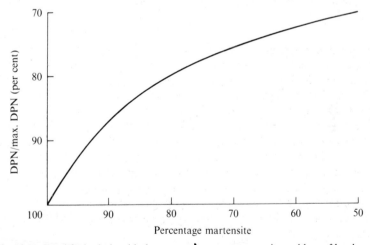

Fig. 12.29. Empirical relationship between percentage martensite and loss of hardness.

12.9 Steels for strength

Iron-based alloys have for long provided cheap and strong materials for engineers and their development continues on both fronts. New production techniques giving cleaner and cheaper steels have already been described (§ 12.2). As regards strength, at one end of the scale is zone refined iron which has a tensile strength of 40 MN/m² (6000 psi) as a single crystal and 100 MN/m² (15 000 psi) as polycrystalline material. At the other end of the scale, the theoretical strength of iron is $10\,GN/m^2$ (1–2×10^6 psi), a value which has been obtained in $2\,\mu m$ diameter iron whiskers. Besides strength, there is always the need for ductility and notch, or fracture, toughness without which the maximum stress obtained in an idealised laboratory tensile test cannot be safely utilised in practical structures. In the development of iron-based materials the problem is thus to obtain progressively higher strengths whilst retaining adequate ductility and notch toughness. Iron whiskers have, for example, negligible elongation to fracture. The minimum acceptable values for these parameters are very debatable (see Chapter 10) but in the first instance an elongation of 10 per cent and an Izod or Charpy value of 20 J (15 ft lb) can be taken; very much higher values are usually obtained except in very high strength steels. In addition, the fatigue strength is important in many applications and it is found that changes which increase the tensile strength frequently do not raise, or may even lower, the fatigue limit, particularly when any stress concentrators are present, such as notches or sharp corners (see § 5.11). Finally, the economic aspect remains dominant in most circumstances.

The lower part of the strength range, up to $550\,MN/m^2$ (8×10^4 psi), is met by plain carbon steels of increasing carbon content, used in the normalised

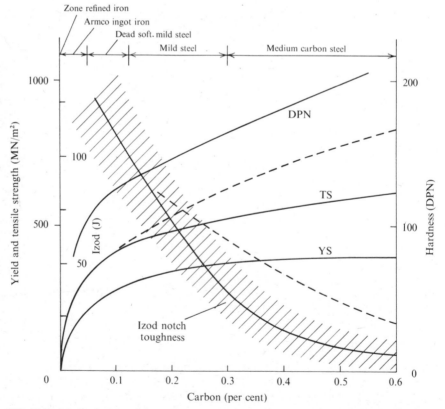

Fig. 12.30. Mechanical properties of plain carbon steels. (1 J = 0.74 ft lb, 1 MN/m² = 145 psi.)
—— normalised, – – – quenched and tempered (small specimens).

condition. Above this strength, the ductility and notch toughness are inadequate. In small specimens medium carbon steels (0.3–0.6 per cent carbon) can be used in the quenched and tempered state to give strengths up to 800 MN/m² with adequate fracture toughness. Typical property values are summarised in Fig. 12.30.

From 0.7 to 1.4 GN/m² (1–2 × 10⁵ psi), alloy steels of the machine steel group are available, used in the heat-treated condition. As regards the parameters measured in the tensile test at room temperature, it has been found that there is very close similarity between the various steels as long as the fully martensitic structure is developed on quenching, that is, as long as the hardenability is adequate. Figure 12.31 gives the data obtained on many SAE steels; after quenching, the tempering temperature was varied to bring each steel to the required strength level. When it comes to the notch toughness at any given strength, there is a large difference between the steels. Figure 12.32 plots the notch toughness of some nickel alloy steels which have been oil quenched and tempered to various strengths. Selection of a steel should therefore be based, amongst other considerations, on its notch toughness at the required strength level.

Above 1.5 GN/m² (2 × 10⁵ psi) a few steels are available and current development is reaching towards the 3 GN/m² (4 × 10⁵ psi) level. Around

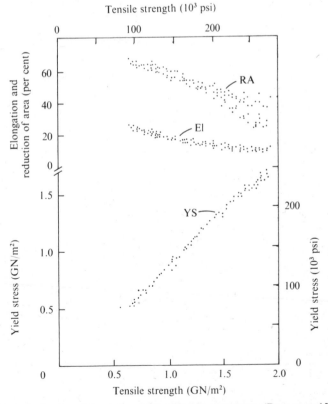

Fig. 12.31. Equivalence of machine steels as regards tensile parameters. (Data cover 15 alloy steels, all fully quenched prior to tempering at various temperatures.) (After E. J. Janitzky and M. Baeyertz.)

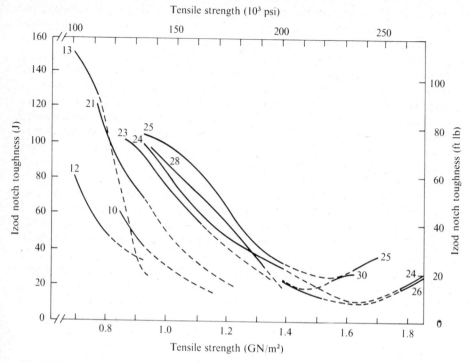

Fig. 12.32. Notch toughness of various nickel alloy steels, 28.6 mm diameter bars, oil quenched and tempered to different strength levels. Solid line indicates recommended tensile range. The numbers refer to En series steels in BS 970.

1.75 GN/m² (2.5 × 10⁵ psi), a few low or medium alloy steels still have adequate ductility and notch toughness: five steels are listed in the BS970 En series in this range. One or two conventional steels go to even higher strengths: a vacuum melted steel, 5 per cent chromium, 0.4–0.5 per cent carbon, develops 2 GN/m² (3 × 10⁵ psi) TS, 10 per cent elongation and 20 J (15 ft lb) Izod. Air melting produced only 11 J (8 ft lb) Izod.

Two new techniques have been invented for producing high strength steels, ausforming (1954) and maraging (about 1960). In ausforming, the steel is austenised, cooled to about 500 °C, strain hardened by working, converted to martensite by quenching to room temperature, and finally tempered. In alloy steels for ausforming there must be a bay between the pearlite and bainite bands on the TTT diagram in which the austenite can be deformed without transforming. The temperature is below the recrystallisation temperature so that the deformation is technically cold work and strain hardening occurs. Although the reasons for the effects have not yet been elucidated it has been shown that the properties are significantly better than those obtained in conventional heat treatment. This is probably related to the small martensite crystallites formed. Results are rather limited but it has been reported that the strength increases from 2 to 3 GN/m² (3 to 4 × 10⁵ psi) without loss of ductility and with even an increase of notch toughness. High temperature and fatigue properties also appear to be enhanced. The process must not be confused with austempering which is the iso-thermal transformation of austenite to bainite, without any working (§ 12.5).

In maraging steels, invented by Bieber (around 1960), iron is alloyed with 18–25 per cent nickel and other elements (titanium, aluminium, cobalt) which form complex precipitates with the nickel in martensitic iron, as described below. No carbon is present (0.03 per cent maximum), although the term steel is, surprisingly, retained. The iron–nickel equilibrium diagram is shown in Fig. 12.33. At high temperatures the fcc γ-iron forms a solid solution with fcc nickel over the full composition range; at low temperatures the bcc α-iron forms a limited solution with nickel. The transformation that occurs on cooling is very dependent on the cooling rate and the nickel content – so much so that the true equilibrium diagram is difficult to determine exactly. Pure iron transforms from its γ to α-form by nucleation and grain growth at, or below, 910 °C on slow cooling. At very fast cooling rates, 5500 deg/s, this mode can be suppressed and a martensitic shear-type transformation to α iron occurs at 545 °C (see Figs. 12.13 and 12.17). With low nickel contents, up to 15 per cent, reconstructive trans-formation to α-iron occurs at cooling rates as high as 5500 deg/s. Between 15 and 18 per cent nickel there appears to be a sharp discontinuity and above 18 per cent nickel martensitic transformation occurs even at a very low cooling rate of 5 degrees per minute. The reason for this rapid change in transformation

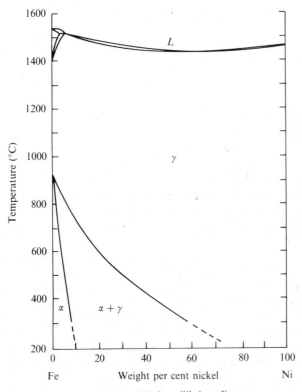

Fig. 12.33. Iron–nickel equilibrium diagram.

modes is not understood. At much higher nickel contents the γ-phase is stable down to room temperature, as shown by Fig. 12.33.

In the range 18–25 per cent nickel, martensite is formed on cooling at about 300 °C. It is a reversible reaction and γ-iron is formed on reheating, but at a very much higher temperature, above 600 °C, as shown in Fig. 12.34. The martensite contains no carbon so it has a bcc crystal structure, rather than the tetragonal structure of iron–carbon martensite, and is soft ($\sim$250 DPN) and ductile. This is the solution-annealed state and is achieved without quenching (cf. solution treatment of precipitation-hardened aluminium alloys). The strength and hardness is developed by precipitating complex compounds of nickel and the other elements during reheating to 500 °C for up to 3 hours, that is, below the temperature at which the martensite reverts to γ-iron.

Maraging was found originally with titanium and aluminium additions to the 18 per cent nickel–iron base. More recently cobalt and molybdenum have been found to be effective and combinations of all four are now being developed, for example, Fe–18Ni–7Co–5Mo–0.4Ti–0.1Al. Strengths are obtained up to 2 GN/m^2 (3 $\times$ 10^5 psi) with good ductility (10 per cent elongation, 50 per cent reduction of area) and exceptional notch toughness (40 J, 30 ft lb, Izod). At the same time the material can be easily machined or formed in the solution-annealed condition and no distortion or scaling occurs on maraging. It is also weldable and corrosion resistant. Applications have already been found in aircraft undercarriages, rocket cases and other parts where the strength/weight ratio is critical and the comparatively high price is not a deterrent.

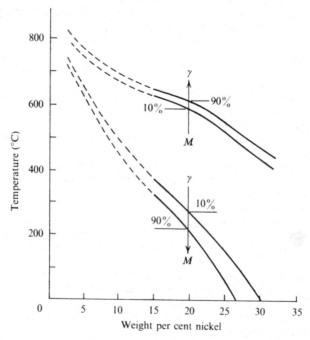

Fig. 12.34. Martensite transformation in iron–nickel alloys on cooling and heating at 5 deg/min. It is now found that below 15 per cent nickel transformation is not martensitic even at very high cooling rates.

Under pressure from the newer metals, titanium and beryllium, steel has recently been shown to be still capable of competing in the race for the maximum strength to weight ratio and it still has potential in hand. It is conceivable that iron will be the basis of future alloys with strengths around $5 \, GN/m^2$ (8×10^5 psi), whilst still retaining ductility and notch toughness.

QUESTIONS

1. Distinguish between (a) cast iron and steel; (b) acid and basic steel; (c) rimming and killed steel; (d) black and bright steel; (e) ingot iron, mild steel and alloy steel.
2. Contrast the state of the carbon in the various types of cast iron: white, grey, malleable and nodular. Indicate briefly the manner of formation, and the effect on mechanical properties.
3. Outline the change in structure and mechanical properties of steel in the normalised state as the carbon content is increased. Why is the content limited to about 0.4 per cent in structural applications?
4. (a) Why is a fine grain size desirable in a mild steel?
   (b) Describe two techniques of reducing the grain size in steel.
   (c) Distinguish between inherently fine grained and inherently coarse grained steel.
   (d) What is Widmanstätten structure?
5. (a) Sketch a typical TTT diagram for a eutectoid steel and discuss its meaning.
   (b) Explain the reasons for the knee-shaped curve.
   (c) Discuss the M_s and M_f lines. What is 'retained austenite' and how can it be reduced?
6. Define: austempering, ausforming, martempering, maraging.
7. Contrast the structure and manner of formation from austenite of: pearlite, bainite, martensite and sorbite.
8. (a) Sketch a continuous cooling transformation diagram and indicate the critical cooling rate and a cooling rate leading to split transformation.
   (b) What is their significance?
   (c) What are the difficulties in deriving such a diagram from an isothermal transformation diagram?
9. Describe the objective and process of conventional heat treatment of steel in terms of the microstructure and properties of the principal phases formed.
10. (a) What is the object of tempering martensite?
    (b) Discuss the various stages of tempering.
    (c) Compare a slack quenched steel and a fully quenched steel which have been brought to the same hardness by adjusting the tempering temperature.
11. Distinguish between the following temperature ranges: brittle temper, blue brittle and temper brittle.
12. (a) Define hardenability, mass effect, limiting ruling section.
    (b) What is the purpose of alloying elements in a machine steel?
    (c) Is hardenability the only criterion for selecting a machine steel?
    (d) Why are several limiting sections quoted for each steel in the BS970?
13. (a) Describe the Grossmann hardenability test.
    (b) A steel had a critical diameter (50 per cent martensite) of 30 mm when oil quenched with moderate agitation. Determine the critical diameter using a water quench with moderate agitation.

(c) What is the objection to using even more severe quenching to achieve a martensitic structure throughout larger specimens?

14. (a) Describe the Jominy hardenability test.

   (b) Given the results of end-quench tests on a nickel steel and a nickel-chromium steel (see Fig. 12.27), plot the hardness across a section of 25 mm and 50 mm diameter bars for each material after oil quenching.

15. Discuss steels for the strength range 300 MN/m^2 (ingot iron) to 3 GN/m^2 (maraging steel).

FURTHER READING

Transformation Characteristics of Nickel Steels. The Mond Nickel Company, London (1952).

G. V. Kurdjumov: Phenomena occurring in the quenching and tempering of steels. *J.I.S.I.* **195**, 26 (1960).

M. Cohen: The strengthening of steel. *Trans. A.I.M.E.* **224**, 638 (1962).

High-strength steels. Conf. Rep. No. 76 B.I.S.R.A., London (1962).

J. S. Kirkaldy and R. G. Ward (eds.): *Aspects of Modern Ferrous Metallurgy.* Blackie (1964).

G. N. T. Gilbert: The mechanical properties of cast iron. *Chart. Mech. Engr.*, 316 (1965).

E. Bull Simonsen and J. M. Dossin: Mechanical properties of zone-refined polycrystalline, high purity iron. *J.I.S.I.* **202**, 380 (1965).

V. F. Zackay (ed.): *High Strength Materials.* Wiley (1965).

G. P. Contractor: The marvel of maraging. *J. Metals,* **18**, 938 (1966).

H. Morrogh: The status of the metallurgy of cast irons. *J.I.S.I.* **206**, 1 (1968).

13. Electrical properties 1: metals

13.1 Introduction

The importance of the study of the electrical properties of solids is due to the increasing use of electricity, as a form of power, for computation, communication and control. This and the following chapter give a theoretical introduction to the properties, and describe briefly the applications, of the three types of electrical materials. These materials, classified according to their function, are:

(a) Conductors, which serve to move quantities of electricity from one region in space to another. The requirement here is for high conductivity, and this is fulfilled by *metals*.

(b) Semiconductors, which, because of the fact that the density and mobility of their conduction electrons are extremely sensitive to purity, temperature and electric field, are important in communications and control engineering. Semiconductors are weakly covalent.

(c) Insulators, the opposite of conductors, used to prevent electricity moving from one region in space to another. Insulators must have very low conductivity, and are usually strongly covalent or ionic, as are many ceramics, glasses and polymers. A class of insulators known as dielectrics, which, due to their high polarisability, enhance electric fields, are used where high capacitance is required.

The division of solids into metals, semiconductors and insulators in terms of the zone structure of their electron energy levels has been described in Chapter 3. This classification can also be achieved by a consideration of electrical resistivity. The resistivities of some common solids are listed in Table 13.1, and it can readily be seen that the division is a valid one. Pure metals have resistivities from 10^{-8} to 10^{-7} ohm m, metallic alloys from 10^{-7} to 10^{-5} ohm m, semiconductors range from 10^{-3} to 10^5 ohm m, and insulators from 10^9 to 10^{17} ohm m. Each class is separated from the next by changes in resistivity of several powers of ten. There is a difference in kind between metallic conduction and conduction in semiconductors and insulators. The resistivity of metal increases with increasing temperature due to the thermal scattering of electrons, but the resistivity of semiconductors and insulators decreases with increasing temperature over most of the temperature range due to thermal excitations of electrons over the forbidden energy levels. Insulators are less sensitive to temperature than semiconductors because of their larger energy gaps. There are a few exceptions to this classification, notably the semimetals, and anomalies such as graphite.

The semimetals, arsenic, antimony, bismuth, selenium and tellurium, have resistivities in the semiconductor range, but their behaviour with respect to temperature is metallic. Like metals, they do not have forbidden energy levels, but their density of conduction electrons is low, and the electrons have low mobilities. Graphite has a simple hexagonal crystal structure (Fig. 13.1) in which close-packed layers (the basal planes) are separated by rather large

Table 13.1. *Room temperature resistivities of some common solids*

Metal	Resistivity (Ωm)	Semiconductor	Resistivity (Ωm)	Insulator	Resistivity (Ωm)
Silver	1.6×10^{-8}	Germanium	0.47	Glass (various)	10^9–10^{15}
Copper	1.7×10^{-8}	Silicon	5×10^3	Mica	9×10^{14}
Aluminium	2.8×10^{-8}	Magnetite	0.01	Diamond	10^{14}
Iron (pure)	10×10^{-8}	InSb	2×10^4	Al_2O_3	10^{13}
Lead	20×10^{-8}			Bakelite	10^{11}
				Polyethylene	10^{15}–10^{17}
				Polymethyl methacrylate	10^{14}–10^{17}

distances from one another. (The weak binding between the layers which is responsible for this separation also enables the layers to slide easily over one another, giving rise to the good lubricating properties of graphite.) Graphite shows metallic conduction in the basal planes, but semiconduction across them.

This chapter gives a brief introduction to conduction processes in metals and alloys, and indicates how they are affected by the atomic structure and

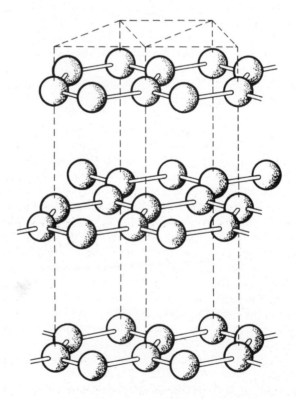

Fig. 13.1. The crystal structure of graphite.

microstructure. In the following section the electrical conductivity of a crystal is related to its zone structure and to a parameter, the mean free path, which is microstructure dependent. The approach is initially in terms of classical mechanics, and the results so obtained are then reconciled with wave mechanics. Finally mention is made of the phenomenon of superconductivity, with an indication of possible commercial applications.

13.2 Electrical conductivity

Electrical conductivity is defined by Ohm's law, which states that, when an applied field causes a current to flow, the current density J is proportional to the field strength $\mathscr{E}$. The constant of proportionality is called the conductivity σ;

$$J = \sigma\mathscr{E} \tag{13.1}$$

(J and $\mathscr{E}$ are both vector quantities, and for a non-isotropic material σ is a tensor). The resistivity ρ is defined as the reciprocal of the conductivity. The electronic interpretation of Ohm's law can be achieved by treating electrons as particles which obey the classical laws of motion, but wave mechanics is required to determine the magnitude of various parameters.

Consider the motion of a single electron of charge e in a field $\mathscr{E}$. The force on the electron is $e\mathscr{E}$, and from Newton's second law it equals the rate of change of momentum of the electron:

$$e\mathscr{E} = \frac{\mathrm{d}p}{\mathrm{d}t}.$$

In a solid with N conduction electrons the total change of electronic momentum due to the application of the field is:

$$\mathrm{d}P = N e\mathscr{E}\,\mathrm{d}t. \tag{13.2}$$

This suggests that as long as the field is applied electrons should continue to accelerate and the electronic current to increase. However, it is known from experience that as soon as a field is applied the current rises almost instantaneously to the value given by Ohm's law; and that while the current flows heat is developed, indicating that the moving electrons are losing energy to the lattice by scattering. As mentioned in Chapter 3, it was held in the free electron theory that scattering was caused by collisions with the ions of the lattice. It is now known that the electrons are scattered by interaction with irregularities in the lattice, the most numerous of which are disturbances in the lattice perfection due to thermal vibrations.

It is assumed in the scattering process that:

(i) The probability of an electron being scattered in a time $\mathrm{d}t$ is given by $\mathrm{d}t/\tau$, where τ is the mean interval between scattering events.

(ii) The speed with which the electron is travelling is not changed by the scattering process.

(iii) The electron has an equal probability of being scattered in any direction. The average momentum of the electrons after scattering must therefore be zero and hence the average change in momentum due to a scattering event is equal to the mean momentum of the electron in the forward direction before it was scattered:

$$\Delta \bar{p} = -\bar{p}.$$

The number of electrons scattered in a time dt is:

$$\mathrm{d}v = N \, \mathrm{d}t/\tau. \qquad (13.3)$$

The total change in forward momentum in a time dt due to scattering is then:

$$\Delta \bar{p} \, \mathrm{d}v = -\bar{p}N \, \mathrm{d}t/\tau. \qquad (13.4)$$

The application of an electric field increases the total electron momentum according to (13.2); scattering decreases it according to (13.4). For a steady state condition the total change in momentum must be zero:

$$Ne\mathscr{E} \, \mathrm{d}t - \bar{p}N \, \mathrm{d}t/\tau = 0,$$

and the average electron momentum $\bar{p} = e\mathscr{E}\tau$.
Putting $\bar{p} = m\bar{v}$ where m is the mass of an electron, and $\bar{v}$ its average velocity in the direction of the applied field gives:

$$\mathscr{E} = m\bar{v}/(e\tau).$$

The current density J is given by the net charge crossing unit area in one second, that is all the electrons contained in a depth $\bar{v}$ on one side of a unit element of area:

$$J = en\bar{v}, \qquad (13.5)$$

where n is the number of conduction electrons per unit volume. Substituting for $\bar{v}$ in (13.5) gives:

$$\mathscr{E} = \frac{m}{ne^2\tau}J.$$

By comparison with Ohm's law (13.1) the conductivity is

$$\sigma = ne^2\tau/m. \qquad (13.6)$$

If the mean free path of the conduction electrons is l, $\tau = l/\bar{v}$ and:

$$\sigma = \frac{ne^2 l}{m\bar{v}} \qquad (13.7)$$

and the resistivity

$$\rho = \frac{1}{\sigma} = \frac{m\bar{v}}{ne^2 l}. \qquad (3.1a)$$

This result has been achieved by the use of classical mechanics; strictly it is applicable only to metals, and then only if interpreted in the light of the wave-mechanical theory developed in Chapter 3. A rigorous derivation produces a formula similar to (3.1a), but the quantities n, $\bar{v}$, l and m are differently defined. n is the number of electrons per unit volume which are able to take part in the conduction process. Wave mechanics shows that these are the electrons close to the Fermi level, which can be excited into empty states. n is thus given by the density of states at the Fermi level, $N(0)$. Their average velocity $\bar{v}$ is obtained by equating their average kinetic energy $\frac{1}{2}m\bar{v}^2$ to the Fermi level ε_f. This velocity is known as the Fermi velocity, v_f. The mean free path, as already mentioned, is determined by interruption of wave propagation caused by irregularities in lattice periodicity, which will be considered in the next section.

A further consequence of wave mechanics is that electrons do not always behave (as judged by classical mechanics) as if their mass were that of an isolated electron. Electrons must be assigned an 'effective mass' m^*, which close to a zone boundary may be quite different to m. The effective mass is derived as follows.

The wave motion equivalent of the velocity of a particle is called the 'group velocity' of the wave, and is defined as $d\omega/dk$, where ω is the angular frequency, and k is the wave number, of the wave. Since according to Planck, $E = \hbar\omega$, the group velocity $V_g = d\omega/dk = (1/\hbar)(dE/dk) = $ velocity v of an electron. The force on an electron moving in an electric field, $e\mathscr{E}$, must be equal to the rate of change of energy with distance, dE/dx. Thus

$$dE = e\mathscr{E}\,dx = e\mathscr{E}v\,dt = \frac{e\mathscr{E}}{\hbar}\frac{dE}{dk}\,dt,$$

from which it can be seen that

$$\frac{dk}{dt} = \frac{e\mathscr{E}}{\hbar}. \tag{13.8}$$

The acceleration, a, of the electron equals

$$\frac{dv}{dt} = \frac{1}{\hbar}\frac{d}{dt}\frac{dE}{dk} = \frac{1}{\hbar}\frac{dk}{dt}\frac{d^2E}{dk^2}$$

and, substituting (13.8) into this equation

$$a = \frac{e\mathscr{E}}{\hbar^2}\frac{d^2E}{dk^2}; \tag{13.9}$$

and this is the wave-mechanical equation for the acceleration of an electron.

The corresponding equation in classical mechanics is

$$a = \frac{e\mathscr{E}}{m}$$

from which it can be seen that the effective mass

$$m^* = \hbar^2\left(\frac{d^2E}{dk^2}\right)^{-1}. \tag{13.10a}$$

Note that for 'free' electrons, for which (see (3.9))

$$E = \frac{\hbar^2 k^2}{2m},$$

$$\frac{d^2E}{dk^2} = \frac{\hbar^2}{m} \quad \text{and} \quad m^* = m.$$

In fact m^* differs from m when the relation between E and K departs from the simple parabola, (3.9), as the zone boundaries are approached (§ 3.8).

For values of E just above a zone boundary, that is at the lower energy levels in a band, d^2E/dk^2 is positive, and m^* is positive. For energy levels at the top of a band, just below the zone boundary, d^2E/dk^2 is negative. In order to avoid the mental gymnastics which are thought necessary to comprehend the notion of a negative mass, the effective mass is redefined as:

$$m^* = \hbar^2 \left| \left(\frac{d^2E}{dk^2} \right)^{-1} \right|,$$

(13.10b)

and an 'effective charge' e^* is defined:

$$e^* = e \frac{d^2E}{dk^2} \bigg/ \left| \frac{d^2E}{dk^2} \right|.$$

(13.11)

Thus e^* always has the same magnitude as e, and its sign is the same as e (i.e. negative) when d^2E/dk^2 is positive, and is equal to $-e$ (i.e. positive) when d^2E/dk^2 is negative. When an electron has a positive charge, i.e. when d^2E/dk^2 is negative as at the top of a band, it is called a 'hole'. Holes are discussed further when dealing with semiconductors in the next chapter.

It is the effective mass and the effective charge which must be used in the formula for resistivity

$$\rho = \frac{m^*v_f}{N(0)e^{*2}l},$$

(3.1b)

where

$$v_f = \left(\frac{2E_f}{m^*} \right)^{\frac{1}{2}}.$$

This formula is applicable when only one type of charge carrier is present. When both electrons and holes are present together, as frequently occurs in semiconductors, then a rather more complicated version of the formula, given in the next chapter, is required.

13.3 Scattering centres and Matthiessen's rule

Of the terms in (13.7), only the mean free path is strongly dependent upon the details of the microstructure. The deviations from periodicity in the crystal lattice which lead to scattering are thermal vibrations, impurity atoms, vacancies and interstitials, dislocations, stacking faults, grain boundaries and particles of a second phase. Each type of scattering centre is assumed to act independently of the presence of any other type and each is characterised by its own scattering probability, $1/\tau$. Resistivity ρ is directly proportional to scattering probability, (13.6), and as probabilities are additive the contributions to the total resistivity from each type of scattering centre are additive:

$$\rho = \frac{m}{ne^2} \left(\frac{1}{\tau_1} + \frac{1}{\tau_2} + \frac{1}{\tau_3} + \dots \right) = \frac{m\bar{v}}{ne^2} \left(\frac{1}{l_1} + \frac{1}{l_2} + \frac{1}{l_3} + \dots \right).$$

Matthiessen first suggested, in 1864, the separation of resistivity into a temperature independent portion ρ_i and a temperature dependent portion ρ_t

$$\rho = \rho_i + \rho_t.$$

The only scattering which is strongly temperature dependent is that due to thermal vibrations, and theory indicates that it is directly proportional to temperature except at very low temperatures (§ 13.4). Hence Matthiessen's rule may be written as:

$$\rho = \rho_i + aT \tag{13.12}$$

where a is a constant of the material. This relationship is confirmed by experiment (Figs. 13.2 and 13.3).

Thermal vibrations give a mean free path at room temperature of about one hundred atom spacings. In order to have comparable influence defects contributing to ρ_i should be not more than one hundred atoms apart. Most important are impurities, which must have a concentration of at least 10^{-3} atomic per cent to be effective. Dislocations also act as scattering centres, and the equivalent density of 10^9 lines mm² is only found in heavily cold-worked material. Concentrations of vacancies are sufficiently high only in a material near to its melting point or after severe neutron irradiation. Stacking faults, grain boundaries and particles of a second phase are normally two widely separated to have much effect.

At low temperatures, the resistivity consists almost entirely of ρ_i, and the resistivity of high purity metals at liquid helium temperature (4.2 °K) is very sensitive to the presence of trace impurities. The resistivity ratio, defined as the ratio of the difference between room temperature resistivity and helium temperature resistivity to helium temperature resistivity, is used to indicate impurities present in quantities lower than can be detected by normal chemical means. Resistivity ratios are typically one for alloys, ten for commercially pure metals, a few hundred for floating zone-refined refractory metals, and up to several thousand for metals such as tantalum heated for several hours at 3000 °C *in vacuo* of 10^{-10} torr.

The sensitivity of resistivity to the presence of impurities is important if it is wished to improve the strength of a metal without impairing its electrical properties. Second-phase hardening must be employed rather than solid-

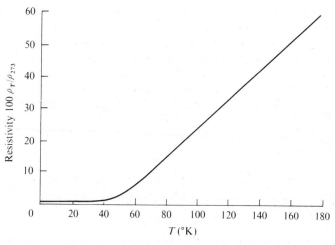

Fig. 13.2. The resistivity of pure copper as a function of temperature, expressed as a percentage of its value at 273 °K.

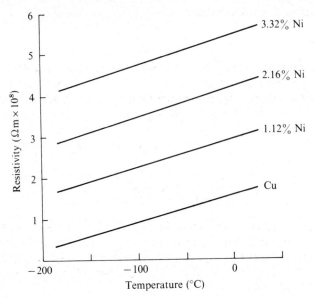

Fig. 13.3. Resistivity versus temperature for a series of Cu–Ni alloys (after J. O. Linde, *Ann. Phys.* **15**, 219 (1932)). These data can be seen to fit Matthiessen's rule, (13.12).

solution hardening in such cases. Even so the resistivity is increased : precipitation-hardened copper–beryllium alloys have a resistivity of the order of ten times greater than pure copper.

13.4 Temperature dependent resistivity

The temperature dependent part of the resistivity arises because electrons are scattered by the atoms, which, as a result of thermal vibrations, spend very little of the time in their equilibrium positions. Scattering theory shows that the scattering probability is proportional to the mean square displacement of the atoms from their equilibrium position, $\overline{x^2}$, where x is the displacement. The equation of motion for each atom is

$$\frac{M\,\mathrm{d}^2 x}{\mathrm{d}t^2} + \alpha x = 0$$

where M is the mass of the atom, and α is the elastic restoring force on the atom. This equation describes a simple harmonic vibration with frequency

$$\omega = \left(\frac{\alpha}{M}\right)^{\frac{1}{2}}.$$

The average frequency of vibration as mentioned in §2.9, for temperatures greater than the Debye temperature, is given by:

$$\omega = \frac{k\theta}{\hbar} \tag{13.13}$$

where k is Boltzmann's constant and θ is the Debye frequency of the crystal lattice. Equating these two gives

$$\alpha = \frac{Mk^2\theta^2}{\hbar^2}.$$

The potential energy stored when an atom is displaced an amount x is αx^2, and the average potential energy per mode of vibration is thus $\alpha \overline{x^2}$. This is also equal to one half of the total energy per mode of vibration, kT. Thus:

$$\alpha \overline{x^2} = \tfrac{1}{2}kT$$

and

$$x^2 = \frac{\hbar^2 T}{2Mk\theta^2}.$$

Since

$$\rho_T \propto \overline{x^2} \propto \frac{T}{M\theta^2},$$

the resistivity, as stated in the previous section, shows a linear variation with temperature. This is only true for temperatures greater than the Debye temperature. At lower temperatures not all of the modes of vibration are excited, and the average frequency is less than that given by (13.13). In this region the resistivity is found to be proportional to T^5.

13.5 Resistivity of alloys

Alloying elements in solid solution introduce scattering centres which reduce the mean free path and increase resistivity. The mean free path is a function of the distance between the atoms of the alloy, and the effectiveness of each alloy atom in producing a scattering event. The distance between alloy atoms is a function of the concentration c, and the resistivity may be approximated by (Nordheim, 1931; see Fig. 13.4):

$$\rho_{\text{alloy}} = \rho_{\text{matrix}} + Ac(1 - c).$$

ρ_{matrix} is the resistivity of the matrix, due to thermal and other defects, in the absence of alloying elements. A is a constant which is determined by the

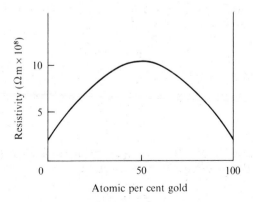

Fig. 13.4. Room temperature resistivity of silver–gold alloys as a function of composition (after Beckman, thesis, Uppsala 1911). The curve follows closely the Nordheim relation.

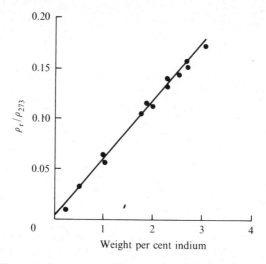

Fig. 13.5. The residual resistivity of dilute solutions of indium in tin, expressed as a fraction of the resistivity at 273 °K (after A. B. Pippard, *Proc. Roy. Soc. A***248**, 97 (1955)). The predicted linear relationship is observed.

effectiveness of the foreign atom as a scatterer. At concentrations below about 5 atomic per cent $c(1 - c) \simeq c$, and resistivity is linear with concentration (Fig. 13.5).

The effectiveness of the alloying atom as a scatterer in a given matrix depends upon the extent to which the periodicity of the lattice potential is disrupted. The disturbance can be either a local elastic distortion of the lattice, which is a function of Δa the difference in size between the solvent and solute atoms, or an electronic disturbance due to Δz, where Δze is the difference in positive electronic charge between the nuclei of the atoms. Δz is normally taken as the difference in the valency between solvent and solute. The constant A depends on both the solvent and solute atoms and is a function of Δa and Δz.

Norbury in 1921 and Linde in 1932 investigated the effect of dissolving solutes of differing valency into the noble metals copper, silver and gold and showed that, when Δa was small, for one per cent of solute, the increase in resistivity was proportional to $(\Delta z)^2$. This is shown in Fig. 13.6, and follows from Rutherford's theory of scattering by a charged particle.[†] The Norbury–Linde relationship holds well for simple metal solvents, and solutes of non-transition metals. With transition metals the situation is complicated by the difficulty of ascribing a simple valency to the transition metal, and by the introduction of an additional scattering mechanism. The electronic structure of transition elements is such that it can often give rise to a localised magnetic moment, and the interaction between conduction electrons and localised moments can give rise to an additional contribution to the electrical resistivity. Ferromagnetic materials below their Curie temperature (see Chapter 15) have a higher contribution to their resistivities from this cause than above their Curie temperature, when the magnetic moments are thermally randomised. This property is made use of to produce alloys with resistivities which are invariant over certain temperature ranges; as the temperature increases, the increased thermal scattering is

[†] Rutherford showed in 1911 that the scattering probability of electrons by positively charged ions is $\propto z^2$, where ze is the charge on the ion.

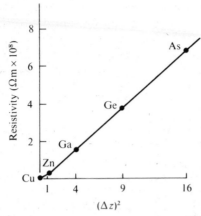

Fig. 13.6. Resistivity of copper alloys containing one atom per cent of elements in the same period, versus $(\Delta z)^2$ (after J. O. Linde, *Ann. Phys.* **15**, 219 (1932)). The linear relationship predicted by Norbury and Linde is observed.

counteracted by a decrease in magnetic scattering as the moments are randomised. Manganin, an alloy of copper and manganese, is an example of this.

The influence of solute size on the resistivity of alloys may be illustrated by reference to Fig. 13.7(a) and (b) which show the effect of various additions to niobium and tantalum respectively. For a given concentration of solutes of the same valency (hafnium and titanium) the one having the larger atomic radius (hafnium) has the greater effect on the resistivity. Also shown is the effect of

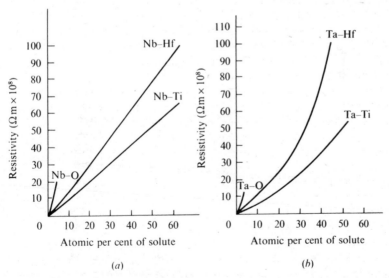

(a)   (b)

Fig. 13.7. Resistivity of alloys of (a) niobium and (b) tantalum with hafnium and titanium (at 1.2 °K) and with oxygen (at 10 °K). The effect of difference in atom size is obvious. (From W. DeSorbo, *Phys. Rev.* **130**, 2177 (1963).)
(a) Nb = 0.294 nm, Hf = 0.317 nm, Ti = 0.293 nm.
(b) Ta = 0.294 nm, Hf = 0.317 nm, Ti = 0.293 nm.

the interstitial solute oxygen, which, at low concentrations, is very much greater than that for a similar quantity of substitutional solute. The effect of solute size was attributed by Blatt, in 1957, to an effective valence of the solute z^*, which is the true valence z plus a volume correction term. According to this theory

$$z^* = z - \left(\frac{\delta V}{V}\right) z_0$$

where $\delta V/V$ $(=3\Delta a/a)$ is the relative change in volume due to the presence of the solute and z_0 is the valence of the solvent. The volume correction term arises as follows. The presence of the solute atom causes a local dilatation of the lattice. If the solute atom is now replaced by a solvent atom, without allowing the lattice to relax, then the charge on the solvent atom, z_0, must be spread over a larger volume, $V + \delta V$, and hence the charge density is reduced by $(\delta V/V)z_0$. A smaller solute would produce an increase in charge density. The valency difference to be used in the Norbury–Linde relationship should be:

$$\Delta z^* = z^* - z_0 = z - z_0 - \left(\frac{3\Delta a}{a}\right) z_0.$$

and for solutes and solvents of equal valence, resistivity should be proportional to Δa^2.

The above treatment for size and valence differences is simple, but seems to hold fairly well if solvent and solute are not too far apart in the periodic table. A rigorous treatment of the resistivity of alloys must, however, take into account changes in phase of the electrons upon scattering, and can only be done by using advanced quantum mechanical concepts.

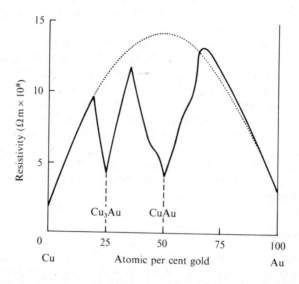

Fig. 13.8. Room temperature resistivity of copper–gold alloys as a function of composition. Annealed samples (——) show ordering at compositions corresponding to Cu_3Au and CuAu. Quenched samples (· · · ·) are fully disordered, and their resistivity obeys the Nordheim relation.

An alloy which undergoes an order–disorder transformation will have a much lower resistivity in the ordered than in the disordered state. In the ordered state a new, super-lattice of high regularity is formed and scattering is reduced. This effect is shown for copper–gold alloys in Fig. 13.8; on slow cooling ordering occurs at the Cu_3Au and $CuAu$ compositions. The dotted line, for quenched alloys, in which the ordering reaction is suppressed, corresponds very closely to the function $c(1 - c)$. The solid line, for the slow-cooled alloys, shows very sharp minima at the ordering compositions.

13.6 Resistivity of the transition metals

The resistivity of the transition metals, that is, those with unfilled d-bands is anomalously large. The overall density of states at the Fermi level is high, of the order of ten times that of a normal metal. Equation (13.6) suggests that the electrical conductivity will therefore be high. This, however, is not the case: a reference to Table 13.1 shows that the transition metal iron has a resistivity greater, and hence a conductivity lower, by a factor of six than the non-transition metal copper. This is due to two facts. The d-electrons have a very low mobility which more than outweighs their high density of states, and as a result they contribute very little to the conduction process. The 4s and 4p-electrons should still ensure a conductivity comparable with non-transition metals, but they do not do so because their mean time between collisions (τ) is short.

Previously τ has been assumed to depend only upon the velocity of the electrons and the distance between the scattering centres. However there is a further factor, the probability that an electron will interact and be scattered by any given centre. The velocity of the electrons is determined by their energies, in this case very near to the Fermi energy. This does not change very much for metals in the same period of the Periodic Table and could not explain the difference already quoted between copper and iron. Likewise, for metals of similar purity the distance between scattering centres will not be very different. The only quantity which can vary is the effectiveness of the scattering centres, which depends, among other things, upon the availability of empty states into which the electrons can be scattered. The high density of states in the transition metal's d-band, into which d-electrons are rarely scattered because of their high effective mass, is available to receive the scattered s–p-electrons which are carrying the current, making the probability of the electrons being scattered high and the mean free paths low. Thus a high density of electrons at the Fermi level produces two conflicting effects: a large number of conduction electrons and a high probability of their being scattered by any centres present. The net result in transition metals is a high resistivity.

13.7 Commercial conductors

The material normally used for applications where high electrical conductivity is required, such as cables, bus-bars, circuit connections etc., is copper. Silver, with a slightly higher conductivity, is normally not used on account of its high price, though it may be in certain applications such as heavy switch gear in which the cost of the material is only a very small fraction of the total cost.

Though the conductivity of aluminium is only about half that of copper it is finding increasing application in cables and overhead conductors. This is because the density of aluminium is one third that of copper, and weight for weight an aluminium conductor shows a higher overall conductivity than does a copper conductor, though its cross-section will be considerably larger.[†] Railway electrification is now proceeding with overhead conductors of steel-cored aluminium. The steel core provides strength. The lower weight of this conductor compared with copper allows of considerable simplification in the support structures.

High conductivity copper must be very pure and free from inclusions. The presence of oxygen has a serious effect on conductivity, as do all the elements normally used as deoxidisers. Oxygen-free high conductivity copper (OFHC) is produced using most stringent precautions against contamination and is melted under a protective layer of charcoal. Deoxidation by elements such as phosphorus is undesirable as residual phosphorus will go into solid solution and lower the conductivity.

The conductivity of electrical materials is usually measured as a percentage of the International Annealed Copper Standard (per cent IACS). This standard is a material of which the resistance of a wire one metre in length and weighing one gramme is 0.153 28 ohm at 20 °C. This is equivalent to a resistivity of 1.7×10^{-8} ohm m, the resistivity of the purest copper commercially available at the time the standard was set up. Currently high conductivity copper is available with a conductivity of 103 per cent IACS. Pure silver is 106 per cent IACS and pure aluminium 60 per cent IACS.

For some applications, such as resistors for dissipating energy or for heating elements, a high resistivity is desirable. This is usually achieved with alloys of transition elements, often in the form of cast iron alloys. Heating elements for furnaces are made from alloys such as nichrome, a nickel chromium alloy, which combines a high electrical resistance with resistance to oxidation at high temperatures.

13.8 Hall effect

If a conductor carrying an electric current has a magnetic field placed across it at an angle to the current, as shown in Fig. 13.9, a potential is developed across the conductor normal to both the current and field. This effect (discovered in 1879) is called after Hall and the developed potential is the Hall potential. If the current density is J, the magnetic field is H and the angle between the directions of J and H is θ, the Hall potential is given by:

$$\mathscr{E}_H = RJH \sin \theta \tag{13.14}$$

in a direction at right angles to both J and H, given by the right-hand screw rule. The constant of proportionality R is the Hall coefficient.

The Hall effect arises from the (Lorentz) force experienced by a moving charge in a magnetic field, which acts in a direction normal to both the direction of motion of the charge and the magnetic field. (An electric motor operates on

[†] Aluminium is now cheaper than copper.

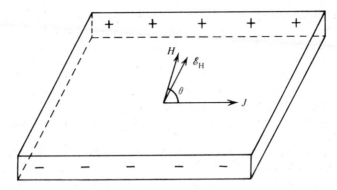

Fig. 13.9. The Hall effect. The magnetic field H deflects the moving electrons to the edges of the conductor, where the resulting accumulation of charge produces the Hall potential $\mathscr{E}_H$.

this principle. It is interesting to note that, prior to Hall's experiment, Maxwell had written that the force acted on the conductor, not on the electrons.) The force on a moving electron is:

$$F_H = -evH \sin \phi$$

where v is the velocity, ϕ is the angle between H and v. The path of an electron moving in a magnetic field is curved until a charge is built up on the faces of the conductor normal to this force, as shown in the figure. The charge induces an electric field (the Hall potential) $\mathscr{E}_H$ which produces a force $F_E = e\mathscr{E}_H$ on the electrons such as to counteract the force due to the magnetic field. For a steady current flow, these two opposing forces must cancel:

$$e\mathscr{E}_H = e\bar{v}H \sin \phi.$$

Using the value of $\bar{v}$ given by (13.5), and substituting in the above:

$$\mathscr{E}_H = \frac{1}{ne}JH \sin \phi.$$

Now J and $\bar{v}$ are in the same direction, so ϕ is equal to θ, and a comparison with (13.14) gives the Hall coefficient

$$R = \frac{1}{ne} \tag{13.15}$$

where n is the number of conduction electrons per unit volume. n is large ($\sim 10^{29}$ electrons/m³) for most metals, and the Hall coefficient is small. It is, however, of technological importance. In the Faraday disc experiment, a copper disc is rotated between the poles of a magnet. The electrons on the copper atoms are moved at right angles to the field by the motion of the disc, and a voltage is developed between the axis and the periphery of the disc. Large homopolar devices have been developed on this principle, to act as DC current generators. Conversely the application of a DC voltage to such a device causes it to act as a motor. Magnetic pumps for circulating liquid metals and other conducting fluids are also based on the Hall effect, as is magnetohydrodynamic generation of power. In this latter, a highly ionised, and therefore conducting, gas (or plasma) is passed at high velocity through a transverse magnetic field.

The Hall voltage developed across the plasma is sufficiently large to act as a source of electric power.

It will be seen in the next chapter that the Hall effect is also important in semiconductors, where the effect is many orders of magnitude larger, due to the much smaller number of charge carriers in a semiconductor.

The application of a magnetic field has an effect additional to the Hall effect; the electrical resistance is found to increase. This is called *magnetoresistance*. Its explanation, which can only be attempted by the use of advanced quantum mechanics, will not be given here.

Experimentally it is found that the proportional increase in resistance due to the magnetic field H is given by

$$\frac{\Delta\rho}{\rho} = aH^2,$$

valid for fields up to ~ 10 T. Beyond this field $\Delta\rho/\rho$ varies linearly with H. Values of a for metals are typically $\sim 10^{-16}$ m^2/A^2. For a metal with a perfectly spherical Fermi surface there should be no magnetoresistance, $a = 0$. The magnitude of a is a measure of the asymmetry, or departure from perfect sphericity, of the Fermi surface. In single crystals, the magnetoresistance is anisotropic, depending upon the crystallographic direction along which the field is applied, as well as that along which the resistance is measured.

13.9 Thermoelectricity

When two dissimilar metals are placed in contact, a potential difference is set up between them. If a second junction is made by joining their opposite ends, so as to establish a closed circuit, the contact potential difference at the second junction is just equal and opposite to that at the first junction, provided that the two junctions are at the same temperature, and no net e.m.f. appears in the circuit. If the junctions are not at the same temperature, the two contact potentials are not equal, and an e.m.f. will exist which can drive current around the circuit. This thermoelectric e.m.f. was first discovered by Seebeck in 1821, and the effect is called the Seebeck effect. Its magnitude is of the order of 10^{-6} V per degree temperature difference between the two junctions.

The converse of the Seebeck effect is the Peltier effect (1834), in which current passed through a junction between dissimilar metals causes either the absorption or the emission of heat. If the sign of the Seebeck effect is such as to cause the current to flow from metal A to metal B at the hot junction, then the passage of a current in this direction by means of an external e.m.f. will produce a cooling effect at this junction by the Peltier effect.

Thomson, as a result of a thermodynamic analysis of these two effects, predicted (1854) a third effect, an e.m.f. resulting from a temperature gradient along a conductor. The existence of the Thomson effect was confirmed experimentally in 1856.

The qualitative explanation in terms of electron theory is quite simple. A quantitative description will not be attempted. The energy required to remove an electron from the top levels of an energy distribution in a crystal to infinity is called the work function, ϕ, and is of the order of a few electron volts. It is different for different metals, and varies with temperature. If two metals, A and B, with work functions ϕ_A and ϕ_B such that $\phi_A < \phi_B$, are placed in contact, electrons will flow from A to B. This flow of electrons creates a potential difference between A and B, causing the Fermi level of B to be raised to equal that of A. When the two Fermi levels are equal, the flow of electrons ceases, and the resulting potential difference, $\phi_B - \phi_A$, is called the contact potential. Obviously, if the two pieces of metal are joined at each end, the contact potential will be the same, provided the temperature is the same, and the potential differences will be equal and opposite. Since the work function varies with temperature, and since this temperature variation is different for different metals, the contact potential also varies with temperature. It is this variation of contact potential which is responsible for the Seebeck effect.

When a current is forced, by an external e.m.f., against the potential difference at a junction, the electrons comprising the current must receive additional energy to overcome the potential. This energy can be supplied thermally, causing the junction to cool. This is the Peltier effect. When a temperature gradient exists along a conductor the average kinetic energy of those electrons with energies greater than the Fermi energy is higher at the hot end than at the cold end. As a result more electrons travel from the hot to the cold end than *vice versa*, until the resulting potential along the conductor prevents further diffusion of electrons. This is the origin of the Thomson effect.

The Thomson coefficient σ_A is defined such that the potential difference in conductor A over an element of length with a temperature difference dT is $\sigma_A\,dT$, and is positive if the direction of the e.m.f. is from cold to hot. The Peltier coefficient Π_{AB} is defined by stating that when a quantity of charge q coloumbs passes from metal A to metal B, Πq joules of heat are absorbed. The thermoelectric power of a circuit comprising two junctions between metals A and B is dV/dT, where dV is the thermoelectric e.m.f. generated by a temperature difference dT between the two junctions. It can be shown thermodynamically that

$$\frac{dV}{dT} = \frac{\Pi_{AB}}{T} = S_B - S_A,$$

where S is the absolute thermoelectric power for the metal, at the temperature T, and is given by

$$S = \int_0^T \frac{\sigma}{T}\,dT.$$

Values of S for metals are of the order of 10^{-6} V/deg, but are higher for semiconductors (e.g. S for germanium at room temperature is $\sim 10^{-3}$ V/deg).

The main use of thermoelectricity is in temperature measurement. If one junction of a circuit between dissimilar metals is maintained at a standard temperature (usually $0\,°C$) then the e.m.f. in the circuit can be used to determine the temperature of the other junction. Such a device is called a thermocouple. Different combinations of metals are used over different temperature ranges; e.g. copper–constantan (60% Cu + 40% Ni) from -200 to $+400\,°C$, chromel (90% Ni + 10% Cr)–alumel (95% Ni + 5% Al) from 0 to $1000\,°C$, platinum–platinum + 13% rhodium up to $1700\,°C$. Some very dilute alloys of transition metals in noble metals show what is called 'superthermopower' at very low temperatures of a few °K (e.g. for Au + 0.03% Fe, S $\simeq 10^{-5}$ V/deg) and are used for thermometry at cryogenic temperatures. The thermoelectric properties of metals are affected by metallurgical variables in a poorly understood fashion. Once a thermocouple has been calibrated, great care must be taken not to subject it to mechanical strain.

The Peltier effect can be made use of to provide refrigeration. Maximum efficiency requires a high thermopower and low thermal conductivity, a combination only to be found in semiconductors. Even the best efficiencies are too low for normal use, and thermoelectric refrigeration is only resorted to when cooling of a small volume by the simplest possible means is required.

13.10 Superconductivity

While investigating the electrical resistivity of metals at very low temperatures, H. Kammerlingh Onnes in 1911 discovered that, at 4.1 °K, the resistivity of mercury dropped suddenly to zero. He was able to demonstrate that the resistance was truly zero by the fact that a current induced in a ring of mercury did not decay with time, up to a period of several months. He called this phenomenon 'superconductivity', and was soon to show that it existed in several metallic elements. Each superconductor is characterised by a critical temperature, T_c. Above T_c, superconductors behave exactly as normal metals, their resistivity increasing with increasing temperature. At T_c, the resistance drops sharply to zero and remains zero at all temperatures below T_c.

At any temperature below T_c, the application of a magnetic field greater than a certain value H_c will destroy superconductivity and return the material to the normal state. The critical field varies with the material and also with temperature, falling from a value $H_c(0)$, usually a few hundred oersted, to zero at the critical temperature. The temperature variation of the critical field approximates closely to a parabolic law:

$$\frac{H_c}{H_c(0)} = 1 - \left(\frac{T}{T_c}\right)^2.$$

The ability of a superconductor to carry current without resistance, and hence transmit power without loss, is naturally of great interest to electrical

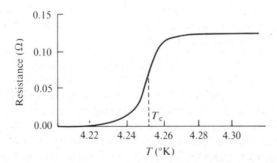

Fig. 13.10. Electrical resistance of mercury, showing the superconducting transition (after H. K. Onnes, *Leiden Comm.* **1226** (1911)).

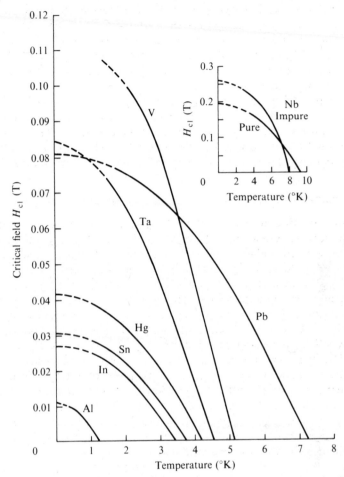

Fig. 13.11. Critical field versus temperature for some elemental superconductors. (After V. L. New-house, *Applied Superconductivity*, Wiley, 1964.)

engineers. Unfortunately a superconductor can carry only a limited current without resistance being restored; the Silsbee hypothesis states that the critical current I_c is that which will create the critical field at the surface of the super-conductor. For a circular wire of radius a,

$$I_c = \tfrac{1}{2}aH_c.$$

A further effect in superconductors is that a magnetic field is completely excluded from the body of a superconducting material. This is known as the Meissner effect. The magnetic induction does not fall abruptly to zero at the surface, but exponentially decreases over a characteristic distance λ, the penetra-tion depth, from the surface. λ is typically 5×10^{-8} m. The field is excluded by supercurrents which flow in the penetration layer so as to produce a field within the superconductor which cancels exactly the applied field.

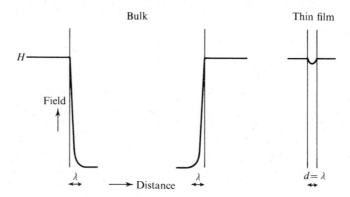

Fig. 13.12. Penetration of an applied magnetic field into bulk and thin film superconductors.

Below T_c the free energy of the superconducting state is less than that of the normal state by an amount Δg_{ns} per unit volume, the superconducting condensation energy. The exclusion of a field H raises the free energy by an amount $\mu_0 H^2/2$ per unit volume. The critical field is the field at which the two terms balance: $H_c = (2\Delta g_{ns}/\mu_0)^{\frac{1}{2}}$. This is only accurate if the dimensions of the superconductor are large compared with λ. The field will penetrate extensively into a thin film or filament of thickness d equal to or less than λ. Superconductivity will then persist to a higher critical field H_d given by

$$H_d \simeq H_c \frac{\lambda}{d}.$$

13.11 Theories of superconductivity

A proper explanation of superconductivity cannot be attempted without the use of advanced quantum mechanics. In the superconducting state an attractive interaction causes some of the valence electrons to condense into pairs. The loss of resistance is not due to a disappearance of the normal scattering processes, but to the reaction of the electron pairs to scattering. If one member of a pair is scattered and suffers a change in momentum, the interaction which binds the pair is such that the other member changes its momentum so as to keep the total momentum of the pair constant. The net electron momentum remains constant despite scattering, and current flows without resistance.

The Bardeen, Cooper, Schrieffer (BCS) theory (1956), now the accepted theory of superconductivity, is able to treat the formation and behaviour of these pairs in quantum mechanical formulation. It predicts *inter alia* that the condensation of electrons into pairs causes an alteration of the energy levels near the Fermi surface, and an energy gap appears. The value of the energy gap at $0\,^\circ$K is given by the theory as $2\Delta(0) = 3.53kT_c$, and is approximately equal to 10^{-22} J. The existence of this gap has been demonstrated and many superconductors are found to agree fairly well with BCS, though there are some deviations. The size of the energy gap decreases with increasing temperature, falling rapidly to zero at T_c.

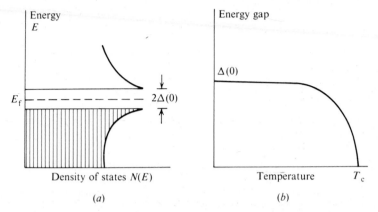

Fig. 13.13. The energy gap in the excitation spectrum of a superconductor, according to the BCS theory.

F. and H. London gave the first phenomenological description of super-conductivity in 1935. The electrodynamics of superconducting electrons differ from those of normal electrons in that there is no destruction of momentum by collisions. Rewriting (13.2) and (13.5) for superelectrons.

$$\frac{\mathrm{d}P_\mathrm{s}}{\mathrm{d}t} = N_\mathrm{s}m\frac{\mathrm{d}V_\mathrm{s}}{\mathrm{d}t} = N_\mathrm{s}e\mathscr{E}$$

and

$$J_\mathrm{s} = n_\mathrm{s}eV_\mathrm{s}.$$

Differentiating with respect to time:

$$\frac{\mathrm{d}J_\mathrm{s}}{\mathrm{d}t} = n_\mathrm{s}e\frac{\mathrm{d}V_\mathrm{s}}{\mathrm{d}t} = \frac{n_\mathrm{s}e^2\mathscr{E}}{m}. \tag{13.16}$$

This equation shows that an electric field only appears in a superconductor when the supercurrent density is changing; when J_s is constant, $\mathscr{E} \equiv 0$.

In a superconductor, just as in any other medium, Maxwell's electrodynamic equations must be obeyed. In particular:

$$\frac{\mathrm{d}B}{\mathrm{d}t} = -\nabla \times \mathscr{E}, \tag{13.17a}$$

$$\nabla \cdot B = 0, \tag{13.17b}$$

$$\nabla \times B = \mu_0 J_\mathrm{s}. \tag{13.17c}$$

∇ is a vector operator;

$$\nabla = \frac{\partial}{\partial x}i + \frac{\partial}{\partial y}j + \frac{\partial}{\partial z}k.$$

From (13.16) and (13.17a):

$$\frac{\mathrm{d}B}{\mathrm{d}t} = -\frac{m}{n_\mathrm{s}e^2}\nabla \times \frac{\mathrm{d}J_\mathrm{s}}{\mathrm{d}t},$$

and from (13.17c)

$$\frac{d\boldsymbol{J}_s}{dt} = \frac{1}{\mu_0}\frac{d}{dt}(\nabla \times \boldsymbol{B}).$$

Hence

$$\frac{d\boldsymbol{B}}{dt} = -\frac{m}{\mu_0 n_s e^2}\nabla \times \frac{d}{dt}(\nabla \times \boldsymbol{B}). \tag{13.18}$$

Now, if the space and time dependence of B are independent of one another, (13.18) can be rewritten

$$\frac{d\boldsymbol{B}}{dt} = -\frac{m}{\mu_0 n_s e^2}\frac{d}{dt}(\nabla \times \nabla \times \boldsymbol{B}),$$

and integration with respect to time yields:

$$\boldsymbol{B} = -\lambda_L^2 \nabla \times \nabla \times \boldsymbol{B}$$

where

$$\lambda_L = \left\{\frac{m}{\mu_0 n_s e^2}\right\}^{\frac{1}{2}}. \tag{13.19}$$

Now, it is a vector identity that

$$\nabla \times \nabla \times \boldsymbol{B} \equiv \nabla(\nabla \cdot \boldsymbol{B}) - \nabla^2 \boldsymbol{B}$$

and since

$$\nabla \cdot \boldsymbol{B} = 0 \tag{13.17b}$$

then

$$B = \lambda_L^2 \nabla^2 B.$$

This is a differential equation with a solution of the form:

$$B(x) = B(0)\exp(-x/\lambda_L).$$

If the value of the magnetic induction at the surface of a superconductor is $B(0)$, it falls exponentially with increasing depth x into the superconductor, and is effectively zero for $x > \lambda_L$. This treatment predicts the Meissner effect, with λ_L the penetration depth.

This theory is a 'local' theory, in that it supposes that the current density at a point is related to the magnetic field at that point. Pippard pointed out in 1950 that the superconducting electrons cannot change their behaviour instantaneously, but that any change must take place over a 'range of coherence', ξ. The range of coherence is interpreted on the BCS theory as the average distance between paired electrons, and for pure metals,

$$\xi_0 = \frac{0.18\,hv_f}{kT_c},$$

and has a value of about 10^{-6} m. The current density at a point is now related to the value of the magnetic field averaged over a sphere of radius ξ centred on that point. Further development of this 'non-local' theory shows that for 'dirty' superconductors, in which the normal electron mean free path l, as defined in § 13.2, is $<\xi_0$, the actual values of the range of coherence and the penetration depth are modified

$$\xi \simeq (\xi_0 l)^{\frac{1}{2}}$$

and

$$\lambda \simeq \lambda_L \left(\frac{\xi_0}{l}\right)^{\frac{1}{2}}.$$

It can be seen that, as l is reduced by alloying, ξ is also reduced and λ is increased, and for very dirty alloys ξ may be as small as 5×10^{-9} m, and λ as large as 10^{-7} m. Superconductors in which $\lambda > \xi$ behave differently from those in which $\lambda < \xi$; the former are called type II superconductors and the latter type I superconductors. Type II superconductors show an incomplete Meissner effect; magnetic fields are only partially excluded from the body of the superconductor. The magnetic energy term is reduced, and the material can remain superconducting up to very high magnetic fields. A reversible type II material behaves like an ideal superconductor, with a complete Meissner effect, up to a field H_{c_1}. Beyond H_{c_1} magnetic flux enters the specimen as quantised supercurrent vortices each carrying one quantum of magnetic flux $\phi_0 = h/2e$ ($= 2 \times 10^{-15}$ Wb). As the external field is raised, more of these vortices are created within the specimen until at a field H_{c_2} they begin to overlap and the material is no longer superconducting. The radius of the vortex core is $\simeq \xi$, and thus:

$$H_{c_2} = \frac{\phi_0}{2\pi\mu_0\xi^2} \simeq \frac{\phi_0}{2\pi\mu_0\xi_0 l}.$$

Since resistivity is also inversely proportional to l, H_{c_2} is directly proportional to the normal state resistivity.

13.12 Superconducting materials

The majority of applications, actual and proposed, for superconductors requires high critical temperatures, high critical fields, and high critical current densities. The operating temperature of the superconductor should normally be not greater than $\frac{1}{2}T_c$, and a high critical temperature therefore reduces the refrigeration cost. In addition the other two parameters roughly scale with T_c. A high critical field allows the superconductor to operate in a high field environment – important as the main use for superconductors is in the generation of large magnetic fields. High current densities enable compact devices to be constructed, and also reduce the quantity of expensive superconductor required for a particular application.

Twenty-four of the elements are superconductors, their distribution in the Periodic Table and critical temperatures are shown in Fig. 13.14. Of these, niobium has the highest T_c, 9.2 °K, and with vanadium is the only pure metal which is type II. Beryllium, bismuth and germanium when deposited as thin films below 10 °K are superconducting; bismuth is also superconducting when subjected to pressures greater than 2×10^9 N/m². Approximately 1000 alloys and compounds are known to be superconductors, including some materials which are semiconducting at normal temperatures. The highest critical temperatures are found in compounds of the type Nb_3X or V_3X, where X can be Ga, Al, Si, Ge, or Sn. These compounds all have the same cubic crystal structure, and form with difficulty at the stoichiometric proportions which are a requirement for maximum T_c. A ternary compound, $Nb_3(AlGe)$ has recently been prepared with the highest reported T_c of 20.75 °K. There seems at present no likelihood of raising T_c above about 25 °K.

High critical fields can be achieved in type II superconductors by increasing the normal state resistivity by alloying. Alloys based on niobium have values

IA	IIA		IIIB	IVB	VB	VIB	VIIB	VIII	VIII	VIII	IB	IIB	IIIA	IVA	VA	VIA	VIIA	0
H																		He
Li	Be* 6–8.4												B	C	N	O	F	Ne
Na	Mg												Al 1.20	Si	P	S	Cl	A
K	Ca	Sc	Ti 0.39	V 5.30	Cr	Mn	Fe	Co	Ni	Cu	Zn 0.88	Ga 1.09	Ge* 8.4	As	Se	Br	Kr
Rb	Sr	Y	Zr 0.55	Nb 9.13	Mo 0.92	Tc 8.22	Ru 0.49	Rh	Pd	Ag	Cd 0.56	In 3.40	Sn 3.72	Sb	Te	I	Xe
Cs	Ba	La (α) 5.0 (β) 6.3	→Hf	Ta 4.48	W 0.1	Re 1.70	Os 0.66	Ir 0.14	Pt	Au	Hg (α) 4.15 (β) 3.94	Tl 2.39	Pb 7.19	Bi* ~6	Po	At	Rn
Fr	Ra	Ac																

←——— Transition elements ———→

Rare earths	Ce	Pr	Nd	Pm	Sm	Eu	Gd	Tb	Dy	Ho	Er	Tm	Yb	Lu
Actinides	Th 1.37	Pa	U (α) 0.68 (γ) 1.8	Np	Pu	Am	Cm	Bk	Cf	E	Fm	Mv	102	Lw

Fig. 13.14. The occurrence of superconducting elements in the Periodic Table. Each superconductor is outlined, and its critical temperature (in °K) is given. * These elements are superconducting only in the amorphous state (bismuth, under very high pressure, is also a superconductor).

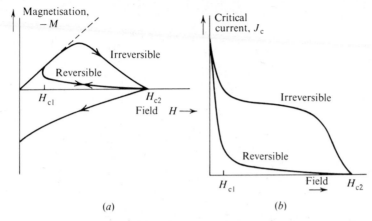

Fig. 13.15. (a) Magnetisation and (b) critical current curves for reversible and irreversible (hard) type II superconductors as a function of applied magnetic field.

of $H_{c_2}(0)$ of 10–15 T; for the high T_c compounds mentioned above $H_{c_2}(0)$ is approximately 20 T, and double this for the ternary $Nb_3(AlGe)$ compound. It is found in practice that increasing the resistivity will not increase $H_{c_2}(0)$ indefinitely. These superconducting materials are all based on transition metals which are strongly paramagnetic; in high fields this paramagnetism reduces the free energy of the normal state to below that of the superconducting state and thus puts an upper limit on the field to which superconductivity can be sustained. Maximum critical fields (in teslas) for the most strongly superconducting materials are 1.5 to 2 times the critical temperature in °K.

Supercurrent vortices in type II superconductors interact with microstructural features such as precipitates or crystal defects introduced by mechanical deformation producing irreversible hysteresis effects in the superconducting properties. For reasons not discussed here, it is these irreversible effects which allow type II superconductors to carry very high currents. These 'hard' superconductors can carry DC currents of nearly 10^{10} amp m^2 in fields of up to 10 T; their commercial importance is obvious. Because of their irreversible properties, hard superconductors are not able to carry large AC currents and the AC applications for superconductors are therefore limited. A serious problem with hard superconductors is that a slight electrical, magnetic, mechanical or thermal disturbance can cause sudden movements of the vortices within the superconductor. These movements, called 'flux-jumps', are capable of rendering the superconductor normal, and could be serious if the superconductor were carrying a heavy current. The material can be 'stabilised' by incorporating it, in the form of fine filaments, in a matrix of normal metal of good conductivity (usually copper). The normal metal acts as a path for the current until the disturbed region of the superconductor has recovered.

Three materials have so far been commercially successful as hard superconductors; the alloys Nb–(25–50 at.%)Zr ($T_c \simeq 11$ °K, $H_{c_2} \simeq 9$ T) and Nb–(40–65 at.%)Ti ($T_c \simeq 10$ °K, $H_{c_2} \simeq 12$ T) and the intermetallic compound Nb_3Sn ($T_c = 18.3$ °K, $H_{c_2}(0) = 24.5$ T). A combination of cold deformation

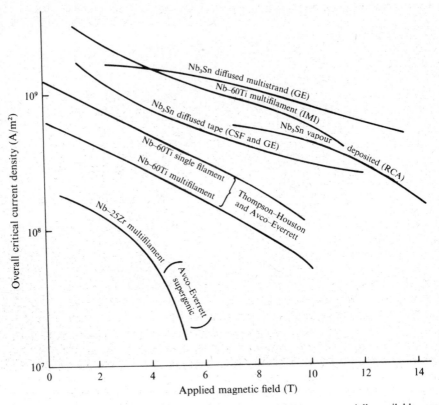

Fig. 13.16. Critical current density versus applied magnetic field, for commercially available super-conducting materials (abstracted from manufacturer's literature). The current density is measured over the overall cross-section of the conductor.

and heat treatment of the alloys produces a microstructure consisting of a finely dispersed precipitate of nitrides, oxides and carbides and dislocation tangles which interact with the vortices to produce the irreversible behaviour necessary for a high current material. The alloys can be co-drawn with copper to provide the necessary stability. The Nb_3Sn compound is very brittle and its technology difficult. Flexible tape may be produced by vapour deposition onto a thin substrate, or by the controlled diffusion of tin into a niobium tape. These processes are more difficult to control, and more costly, than those for producing the alloys, but the currents and fields available are greater.

Superconductors have been suggested for logic and memory circuits in computers, frictionless suspension, motors, generators, transformers, heat switches, radiation detectors and contactless reversing switches. Superconduct-ing amplifiers have been designed, and superconducting devices have been made which can detect very low voltages ($\sim 10^{-14}$ volts), low currents ($\sim 10^{-16}$ amps) and low fields ($\sim 10^{-13}$ T). The only commercial application to date has been of the hard superconductors in solenoids for producing high magnetic fields. A solenoid has been built to produce a field of 4 T in a bore of 0.3 m diameter

Fig. 13.17. A small laboratory superconducting magnet (3 tesla in 40 mm bore) being removed from its cryostat.

over a length of 3 m. Over 10 T has been produced in a 25 mm bore. A conventional copper solenoid would require about 2 megawatts of power and a cooling flow of tens of litres of demineralised water per second to produce this latter field. The power required for a superconducting solenoid is almost zero, and in a well-designed cryostat the liquid helium consumption can be less than one litre per hour. Superconducting solenoids are used to provide high fields for various experimental investigations. They are also used for beam control in particle accelerators, and to provide the magnetic field for homopolar motors. The possibility of power transmission by underground superconducting cables is being actively pursued.

QUESTIONS

1. Copper is an fcc metal with a density of 8.9×10^3 kg/m^3, atomic weight 64. From data given in Chapters 3 and 13, and assuming one free electron per atom, calculate
   (a) The Fermi energy of copper.
   (b) The density of conduction electrons.
   (c) The average velocity of the conduction electrons.
   (d) The electron mean free path.
   (Assume $m^* =$ the free electron mass, 9.1×10^{-31} kg.)

2. A copper strip, 1 mm thick by 10 mm wide, carries a current of 100 A. Using the answers from the above problem, calculate the average drift velocity of the electrons. How does this compare with their Fermi velocity, and the speed with which information could be communicated if each electron were a source of electromagnetic radiation?

3. A sample of pure silver has a resistivity of 1.6×10^{-8} Ωm at 300°K and 3.1×10^{-8} Ωm at 600°K. Nichrome, an alloy of nickel, iron and chromium, has a resistivity of 1×10^{-6} Ωm at 300°K and 1.05×10^{-6} Ωm at 600°K. Explain the difference in behaviour of silver and nichrome between the two temperatures. Assuming Mattheissen's rule is valid for both, what are the contributions to resistivity from impurity scattering in both materials?

4. Describe how you would expect the resistivity of (a) a sample of pure aluminium, (b) a sample of an Al–1.5%Cu alloy, to change both during, and for several hours after, quenching from 500 °C to room temperature.

5. How would you (a) decrease, (b) increase the resistivity of any given metal or alloy? How would you go about producing a metal with the maximum possible resistivity?

6. Discuss the relative economics of copper and aluminium conductors for power transmission.

7. V_3Si is a superconductor with a critical temperature of 17 °K. The slope of the upper critical field versus temperature curve is found to be -2.7 T K at 17 °K. Estimate the value of the upper critical field at 0 °K, assuming it to obey a parabolic variation with temperature.

8. Lead is a superconductor with a critical field at 0 °K of 0.08 T, and a critical temperature of 7.2 °K. What is the maximum supercurrent that can be carried by a 0.3 mm diameter lead wire immersed in liquid helium at atmospheric pressure?

9. Tin is a superconductor; at 2 °K the critical field for bulk tin is 0.022 T, and its penetration depth is 5×10^{-8} m. How thick must a thin film of tin be to remain superconducting in a field of 0.75 T?

10. Why are type II superconductors capable of carrying high current densities called 'hard' superconductors? Compare these materials to mechanically hard solids.

FURTHER READING

N. F. Mott and H. Jones: *The Theory of the Properties of Metals and Alloys.* Oxford University Press (1936).

A. H. Wilson: *The Theory of Metals.* Cambridge University Press (1953).

A. J. Dekker: *Electrical Engineering Materials.* Prentice-Hall (1959).

T. S. Hutchinson and D. C. Baird: *The Physics of Engineering Solids.* Wiley (1963).

J. K. Stanley: *Electrical and Magnetic Properties of Materials.* ASM (1963).

V. L. Newhouse: *Applied Superconductivity.* Wiley (1964).

A. C. Rose-Innes and E. H. Rhoderick: *Introduction to Superconductivity.* Pergamon (1969).

D. Saint-James, G. Sarma and E. J. Thomas: *Type II Superconductivity.* Pergamon (1969).

D. Fishlock (ed.): *A Guide to Superconductivity.* Macdonald (1969).

D. Dew-Hughes: The metallurgical enhancement of type II superconductors. *Reports of Progress in Physics* **34**, 821 (1971).

14. Electrical properties 2: semiconductors and insulators

14.1 Introduction

This chapter is concerned with the physics of semiconduction and insulation, and indicates the effect of material parameters upon properties. Semiconductors and insulators are treated in one chapter because, as has already been mentioned, there is no fundamental physical difference between them. At the absolute zero of temperature all non-metallic materials would be perfect insulators; at finite temperatures thermal activation allows conduction processes to occur. In semiconductors these processes occur readily at normal operating temperatures; in insulators appreciable conduction occurs only at excessive temperatures. The chapter finishes with a consideration of dielectrics, insulators which are highly polarisable; under the action of electric fields the electrons may separate slightly from the positive ions and produce electric dipoles.

No attention is given to photoelectric (photoemission, photoconduction and luminescence) and little to mechanoelectric (piezoelectric) effects. This omission is due entirely to lack of space, and should not be interpreted as meaning that they are not important.

14.2 Semiconductors

Semiconducting materials are the basis of the new electronics industry which has developed over the last twenty years around transistors and other solid state devices. The materials requirements for these devices demand a degree of purity and crystalline perfection previously unknown. Techniques have been developed to produce single crystal materials of very low dislocation densities with impurity levels measured in parts per thousand million. The reasons for this will become apparent after a consideration of the electrical properties of semiconductors.

A semiconductor is a solid in which the highest occupied energy band at absolute zero is completely filled (this is called the valence band) and the next band (called the conduction band) is completely empty of electrons. The lowest energy level in the conduction band is separated from the highest energy level in the valence band by only a small gap of forbidden energy levels, the energy gap being of the order of 10^{-19} J (0.6 eV). This is equivalent to saying that the valence electrons are not very tightly bound to their respective atoms, and may readily be excited by thermal, electromagnetic, or other energy into the conduction band where they are free to move through the crystal under the action of an applied potential. The situation in an insulator is exactly the same, except that the energy gap tends to be much larger, e.g. of the order of 10^{-18} J (6 eV) in diamond, and it is therefore extremely difficult to excite electrons into the conduction band.

A consequence of the excitation of an electron into the conduction band is that an empty state, or 'hole', is left behind in the valence band. The process is illustrated in Fig. 14.1. The hole has an energy equal to that of the electron before excitation, and is localised in the sense that its spatial extension is that of the electron. It is, however, mobile; a succession of electrons may move into the hole from one direction causing motion of the hole in the opposite direction. The motion of negatively charged carriers (electrons) in one direction is electrically equivalent to the motion of positively charged carriers in the opposite direction. The hole may therefore be treated as if it were a positive charge carrier with a charge $+e$ (the electronic charge being $-e$), and may even be assigned an effective mass, m^*. This is the mass it must have were its behaviour governed by the classical laws of motion rather than by quantum mechanics (see § 13.2).

Semiconducting behaviour is determined by the number of electrons in the conduction band and the number of holes in the valence band. The electrons and holes may be created by excitation across the band gap, as described above, or by the introduction of suitable impurity elements into the lattice. These processes are referred to as *intrinsic* or *extrinsic* excitation respectively, and the materials in which they occur are intrinsic or extrinsic semiconductors. Intrinsic excitation occurs in all semiconductors but in an extrinsic semiconductor is overshadowed by impurity effects. Most commercially useful semiconductors are extrinsic.

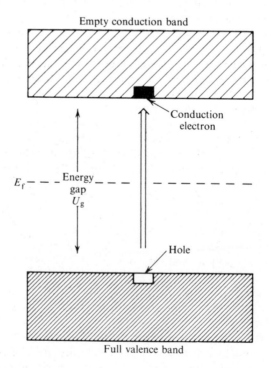

Fig. 14.1. Intrinsic semiconduction: excitation of an electron from the valence band into the conduction band, leaving a hole in the conduction band.

IIIB	IVB	VB	VIB	VIIB
B	C	B	O	F
Al	Si	P	S	Cl
Ga	Ge	As	Se	Br
In	Sn	Sb	Te	I
Tl	Pb	Bi	Po	

Fig. 14.2. The semiconducting elements in the Periodic Table.

Figure 14.2 is a section of the Periodic Table indicating the semiconducting elements. Silicon and germanium are the most frequently used elemental semiconductors. Tin is a semiconductor only in its low temperature modification, grey tin. Bismuth is an interesting material; in bulk it is a semiconductor, but when deposited as a thin film, or under high pressure, it behaves like a superconducting metal. Many compounds are also semiconductors and some of these will be discussed later in the chapter.

The Group IVB semiconducting elements crystallise in the diamond-cubic structure. This has a face-centred cubic lattice, with two atoms per lattice point. Each atom is linked tetrahedrally to four other atoms. Many semiconducting compounds also show this structure or a modification of it. The III–V compounds (see § 14.6) have the zinc-blende structure which is similar to diamond-cubic but the two atoms per lattice point are now of different species. The diamond-cubic and zinc-blende structures are illustrated in Fig. 14.3.

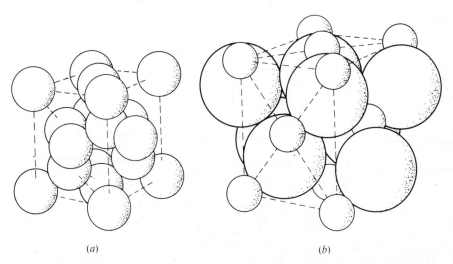

(a)   (b)

Fig. 14.3. (a) Diamond cubic structure; the Bravais lattice is fcc, with two atoms per lattice point. (b) Zinc-blende structure; similar to the diamond cubic, but with atoms of two different species on each lattice point.

14.3 Intrinsic semiconduction

The presence of a small energy gap immediately explains the observed increase of conductivity with temperature in these materials. The excitation energy required to raise electrons from the valence band to the conduction band is supplied thermally and the number of charge carriers, both positive and negative (which will, of course, be equal) increases as the temperature increases, as shown in Fig. 14.4.

The equilibrium number of intrinsic electrons and holes at a given temperature is found in the following way. If n_i electrons per unit volume have been excited into the conduction band then the rate of excitation of more electrons dn/dt is proportional to the number of electrons available for excitation, i.e. the number left in the valence band, times a Boltzmann factor:

$$\frac{dn}{dt} = \alpha(N - n_i) \exp\{-U_g/kT\},$$

where N is the total number of valence electrons per unit volume, U_g is the energy gap and α is a constant. At the same time excited electrons will fall back into the valence band by recombining with the holes; the rate of recombination dn'/dt being proportional to the product of the number of electrons and the number of holes,

$$\frac{dn'}{dt} = -\beta n_i^2$$

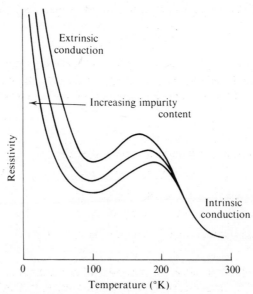

Fig. 14.4. Variation of resistivity with temperature, for several samples of a semiconductor with different amounts of doping (schematic). The resistivity decreases with increasing temperature and impurity content at low temperatures. At $\sim 100\,°K$ all impurity levels are fully ionised, and the resistivity rises due to the temperature dependence of scattering. At $\sim 200\,°K$ intrinsic excitation occurs; resistivity, now independent of impurity content, decreases.

where β is another constant. (The rate of recombination is discussed in § 14.8.) At equilibrium

$$\frac{dn}{dt} + \frac{dn'}{dt} = 0$$

and

$$\frac{n_i^2}{N - n_i} = \frac{\alpha}{\beta} \exp(-U_g/kT). \qquad (14.1a)$$

If, as is usual, $N \gg n_i$, then

$$n_i \simeq \left(\frac{\alpha N}{\beta}\right)^{\frac{1}{2}} \exp(-U_g/2kT). \qquad (14.1b)$$

This treatment is not strictly valid, as not all the electrons have the same energies, from the Pauli Principle. A more accurate estimate of the number of excited electrons, which is equal to the number of holes, is:

$$n_i = 5 \times 10^{15} \, T^{\frac{3}{2}} \exp(-U_g/2kT). \qquad (14.2)$$

For an energy gap U_g of 1.6×10^{-19} J (1 eV) the number of electrons in the conduction band at room temperature ($kT = 4 \times 10^{-21}$ J, 0.025 eV) is $n_i \simeq 5 \times 10^{19}$ electrons per m³, compared to $\simeq 5 \times 10^{28}$ atoms per m³ in most solids. Thus it can be seen that only a small fraction of the valence electrons are excited into the conduction band.

Equation (13.6) gave the conductivity for metals as:

$$\sigma = ne^2\tau/m,$$

where n is the number of electrons per m³, e their charge, m their mass, and τ their mean scattering time. This equation must be modified for semiconductors with two types of charge carrier; the appropriate equation being:

$$\sigma = \frac{n_e e_e^2 \tau_e}{m_e^*} + \frac{n_h e_h^2 \tau_h}{m_h^*}. \qquad (14.3)$$

(The subscripts e and h refer to electrons and holes respectively.) A quantity μ, known as mobility, is defined as the average drift velocity of a charge carrier produced by unit field. From consideration of momentum it is given by:

$$\mu = \frac{e\tau}{m^*}. \qquad (14.4)$$

The formula for conductivity may then be written:

$$\sigma = n_e e_e \mu_e + n_h e_h \mu_h, \qquad (14.5a)$$

or since $e_e = -e_h = e = -1.6 \times 10^{-19}$ coulombs:

$$\sigma = e(n_e \mu_e - n_h \mu_h) \qquad (14.5b)$$

and for an intrinsic semiconductor, where $n_e = n_h = n_i$:

$$\sigma = en_i(\mu_e - \mu_h).$$

The mobility of holes is usually less than that of electrons; for example in pure germanium at room temperature $\mu_e = -0.39$ and $\mu_h = 0.19$ m² V⁻¹ s⁻¹.

Simplified theory indicates that the relaxation time, and hence μ, should be proportional to $T^{-\frac{3}{2}}$. Combining this with (14.2) gives the temperature variation of conductivity:

$$\sigma = \text{const.} \exp(-U_g/2kT).$$

Table 14.1. *Properties of pure Group IVB semiconductors (room temperature).*

Semiconductor	Energy gap (J)	Conductivity $(\Omega^{-1} m^{-1})$	Mobility $(m^2 V^{-1} s^{-1})$	
			electrons	holes
Carbon (diamond)	8.32×10^{-19}	10^{-12}	-0.18	0.12
Silicon	1.93×10^{-19}	5×10^{-4}	-0.14	0.048
Germanium	1.15×10^{-19}	2.2	-0.39	0.19
Tin (grey)	0.128×10^{-19}	5×10^{5}	-0.25	0.24
Lead	no gap	5×10^{6}		

Conductivity is a very sensitive function of the temperature and of the energy gap, as can be seen from the values of room temperature conductivity given in Table 14.1 for elements in Group IVB of the Periodic Table. The strong variation of conductivity with temperature is made use of in the 'thermistor', a device which has many control applications and may be used as a thermometer.

14.4 Extrinsic semiconduction

Charge carriers of either sign may be induced by the presence of certain impurities. If the induced carriers are (negative) electrons the semiconductor is said to be n-type, if (positive) holes, p-type. An impurity concentration of greater than one atom in 10^9 is sufficient to convert germanium from intrinsic to extrinsic behaviour at room temperature. The intentional addition of impurities is known as doping.

If a Group IVB semiconductor such as germanium has a Group VB impurity, say antimony, in substitutional solid solution, then each antimony atom provides one more electron than is required to satisfy the bonding in germanium. The extra electrons are very loosely bound to their antimony atoms, that is, their energy levels are in the germanium energy gap just below the conduction band, by an amount U_d; see Fig. 14.5. The value of U_d is a function only of the host lattice, and is $\sim 1.6 \times 10^{-21}$ J (0.01 eV) for all Group VB elements in germanium, and $\sim 6.4 \times 10^{-21}$ J (0.04 eV) for the same elements in silicon.

These extra energy levels are so close to the conduction band that they can very easily donate their electrons to it leading to n-type semiconduction. The levels are called *donor levels* and impurities which create them are called *donor impurities*. All levels are ionised at normal temperatures, that is all their electrons are donated to the conduction band.

If the antimony in solid solution were replaced by a Group IIIB element, say boron, then each boron atom would have one electron fewer than the four required by the germanium binding. Thus an unfilled energy level or hole, again very loosely bound to its boron atom, is introduced into the energy structure of the material (Fig. 14.6), this time U_a above the valence band. U_a for any Group IIIB element has the same value as U_d for a Group VB element in the same host lattice. Electrons from the valence band are easily excited into these extra, *acceptor*, levels, leaving holes in the valence band to act as positive

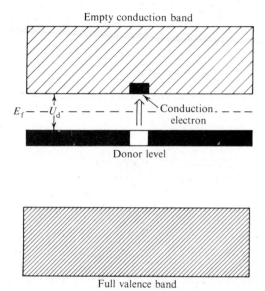

Fig. 14.5. Extrinsic semiconduction; n-type. Electron is excited from donor level into conduction band.

charge carriers. A semiconductor containing *acceptor impurities* is known as *p-type*, and again the levels are fully ionised at normal temperatures. The temperature dependence of the resistivity of extrinsic semiconductors is also shown in Fig. 14.4.

The donor and acceptor levels are localised to the region of the impurity atoms, for low concentration of impurity. This is why electrons or holes in these levels cannot contribute to conductivity. At a sufficiently high concentration of impurities these localised levels begin to overlap, and because of the Pauli Principle they broaden out into bands and the semiconductor is said to be *degenerate*. Degeneracy occurs at an impurity concentration of the order of one hundredth of one atomic per cent.

The number of extrinsic carriers per unit volume is calculated by using exactly the same arguments as for the intrinsic case. However, account must be taken of the intrinsic carriers which will also be present. The total rate of excitation of electrons will be the sum of the rates for intrinsic and extrinsic electrons:

$$\frac{dn}{dt} = \alpha(N - n_i)\exp(-U_g/kT) + \alpha(N_d - n_e)\exp(-U_d/kT)$$

where N_d is the number of donor impurities, and n_e the number of extrinsic electrons in the conduction band, per unit volume.

The rate of recombination

$$\frac{dn'}{dt} = -\beta np = -\beta(n_i + n_e)(n_i + p_e)$$

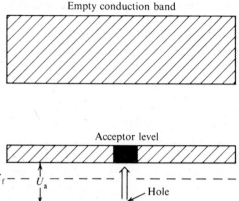

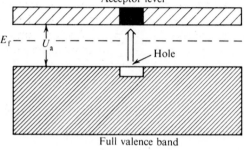

Fig. 14.6. Extrinsic semiconduction; p-type. Electron is excited from valence band into acceptor level, creating a hole in the valence band.

where n, the total number of electrons in the conduction band, is equal to the sum of the intrinsic and extrinsic electrons, and p, the total number of holes in the valence band, is equal to the sum of the intrinsic holes ($= n_i$) and the extrinsic holes, p_e. In the absence of any acceptor impurities, $p_e = 0$, and the condition for equilibrium is:

$$\alpha(N - n_i) \exp(-U_g/kT) + \alpha(N_d - n_e) \exp(-U_d/kT) + \beta(n_i + n_e)n_i = 0.$$

$$(14.6)$$

This expression yields no simple solution for n_e; however, in the temperature range such that $U_d \ll kT \ll U_g$, n_i is small and n_e becomes very nearly equal to N_d. A similar result is found for a semiconductor containing N_a acceptor impurities per unit volume, in which p_e becomes very nearly equal to N_a. It is therefore usual to assume that at temperatures above which $kT > U_d$ (or U_a) (i.e. above liquid nitrogen temperature, 77 °K, for germanium), all of the impurity levels are completely ionised.

At very high temperatures such that $kT \gg U_g$, the number of intrinsic carriers exceeds the extrinsic carriers, and the effect of impurities on carrier density can be ignored.

Extrinsic semiconduction can be obtained by adding any impurity of a valency different from that of the semiconductor. If the valency difference is one, a donor or acceptor level is produced as described above, if it is greater than one then more than one ionisation level is created. The extra levels usually occur deep in the energy gap, and impurities producing these levels are called deep-level impurities. Copper in germanium is a common example of a deep-level impurity, creating an ordinary acceptor level at 6.4×10^{-21} J and a deep acceptor level at 5×10^{-20} J above the top of the valence band.

Deep donor levels are produced in Group IVB semiconductors only by gold and lithium. All other impurities which are soluble in germanium produce acceptor levels. The group VIB elements, which would be expected to produce donor levels, are apparently all insoluble. Deep-level impurities produce very few carriers, and in fact reduce the conductivity by reducing the relaxation time τ through increased scattering.

14.5 Compensation

It has already been stated that intrinsic excitation also occurs in an extrinsic semiconductor. However the presence of extrinsic carriers tends to suppress the production of intrinsic carriers. For example, in a semiconductor containing acceptor impurity atoms the rate of excitation of intrinsic electrons is given by

$$\frac{dn}{dt} = \alpha(N - n_{\mathrm{j}})\exp\left(-U_{\mathrm{g}}/kT\right)$$

where n_{j} is the number of electrons per unit volume excited into the conduction band. The rate of recombination is given by:

$$\frac{dn'}{dt} = -\beta n_{\mathrm{j}}p$$

where $p = N_{\mathrm{a}} + n_{\mathrm{j}}$ is the total number of holes, both intrinsic and extrinsic, assuming complete ionisation of the acceptor levels. At equilibrium:

$$n_{\mathrm{j}}p = (N - n_{\mathrm{j}})\frac{\alpha}{\beta}\exp\left(-U_{\mathrm{g}}/kT\right),$$

which may be compared to the intrinsic case:

$$n_{\mathrm{i}}^2 = (N - n_{\mathrm{i}})\frac{\alpha}{\beta}\exp\left(-U_{\mathrm{g}}/kT\right).$$

With n_{j} and n_{i} much less than N,

$$n_{\mathrm{j}}p = n_{\mathrm{j}}(N_{\mathrm{a}} + n_{\mathrm{j}}) = n_{\mathrm{i}}^2, \tag{14.7}$$

from which it can be seen that n_{j}, the number of intrinsic electrons in a semi-conductor containing acceptor impurities, is less than n_{i}, the number in the same semiconductor, at the same temperature, in the absence of impurities. A similar result is obtained for a semiconductor containing donor impurities.

A more general result, when both acceptor and donor impurities supply extrinsic holes and electrons, is:

$$(n_{\mathrm{e}} + n_{\mathrm{j}})(p_{\mathrm{e}} + p_{\mathrm{j}}) = n_{\mathrm{i}}^2.$$

If the density of donors, N_{d}, and acceptors, N_{a}, is so large that the product $N_{\mathrm{d}}N_{\mathrm{a}} > n_{\mathrm{i}}^2$, then in order that $n_{\mathrm{e}}p_{\mathrm{e}} = n_{\mathrm{i}}^2$, n_{e} and p_{e} must be less than N_{d} and N_{a}. This effect, in which the presence of two types of impurity appears to reduce their degree of ionisation, is known as *compensation*. On account of this effect, measurement of electrical resistivity alone is not a reliable guide to the purity of a semiconductor, but must be combined with Hall effect measurements (see § 14.7).

14.6 Compound semiconductors

In addition to the pure element semiconductors, very many alloys and compounds are semiconductors. These may be formed from semiconducting or non-semiconducting elements. Copper oxide is the oldest semiconductor in commercial use, in copper–copper oxide rectifiers, although their operation is still inadequately understood.

The particular advantage of compounds is that they provide the device engineer with a wide range of energy gaps and mobilities, so that materials are available with properties which match exactly specific requirements. Some compounds are particularly important in that they have high mobilities, without which some of the Hall effect devices mentioned in the next section would not be possible. The compound bismuth telluride is the most successful material for thermoelectric devices.

It is not possible to discuss all of the compounds in detail; the properties of many of them are summarised in Table 14.2. One of the most important classes of compounds is the III–V compounds, formed of elements from Groups IIIB and VB of the Periodic Table. Typical examples are InSb, AlP and GaAs. The latter is one of the most versatile and useful semiconductors, being used in many devices including lasers.

In the III–V materials impurity levels may be produced by deviations from the stoichiometric composition (the ideal composition of the compound, in this case an equal number of atoms of both elements) and control of composition is very important. Group IVB elements also may act as donors or acceptors, depending on whether they replace the Group III or the Group V element in the compound. Similar considerations apply to other compounds.

Table 14.2. *Properties of compound semiconductors*
       (room temperature)

Compound	Energy gap (J)	Conductivity $(\Omega^{-1}\,m^{-1})$	Mobility $(m^2\,V^{-1}\,s^{-1})$ electrons	holes
III–V compounds				
AlP	4.8×10^{-19}			
AlAs	3.7×10^{-19}			
GaP	3.6×10^{-19}		-0.045	0.002
AlSb	2.6×10^{-19}	1000	-0.140	0.020
GaAs	2.1×10^{-19}		-0.85	0.45
InP	2.0×10^{-19}		-0.60	0.016
GaSb	1.1×10^{-19}	500	-0.50	0.085
InAs	0.53×10^{-19}		-2.30	0.010
InSb	0.29×10^{-19}	5000	-7.70	0.075
Other compounds				
CdSb	0.9×10^{-19}			~ 0.10
ZnSb	0.8×10^{-19}	~ 400		0.001
PbS	0.60–0.65×10^{-19}	~ 100	-0.06	0.07
PbTe	0.46–0.51×10^{-19}		-0.18	0.09
PbSe	0.42–0.46×10^{-19}		-0.12	0.10
Bi_2Te_3	0.21×10^{-19}	1.4×10^4	-0.12	0.05
SiC	4.8×10^{-19}		-0.007	0.001

14.7 Hall effect in semiconductors

It was shown in § 13.8 how the application of a magnetic field at right angles to a current flow in a metal conductor causes the electrons to be deflected until the Hall potential is built up which counteracts this deflection. In calculating an expression for the Hall coefficient in semiconductors, account must be taken of both the electron and the hole charge carriers, just as with conductivity, see (14.5). The magnetic field deflects both in the same direction because they are travelling, initially, in opposite directions and have opposite sign charges. Any resulting Hall potential accelerates electrons and holes in opposite directions, and the equilibrium condition is established when the net transverse current is zero, not when the effects of the magnetic and Hall fields balance on both types of carrier individually (which is impossible).

Let the current density in the x-direction of a slab of semiconductor be J_x, the magnetic field H_z be applied in the z-direction, and the Hall potential $\mathscr{E}_y$ be generated in the y-direction (see Fig. 14.7). The current in the x-direction results from the electrons and holes having components of velocity v_{ex} and v_{hx} respectively in the x-direction. The combined action of $\mathscr{E}_y$ and H_z produces velocities of v_{ey} and v_{hy} respectively in the y-direction. Using the concepts and notation of § 14.3, the current density in the x-direction is

$$J_x = e(n_h v_{hx} - n_e v_{ex}) \tag{14.8}$$

and in the y-direction is

$$J_y = e(n_h v_{hy} - n_e v_{ey}).$$

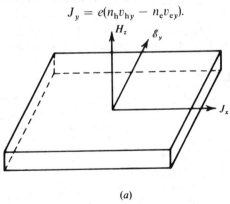

(a)

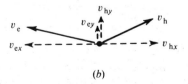

(b)

Fig. 14.7. The Hall effect in a semiconductor. (a) The relation between current, magnetic field and Hall potential. (b) Vector diagram showing the velocities of electrons and holes in the x and y-directions.

As has already been stated, J_y is equal to zero at equilibrium, and

$$n_h v_{hy} = n_e v_{ey}. \tag{14.9}$$

In the y-direction the combined accelerative effect of the Hall potential and the magnetic field on the carriers must be balanced by the deceleration due to scattering. This gives, for electrons:

$$-\mathscr{E}_y e - e v_{ex} H_z = m_e^* v_{ey}/\tau_e,$$

from which

$$v_{ey} = -e(\mathscr{E}_y + v_{ex} H_z)\tau_e/m_e^*$$

and for holes:

$$\mathscr{E}_y e + e v_{hx} H_z = m_h^* v_{hy}/\tau_h$$

from which

$$v_{hy} = e(\mathscr{E}_y + v_{hx} H_z)\tau_h/m_h^*.$$

Substituting these velocities in (14.9) gives

$$-(\mathscr{E}_y + v_{ex} H_z)\frac{\tau_e n_e}{m_e^*} = (\mathscr{E}_y + v_{hx} H_z)\frac{\tau_h n_h}{m_h^*}$$

which may be rearranged:

$$\mathscr{E}_y\left(\frac{\tau_h n_h}{m_h^*} + \frac{\tau_e n_e}{m_e^*}\right) = -H_z\left(\frac{v_{hx}\tau_h n_h}{m_h^*} + \frac{v_{ex}\tau_e n_e}{m_e^*}\right).$$

Introducing mobility as defined in (14.4),

$$\mathscr{E}_y(n_h\mu_h - n_e\mu_e) = -H_z(v_{hx}n_h\mu_h - v_{ex}n_e\mu_e)$$

or

$$\mathscr{E}_y = -H_z\frac{(v_{hx}n_h\mu_h - v_{ex}n_e\mu_e)}{(n_h\mu_h - n_e\mu_e)}. \tag{14.10}$$

The Hall coefficient R is defined by (§ 13.8)

$$\mathscr{E}_y = RJ_x H_z$$

and introducing J_x from (14.8), (14.10) becomes:

$$\mathscr{E}_y = -J_x H_z\frac{(v_{hx}n_h\mu_h - v_{ex}n_e\mu_e)}{e(n_h\mu_h - n_e\mu_e)(n_h v_{hx} - n_e v_{ex})}. \tag{14.11}$$

If the applied potential $\mathscr{E}$ in the x-direction, giving rise to the current J_x, is $\mathscr{E}_x$, then

$$v_{hx} = \mathscr{E}_x\mu_h \quad \text{and} \quad v_{ex} = \mathscr{E}_x\mu_e.$$

Substituting these into (14.11) gives

$$\mathscr{E}_y = -J_x H_z\frac{(\mathscr{E}_x n_h\mu_h^2 - \mathscr{E}_x n_e\mu_e^2)}{e(n_h\mu_h - n_e\mu_e)(\mathscr{E}_x n_h\mu_h - \mathscr{E}_x n_e\mu_e)}.$$

The $\mathscr{E}_x$s now cancel and the Hall coefficient R is given by

$$R = \frac{(n_h\mu_h^2 - n_e\mu_e^2)}{e(n_h\mu_h - n_e\mu_e)^2}. \tag{14.12}$$

The sign of R is determined by the sign of the dominant carrier, that is the carrier for which the product $n\mu^2$ is greatest. For this reason the Hall effect is the most useful experimental quantity for characterising semiconductors. The mobilities of electrons and holes in the various semiconductor materials are now catalogued, and from a simultaneous measurement of resistivity and Hall coefficient the numbers of each type of carrier present can be calculated from (14.5) and (14.12).

Because of the much smaller number of carriers in semiconductors, the Hall coefficient is several orders of magnitude larger in semiconductors than in metals. In intrinsic germanium at room temperature, for example, the densities of electrons and holes are both $\simeq 5 \times 10^{19}/m^3$. Substituting this and the values of electron and hole mobilities from Table 14.1 into (14.12) gives the value of the Hall coefficient $R \simeq -0.37 \, m^3/coulomb$. By comparison, the electron density in a monovalent metal is $\sim 10^{28}/m^3$, and the Hall coefficient $\sim -10^{-9} \, m^3/coulomb$.

Hall effect devices using semiconductors, often compounds, are now finding many applications. They form a simple method for measuring the strength of magnetic fields, the Hall voltage being directly related to the field strength. Other suggested applications include analogue computers, where information can be coded magnetically, and control devices for industrial processes.

14.8 Lifetime and scattering

The carrier lifetime is the average time between the creation of a carrier and its recombination. Through the factor β in (14.1) it determines the number of charge carriers. If it is less than the scattering time τ, then it is this lifetime, rather than τ, which must be used in the expressions for conductivity and carrier mobility. Recombination can only occur if an electron and a hole meet, and when both are mobile the probability of a direct recombination event is small and lifetimes of the order of one second are predicted theoretically. Observed lifetimes are in the range 10^{-6}–10^{-3} seconds, depending on purity, the higher figure being obtained in the most pure and carefully prepared specimens. These values of lifetime can only be explained by assuming that one of the charge carriers is rendered immobile by a trapping process, shown in Fig. 14.8. The carrier which is trapped is usually the electron; the trap being provided by acceptor levels near the middle of the gap created either by deep-level impurities or by the presence of dislocations. An edge dislocation in a semiconductor produces a 'dangling bond' (Fig. 14.9), which can accept an electron

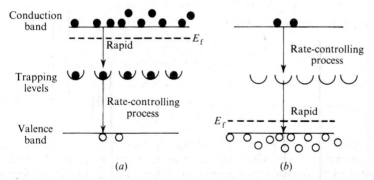

Fig. 14.8. Recombination by trapping. (a) n-type; trapping levels normally full. Electrons in these levels recombine with occasional holes in the valence band. (b) p-type; trapping levels normally empty. Occasional electrons in conduction band are trapped and immediately recombine with holes in the valence band. In both cases the rate-controlling process involves the type of carrier which is present in the minority.

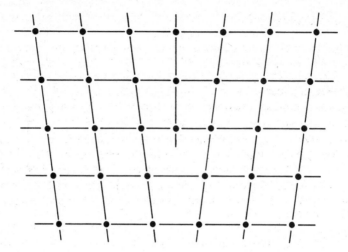

Fig. 14.9. The 'dangling bond' on an edge dislocation in a semiconductor. (Note: a pure edge dislocation does not exist in the diamond cubic lattice. The structure of the 60° dislocation is shown in Fig. 9.41.)

making the dislocations negatively charged. With a dangling bond for every atom length of dislocation, dislocation density of 10^8 lines/m^2 will produce 4×10^{17} traps/m^3; this is to be compared with the intrinsic carrier density of 10^{19}/m^3 in germanium at room temperature. This defines the allowable upper limit on dislocation density if conduction is not to be significantly impaired, and should be compared with a typical value for well-annealed metals of 10^{10} lines/m^2. The lifetime is inversely proportional to the trap density, and is independent of temperature provided the majority carrier density is greater than the trap density.

The electrical properties of semiconductors are much more influenced by the mean time between collisions τ, which is very structure sensitive. Charge carriers in semiconductors, as in metals, are scattered by lattice thermal vibrations, by Rutherford scattering from ionised (charged) impurities and dislocations, and by elastic scattering from un-ionised impurities, dislocations and grain boundaries. Semiconducting materials for solid-state devices must therefore be single crystal, low dislocation density, extremely high purity except for the accurately controlled addition of desired extrinsic impurities (i.e., doping), and very close control must be exercised over the composition of compounds.

The above requirements have been achieved by the development of special techniques, a description of which is beyond the scope of this book. The art and science of crystal growing has been advanced rapidly in recent years by the demands for semiconductor material and extremely high purities have been achieved by the technique known as zone refining. It is possible to produce single crystals of germanium about one inch diameter and six inches long containing no detectable dislocations, and with an impurity content less than one part in 10^9. Similar results are obtainable with silicon and other important semiconductors.

14.9 Uses of semiconductors

The earliest use of semiconducting material, and one that is still of extreme importance, is in rectifiers, some of which are capable of handling very high power levels in small units. The rectification properties of semiconductors were discovered experimentally. It was not until the theory of semiconductors was understood that the transistor was invented in 1948. Transistors, which are made of several junctions between different semiconducting materials, have replaced electronic valves in many applications. They have the advantage of small size, lower power consumption and no 'warm-up' time. Transistors can also be used as extremely fast switches in logic and memory circuits in computers. Transistors based on germanium originally replaced vacuum tubes as individual circuit components; these are now in turn being displaced by complete, miniature, integrated circuits based on silicon.

Semiconductors show much larger thermoelectric effects than metals, as mentioned in § 13.9, and are used in a wide range of devices, for example temperature sensors, such as thermopiles, and thermoelectric refrigerators. The energy for the excitation of charge carriers in semiconductors may also be supplied by electromagnetic radiation, particularly in the infra-red and visible regions. Hence, various photoelectric devices are also possible with semiconductors. Extrinsic superconductors, due to the predominance of one type of charge carrier and the generally higher current levels attainable, are used rather than intrinsic semiconductors, in almost all of the above devices.

This brief treatment of semiconductor uses is concluded with a simplified description of the operation of two devices, the n–p rectifier and the n–p–n junction transistor. The n–p rectifier consists of two extrinsic semiconductors, one n-type and one p-type, in contact. The device is usually made from a single block of semiconductor by diffusing in a donor impurity from one side, and an acceptor from the other. Just as in metals (§ 13.9), when two dissimilar semiconductors are placed in contact, electrons flow from one to the other until a contact potential, which equalises the Fermi levels in both materials, is established. The energy levels are now as shown in Fig. 14.10(*a*). The rectifier is connected in the circuit shown in Fig. 14.10(*b*). With no voltage applied, electrons from the conduction band of the n-type material cannot surmount the barrier into the conduction band of the p-type material, and holes find a similar difficulty in travelling in the opposite direction. When a forward bias is applied, the energy levels in the n-region are raised relative to those in the p-region (Fig. 14.10(*c*)) and a flow of electrons from n to p, and of holes from p to n, occurs; the rectifier is conducting. A negative bias lowers the n energy levels relative to the p levels (Fig. 14.10(*d*)), and the electrons and holes have even larger barriers preventing their motion into the neighbouring material. The rectifier thus becomes even more strongly insulating when negative bias is applied than when no bias is applied. There is always a finite probability of thermal excitation across the barrier, but this is largely cancelled out by a flow of minority carriers in the opposite direction, and is usually ignored.

The n–p–n junction transistor consists of a region of p-type material, the base, sandwiched between two n-type regions, one called the emitter, the other the collector. The device can be constructed quite simply by diffusing a blob

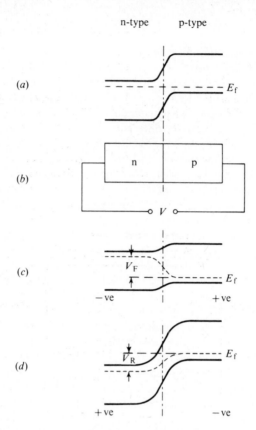

Fig. 14.10. n–p junction diode. (*a*) Energy level diagram for diode with no applied bias. (*b*) Diode in circuit, V is bias voltage of either sign. (*c*) Energy level diagram with forward bias voltage V_F. (*d*) Energy level diagram with reverse bias voltage V_R ($= -V_F$). In (*a*), (*c*) and (*d*) the ordinate is the electron energy.

of arsenic into either side of a slab of p-type germanium (Fig. 14.11(*a*)). The diffusion must be carried out so as to leave the base as narrow as possible without the two n-type regions overlapping. The transistor is connected in the circuit as shown in Fig. 14.11(*b*), a signal is applied between the emitter and the base and a strong forward bias is applied from base to collector. The resulting energy levels are shown in Fig. 14.11(*c*). The emitter–base junction acts exactly as the rectifier described above, and a current flows from base to emitter, modulated by the applied signal voltage. Due to the narrowness of the base region, many of the electrons which flow from the emitter pass right through the base, and are immediately accelerated by the strong forward bias into the collector. The importance of an extremely narrow base region, in ensuring that a very high proportion of electrons pass right through it, is obvious. The signal introduced into the low impedance emitter circuit is reproduced in the high impedance collector circuit. The difference in impedance is such as to cause an amplification of the power of the input signal. The degree of amplification is controlled by the value of the bias voltage between collector and

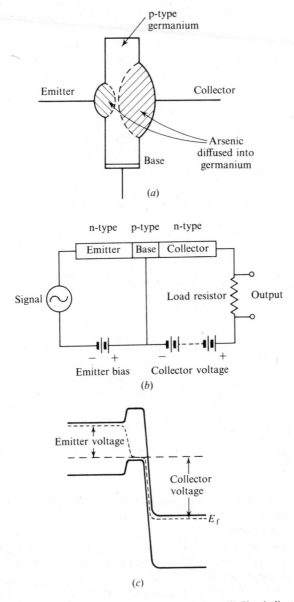

Fig. 14.11. n–p–n junction transistor. (*a*) Fabrication of the transistor. (*b*) Circuit diagram. (*c*) Energy level diagram; the ordinate is the electron energy.

base. As in the previous case of the rectifier, the situation is complicated by the action of the minority carriers, and by thermal currents flowing across the junctions. The latter limit the temperature above which the device cannot operate satisfactorily. Heavy-current devices, in which the operating temperature is high due to Joule heating, are made from silicon which, because of its larger energy gap, is less troubled by thermal effects.

Transistors can be devised to perform almost all of the functions of vacuum tubes, as well as many new functions which cannot be carried out by other devices. For a further description of transistors the reader is referred to a text dealing specifically with solid state electronics.

14.10 Insulators

The requirement of an insulator is that it should conduct little or no electricity ; its resistivity should be as high as possible. A comparatively high resistivity can be achieved by interfering with the motion of electrons by introducing many scattering centres, thus giving a short mean free path, as has been described in Chapter 13 for metals. The resulting resistivities are, however, orders of magnitude smaller than those required for insulating purposes. The resistivity of a good insulator is some 10^{18} times greater than that of the most resistive alloy (see Table 13.1). The resistivity of any solid is given by (3.1), in which the only quantity that can vary by eighteen orders of magnitude is the density of conduction electrons, n, which may be as high as 10^{28} m^{-3} in a pure metal such as copper, $\sim 10^{19}$ m^{-3} in germanium at room temperature, and $\sim 10^9$ m^{-3} in diamond at room temperature. The other quantities in (3.1) will show some variation from metal to insulator, but none sufficiently large to explain the difference in resistivity.

This low density of conduction electrons in an insulator is a consequence of the large energy gap between valence and conduction bands. As mentioned in the introduction to this chapter, the difference between insulators and semi-conductors is purely one of degree. The density of conduction electrons in such materials is an exponential function of the energy gap, (14.2), and a fivefold increase in the energy gap results in a 10^5 decrease of conduction electrons. The gap in semiconductors is 2×10^{-19} J (1 eV) or less (Table 14.1) and in insulators is $\sim 8 \times 10^{-19}$ J (5 eV). This large energy gap is the important characteristic of an insulator. Gaps of this magnitude are found in ionic and strongly covalent solids, in which the valence electrons are firmly bound to the atoms. Crystalline solids which fulfil these requirements are usually ceramic oxides, carbides and nitrides. Many polymeric materials, involving strong covalent bonds, are also insulators. Some values for resistivity of typical insulators are given in Table 13.1.

Insulating materials must be of high purity, as the presence of impurity elements can, as in semiconductors, produce energy levels within the band gap and so increase the number of available charge carriers. This is in contra-distinction to metals, where the presence of impurites invariably increases resistance.

The application of a weak electric field will cause the conduction electrons to move, but because their number is so small, a negligible current will flow. A larger effect can result if charged ions are induced to diffuse under the action of an electric field. This leads to ionic conduction, discussed in § 14.11. Ionic conduction can only occur if the material has some degree of ionic binding, and is usually only important in non-crystalline materials, through which the diffusion of ions is comparatively rapid. Ionic conduction is found principally in glasses and some polymeric materials.

The weak field phenomenon of greatest importance is polarisation. Electrons may be tightly bound to the ions, but under the action of quite small fields may be displaced slightly from them. Such a displacement results in the centres of positive and negative charge no longer coinciding, and small electric dipoles are set up. Polarisability is the ease with which the electron charge distribution can be deformed by a field ; highly polarisable materials are called dielectrics (and are discussed in § 14.12). In a steady field, a small displacement current is set up as the electron distribution distorts, but dies away as the equilibrium configuration for that particular field is reached. In an alternating field the distortion also alternates, and an AC displacement current may therefore flow continuously through a material which is an insulator under DC conditions. The polarisation may be slow in conforming to the changing field, leading to hysteresis and power loss; these effects being dependent upon the nature and microstructure of the material. The behaviour of dielectrics in an alternating field is described in § 14.14.

Dielectricity may be regarded as an electrical analogue of paramagnetism (see Chapter 15). Some materials show permanent electric dipoles which interact with one another, as do the magnetic dipoles in ferromagnetism; such materials are called ferroelectrics, and are discussed in detail in § 14.13.

As the field applied to an insulator is increased, the small current flowing in it, due to the few electrons in the conduction band together with any ionic conduction and displacement effects, will increase in accordance with Ohm's law. At some high field, Ohm's law will cease to be obeyed, the current rising rapidly to a large value indicating a catastrophic reduction in the resistance of the insulator. This is due to several phenomena, known collectively as *breakdown*, which are discussed in § 14.15. As an insulating material is usually used to prevent the passage of electricity, breakdown should occur at as high a field as possible; this field defines the electrical strength of the insulator. Like mechanical strength, it is not a well-defined parameter and depends upon the method of measurement.

It is desirable to know how an insulator behaves in very high fields (breakdown) and how it behaves in weak alternating fields (dielectric properties). Whichever of these is the most important depends upon the application, and a material which is a good insulator may not necessarily be a good dielectric (and *vice versa*).

Finally mention is made of LASER action, an electron–optical phenomenon which occurs in some insulating crystals as a consequence of the presence of impurity levels within the band gap.

14.11 Ionic conduction

Conduction in electrolytes, solutions of ionic solids usually in water, is by the motion of charged ions, rather than by electrons alone. (The term ion is derived from a Greek word meaning wanderer.) In the liquid state the motion of ions is relatively easy. Conduction by ions can also take place in the solid state, but ionic motion is restricted, and takes place by diffusion, see § 2.11. The ions diffuse through the crystal lattice by changing place with lattice vacancies. This motion is normally quite random, and does not produce a net flow of charge and hence an electric current. The application of an electric field will favour motion in a particular direction, and a net current will result.

For an atom or ion to jump from its regular lattice site into an adjacent vacant site requires the expenditure of an activation energy, Δg_m. This is the energy required to squeeze the ion between its neighbouring ions in going from one lattice site to another. The rate at which ions will jump is equal to the probability of there being a vacant site into which to jump times the rate at which a jump is attempted times the probability of a jump being successful. The probability of there being an empty site is given by $\exp(-\Delta g_f/kT)$ where Δg_f is the free energy of formation of vacant sites. The rate at which jumping is attempted is the frequency of vibration of the ions, v, which is related to their mass and to the interionic (lattice) forces. The probability of success is the probability that the ion will have the necessary activation energy; this is given by the Boltzmann term, $\exp(-\Delta g_m/kT)$. Thus the total rate of jumping is given by

$$R_0 = v \exp\left(-\frac{\Delta g_f + \Delta g_m}{kT}\right) = v \exp\left(-\frac{\Delta g}{kT}\right)$$

where $\Delta g = \Delta g_f + \Delta g_m$.

In the absence of an electric field there is an equal probability, once a jump has been made, that the ion will jump back into the site which it has just vacated. The presence of a field $\mathscr{E}$ exerts a force on the ions $\mathscr{E}\beta$, where β is the ionic charge. The direction of this force depends on the sign of the ionic charge and on the direction of the field, but its effect is to lower the activation energy in the 'forward' direction, and raise it in the backward direction, by an amount, $\beta\mathscr{E}d/2$, where d is the distance jumped, i.e. the distance between adjacent lattice sites. The factor of 2 arises because the position of maximum energy occurs midway between lattice sites.

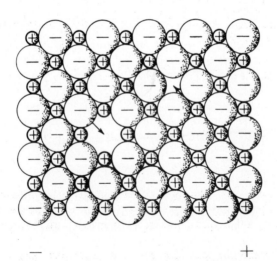

Fig. 14.12. Ionic conduction in a crystal; application of an electric field as shown causes the positive ion vacancy to diffuse to the right and the negative ion vacancy to diffuse to the left.

The rate of forward jumping is now

$$R_{\text{forward}} = v \exp\left(-\frac{\Delta g - \beta \mathscr{E} d/2}{kT}\right),$$

and the rate of backward jumping

$$R_{\text{backward}} = v \exp\left(-\frac{\Delta g + \beta \mathscr{E} d/2}{kT}\right).$$

The net rate of movement $\Delta R = R_{\text{forward}} - R_{\text{backward}}$:

$$\Delta R = v\left\{\exp\left(-\frac{\Delta g - \beta \mathscr{E} d/2}{kT}\right) - \exp\left(-\frac{\Delta g + \beta \mathscr{E} d/2}{kT}\right)\right\}$$

$$= R_0\left[\exp\left(+\frac{\beta \mathscr{E} d}{2kT}\right) - \exp\left(-\frac{\beta \mathscr{E} d}{2kT}\right)\right] = 2R_0 \sinh \frac{\beta \mathscr{E} d}{2kT}.$$

The ionic charge β will be one, two or three times the electronic charge, d will be $\simeq 3 \times 10^{-10}$ m, and for all reasonable fields at room temperature $\beta \mathscr{E} d \ll 2kT$. For small values of x, $\sinh x \simeq x$, and therefore

$$\Delta R = \frac{R_0 \beta \mathscr{E} d}{kT}.$$

To relate this to ionic conductivity, Ohm's law is used: $J = \sigma \mathscr{E}$. The current density J is made up of both positive and negative ions flowing in opposite directions:

$$J = n^+ \beta^+ v^+ + n^- \beta^- v^-$$

where n^+ (n^-) is the number of positive (negative) ions per unit volume, and v^+ (v^-) their mean velocity, given by $v = \Delta R d$.

Therefore
$$J = \frac{n^+ \beta^{+2} R_0^+ \mathscr{E} d^2}{kT} + \frac{n^- \beta^{-2} R_0^- \mathscr{E} d^2}{kT}.$$

For simple ionic structures, $n^+ = n^- = n$, and $\beta^+ = -\beta^- = xe$, where x is some small integer:

$$J = \frac{x^2 n e^2 d^2 \mathscr{E}}{kT}(R_0^+ + R_0^-)$$

and
$$\sigma = \frac{nx^2 e^2 d^2}{kT}(R_0^+ + R_0^-),$$

but
$$d^2(R_0^+ + R_0^-) = D, \text{ the diffusion coefficient,} \qquad (2.17a)$$

hence
$$\sigma = \frac{nx^2 e^2}{kT} D.$$

This is Einstein's equation relating ionic conductivity to the diffusion coefficient. Ionic conductivity is therefore influenced by the lattice properties which govern ordinary diffusion, rather than by electronic properties as is the case for electronic conductivity.

At high field strengths ionic currents can cause local ohmic heating and produce thermal breakdown of insulation. High polymers with ionic radicals or catalyst residues can show some ionic conductivity. Soda-glasses and other

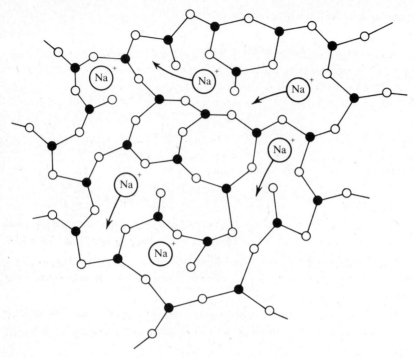

Fig. 14.13. Ionic conduction in a glass; schematic two-dimensional representation of a SiO_2–Na_2O (soda) glass. The electric field causes the positive sodium ions to diffuse to the left.

glasses containing network modifying ions may show ionic conduction through the migration of these ions. In both of these last two cases ionic conduction can be particularly important as diffusion rates are high in non-crystalline structures.

14.12 Dielectrics: weak field effects

It is necessary to begin by defining some quantities. $\mathscr{E}$ is the electric field strength, and in the SI system of units is measured in volts per metre. It may be an applied external field $\mathscr{E}_a$, or it may be a local field, $\mathscr{E}_{loc}$, due to the sum of the external field, the depolarisation field $\mathscr{E}_d$ created by free charges induced on the surface of the specimen, and the internal field $\mathscr{E}_i$ due to electric dipoles on the atomic or molecular scale. D, the electric displacement or flux density, measured in coulomb/m², is related to $\mathscr{E}$ by

$$D = \varepsilon_0 \varepsilon_r \mathscr{E}, \qquad (14.13)$$

where ε_r is the (dimensionless) *relative permittivity* (dielectric constant) of the material ($\varepsilon_r = 1$ for a vacuum) and ε_0 is the *permittivity of free space* ($= 8.854 \times 10^{-12}$ farad/m). ε_0 has no physical meaning, it is merely a consequence of choosing to work with the SI system of units. The dipole moment per unit volume of a material is called the polarisation, P, and is expressed in coulomb/m².

P is related to $\mathcal{E}$ by

$$P = \varepsilon_0(\varepsilon_r - 1)\mathcal{E}. \qquad (14.14)$$

ε_r is a constant for simple materials in which the polarisation is produced by the applied field alone. In ferroelectrics, polarisation exists in zero field, the above relationships do not hold, and ε_r has no unique value.

The dipole moment μ induced on one molecule by a local field $\mathcal{E}_{loc}$ is given by

$$\mu = \alpha\mathcal{E}_{loc} \qquad (14.15)$$

where α is the *polarisability* of the molecule. This microscopic quantity α is related to the experimentally measured property ε_r as follows:

If there are N dipoles per unit volume, then

$$P = N\mu = N\alpha\mathcal{E}_{loc},$$

and from (14.14)

$$\varepsilon_r - 1 = \frac{P}{\varepsilon_0\mathcal{E}} = \frac{N\alpha\mathcal{E}_{loc}}{\varepsilon_0\mathcal{E}}.$$

This now introduces the problem of determining the local field. As mentioned above, this consists of the applied external field plus the depolarisation and

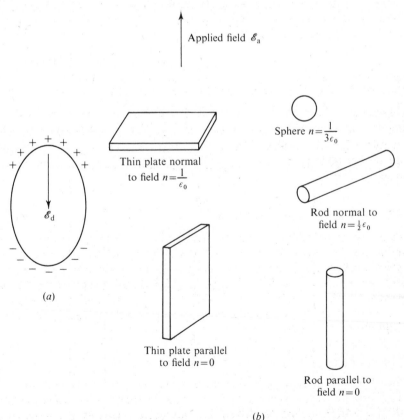

(a)

(b)

Fig. 14.14. (a) Depolarisation field $\mathcal{E}_d$ resulting from charges induced on the surface of an ellipsoid in an applied field $\mathcal{E}_a$. (b) Values of the depolarisation factor n, for dielectrics of various shapes.

internal fields. The depolarisation field $= -nP$, where n is the *depolarisation factor*; $n = 0$ for a thin rod parallel to $\mathscr{E}$, and $= 1/3\varepsilon_0$ for a sphere, see Fig. 14.14. The internal field at the particular molecular dipole due to all of the other dipoles can be estimated by the following argument. Suppose the particular molecule occupies a spherical cavity in the material from which it is removed, leaving a vacuum. Charges will appear on the surface of the cavity, and will produce a depolarisation field (the Lorentz field) in the cavity equal to $P/3\varepsilon_0$, where P is the polarisation of the surrounding material (Fig. 14.15). This is then the field due to all of the other dipoles, and

$$\mathscr{E}_{\text{loc}} = \mathscr{E} - nP + \frac{P}{3\varepsilon_0}.$$

This derivation is only strictly correct if the material is isotropic, and if the imaginary spherical cavity has a radius greater than the size (that is the separation of the charges) of the dipole. A more general expression is

$$\mathscr{E}_{\text{loc}} = \mathscr{E} - nP + \frac{\gamma P}{\varepsilon_0}$$

where γ is referred to as the internal field constant, and has a value $\approx \frac{1}{3}$. Note that for a spherical specimen, $\mathscr{E}_{\text{loc}} \approx \mathscr{E}$, and $\varepsilon_r - 1 \approx N\alpha/\varepsilon_0$. For a rod specimen with $n = 0$, the *Clausius–Mosotti* expression

$$\frac{\varepsilon_r - 1}{\varepsilon_r + 2} = \frac{N\alpha}{3\varepsilon_0}$$

is found.

There are three mechanisms by which a material can be polarised, and hence three contributions to the polarisability. They are illustrated in Fig. 14.16.

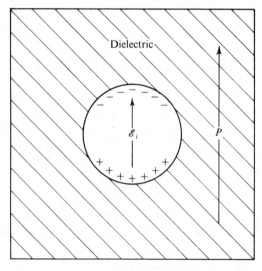

Fig. 14.15. Lorentz field $\mathscr{E}_i = P/3\varepsilon_0$ produced in a spherical cavity within a dielectric of polarisation P.

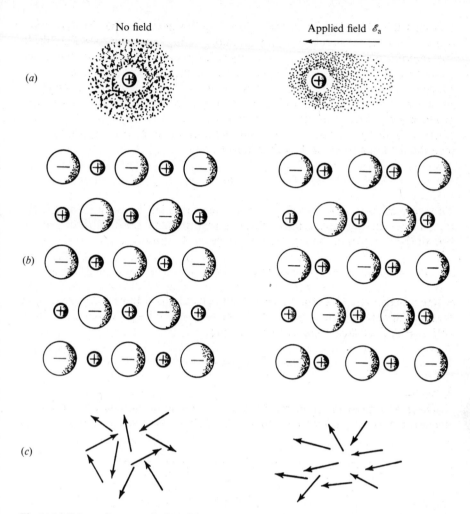

Fig. 14.16. Schematic representation of the three sources of polarisation; (a) electronic, (b) ionic and (c) orientation polarisation.

Electronic polarisation arises when the application of an electric field tends to displace the centre of gravity of the electrons away from the centres of the positive nuclei. This displacement is counteracted by the Coulombic attraction between the electrons and the nucleus, and the net displacement is

$$d = \frac{4\pi\varepsilon_0 r^3 \mathscr{E}}{Ze}$$

where r is the atomic radius and Z the number of electrons. Charges $+Ze$ and $-Ze$ separated by a distance d produce a dipole moment

$$\mu = Zed = 4\pi\varepsilon_0 r^3 \mathscr{E}$$

and from (14.15), the electronic polarisability

$$\alpha_e = 4\pi\varepsilon_0 r^3.$$

Ionic polarisation results from a shift of positive ions relative to negative ions, in a material with some degree of ionic binding. This polarisation cannot be treated in so simple a manner as can electronic polarisation. Part of the difficulty is due to the fact that the local field experienced by a positive ion may be different from that experienced by a negative ion. In simple ionic salts such as the alkali halides, the ionic polarisability α_i is from 2 to 5 times the electronic polarisability.

As this ionic polarisation involves relative motion of opposite ion lattices, it is obvious that the elastic properties of the material are involved. The detailed relationship between ionic polarisability and elastic moduli is difficult to deduce, but experimentally ionic solids with low moduli have high values of polarisability, and *vice versa*, as would be expected.

If an ionic solid is subjected to mechanical deformation, for example a uniform tension along some axis, then all the interionic distances along this direction are increased. This results in an increase in all of the local dipole moments between pairs of ions. For a crystal with a centre of symmetry, for each dipole of one sign there will be a dipole of opposite sign, and the total dipole moment of the crystal will remain zero. If the crystal does not have a centre of symmetry (of the 32 crystal classes only 12 have a centre of symmetry) dipoles of opposite sign will not be balanced after deformation and the crystal will exhibit a net dipole moment. Conversely the application of an electric field will favour an excess of dipoles with a component lying in the field direction, and will cause mechanical deformation of the crystal. This coupling of electrical and mechanical effects is known as piezoelectricity. Piezoelectric crystals such as quartz and cadmium sulphide can be caused to vibrate by the application of an AC field and act as transducers, transforming electrical into mechanical vibration. The amplitude of vibration becomes large when the frequency of the field is the same as the natural frequency of vibration of the crystal, determined by the dimensions and elastic constants, and such crystals are used as oscillators in many electronic applications.

Orientation polarisation occurs in substances the molecules of which have permanent dipole moments, and are free to rotate such as to bring these dipoles into alignment with an applied field. This freedom to rotate is found in gases and liquids, rarely in solids. The tendency for dipoles to rotate into alignment with the field is resisted by thermal disordering. This leads to a temperature dependence of polarisability

$$\alpha_0 = \frac{\text{const.} \, \mu_0^2}{T}$$

where μ_0 is the permanent dipole moment. This equation is similar to that for paramagnetic susceptibility, derived in the next chapter.

Orientation polarisation is found in certain high polymers, due to asymmetrical distribution of charge in the structure. Such materials show an additional temperature effect to the one given above. At low temperatures the viscosity of the polymer is too high to allow for the bond rotation which would enable dipoles to align themselves parallel to the field. As the viscosity decreases at higher temperatures alignment becomes more easy, but has to contend with a greater thermal randomisation. The overall relation between polarisability and temperature can be complicated, usually passing through a maximum at some temperature below the flow temperature. Increasing degree of crystallinity will make bond rotation more difficult and reduce the polarisability. Dielectric relaxation peaks are described in § 8.11.

14.13 Ferroelectrics

Some crystalline materials have permanent dipoles which interact with one another in such a way that adjacent dipoles are aligned in the same direction. The crystal is divided into domains, the direction of dipole alignment is different in different domains, and in the absence of a field there is no net dipole moment. The application of an electric field favours the growth of those domains which are aligned in the direction of the field, at the expense of the other domains, and a net polarisation is produced. Unlike other dielectric materials the polarisation is not a linear function of the applied field, but depends on the previous history of the sample. These materials exhibit hysteresis effects, and their behaviour is reminiscent of ferromagnetic materials described in the next chapter; for this reason they are known as *ferroelectric* materials.

The polarisation of a typical ferroelectric as a function of field is shown in Fig. 14.17. When a field is applied to an initially depolarised specimen, the polarisation increases along the sigmoidal curve $OABC$. Reduction of the field to zero leaves the specimen with a remanent polarisation P_r. A coercive field $\mathscr{E}_c$ must be applied in the reverse direction to reduce the polarisation to zero. Increasing the reverse field causes polarisation in the opposite direction, and a complete hysteresis loop can be constructed, as in Fig. 14.17.

As the applied field is increased, the growth of favourably oriented domains occurs, not, as in the case of ferromagnetics by the sideways motion of domain boundaries, but by the growth of thin needle-like domains of favourable orientation within unfavourably oriented grains. These needle domains, about one

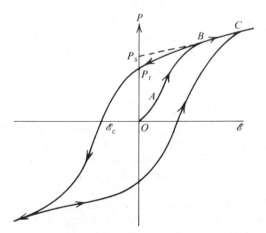

Fig. 14.17. Hysteresis curve for a ferroelectric material (schematic). P_s is the spontaneous polarisation of a single domain in the absence of an applied field, P_r the remanent polarisation, and $\mathscr{E}_c$ the coercive field.

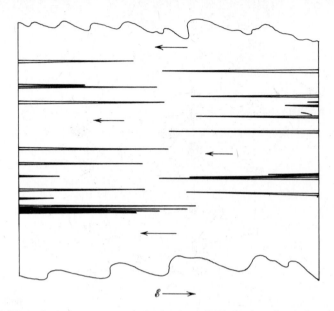

Fig. 14.18. Reversal of spontaneous polarisation by the polarisation of new domains in a ferro-electric crystal. Each needle is a domain with its direction of polarisation parallel to the applied external field.

micron in width, are nucleated at domain boundaries and grow in the forward direction only, their width remaining essentially constant. The domains, which can be rendered visible in polarised light, are shown schematically in Fig. 14.18. The nucleation and growth of these needles continues until the entire specimen has been converted to a favourable orientation. Hysteresis arises from the resistance to nucleation and growth of these needles. The spontaneous polarisation vanishes above a certain temperature, the ferroelectric Curie temperature θ_f, when thermal agitation is sufficient to overcome the forces which align neighbouring dipoles. Above the Curie temperature a ferroelectric behaves as a normal dielectric, with a permittivity given by

$$\varepsilon_r = \frac{\text{const.}}{T - \theta}$$

where θ is a characteristic temperature usually a few degrees below θ_f.

The phenomenon of ferroelectricity was discovered in 1921 by Valasek in Rochelle salt ($NaKC_4H_4O_6.4H_2O$). In 1935 potassium dihydrogen phosphate KDP (KH_2PO_4) was found to be ferroelectric. Other alkali dihydrogen phosphates and arsenates also show the phenomenon. Barium titanate, $BaTiO_3$, a representative of the oxygen octahedron group, is probably the best known ferroelectric.

Above its Curie temperature of 120 °C, barium titanate has the cubic structure shown in Fig. 14.19. The Ba^{2+} ions are located at the centre of the cube, the O^{2-} ions at the centres of the cube edges, and the Ti^{4+} ions at the cube corners, inside an octahedron formed by the oxygen ions. The material has a high ionic polarisability, partly because the titanium ion has a charge of $4e$, and partly

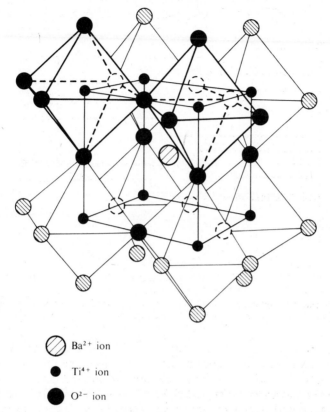

Ba²⁺ ion
Ti⁴⁺ ion
O²⁻ ion

Fig. 14.19. The crystal structure of barium titanate, BaTiO₃, above its Curie temperature.

because it can be displaced over a relatively large distance due to the fact that it is much smaller than the space available for it within the oxygen octahedron.

Below 120 °C barium titanate is spontaneously polarised along a ⟨100⟩ direction, and the crystal structure undergoes an expansion in this direction, becoming tetragonal. As the direction of polarisation may be along any one of six ⟨100⟩ directions, different domains have a different tetragonal axis, and internal stresses are developed in the material. These may contribute to the hysteresis. At 5 °C the direction of polarisation changes to ⟨110⟩, and at −80 °C changes again to ⟨111⟩. Each of these changes involves a change in the crystal structure. The dielectric constant as a function of temperature is shown in Fig. 14.20.

Barium titanate is used in high capacity condensers in electronic circuits.

14.14 Dielectrics: alternating field effects

The polarisability of a dielectric in an alternating field depends upon the ability of the dipoles to reverse their alignment with each reversal of the field. The time taken for any disturbed system to return to equilibrium is known as the *relaxation time,* and the reciprocal of the relaxation time is known as the relaxation

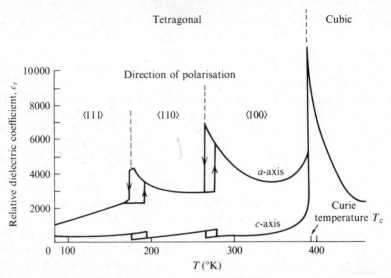

Fig. 14.20. Relative dielectric coefficient, ε_r, for a sample of $BaTiO_3$, measured with the field parallel to the a and to the c-axes, in both increasing and decreasing temperature.

frequency. Relaxation frequencies differ for each type of polarisation mechanism discussed in § 14.12. If polarisability is measured as a function of frequency it will be found to vary in a complicated manner, as shown schematically in Fig. 14.21. At low frequency the period of the field cycle is greater than the relaxation time of the dipoles. The dipoles are always in phase with the field, and the polarisability is the same as for a steady field. At high frequencies the period is much shorter than the relaxation time, the dipoles are quite unable to follow the field and the dielectric properties of the material are not in evidence. At intermediate frequencies the dipoles are able to follow the field, but due to the finite time re-

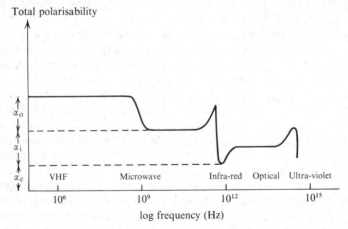

Fig. 14.21. Polarisation versus frequency for a solid which exhibits electronic, ionic and orientation polarisation.

quired for adjustment, they will be slightly out of phase with it. Resonance effects may be observed when the applied frequency coincides with a relaxation frequency.

By the time the system of dipoles has responded to a particular value of the field, the field will have changed to a different value. This difference in phase leads to hysteresis and power loss within the dielectric, which appears in the form of heat. This power loss is of great importance. The use of dielectrics is either to increase the capacitance of condensers in electronic circuits, or as insulation. In either application, power is removed from the system, and the dielectric must be cooled. The heating effect can be utilised, for warming polymers prior to forming operations, or in a rapid form of cooking where moisture in the food is the dielectric.

In a material which is composed of only one kind of atom, such as silicon or germanium, the only component of polarisability present is electronic. The momentum of an individual electron is small and the relaxation frequency is about 10^{15} Hz. This is at the top end of the optical frequency range, in the ultra-violet region. According to Maxwell's theory of electromagnetic radiation, in materials for which the magnetic permeability is equal to that of a vacuum, $\varepsilon_r = n^2$, where n is the optical index of refraction. Thus electronic polarisability can be measured by a simple refraction experiment.

The relaxation frequency for ionic polarisation is much lower, generally lying in the infra-red region ($\sim 10^{13}$ Hz). This is because the heavy ions cannot follow field variations as rapidly as can electrons. At optical frequencies, only the electronic polarisability is measured. Measurements at lower frequencies give both electronic and ionic polarisabilities, and the difference between the two measurements enables the ionic polarisability to be deduced. Electronic dielectric constants for the Group IVB elements, diamond, silicon and germanium are 5.68, 12 and 16 respectively. Total (electronic plus ionic) dielectric constants for the alkali halides, are 2–5 times greater than their electronic dielectric constants, which have values of 2–4. Electronic and ionic polarisabilities are related to atomic size and elastic constants; these quantities vary little with temperature, and the polarisabilities may be regarded as independent of temperature. This does not hold true for orientation polarisation.

The most important source of polarisation in glasses and high polymers is orientation polarisation, and these are the materials frequently found in practical applications. Orientation processes are thermally activated, and therefore relaxation frequencies and polarisability are strongly temperature dependent. Frequencies vary from a few cycles per second to microwave frequencies ($\sim 10^{10}$ Hz), and cover the range of power and communication frequencies. Orientation polarisation, and resulting dielectric losses, are therefore of great technological importance.

The fact that the polarisation lags behind the field leads to a complex dielectric constant given by

$$\varepsilon_{ro}^* - 1 = (\varepsilon_{ro} - 1)\left[\frac{1}{1 + \omega^2\tau^2} - i\frac{\omega\tau}{1 + \omega^2\tau^2}\right]$$

where ε_{ro} is the orientation dielectric constant in a steady field, τ is the relaxation time, and ω is the frequency of the applied field. The first term, the real part,

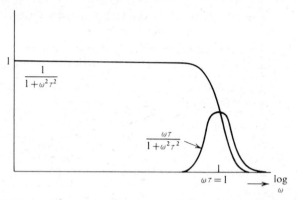

Fig. 14.22. Complex dielectric constant as a function of frequency showing a resonance peak at $\omega\tau = 1$. Real part $\varepsilon_r' \propto 1/(1 + \omega^2\tau^2)$, imaginary part $\varepsilon_r'' \propto \omega\tau/(1 + \omega^2\tau^2)$. ($\tau$ is the relaxation time for the resonance.)

is in phase with the field; the second, imaginary, term lags 90° behind the field. It is this imaginary part of the dielectric constant, ε_r'', which gives rise to power losses. The power loss per m³ per second,

$$W = \tfrac{1}{2}\omega\varepsilon_0\varepsilon_r''\mathscr{E}_0^2,$$

where $\mathscr{E}_0$ is the maximum amplitude of the applied field. The losses in a dielectric are usually characterised by quoting the loss tangent

$$\tan \delta = \varepsilon_r''/\varepsilon_r'$$

where ε_r' is the real part of the dielectric constant (Fig. 14.22).

Orientation polarisation arises from a non-uniform charge distribution giving rise to permanent dipoles, and is to be expected in many asymmetrical polymers, condensation polymers in particular. Several processes may be involved in orientation polarisation in any given polymer, and a broad spectrum of relaxation times is expected. The loss tangent will depend upon the ease with which these processes can occur, and any factor which interferes with the motion of polar groups will alter the relaxation frequency and loss tangent, as discussed in § 8.11.

The relaxation time (or times) varies with temperature according to a Boltzmann relation, indicating that the processes are thermally activated. Loss tangent at constant frequency plotted versus reciprocal temperature, or at constant temperature plotted versus frequency, are similar in form and show maxima when the frequency is equal to the relaxation frequency.

Table 14.3 gives the real dielectric constant and the loss tangent at a frequency of 1 kHz for a number of dielectric materials.

14.15 High field effects: breakdown

When the electric field applied to an insulator exceeds a critical value the current suddenly increases; the resistance catastrophically decreases to a very low value. This is known as *breakdown*, and can be due to one of several phenomena:

    (*a*) Intrinsic breakdown,
    (*b*) Thermal breakdown,

Table 14.3. *Dielectric constant* (ε_r) *and dielectric
loss* $(tan\ \delta)$ *at 1 kHz*

Material	ε_r	$tan\ \delta$
Linear polyethylene	2.3	<0.0002
Branched polyethylene	2.3	<0.0005
Polypropylene	2.0	<0.0003
Polycarbonate	3.0	0.0011
PTFE	2.0	<0.0002
Polystyrene	2.5	0.0001–0.0003
HIPS	2.4–4.5	0.0004–0.002
SAN	2.5	0.007–0.010
ABS	2.7–4.8	0.002–0.012
PMMA	3.0–3.5	0.03–0.05
Cellulose acetate	3.5–7.0	0.01–0.06
Nylon-6,6	4.0–4.5	0.02–0.04
Glass (Corning 0010)	6.6	~0.005
Porcelain no. 4462	9.0	~0.0015

(*c*) Electrochemical breakdown,
(*d*) Breakdown by discharges,
(*e*) Breakdown by tracking.
These are discussed briefly in this section.

The intrinsic electrical strength of a material is achieved only when various factors which reduce the electrical strength, discussed later in this section, are absent. Its value varies from material to material, but is of the order of 10^9 volts/m. The intrinsic strength is limited by one of two processes: electronic or electro-mechanical.

The high field accelerates the few electrons present in the conduction band, and, as in a metal, these accelerated electrons collide with imperfections in the structure and lose energy gained from the field. The energy of an individual electron before a collision can be very high, and if this energy is given up to an atom in the collision, it may be sufficient to ionise that atom and release another electron to the conduction band. This electron in turn is accelerated by the field. At some field a stage is reached when the intrinsic and ionised electrons are receiving more energy from the field than they lose by collision, and the current increases catastrophically; intrinsic electronic breakdown has occurred. The phenomenon is similar to the ionisation of a gas by electrons in a discharge tube.

In materials which are mechanically weak, the mechanical stresses resulting from the application of large fields can cause compressive deformation. The thickness of the material is decreased, and at a constant voltage the field increases, thus increasing the mechanical forces and causing further deformation. Above a critical field the situation becomes unstable and intrinsic electromechanical breakdown ensues. The ultimate breakdown may be electronic, the actual field having exceeded the electronic breakdown field as the material is compressed, or it may be completely mechanical, if the stresses are so high as to cause the material to fracture. This latter type of electromechanical breakdown is more likely to occur in brittle materials.

Thermal breakdown occurs as a result of the ohmic heating due to the, initially, small currents which flow, either from the intrinsic electrons or from ionic conduction. Insulators are generally poor thermal conductors, and any local temperature fluctuation takes a long time to diffuse away. Local temperature rises due to ohmic heating cause the thermal excitation of more electrons into the conduction band and with ionic conduction cause the diffusion coefficient to increase. The increase in temperature gives rise to an increase in current, which in turn produces a further rise in temperature. Above a certain field the rate of increase of temperature is greater than the rate of decrease due to thermal diffusion and cooling from surfaces; the process becomes catastrophic, and thermal breakdown occurs. The situation in polymers can be very serious, as the local heating can cause thermal degradation of the polymer, leading to both electrical and mechanical collapse.

Impurities which give rise to energy levels within the band gap will reduce intrinsic and thermal breakdown strengths, by providing more readily ionised electrons to swell the breakdown current. Close control over the purity must be exercised in insulator manufacture.

Electrochemical breakdown occurs at fields sufficiently large to cause separation of oppositely charged ions. This leads to exaggerated ionic conduction and a separation of the constituent ions or radicals of the material, as happens when a current is passed through an electrolyte. Weakly bound ionic materials are susceptible to this form of breakdown.

Electrical discharges across gas-filled cavities within the body of an insulator, or at the edges of electrodes adjacent to the material, can lead to breakdown (Fig. 14.23). The actual processes involved in breakdown by discharges are

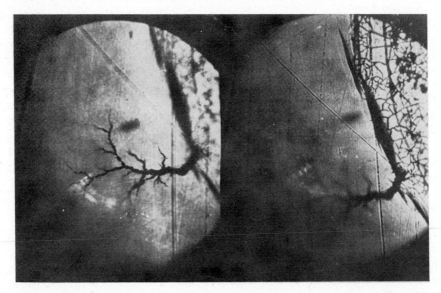

Fig. 14.23. Breakdown in polystyrene initiated at surface crazes caused by discharges. Breakdown results from the high fields which arise due to the sharp radii of curvature at the limits of the surface cracks. (From N. Parkman, in P. D. Ritchie, editor, *The Physics of Plastics*, Iliffe Books Limited (1965).)

Table 14.4. *Room temperature breakdown strength of various insulators*

Insulator	Breakdown strength MV/m
LiF	310
NaF	240
KF	190
NaCl	160
KCl	100
NaBr	80
KBr	70
KI	50
Quartz	600
Glass	700
Mica	1000
Polymethylmethacrylate	1200
Polyethylene	750
Polystreyene	700
Polyisobutylene	100

Note: these values of breakdown strength are averaged from a number of sources. The true value depends upon specimen thickness, rate of application of electric field, and geometry of test.

complex, the electrical, mechanical and thermal stresses at the surfaces causing the material to degrade. Cavities may be punched right through the material by such discharges. This form of breakdown is minimised by ensuring homogeneity of the insulator, with absence of fissures or porosity. The breakdown of brittle materials such as mica is facilitated by fissures which follow the fracture planes.

Surface breakdown occurs when moisture and other impurities adsorbed onto a free surface provide conducting paths. The heat generated by surface currents on organic insulators may cause thermal degradation of the surface, and produce carbonised 'tracks' of high conductivity. Once tracking has occurred the insulator is useless as such until the surface layer has been removed. High tension insulators are fashioned in typically corrugated shapes to increase the length of any surface path, and hence reduce the voltage gradient along it. Surfaces are also glazed, to prevent adsorption of impurities which aid breakdown.

The electrical strengths of various insulating materials are shown in Table 14.4. It must be emphasised that these values are sensitive to temperature, environment, geometry of test specimens and method of testing.

14.16 Lasers

The LASER, named from the initials of 'light amplification by the stimulated emission of radiation' is an interesting example of an electro-optical phenomenon. Laser action was discovered by Maiman in a ruby crystal, and depends upon the presence of localised energy levels within the band gap. These energy levels arise because of the presence of impurities, just as in extrinsic semiconductors. There is, however, an important difference in that laser crystals are insulators, with a large band gap; the impurity levels are far below the conduction band.

Ruby is aluminium oxide (Al_2O_3) in which a few aluminium ions are replaced by chromium ions, which introduce several localised energy levels into the band-gap, as shown in Fig. 14.24. Normally the electrons are in the ground state, E_0. Irradiation by light with wavelengths in the green part of the spectrum excites electrons into one of two energy levels E_2 and E_3. From these levels the electrons fall into the metastable level E_1, from which they fall back into the ground state, emitting radiation in the red, wavelength 0.6943 μm.

This last transition, from E_1 to E_0, is normally slow, and the population of electrons in the E_1 level can be built up to a high density before one electron falls to E_0, emitting a photon of frequency v given by $hv = E_1 - E_0$. The peculiarity of laser action is that this photon can now stimulate other electrons in E_1 to undergo the transition to E_0 much earlier than they would normally do. As each transiting electron produces a photon, more and more transitions are stimulated in a very short period of time by an avalanche process, and an intense burst of radiation of single wavelength is emitted. A characteristic of the stimulation process is that the emitted photon is exactly in phase with the photon which stimulated its emission. The radiation is therefore *coherent.*

The ruby crystal is in the form of a rod, the ends of which are optically polished and semisilvered. Photons are reflected back and forth along the rod many times in order to stimulate as many transitions as possible. Electrons are pumped into the higher energy levels by exposure to radiation, usually provided by a xenon flash tube. Above a certain intensity of pumping radiation, laser action occurs and an intense red beam flashes from the end of the rod. The flash lasts for about 5×10^{-4} s with a power of 10^4 watts over a cross-section of 10^{-4} m^2, and with an angular spread of less than one degree. The very high power density allows only of discontinuous operation of a ruby laser.

Laser action has been found in various doped insulating crystals, in some degenerately doped semiconductors (mainly III–V compounds), in solutions of

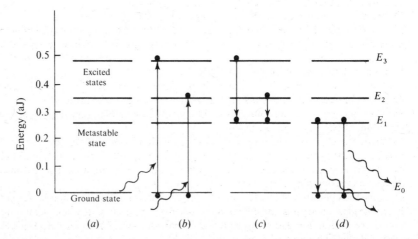

Fig. 14.24. The operation of a ruby laser. (a) Energy levels for an excited electron due to chromium ions in Al_2O_3. (b) Optical excitation of electrons from ground state to excited state (pumping). (c) Excited electrons falling into metastable state. (d) Stimulated transition from metastable state to ground state with the emission of coherent photons.

fluorescent dyes, and in certain gas mixtures. Gas and dye lasers can generally be operated in a continuous mode.

Laser crystals must be of high perfection; flaws, crystal defects and unwanted impurities can reduce efficiency by scattering photons out of the main beam. A defect in insulating crystals which has a most serious effect as a scatterer is the 'negative crystal'. This is a void, formed during crystal growth, within the crystal which has sides corresponding to crystal facets. It is a particularly stable defect, due to being formed of low energy surfaces, and much effort is being put into the growth of crystals without these defects.

Lasers find several uses. Laser beams, because of their coherence and closely controlled wavelength are ideal for information transmission, and are used in satellite communication. The intense energy carried in a fine beam can be used for micromachining and welding, and also for delicate eye and brain surgery. The coherence of their radiation can be used for information storage by holography.

QUESTIONS

1. (a) What are the differences between semiconductors and insulators?
   (b) What is a 'dielectric'?
   (c) Describe the various mechanisms which are responsible for electrical conduction in crystalline solids.
2. Silicon has the diamond cubic structure, with a lattice parameter of 0.543 nm. Calculate the number of atoms per lattice point, and the number of atoms per cubic metre. Assuming the silicon atoms to be perfect spheres in contact, calculate their diameter.
3. Explain what is meant by intrinsic and extrinsic semiconductors. Give three examples of each. List the important parameters which determine the electrical properties of a semiconductor.
4. Calculate the intrinsic carrier density of pure silicon from its room temperature intrinsic resistivity and the values of mobility given in Table 14.1. How does this compare with the value calculated from the energy gap?
5. Determine the concentration at room temperature (300 °K) of:
   (a) Donor impurities necessary to make germanium an extrinsic n-type semiconductor;
   (b) Acceptor impurities necessary to make germanium an extrinsic p-type semiconductor.
   Use the values of quantities given in Table 14.1.
6. Germanium is doped with 10^{-2} atomic per cent aluminium. Calculate the room temperature resistivity of this material, using quantities given in Table 14.1.
7. How do you expect the resistivity of the material in Question 6 to vary, as progressively larger quantities of arsenic are added?
8. Calculate the Hall coefficient for the doped germanium sample of Question 6.
9. What is meant by: carrier lifetime, trapping? Describe qualitatively how the resistivity of an intrinsic semiconductor varies with (a) increasing impurity concentration, (b) with increasing temperature from 4.2 °K.
10. Describe the operation of two semiconductor devices, other than the simple rectifier.
11. In simple glass structures, the diffusion coefficient of the potassium ion is smaller than that of the sodium ion. How do you think the substitution of K_2O for Na_2O in soda-glass will affect its electrical properties?
12. Define polarisation, displacement, relative permittivity. Describe the three possible contributions to the polarisation of a substance. Calculate the relative permittivity of argon at NTP, given that the polarisability of an argon atom is 1.43×10^{-40} farad m^2.

13. Derive the Clausius–Mosotti relation for a material containing N identical atoms/m^3 and polarisability α. If $N = 5 \times 10^{28}$ m^{-3}, and $\alpha = 2 \times 10^{-40}$ farad m^2, calculate its relative permittivity.

14. What is piezoelectricity, ferroelectricity? Would you expect to find piezoelectricity exhibited by a ferroelectric crystal?

15. The polarisation of a sample of barium titanate is found to be 0.24 C/m^2. Assuming this to be entirely due to relative motion of the titanium atoms, by what distance must these atoms have moved to produce the observed polarisation. The lattice parameter is 0.4 nm.

16. The index of refraction of sodium iodide is 1.7; its static dielectric constant is 6.60. What fractional contribution does ionic polarisability make to the total polarisability?

17. Explain the origin of dielectric loss. What uses may be made of this loss?

18. What is meant by a 'loss peak'? How would you measure the activation energy of a particular relaxation leading to a loss process? How would you go about producing a low-loss polymeric insulator?

19. What is meant by 'dielectric strength'? Is this a unique measure of the properties of a material? Describe the various processes contributing to the breakdown of dielectrics.

20. What do the initials LASER stand for? Describe the action of a laser. How does it differ from that of a maser? Give some possible applications for lasers.

FURTHER READING

A. R. von Hippel: *Dielectric Materials and their Applications.* Wiley (1954).

W. C. Dunlap: *An Introduction to Semiconductors.* Wiley (1957).

N. Cusack: *The Electrical and Magnetic Properties of Solids.* Longmans (1958).

R. A. Smith: *Semiconductors.* Cambridge University Press (1959).

T. S. Hutchinson and D. C. Baird: *The Physics of Engineering Solids.* Wiley (1963).

J. C. Anderson: *Dielectrics.* Chapman and Hall (1964).

J. J. O'Dwyer: *The Theory of Dielectric Breakdown of Solids.* Oxford University Press (1964).

N. F. Mott and R. W. Gurney: *Electronic Processes in Ionic Crystals*, 2nd edition. Dover (1964).

S. Wang: *Solid State Electronics.* McGraw-Hill (1966).

G. C. Jain: *Properties of Electrical Engineering Materials.* Harper and Row (1967).

N. G. McCrum, B. E. Read and G. Williams: *Anelastic and Dielectric Effects in Polymeric Solids.* Wiley (1967).

N. E. Hill, W. E. Vaughan, A. H. Price and Mansel Davis: *Dielectric Properties and Molecular Behaviour.* Van Nostrand (1969).

15. Magnetic properties of solids

15.1 Introduction

This chapter begins with the basic phenomenon and theory of magnetism and concludes with a description of useful magnetic materials.

Some symbols used in this chapter are defined as follows. H is the magnetic field strength; in the SI system of units it is measured in amperes per metre. Where a magnetised specimen has free poles induced on its surface H is the net field strength due to the externally applied field minus the demagnetising field created by the free poles. M is the induced magnetic moment per unit volume, called the intensity of magnetisation or, simply magnetisation. The dimensionless ratio M/H is called the magnetic susceptibility, χ. B is the (magnetic) induction or flux density, measured in teslas or webers per metre^2. $B = \mu_0\mu_r H$, where μ_r, the relative permeability of the material ($\mu_r = 1$ for vacuum), is dimensionless, and μ_0 is the permeability of free space ($= 4\pi \times 10^{-7}$ henry/m). μ_0 has no physical meaning, it is merely a consequence of choosing to work with the SI system of units. $B = \mu_0(H + M)$ and $\chi = \mu_r - 1$. χ and μ_r are constants for simple materials, but vary with H in a complicated way for ferromagnetics, as will be seen below.

The magnetic behaviour of materials can be classified into five types, two of which, ferromagnetic and ferrimagnetic, are usually regarded as magnetic. In all materials, the application of a magnetic field causes the angular frequency of the electrons in their orbits to change so as to create a magnetic field in opposition to the applied field. The susceptibility is negative and for all materials is $\simeq -10^{-5}$ in agreement with theory. The effect is known as *diamagnetism* and a material which displays no other magnetic behaviour is said to be diamagnetic. Examples are the rare gases, liquids such as water, mercury and benzene, and solids such as bismuth and magnesium. Values of the susceptibilities of the more common diamagnetics are given in Table 15.1. Diamagnetism is of little engineering importance.

Some materials, even in the absence of a magnetic field, have atoms that are permanent magnetic dipoles, due to electron spin. Every electron spins on its own axis as it travels in orbit around the nucleus, and acts as a magnetic dipole (see Chapter 1). The magnetic moment of a single spinning electron is called the Bohr magneton μ_B and equals $\pm eh/4\pi m$ ($= 9.27 \times 10^{-24}$ A m^2.) The electron spin is in one of two senses such that the dipole supports or opposes any applied magnetic field. Usually the electron spins are paired and their magnetic moments cancel, but unbalanced spins leave the atom as a permanent dipole. From this it might be assumed that an atom with an even number of electrons would have no dipole moment, and that an atom with an odd number of electrons would have a dipole moment of one Bohr magneton. This is in general true for simple atoms, but the elements with incomplete inner shells such as the transition metals and the rare earths, are more complicated. The interaction which produces

587

Table 15.1. *Susceptibilities of some diamagnetic and para-magnetic materials at 20 °C*

	Diamagnetic		Paramagnetic
Material	$\chi = \mu_r - 1 \, (\times 10^5)$	Material	$\chi = \mu_r - 1 \, (\times 10^3)$
Cu	−0.96	Li	0.0228
Ag	−2.5	Na	0.0074
Au	−3.4	K	0.0049
Zn	−1.25	Mg	0.0055
Hg	−2.9	Ca	0.0213
Graphite	−61.0	Al	0.0202
Diamond	−2.1	Ti	0.192
Si	−3.2	V	0.45
Ge	−7.1	Mn	0.90
Pb	−1.57	Nb	0.25
Se	−1.73	Pd	0.80
Te	−2.42	$CrCl_3$	1.5
Al_2O_3	−0.5	CrO_3	1.7
$BaCl_2$	−2.0	$MnSO_4$	3.6
NaCl	−1.2	Fe_2O_3	1.4
		$Fe_2(SO_4)_3$	2.2
		$FeCl_2$	3.7
		$FeSO_4$	2.8
		CoO	5.8
		$CoSO_4.H_2O$	2.0
		$NiSO_4$	1.2

Table 15.2. *Configuration of 3d-electrons and resultant spin for elements in the first transition series*

Element (free atom), ion (dilute) or metal	Atomic number	No. of 3d-electrons	Resultant spin in Bohr magnetons	
Calcium	20	0	0	
Scandium	21	1	1	↑
Titanium	22	2	2	↑↑
Vanadium	23	3	3	↑↑↑
Chromium and Cr^{2+}	24	4	4	↑↑↑↑
Cr^{3+} (chromic)		3	3	↑↑↑
Manganese and Mn^{2+}	25	5	5	↑↑↑↑↑
Mn^{3+} (manganic)		4	4	↑↑↑↑
Iron and Fe^{2+}	26	6	4	↑↑↑↑↑↓
Fe^{3+} (ferric)		5	5	↑↑↑↑↑
metal		7.4	2.2	4.8↑ 2.6↓
Cobalt	27	7	3	↑↑↑↑↑↓↓
metal		8.3	1.7	5.0↑ 3.3↓
Nickel	28	8	2	↑↑↑↑↑↓↓↓
metal		9.4	0.6	5.0↑ 4.4↓
Copper and Cu^+	29	10	0	↑↑↑↑↑↓↓↓↓↓
Cu^{2+} (cupric)		9	1	↑↑↑↑↑↓↓↓↓

ferromagnetism, to be described below, favours the situation where several of the electrons in the incomplete shell do not pair off with opposite spins, and the resulting magnetic dipole is greater than one Bohr magneton. The dipole strengths for the elements in the first transition period are given in Table 15.2. The dipole strength in a chemical compound depends upon the valency displayed by the atoms in the compound. The values for ferrous and ferric ions are also shown in the table.

Materials with permanent dipoles are of four types, depending on the nature of any interaction between the dipoles (Fig. 15.1). In a *paramagnetic* material there is no interaction and in the absence of a magnetic field the energy of a dipole is constant, irrespective of the direction of its magnetic vector. The dipoles are randomly oriented and there is no net magnetic moment over a large number of atoms. When an external magnetic field is applied the energy of the dipoles which are aligned parallel to the field, i.e. with their magnetic vectors in the same direction as the field, is lowered by an amount $\mu_0\mu_B H$, and the energy of those dipoles which are aligned antiparallel, i.e. with vectors in the opposite direction to the field, is raised by an equal amount. The application of the field has produced two energy levels, differing by $2\mu_0\mu_B H$. At the absolute zero of temperature, all dipoles will assume the lowest energy and produce a magnetisation $M = N_0\mu_B$ where N_0 is the total number of dipoles per unit volume. At any finite temperature some fraction of the dipoles will be excited to the higher energy state. The ratio of antiparallel to parallel dipoles is given by Boltzmann statistics:

$$\frac{N_a}{N_p} = \exp\left(-\frac{2\mu_0\mu_B H}{kT}\right). \tag{15.1}$$

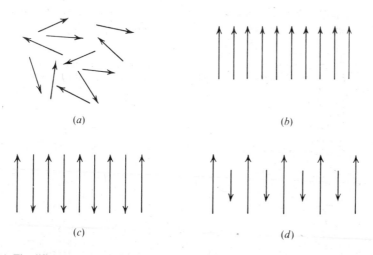

(a)

(b)

(c)

(d)

Fig. 15.1. The different types of magnetic material. (*a*) Paramagnetic; dipoles randomly arranged, in the absence of a field. (*b*) Ferromagnetic; dipoles permanently aligned below the Curie temperature. (*c*) Antiferromagnetic; neighbouring dipoles are antiparallel, below the Néel temperature. (*d*) Ferrimagnetic, two groups of dipoles of unequal strengths, aligned antiparallel, below the Curie temperature.

The net magnetisation is given by:

$$M = \mu_B(N_p - N_a) = N_0\mu_B \tanh\left(\frac{\mu_0\mu_B H}{kT}\right). \tag{15.2}$$

This relation is plotted in Fig. 15.2. For $x \ll 1$, $\tanh(x) \simeq x$, and for $x \gg 1$, $\tanh(x)$ approaches unity. Thus at low temperatures and high fields the magnetisation approaches the value $N_0\mu_B$, that is when all the dipoles are aligned parallel to the field. In practice, $\mu_0\mu_B H$ is generally $\ll kT$, and at low fields

$$M = \frac{N_0\mu_0\mu_B^2 H}{kT}, \tag{15.3}$$

and thus the initial susceptibility is given by

$$\chi = \frac{N_0\mu_0\mu_B^2}{kT} = \frac{C}{T}. \tag{15.4}$$

This temperature dependence of the susceptibility was found experimentally by Curie in 1895 and is known as the Curie law. $C = N_0\mu_0\mu_B^2/K$ is called the Curie constant. This result is only strictly accurate for a gas or dilute solution. The interaction between the closely packed atoms of a solid causes the susceptibility to obey a modified law, proved theoretically by Weiss in 1907. This Curie–Weiss law is

$$\chi = \frac{C'}{T - \theta} \tag{15.5}$$

where θ, the paramagnetic Curie temperature, is a constant of the material. For most paramagnetic materials, θ is only a few degrees Kelvin or less, and the Curie law may be considered to be quite accurate, except at very low temperatures. In the next type of material to be described, the ferromagnetic, the value of θ is sufficiently high that the Curie–Weiss law must be used.

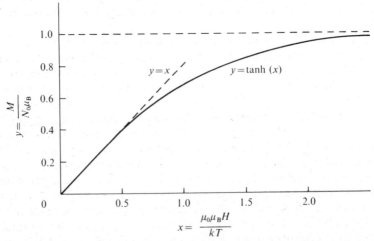

Fig. 15.2. Magnetisation as a function of H/T for a paramagnetic, (15.2).

The average value of N_0, reasonable for almost all solids, is 5×10^{28} m^{-3}. Inserting the appropriate values of the other constants, assuming an unbalanced spin of one Bohr magneton, gives the Curie constant C a value of ~ 0.4 and at room temperature $\chi \simeq 10^{-3}$. This is the right order of the experimentally observed values for many paramagnetic salts. It is small but sufficient to overcome the opposing diamagnetism which is always present. The unbalanced spin in paramagnetic metals, due to band structure effects, is usually only a small fraction of one Bohr magneton, and $\chi \simeq 10^{-5}$. The susceptibilities of some paramagnetics are given in Table 15.1.

In a *ferromagnetic* material, a very important magnetic type, there is an interaction between the dipoles such that even without an external magnetic field they tend to align parallel to and in the same sense as each other, a phenomenon called spontaneous magnetisation. The direction of alignment is constant only across microscopic regions, called domains, and a bulk specimen will not normally show a net magnetisation. The all important influence of domains on the magnetic behaviour is described later. The ferromagnetic elements are bcc iron, fcc nickel, cph cobalt and gadolinium.

In an *antiferromagnetic*, on the other hand, the neighbouring spins tend to align themselves in opposite directions (antiparallel), so there is no magnetisation even within a domain. Examples of antiferromagnetics are FeO, NiF$_2$ and various salts of manganese.

Finally, a *ferrimagnetic* material has an interaction such that neighbouring spins are again antiparallel but they are of unequal strength and there is a net magnetisation. The original materials of this class, from which the name is taken, were the ferrites. These are a group of materials based on iron oxide, of which a naturally occurring example is magnetite Fe$_3$O$_4$, the lodestone of the ancients. The terminology is not perhaps happily chosen since antiferromagnetism is a special case of ferrimagnetism. Ferrites do not usually have as high magnetisation as ferromagnetics but they are non-conductors, which is important for the elimination of eddy current losses in high frequency applications. Reliable methods of controlling the composition during manufacture were discovered by Snoek in 1940.

The theory of the interaction between electrons was given in quantum mechanical terms by Heisenberg, in 1928, and later by Bethe and others. The term responsible for the interaction between dipoles, known as the exchange interaction, is a function of the ratio of the interatomic distance to the radius of the incomplete shell, see Fig. 15.3. At large distances the exchange forces are small and the material is paramagnetic. As the ratio is decreased, a positive interaction develops, leading to parallel alignment of adjacent dipoles and ferromagnetism. Only four elements come into this category: gadolinium, nickel, cobalt and iron. Finally, at small ratios of interatomic distance to shell size, the exchange becomes negative and, in appropriate crystal structures, causes antiferromagnetism or ferrimagnetism. Manganese in the pure state has a slight negative interaction and is antiferromagnetic, but it can be made ferromagnetic by increasing the interatomic distance by alloying; for example, MnCuAl$_2$ one of a series of manganese alloys investigated by Heusler in 1903, and MnBi (Bismanol). Gamma iron (austenite) has a smaller interatomic distance

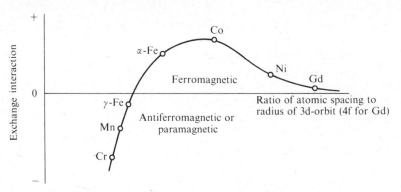

Fig. 15.3. Variation of exchange interaction with atomic spacing for some transition metals (after Bethe).

than ferrite and the exchange interaction is negative; but it is not antiferromagnetic on account of the structure (see the next section).

The temperature dependence of the magnetic properties for each of the classes of magnetic materials is quite distinct. Diamagnetism is independent of temperature, but in the other magnetic types the thermal agitation tries to randomise the ordering of the spins created by the exchange interaction and applied field. The temperature dependence of susceptibility for the magnetic types (other than diamagnetism) is shown in Fig. 15.4.

The magnetic behaviour of a ferromagnetic is dependent on the temperature in a complex manner. The variation of the susceptibility at fixed field strengths is shown in Fig. 15.4(*b*). At low field strengths it increases at first with rising temperature, but at high strengths it remains approximately constant. As the temperature approaches a critical value, the susceptibility drops sharply due to the effect of thermal agitation and approaches zero at the *ferromagnetic Curie temperature*. Above this temperature the material behaves as a paramagnetic and the susceptibility, now small, obeys the Curie–Weiss law given previously for paramagnetics. The paramagnetic Curie temperature, θ is slightly above the ferromagnetic Curie temperature, θ_f. Typical values of Curie temperatures are given in Table 15.3.

The initial susceptibility of an antiferromagnetic increases at first with temperature (Fig. 15.4(*c*)), but above the Néel temperature, θ_n, the antiferromagnetic equivalent of the Curie temperature, it also exhibits paramagnetism. The paramagnetic susceptibility of an antiferromagnetic obeys a modification of the Curie–Weiss law:

$$K = C''/(T + \theta_n). \tag{15.6}$$

The behaviour of a ferrimagnetic is similar to that of a ferromagnetic.

15.2 Crystal structure

In an antiferromagnetic the dipoles of neighbouring atoms are aligned in opposite directions and the crystal structure must therefore consist of two interpenetrating lattices such that the spins in either lattice are all parallel but are

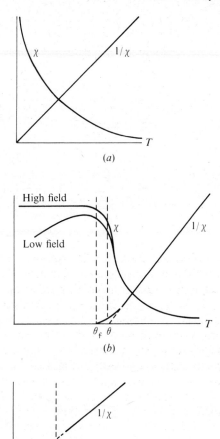

Fig. 15.4. Temperature variation of the susceptibility χ (schematic). (a) Paramagnetic. (b) Ferromagnetic. (c) Antiferromagnetic; ferrimagnetic is similar.

Table 15.3. *Ferromagnetic and paramagnetic Curie temperatures ($^{\circ}K$) of the ferromagnetic elements*

Element	θ_f	θ
Iron	1043	1101
Cobalt	1394	1408
Nickel	631	650
Gadolinium	293	317

antiparallel to the spins in the other lattice. Crystal structures which satisfy this arrangement are body-centred cubic, sodium chloride and caesium chloride (see Fig. 2.6 and Fig. 3.2). A face-centred cubic material cannot be antiferromagnetic, as the structure will not split into two interpenetrating simple cubic lattices although it will into four; the close-packed hexagonal lattice is in the same category. This is the reason why neither austenite (fcc) nor chromium (cphex), both of which have negative exchange interactions, are antiferromagnetic.

The behaviour of ferrites is also explained by reference to their crystal structures. Magnetite, Fe_3O_4, for example, may be thought of as a mixture of ferrous and ferric oxides, FeO and Fe_2O_3, though in fact it cannot be formed merely by mixing the two oxides. A ferrous ion Fe^{2+} has a moment of 4 Bohr magnetons and a ferric ion Fe^{3+} of 5 Bohr magnetons. If the interaction were entirely ferromagnetic, magnetite would have a total spin of 14 Bohr magnetons per molecule. However, the saturation magnetisation of magnetite corresponds to only 4 Bohr magnetons per molecule. The difference can be explained by considering the crystal structure of magnetite.

Magnetite has the spinel structure (Fig. 15.5), which is a complex arrangement of relatively large oxygen ions in an fcc lattice with the much smaller metal ions in interstitial positions. There are two types of interstitial site. In the *A* sites the metal ion is surrounded tetrahedrally by four oxygen ions and in the *B* sites octahedrally by six oxygen ions. The antiferromagnetic type of interaction exists between metal ions in the *A* sites and metal ions in the *B* sites. In magnetite half the Fe^{3+} ions are in *A* sites and half in *B* sites; their moments therefore

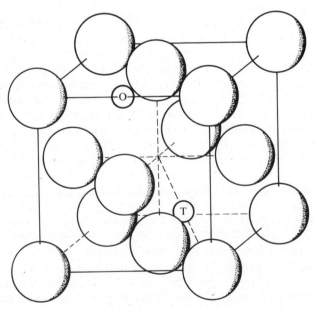

Fig. 15.5. Crystal structure of magnetite, Fe_3O_4, showing one eighth of a unit cell. ○ 32 oxygen ion sites (all occupied). ◎ 32 octahedral sites, 8 of which are occupied by Fe^{3+} ions. ⊕ 64 tetrahedral sites, 8 of which are occupied by Fe^{2+} ions and 8 by Fe^{3+} ions.

cancel. The Fe^{2+} ions all go into B sites and are responsible for 4 Bohr magnetons per molecule.

Other ferrites are obtained by replacing the Fe^{2+} ions by other divalent metallic ions. Although chemically equivalent they do not physically replace the Fe^{2+} ions in the crystal structure. If they did ions such as Zn^{2+} or Cd^{2+} would be expected to reduce the value of the saturation magnetisation, but this is found to be not so. These ions preferentially occupy A sites, forcing more Fe^{3+} ions into B sites (which are not fully occupied) and their addition, at least up to several per cent, results in an increase in saturation magnetisation. It is possible, by adjusting the composition, to produce ferrites with a wide range of magnetic properties. There are also several materials other than those based on magnetite which show ferrimagnetic behaviour.

15.3 Ferromagnetic magnetisation and hysteresis

For a ferromagnetic[†] the magnetic moment is not proportional to the applied field but exhibits hysteresis, see Fig. 15.6. If the specimen is initially unmagnetised, either by cooling it from above the Curie temperature in zero field or by applying an alternating magnetic field of decreasing strength, the magnetisation curve is given by OA. The slope initially increases and later decreases until saturation is achieved. The concept of permeability B/H is therefore physically meaningless as it varies with the field, but an initial permeability μ_i is defined as the slope of the curve at the origin and a maximum differential permeability μ_{max} by the slope of that tangent passing through the origin with maximum inclination. A mean permeability is used in engineering design calculations, for soft magnetic materials.

When the applied field is reduced, the induction does not fall back along the original curve OA but decreases at a much slower rate, along the line AC. At zero field, a flux density B_r, the remanence or retentivity, remains. In order to reduce the induction to zero a reverse field $-H_c$, the coercivity or coercive force, must be applied. On increasing the reverse field the induction follows the curve CDE to saturation in the opposite sense. If the field is now returned to the positive sense the induction follows along $EFGA$, symmetrical curve to $ACDE$, and a hysteresis loop is described. Other loops may be generated at any part of the cycle upon reduction or reversal of the field, e.g. loop LM in the figure. The curve $ACDEFG$ is the envelope of all such minor loops and is achieved only by taking $\pm H$ above the value which produces saturation of magnetisation.

Materials have various shapes of loop, ranging from the very narrow, with low remanence and coercivity (Fig. 15.6(*b*)), to the almost square, with high remanence and coercivity (Fig. 15.6(*c*)). Some typical values are given for common magnetic materials in Tables 15.5 and 15.6. The former are known as magnetically soft and the latter as magnetically hard, the names generally corresponding with mechanical properties. Thus pure annealed iron and hardened steel are the original soft and hard magnetic materials. Soft magnetic material is used in AC applications such as transformer laminations and alternator cores, where

---

[†] In this and subsequent sections on domain theory, ferrimagnetics are to be included in the term ferromagnetic, since both are spontaneously magnetised into domains.

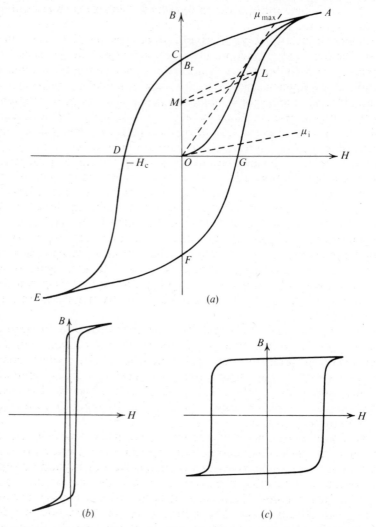

Fig. 15.6. B–H loops for ferromagnetic materials. (*a*) Hysteretic behaviour of induction B as a function of applied field H. (*b*) 'Soft' ferromagnetic with narrow loop. (*c*) 'Hard' ferromagnetic with square loop.

hysteresis energy losses, proportional to the loop area, must be minimised. Hard magnetic material is required for permanent magnets and, in recent years, computer stores, since magnetisation can be rapidly switched between two states. The detailed requirements of soft and hard magnets are discussed later.

15.4 Domain theory

The domain theory of magnetisation was postulated in 1907 by Weiss and developed by many workers, recently by Néel. An unmagnetised ferromagnetic is divided into microscopic regions of spontaneous magnetisation, the domains.

The spins are aligned within a domain but the direction of magnetisation changes from one domain to the next. Within a domain the magnetisation vector corresponds with certain crystallographic directions in which magnetisation is easiest. The observed magnetisation curve results from the increasing volume of material with its direction of magnetisation favourably oriented in the direction of the applied field, either by domain boundary movement, which increases the size of some domains at the expense of others, or by rotation of the magnetisation vectors in the domains, as the externally applied field is increased. These processes may be reversible or irreversible as described below. The local intensity of magnetisation remains constant at the saturated value.

The first stage of magnetisation determining the initial permeability, is caused by reversible boundary movement enlarging the favourably oriented domains (and to small extent, reversible rotation of the magnetisation vectors). The second stage, during which the induction increases rapidly is due to irreversible domain wall motion and, if that is prevented, to irreversible rotation of the magnetisation vectors which suddenly flip over from an unfavourably oriented direction to an easy direction of magnetisation which is closer to the applied field. The final stage, approaching saturation, is again reversible, caused by the gradual rotation of the magnetisation vectors away from the easy direction towards the applied field. The stages on the magnetisation curve are shown in Fig. 15.7.

On reducing the field, the magnetisation vectors return to the nearest easy direction, thus determining the drop in the magnetisation from saturation to remanence. From the remanence point, through the coercivity point to approaching reverse saturation, the mechanism is irreversible boundary motion or, if prevented, irreversible rotation of the magnetisation vectors. The coercivity is dependent on the field strength required. Finally there is again reversible rotation of the vectors away from the easy directions.

The ease of domain boundary movement is the primary variable in determining the shape of the *B–H* loop and it is extremely structure sensitive. In contrast, the saturation magnetisation is a structure insensitive property and is determined by the overall composition. If movement is easy, the loop is small and the material magnetically soft. If movement is restrained, the loop is broad and the material magnetically hard. The influence of microstructure on domain boundary movement is the proper study of the physical metallurgy of magnetic materials and will be considered in § 15.6.

Domains are formed so as to minimise the sum of several energy terms, which are due to exchange forces, magnetocrystalline anisotropy, magnetostatic effect and magnetostriction. The exchange forces, already discussed, are responsible for an energy term which would be minimised if all spins were parallel throughout a specimen.

Magnetocrystalline anisotropy describes the variation of magnetic properties with crystal direction. Brillouin zone theory, discussed in § 3.8, shows how the energy of electrons varies with direction in a crystalline solid. Electrons are responsible for magnetism, and it is therefore not surprising that magnetic properties are influenced by crystallographic direction. It is found that the energy of a magnetic system is lower if all of the spins are aligned along a particular direction in the crystal. This direction is known as the easy direction of

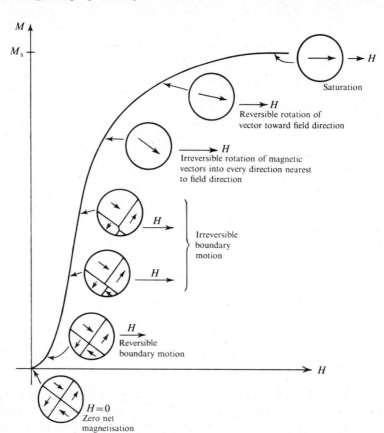

Fig. 15.7. Domain behaviour during various stages of magnetisation.

magnetisation; initial permeability is highest and saturation magnetisation is reached soonest, along this easy direction. Fig. 15.8 shows the magnetisation curves for single crystals of iron in three directions $\langle 100 \rangle$, $\langle 110 \rangle$ and $\langle 111 \rangle$. The $\langle 100 \rangle$ direction can be seen to be the easy direction. In nickel, also cubic, the easy direction of magnetisation is $\langle 111 \rangle$, and in cobalt it is along the hexagonal axis $\langle 0001 \rangle$. In the analysis of this effect crystal anisotropy constants are defined which relate the magnetisation energy to the direction of magnetisation (with respect to the crystal axes). The ratio of the energies in the hard and soft directions of magnetisation is called the anisotropy constant, K. The value of K is strongly temperature dependent, and may be as high as 1000, in for example, cobalt. A calculation, from first principles of the value of K, or even a prediction of the easy direction, requires detailed quantum mechanical considerations, and is quite beyond the scope of this book.

Free poles are created at a boundary whenever the component of magnetisation normal to the boundary changes across the boundary. The boundary may be a free surface or a domain wall. The magnetostatic energy term is due to the self-demagnetising field which results from these free poles in opposition to the local magnetisation. It can be calculated for some simple shapes. The

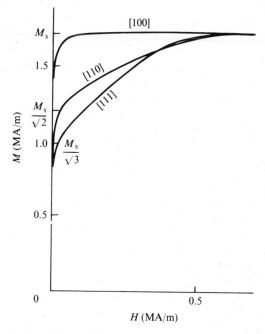

Fig. 15.8. Magnetisation of iron single crystals along different crystallographic directions (after Honda and Kaya).

demagnetising field and energy per unit volume are given by NM_S and $\mu_0 NM_S^2$ respectively where M_S is the intensity of magnetisation and N a numerical factor. For an iron sphere $N = 4\pi/3$, and $M_S = 1.7 \times 10^6$ A/m, giving a demagnetising field of 7.1×10^6 A/m, and an energy per unit volume of 15.6 MJ/m^3. For a fibre with a length/diameter ratio of ten, $N = 4\pi/30$. Hence a specimen tends to break down into fibre-shaped domains with opposing magnetisation vectors in order to minimise the magnetostatic energy, see Fig. 15.9.

Lastly, the magnetostriction term is the elastic strain energy arising from the change of dimensions occurring on magnetisation. With different directions of magnetisation in adjacent domains in a crystal internal stresses and strains are developed. The (Joule) magnetostriction λ is defined as the fractional change in length in the direction of magnetisation. λ is a function of the field strength, the variation for some typical materials being shown in Fig. 15.10. It may be positive or negative, i.e. the material may increase or decrease in length upon magnetisation. The saturation magnetostriction, λ_s, is the fractional change in length upon magnetisation from zero to saturation. With an increase in length λ in the direction of magnetisation, for there to be no change in volume, requires a decrease in length $\lambda/2$ in directions normal to the direction of magnetisation.

15.5 Domain boundaries

Domains were first observed in 1931 by Bitter and, independently, by Hamos and Thiessen. Bitter applied a suspension of iron oxide to a polished magnetised surface and observed under a microscope patterns which he attributed to the

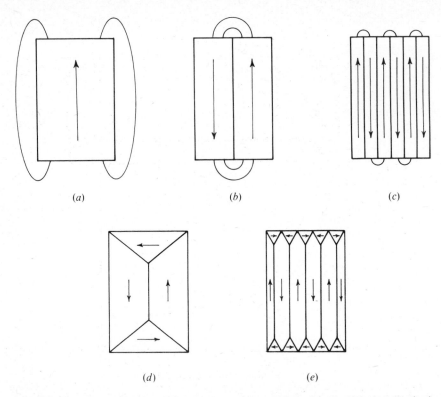

Fig. 15.9. The formation of domains in a ferromagnetic. (*a*) Single domain, high demagnetisation field. (*b*) Two domains, demagnetisation field considerably reduced. (*c*) Several domains, demagnetising field almost zero. (*d*) Closure domains, demagnetisation field now zero, but anisotropy and magnetostriction contribute to energy of closure domains. (*e*) Many small closure domains reduce total volume with unfavourable magnetocrystalline energy and magnetostriction.

inhomogeneous field at domain boundaries. Later workers developed finer and more stable colloids. It has also been shown that the original maze-like patterns were due to the distorted layer produced by mechanical polishing and that electropolishing is necessary to reveal the true domain structure of the original material. The two types of pattern are contrasted in Fig. 15.11. Other recent techniques use polarised light, which may be reflected or transmitted (through thin films or transparent specimens), transmission electron microscopy and X-ray microscopy.

The direction of magnetisation in a domain can be revealed in various ways, the simplest being to lightly scratch the surface and observe if any free poles attract the colloid. No poles are formed if the scratch is parallel to the magnetisation vector. The sense of the vector can be found by applying a weak magnetic field perpendicular to the surface.

Detailed studies of domains in single crystals of iron–3.8 per cent silicon were made in 1949 by Williams and co-workers. They showed for the first time the shape of domains and the effects due to stress and fields. Their experiments strikingly confirmed the quantitative predictions of domain theory. Figure

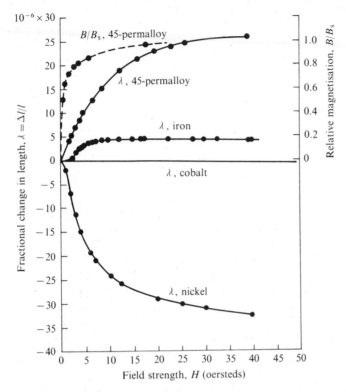

Fig. 15.10. Magnetostriction of some common materials. (From R. M. Bozorth, *Ferromagnetism*. © 1951 by Litton Educational Publishing, Inc. Reprinted with permission of Van Nostrand Reinhold Company.)

Fig. 15.11. Bitter patterns: (*a*) mechanically polished surface; (*b*) same surface after electropolishing. (From R. M. Bozorth, *J. Phys. Radium*, **12**, 308 (1957).)

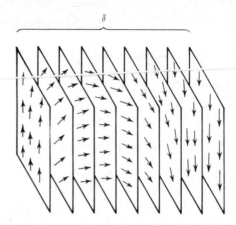

Fig. 15.12. Schematic representation of rotation of spin in a 180° Bloch wall.

15.11(b), shows the typical tree pattern of domains and how they are modified by stress: with positive magnetostriction, the domains magnetised along the tension axis grow at the expense of the cross-magnetised domains.

The boundary between domains is sometimes called a Bloch wall. Bloch showed in 1932 that the direction of magnetisation could not change abruptly at a boundary, but that the change must be gradual. The exchange forces allow of only a minute change in direction of spin from one atom to the next. Thus to achieve a change in direction of, usually, 90° or 180° the wall must be many atoms thick. The structure of a boundary is shown schematically in Fig. 15.12.

If the exchange energy were the only term to be considered, the wall would be infinitely thick. However, magnetocrystalline anisotropy energy and magneto-strictive energy, which are both a function of the volume of the wall, tend to restrict its thickness. Magnetostatic energy is generally not important, as the wall will try to align itself so that there is no change in the normal component of H, and hence no free poles, on the wall.

A simple estimate of the thickness δ and energy U of a Bloch wall can be made as follows. The exchange energy between a pair of atoms whose spins differ in direction by an angle ε is given by $A(1 - \cos \varepsilon)$, where A is the exchange energy per atom. For a 180° boundary, $\varepsilon = \pi a/\delta$ where a is the interatomic distance. The total number of atoms per unit area of wall is δ/a^3 and for small ε, $1 - \cos \varepsilon$ can be approximated to $\frac{1}{2}\varepsilon^2$. The exchange energy per unit area of wall is thus given by $\pi^2 A/2a\delta$. The magnetocrystalline energy term should be obtained by evaluating for each angle away from the easy direction of magnetisation, but can be approximated by assuming that as soon as spin deviates from an easy direction it takes on the maximum anisotropy energy, i.e. all of the atoms in the wall have maximum anisotropy. The anisotropy energy per unit area is then $K\delta$, where K is the magnetocrystalline anisotropy energy per unit volume. The magneto-strictive energy term, from simple elasticity theory, is $\frac{1}{2}E\lambda^2\delta$ per unit wall area (E is Young's modulus). The total wall energy per unit area is then:

$$U = \frac{\pi^2 A}{2a\delta} + (K + \tfrac{1}{2}E\lambda^2)\delta. \tag{15.7}$$

The actual wall thickness is that value of δ which makes U a minimum:

$$\delta = \left(\frac{\pi^2 A}{2a(K + \frac{1}{2}E\lambda^2)}\right)^{\frac{1}{2}} \tag{15.8}$$

and the corresponding value of the wall energy is:

$$U = \left(\frac{2\pi^2 A}{a}(K + \frac{1}{2}E\lambda^2)\right)^{\frac{1}{2}}. \tag{15.9}$$

The value of A can be estimated since at the Curie temperature thermal energy overcomes the exchange interaction and hence $A \simeq k\theta_f$. For iron, $\theta_f \simeq 10^3 \, °K$, $a = 0.4 \, nm$, and $K = 4 \times 10^4 \, J/m^3$ ($E \simeq 2 \times 10^{11} \, N/m^2$ and $\lambda = 2 \times 10^{-5}$; the magnetostrictive energy term is small compared to the anisotropy term and may be neglected). The values obtained are: $\delta \simeq 0.1 \, \mu m$ and $U \simeq 5 \, mJ/m^2$. These results are confirmed by more sophisticated calculations and are in remarkably good agreement with experimental results.

15.6 Domain boundary motion

The connection between the shape of the magnetisation curve and domain boundary motion has already been mentioned (§ 15.4); in particular, the greater the resistance to motion the wider the hysteresis loop becomes. Boundary motion only occurs because magnetocrystalline anisotropy makes rotation of the spin moments energetically difficult. Boundary motion is then inhibited by stress and magnetic inhomogeneity.

Since the direction of magnetisation changes across a domain boundary, the direction of the magnetostrictive strain will also change (except for a 180° boundary). A rather complicated stress pattern develops at a domain boundary, as a result of which a boundary interacts with any non-homogeneous stresses in the material. The interaction is stronger, the greater the value of the magnetostrictive coefficient.

The presence of magnetic inhomogeneities, such as non-magnetic inclusions, in the middle of a domain, causes the appearance of free poles. These are suppressed when a domain boundary intersects the inclusion and closure domains are formed, see Fig. 15.15(d). The attraction of the boundary to the inclusion is stronger the greater the value of the magnetocrystalline anisotropy. Both types of interaction will be studied in greater detail below.

The width of a domain boundary affects its ease of movement. In a wide boundary, for a small movement each moment rotates through a small angle, and motion is easier than in a narrow boundary. As can be seen from the previous section, a wide boundary is favoured by low values of K and λ. Through all of these causes boundary motion is therefore favoured by low values of K and λ.

The initial permeability is determined by reversible motion and Becker in 1930 has estimated its theoretical maximum value in the following way. Consider a strip of cross-sectional area A in a ferromagnetic crystal with positive magnetostriction. Assume there are longitudinal residual stresses varying in a sinusoidal manner, see Fig. 15.13. Below the Curie temperature and under no applied magnetic field the domain boundaries will form at the stress nodes and the direction of magnetisation will vary along the strip so as to relieve the original

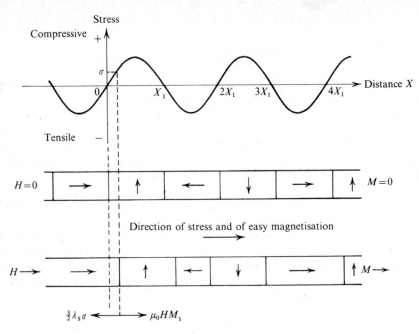

Fig. 15.13. Initial permeability: movement of domain boundaries in the presence of uniformly varying internal stress (after Becker).

strains, i.e. along the strip axis in the regions under tension and perpendicular to the strip axes in the regions of compression.

Let the length of each domain be initially X_1, and the variation of stress with distance be:

$$\sigma = \sigma_m \sin \pi \frac{X}{X_1}. \tag{15.10}$$

If a field H is applied along the strip, the boundaries will move so as to increase the length of those domains magnetised in the direction of the field, and to decrease the length of those magnetised in opposition to H. The change in the normal component of magnetisation across a boundary is the saturation intensity M_S and hence a density M_S of free poles is established at the boundary. The magnetic pressure on the boundary is $\mu_0 H M_S$ per unit area. The motion of the boundary is resisted by elastic forces. Consider a boundary, having moved a distance X, being displaced a further distance dX. The swept volume element $A\, dX$ changes its strain from $-\frac{1}{2}\lambda_s$ to λ_s, a change of $\frac{3}{2}\lambda_s$, and the change in elastic strain energy is $dE = \frac{3}{2}\lambda_s \sigma A\, dX$. The elastic pressure on the boundary resisting its motion is:

$$\frac{1}{A}\frac{dE}{dX} = \frac{3}{2}\lambda_s \sigma = \frac{3}{2}\lambda_s \sigma_m \sin \pi \frac{X}{X_1}. \tag{15.11}$$

The boundary reaches a stable position X' when the magnetic pressure just balances the elastic pressure:

$$\mu_0 H M_S = \frac{3}{2}\lambda_s\sigma_m \sin \pi \frac{X'}{X_1}. \tag{15.12}$$

Each boundary undergoes a similar displacement and the net magnetisation is $M = M_S X'/X_1$ in the direction of H. The susceptibility

$$\chi = \frac{M}{H} = \frac{2\mu_0 M_S^2 X'/X_1}{3\lambda_s\sigma_m \sin (\pi X'/X_1)}. \tag{15.13}$$

The initial permeability $\mu_i = 1 + \chi_i$, and for ferromagnetics $\chi_i \gg 1$, and therefore $\mu_i \simeq \chi_i$. When X' is very small, the initial susceptibility

$$\chi_i = \frac{2\mu_0 M_S^2}{3\pi\lambda_s\sigma_m} \simeq \mu_i. \tag{15.14}$$

This equation is in agreement with the observation that maximum initial permeability is obtained in material with low internal stresses and low magnetostriction. The maximum value of this expression will be obtained when the internal stresses are due solely to the formation of the domains themselves, that is, σ_m is of the order of $\lambda_s E$. Then

$$\mu_i = \frac{2\mu_0 M_S^2}{3\pi\lambda_s^2 E}.$$

For iron at room temperature, $M_S = 0.17\ T$, $\lambda_s = 2 \times 10^{-5}$ and E is $20 \times 10^{10}\ \text{N/m}^2$, giving $\mu_i \simeq 10\,000$, a value which is close to the best experimental one.

In real materials the stress distribution will be more complicated than in the example just described. For small fields the boundary movement will still be reversible, but for large fields it will become irreversible. This is illustrated qualitatively by reference to Fig. 15.14 which shows a boundary between two domains moving in a non-uniform stress field. In the absence of any externally applied field the boundary will sit at the point of zero stress. The application of a field in the direction of the magnetisation of the left-hand domain will cause the boundary to move to the right against the resistance of the stress field (until the point A is reached), the magnetisation increasing with increasing field. The motion is reversible up to the point A, at which the boundary can suddenly jump to B for no increase in field, giving a jump in magnetisation. This is followed by a further steady rise to C. Upon reduction of the field, the motion is smooth until the point D at which a jump to E occurs, this time with a sudden reduction in magnetisation. The motion of this one boundary has given rise to a small hysteresis loop, and the complete B–H curve is the sum of many such small loops.

Magnetic inhomogeneities, the extreme case of which is the presence of a non-magnetic inclusion, can similarly cause irreversible boundary motion and hysteresis loops. Free poles are produced at any surface across which the normal component of magnetisation changes. For example, if a non-magnetic cube were embedded in a domain with the direction of magnetisation normal to one pair of cube faces, free poles would be produced at these faces (Fig. 15.15(a)). The resulting magnetostatic energy is reduced by the formation of

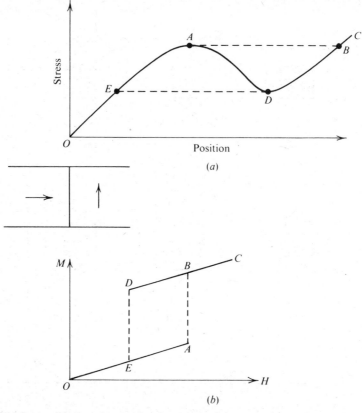

Fig. 15.14. Irreversible boundary motion due to stress inhomogeneity. (*a*) Domain boundary and non-uniform internal stress distribution. (*b*) Hysteresis loop resulting from motion of boundary through the internal stress distribution.

supplementary domains (Fig. 15.15(*b*)). If a 180° boundary now moves across the domain, it may interact with these supplementary domains and completely remove all of the free poles (Fig. 15.15(*c*)) with a sudden lowering of energy. The energy is further lowered when the boundary bisects the inclusion (Fig. 15.15(d) – note the closure domains) as the total boundary area is minimised. Further movement of the boundary away from the inclusion causes an increase in energy (Fig. 15.15(*e*)) with a sudden fall as the boundary breaks away from the supplementary domains (Fig. 15.15(*f*)). The energy of the boundary as a function of position and the force necessary to move the boundary is also shown in the figure. An electron micrograph of a domain boundary and supplementary domains around a non-magnetic inclusion is given in Fig. 15.16.

It will be seen from the above that irreversible wall motion does not take place smoothly but involves a series of jerks. These jerks are known as Barkhausen jumps and can be detected as a rustling or crackling noise in a loudspeaker connected to a small coil wound around a sample of the ferromagnetic material as the external field is changed.

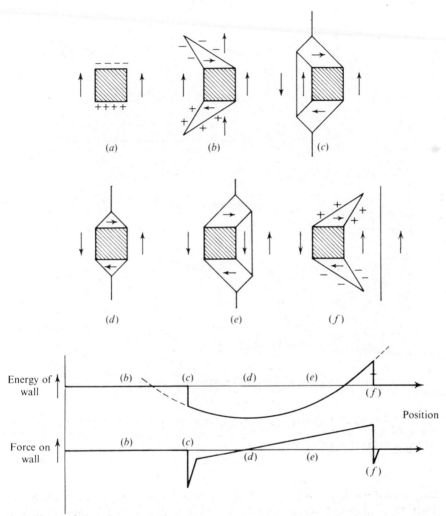

Fig. 15.15. Domain boundary movement past a non-magnetic inclusion (schematic). (*a*) Free poles on surface of inclusion. (*b*) Energy reduced by supplementary domains. (*c*) Domain boundary approaches and interacts with inclusion. (*d*) Inclusion bisected by boundary. (*e*) Boundary moving away from inclusion. (*f*) Boundary separated from inclusion. (+ and − indicate north- and south-seeking free poles respectively.) (After K. H. Stewart, *Ferromagnetic Domains*, Cambridge University Press (1954).)

15.7 Fine particles

The bulk magnetic properties have been shown in the preceding sections to be due to the domain structure and the movement of the boundaries. Below a critical size of specimen, domains become energetically undesirable and fine particles exist as single domains. This causes a strikingly different magnetic behaviour, in particular the coercive force, which is a measure of the ease of boundary movement, is raised sharply. Iron fibres with 15 nm diameter have a coercive force ten thousand times higher than the bulk material. At even smaller

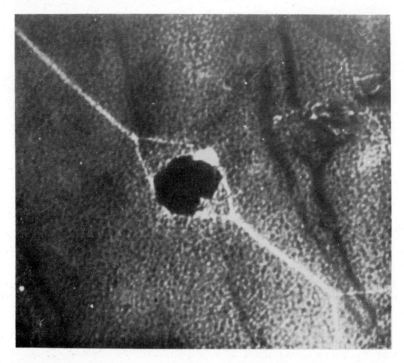

Fig. 15.16. Closure domains at a carbide particle in a rimming steel, × 45 000. This photograph was taken with an electron microscope. (By courtesy of I.R.D. Co. Ltd.)

sizes the ferromagnetic properties are lost due to thermal fluctuation overcoming the exchange forces. This state is called superparamagnetism. The variation of coercive force as the particle size is reduced is shown in Fig. 15.17. A whole new class of permanent magnets has been developed based on the use of fine particles.

The critical particle size below which a single-domain structure is formed can be roughly calculated, following Kittel (1949). The magnetostatic energy of a single-domain spherical particle, diameter d, is $2\pi\mu_0 d^2 M_S^2/9$ (§ 15.4). This is halved if the particle forms two opposing domains, as in Fig. 15.18. But this introduces a domain boundary energy, equal to $\pi d^2 U/4$ where U is a boundary energy per unit area. A single domain will exist when the energy of a domain boundary would exceed the reduction in magnetostatic energy, that is, below a critical particle diameter

$$d \simeq U/\mu_0 M_S^2. \tag{15.15}$$

For iron, $U \simeq 3 \times 10^{-3}$ J/m^2 and $M_S = 1.7 \times 10^6$ A/m, giving $d \simeq 10$ nm which is in agreement with the evidence in Fig. 15.17. This calculation is valid only if the particle's size is large compared with the boundary width, that is, in materials with high magnetocrystalline anisotropy.

In the absence of domain boundaries, a particle can only change its magnetisation by simultaneous rotation of the spin vectors, a difficult process because it is

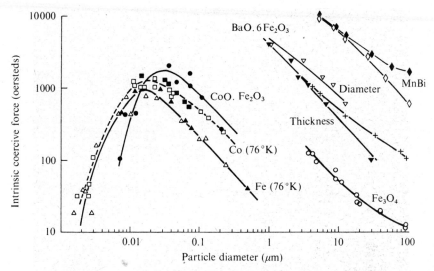

Fig. 15.17. Observed relation between coercive force and particle diameter for small ferromagnetic particles. The coercive force is desired mainly from crystal anisotropy energy (From F. E. Luborsky, *J. Appl. Phys.* (suppl) **32**, 171s (1961). Published by the American Institute of Physics.)

opposed by magnetocrystalline anisotropy. A reverse field of $2K/\mu_0 M_S$ is required to overcome it, where K is the magnetocrystalline anisotropy. In addition, fibres will have shape anisotropy opposing rotation due to the magnetostatic effect. The maximum coercivity due to shape is that for a long, thin, needle-shaped single domain particle and is given by $M_S/2$. The predicted and experimental values of the coercive force for various materials are listed in Table 15.4, from which it can be seen that the highest coercivities are to be obtained from shape effects and that the most important route to improving permanent magnet properties is to find new materials with higher saturation magnetisations.

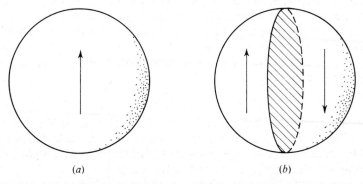

Fig. 15.18. Division of spherical particle into two domains. (*a*) Single domain sphere, magnetostatic energy $2\pi\mu_0 d^3 M_S/9$. (*b*) Sphere bisected by domain boundary of area $\pi d^2/4$.

Table 15.4. *Calculated and observed coercivities for single-domain particles.*
*(Data from F. E. Luborsky, J. Appl. Phys. **32** (suppl.)171S (1961).)*

Material	Crystal anisotropy constant, K (MJ/m³)	Saturation magnetisation, M_s (MA/m)	Anisotropy field, $2K/\mu_0 M_s$ (MA/m)	Shape field, $M_s/2$ (MA/m)	Maximum observed coercivity, H_c (MA/m)
MnBi	1.16	0.62	3.0	0.31	0.96
MnAl	~1.0	0.495	~3.2	0.25	0.48
BaO . 6Fe₂O₃	0.33	0.38	1.4	0.19	0.36
CoO . Fe₂O₃	0.25	0.425	0.94	0.21	0.34
Cobalt	0.40	1.4	0.46	0.70	0.17
Iron	0.04	1.7	0.037	0.85	0.08
Fe₃O₄	0.11	0.48	0.037	0.24	0.01

15.8 Soft magnetic materials

Soft magnetic materials are required as field amplifiers, that is, to respond immediately to an applied field and themselves produce a larger one, and as shields to prevent magnetic fields from penetrating regions where they are not wanted. As field amplifiers they are employed as cores in transformers, motors, dynamos and similar electromagnetic devices.

The requirements for soft magnetic materials are a high saturation intensity, high initial permeability and low hysteresis. These properties enable a material to reach a high magnetisation in a low applied field with little absorption of energy. The high saturation intensity is achieved by using an appropriate composition. Low hysteresis and high permeability call for a material in which the resistance to domain wall movement is a minimum. Eddy current losses must also be small for materials used in AC applications. These losses are due to the currents (eddy currents) which are induced in any electrical conductor which is exposed to a changing magnetic field. Energy is absorbed from the magnetic field in order to create these currents, and is dissipated in the form of heat by the resistance of the material. The rate of energy loss, due to eddy currents, in a thin strip of conductor, thickness d, and resistivity ρ, subjected to a sinusoidally varying magnetic field of frequency f, producing a maximum induction in the material of B_{max}, can be shown to be proportional to:

$$\frac{(df\,B_{max})^2}{\rho}.$$ (15.16)

Hence soft ferromagnetics are usually used in the form of thin laminations insulated from one another, and, particularly for high frequency applications, should have high electrical resistivity. Table 15.5 lists the properties of some common soft magnetic materials.

Pure iron has been prepared with an initial permeability of 1.4×10^4 and a maximum differential permeability of 8.8×10^4. Stress and magnetic inhomogeneities are small and there is little to hinder boundary motion, but the resistivity is comparatively low. If the situation is to be improved by alloying, the material must remain single phase and the saturation intensity of magnetisation should not be seriously affected. Silicon and aluminium have the greatest effect on the resistivity of iron, four per cent of either raising the resistivity by a factor of five, but the saturation magnetisation is reduced by about ten per cent. The addition of either of these elements also raises the permeability, presumably because they are both good deoxidisers for iron and thus reduce the inclusion content. Both elements close the gamma-loop in iron (see Chapter 12) and must be used in excess of about $2\frac{1}{2}$ per cent to preserve a single-phase structure at all temperatures. The upper limit to silicon content is $4\frac{1}{2}$ per cent as above this amount the alloy becomes too brittle; the usual composition is $3\frac{1}{2}$ per cent silicon. The aluminium alloys do not embrittle, and up to 16 per cent aluminium has been employed. The high aluminium alloys have properties similar to, and on grounds of cost may replace, some of the iron–nickel alloys described below.

An ideal soft magnetic material, in addition to the factors described above, would be single crystal with an easy direction of magnetisation (§ 15.4) always aligned parallel to the applied field. This is technically feasible but economically

Table 15.5. *Some soft magnetic materials*

Material	Composition (weight per cent)	Initial permeability μ_i	Maximum differential permeability	Saturation magnetisation intensity M_s (MA/m)	Resistivity (Ωm)	Energy loss at $B_{max} = 0.5$T (Jm^{-3} cycle^{-1})
Iron (pure)	99.95	10 000	200 000	1.7	10×10^{-8}	
Dynamo sheet	99.8	150–250	5000	1.7	14×10^{-8}	70
Silicon-iron	96Fe, 4Si	500	7000	1.57	60×10^{-8}	40
Silicon-iron (Goss texture)	97Fe, 3Si	1500	40 000	1.59	47×10^{-8}	
Silicon-iron (cube texture)	97Fe, 3Si		116 000	1.59	47×10^{-8}	
Aluminium iron	96.5Fe, 3.5Al	500	19 000	1.51	47×10^{-8}	
Alfer	87Fe, 13Al	700	3700	0.96	90×10^{-8}	
Alperm	84Fe, 16Al	3000	55 000	0.64	140×10^{-8}	
Permalloy A	21.5Fe, 78.5Ni	8000	100 000	0.86	16×10^{-8}	5
Mo-Permalloy	17.7Fe, 78.5Ni, 3.8Mo	20 000	75 000	0.68	55×10^{-8}	
Cr-Permalloy	17.7Fe, 78.5Ni, 3.8Cr	12 000	62 000	0.64	65×10^{-8}	
Supermalloy	15Fe, 79Ni, 5Mo, 0.5Mn	100 000	1 000 000	0.63	60×10^{-8}	<0.5
Mumetal	16Fe, 77Ni, 5Cu, 2Cr	20 000	100 000	0.52	62×10^{-8}	4.5
Rhometal	64Fe, 36Ni	250–2000	1200–8000	0.72	90×10^{-8}	45
Permendur	50Fe, 50Co	800	5000	1.95	7×10^{-8}	
Ferroxcube A	{48MnFe$_2$O$_4$, 52ZnFe$_2$O$_4$; 79MnFe$_2$O$_4$, 21ZnFe$_2$O$_4$}	1400 / 700		0.26 / 0.41	2000 / 8000	
Ferroxcube B	{36NiFe$_2$O$_4$, 64ZnFe$_2$O$_4$; 100NiFe$_2$O$_4$}	650 / 17		0.29 / 0.18	10^7 / 10^7	

prohibitive. A compromise solution is to have a multigrained material, with the grains so oriented that a majority of them have an easy direction of magnetisation within a degree or two of some given direction in the material (usually the rolling direction in the case of sheet). Such a structure is called a *texture*, and can result from cold working (deformation texture) or annealing (annealing texture) or a combination of both. Textures are common in heavily deformed materials. If a preferred orientation texture can be developed such that an 'easy' direction is always parallel to the applied field direction, saturation will be reached for lower applied fields than for a randomly oriented sheet. Goss in 1933 developed a process for producing a (110)[001] texture in silicon-irons. The process consists of hot breaking down of the ingot followed by a 60 per cent cold-reduction, an intermediate anneal and a further 60 per cent cold-reduction and a final anneal at 1100 °C in a hydrogen atmosphere. The resulting product, called grain-oriented silicon-iron, has the majority of crystals with a (110) plane in the plane of the sheet, and a [001] direction parallel to the rolling direction (Fig. 15.19(*a*)). The magnetisation curves for random-oriented and grain-oriented silicon–iron are shown in Fig. 15.19(*b*). A cube texture, (100)[001], has recently been developed for silicon–irons which has some advantages over the Goss texture.

In iron–nickel alloys the magnetostrictive coefficient λ and the magneto-crystalline anisotropy K vary with composition as shown in Fig. 15.20. In the alloy range 76–82 per cent nickel both λ and K go to zero, leading to easy domain boundary movement. This composition is the basis of 'Permalloy'. Often a third element is added to increase the resistivity. The high permeability is obtained on quenching (cooling in air from 600 °C is sufficient), since slow cooling forms the ordered $FeNi_3$ in which K is not equal to zero. There is a maximum in the resistivity at 30–40 per cent nickel, and alloys based on this composition, 'rhometal', are used where cost is of greater importance than the highest permeability. As with the silicon and aluminium alloys, these alloys can be improved by suitable cold rolling and annealing to given grain-oriented sheet.

Permalloy, in the form of thin films, is used for rapid switching memory units in fast computers. A magnetic material with a 'rectangular' hysteresis loop (Fig. 15.21) can have two possible remanent states, which can be used to represent the binary digits '0' and '1'. An array of pieces of such material can be used to build up a complete memory system, each piece of magnetic material storing one 'bit' of information. Among the requirements for a magnetic memory are: (*a*) it should be stable, switching only above a particular field, and remaining switched until it experiences equal but opposite field; (*b*) the switching operation should be extremely rapid; (*c*) the device must be compact. Requirement (*a*) is met by the material having a rectangular magnetisation loop. All three requirements can be satisfied by thin films of a material such as permalloy with the magnetisation direction lying in the plane of the film. In such a configuration the demagnetising field is low, and the film is a single domain (§ 15.7). Switching takes place by magnetisation vector reversal, not by domain boundary motion. The rate of switching is controlled by damping resulting from induced eddy currents in the film. Thus rapid switching is promoted by thin films of high resistivity material, (15.16). Switching times of 10^{-8} s have been achieved in 1 μm thick permalloy films. Thin films also allow of high packing density, thus satisfying (*c*) above. Films must be of high perfection, as any irregularities will

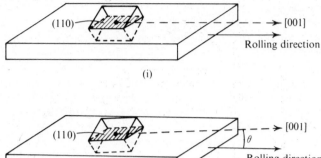

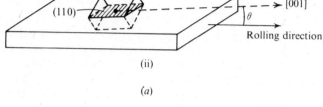

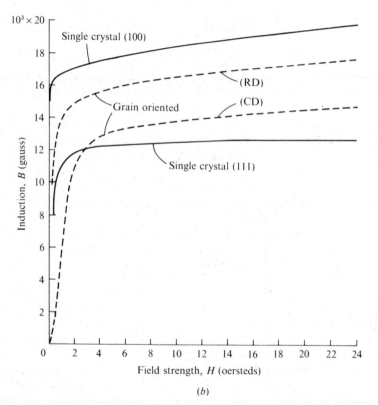

Fig. 15.19. (a) Goss texture in silicon–iron sheet. (i) Ideal: (110) parallel to rolling plane, [001] parallel to rolling direction. (ii) In practice (110) and [001] make a random angle θ (usually not greater than 10°) with the rolling direction.

(b) Magnetisation curves of textured sheet measured parallel to rolling and cross directions, compared with single crystal curves, of silicon–iron. (From R. M. Bozorth, *Ferromagnetism*, © 1951 by Litton Educational Publishing Inc. Reprinted by permission of Van Nostrand Reinhold Company.)

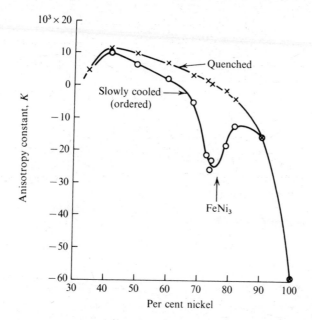

Fig. 15.20. Crystal anisotropy constant of iron–nickel alloys after quenching and after slow cooling to produce ordered FeNi$_3$. (From R. M. Bozorth and J. G. Walker, *Phys. Rev.* **89**, 624 (1953).

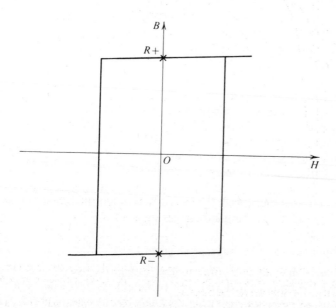

Fig. 15.21. Material with rectangular B–H loop, suitable for magnetic memory, showing two remanent points, R+ and R−, which represent the two stable states of the memory.

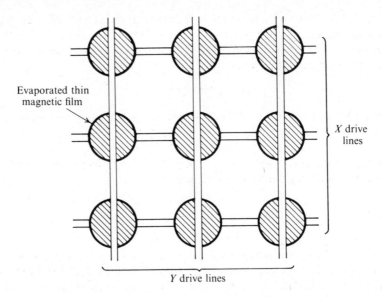

Fig. 15.22. Part of a magnetic memory matrix using thin film spots (schematic).

slow down the switching time. Each 'bit' consists of a small disc of material, many hundreds of which are evaporated into an insulating substrate. The arrangement is shown schematically in Fig. 15.22.

Magnetically soft ferrites, usually known under the trade name of Ferroxcube, are based on the magnetite structure. Electrically they are semiconductors, and their DC resistivity is several orders of ten higher than that for iron. Ferrites are therefore very useful in cutting down eddy current losses at high frequencies, for example in microwave transformers. Ferrites can also be made to give hysteresis loops of square form but narrow width which are required in computer circuits, particularly in memory units.

Two types of soft ferrite are commonly used, Ferroxcube A, manganese zinc ferrite, in which the Fe^{2+} ions are replaced by Mn^{2+} and Zn^{2+}; and Ferroxcube B, nickel zinc ferrite, Fe^{2+} being replaced by Ni^{2+} and Zn^{2+}. The compositions are adjusted to give differing saturation magnetisation (§ 15.2) and resistivity, according to the frequency range in which they are to be used. The properties of some ferrites are included in Table 15.5.

Ferroxcube is a hard black substance, chemically very inert and unaffected by humidity. It is produced by powder metallurgical techniques from inexpensive raw materials, and is pressed and sintered to final shape. It can be machined only by grinding after sintering.

15.9 Requirements for permanent magnets

A permanent magnet is required to have as high as possible a field in its air gap for a given volume of material, and to be able to resist demagnetisation, not only by its own demagnetising field but also by any external fields. The material from which the permanent magnet is constructed should therefore have both a high remanence, B_r, and a high coercivity, H_c.

The highest field is produced from a given volume of magnet when the induction B in the material is such that the product BH has a maximum value. In the absence of any externally applied fields, the integral of the magnetomotive force around the magnet must be zero:

$$\oint H \, dl = 0.$$

Referring to Fig. 15.23, where H_g is the field in the gap and H the field in the magnet:

$$H l_m + H_g l_g = 0.$$

Assuming that the gap is sufficiently small, so that there is no flux leakage, the flux must be continuous around the magnet and

$$B A_m = H_g A_g.$$

Combining these two equations gives

$$H_g^2 = -BH(V_m/V_g)$$

where V_m and V_g are the volumes of the material and of the gap respectively. For a given volume of magnet and air gap, the maximum value of H_g is given by the maximum value of BH, i.e. BH_{max}. BH_{max} is called the energy product, as it is proportional both to the energy per unit volume of the gap field, $\frac{1}{2}\mu_0 H_g^2$, and the magnetic energy per unit volume stored within the magnet $\frac{1}{2}BH$. The energy product can be found by the empirical construction shown in Fig. 15.24, which seems to hold almost exactly for all hard magnetic materials. The value is usually quoted to indicate the quality or efficiency of a permanent magnet material.

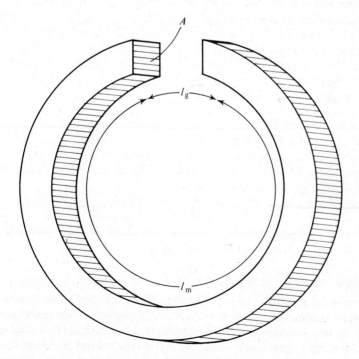

Fig. 15.23. Schematic representation of a permanent magnet, showing the air gap.

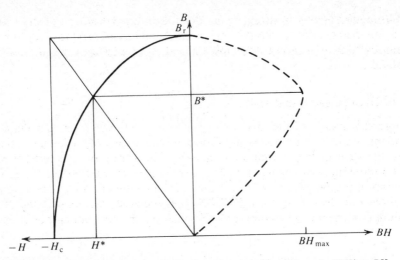

Fig. 15.24. Empirical construction for determining BH_{max} (after Watson). $B^{*}H^{*} \cong BH_{max}$ for almost all ferromagnetics.

The energy product can be increased by raising B_r or H_c, but it is rarely possible to raise both together. To achieve a given energy product, the application will determine whether a material with a high coercivity or one with a high remanence is required. If the magnet must be used in an environment influenced by other magnetic fields, particularly varying fields, then a high coercivity is most important. Often the shape of the magnet is predetermined. A short, fat magnet will have a large demagnetising field and calls for high coercivity. The low working B can be tolerated as, due to the large cross-section area, the total flux is high; a high gap field can be produced by concentrating this flux with soft-iron pole-pieces. A long thin magnet on the other hand will require a high remanence, otherwise the small area will give a small flux. The demagnetising field in this case will be low, so the value of the coercivity is not critical. The material of the magnet for a given geometry must be chosen so that it operates at the point on the demagnetisation curve which corresponds to BH_{max}.

High values of B_r, H_c and BH_{max} in a ferromagnetic are obtained by severely limiting domain boundary movement, and by ensuring that any movement which does occur is highly irreversible and only takes place under the influence of very large applied fields. Permanent magnet materials should therefore have high values of magnetocrystalline anisotropy and magnetostrictive coefficients and contain a high degree of magnetic and stress inhomogeneity. Materials which are magnetically hard are therefore usually mechanically hard. Single-domain particles, as described in § 15.7 are also magnetically hard.

There are two metallurgical approaches to the production of magnetically hard materials. One is precipitation; either non-magnetic precipitates in a magnetic matrix to obstruct domain boundary movement, or magnetic precipitates in a non-magnetic matrix to produce single-domain particles. Any internal stresses resulting from this precipitation are also advantageous. The inhomogeneities introduced to increase coercivity normally decrease the remanence, both because less of the volume is now magnetic and because most alloy additions tend to decrease the saturation magnetisation of the matrix. The other approach is that of fine particle materials produced by powder methods. This is a recent development and its possibilities are not yet fully realised.

15.10 Hard magnetic materials

Before the middle of the last century the only permanent magnet materials available were plain carbon steels. Optimum magnetic properties are obtained with the steel in the martensitic condition. The variation in properties with carbon content is shown in Fig. 15.25. The remanence decreases, due to a drop in saturation magnetisation, with increasing carbon content, to a constant value at above one per cent carbon. The coercivity increases rapidly with increasing carbon content up to the eutectoid composition, 0.8 per cent carbon; beyond

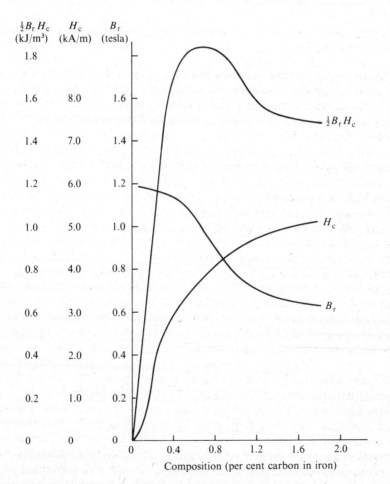

Fig. 15.25. Remanence B_r, coercivity H_c, and $\frac{1}{2}B_rH_c$ ($\approx$ energy product) for as-quenched plain carbon steels, as a function of carbon content (after Gurnlich).

this the increase is less rapid. BH_{max} has a maximum value of about 1.6×10^3 AW/m³ at from 0.9 to 1.0 per cent carbon, depending upon the exact heat treatment.

The magnetic hardness is due to magnetic inhomogeneities resulting from the presence of, non-magnetic, retained austenite. Improvements on the plain carbon steels are usually achieved by alloying with elements which increase the quantity of retained austenite.

The earliest use of an alloy steel for a permanent magnet was in 1895, when both tungsten and chromium steels were tried. The tungsten steels, containing about 6 per cent tungsten and 0.7 per cent carbon, were in general use before 1914. During the First World War, due to the shortage of tungsten, they were replaced by the chromium steels, 3.5 per cent chromium and 0.9 per cent carbon. Values of BH_{max} of 2.4×10^3 AW/m³ were obtained. A typical heat treatment for the latter is austentising at 850 °C followed by an oil quench. Tempering for 24 hours at 100 °C increases the stability of the martensite for a slight loss in magnetic properties. The ultimate development of magnet steels was the introduction in 1917 by Honda of the cobalt steels. The addition of cobalt to chromium tungsten steels, called KS magnet steel, increases the saturation magnetisation and with it the remanence and coercivity. The composition range of the KS steels is 0.7–1 per cent carbon, 6–8 per cent tungsten, 1–4 per cent chromium and 15–42 per cent cobalt. A steel has been made with an energy product of eight thousand AW/m³ and a coercivity of 2×10^4 A/m. These values have never been exceeded in a steel. The properties of these steels, and other hard magnetic materials, are summarised in Table 15.6.

The KS magnet steels reigned supreme until investigation into the iron–cobalt–tungsten and iron–cobalt–molybdenum systems in 1931 produced Remalloy. This is a carbon-free alloy of iron with about 15 per cent cobalt and 20 per cent tungsten or molybdenum and represents the first real improvement over the cobalt steels. Another alloy of 10–25 per cent nickel, 8–25 per cent titanium and 15–36 per cent cobalt was reported by Honda in 1934 to have an energy product of 1.6×10^4 AW/m³. The hardening mechanism in these alloys is precipitation of non-magnetic particles.

Another Japanese, Mishima, also in 1931, during an investigation of the iron–nickel–aluminium alloys, discovered the remarkable properties of the alloy corresponding to Fe₂NiAl (58 per cent iron, 24 per cent nickel, 13 per cent aluminium). This composition is the basis of the Alni, Alnico, Alcomax and Ticonal series of diffusion hardening magnet alloys, which have undergone continuous development since their first discovery. They rely on the formation of iron rich clusters or precipitates sufficiently small to contain single domains. At high temperatures the alloy consists of a single bcc phase corresponding to Fe₂NiAl. Upon cooling this phase undergoes a decomposition into two other bcc phases:

$$Fe_2NiAl \rightarrow Fe + FeNiAl.$$

The phenomenon is similar to age hardening of aluminium–copper alloys. The alloy is solution treated at 1250 °C, rapidly cooled to 900 °C (to avoid an undesirable phase change), slow cooled to 600 °C and held at this temperature for several hours. The precipitation and growth phenomena during ageing are

Table 15.6. Some hard magnetic materials

Type of Hardening	Material	Composition (weight percent, remainder iron)	Remanence B_r (Tesla)	Coercivity H_c (kA/m)	$(BH)_{max}$ KJ/m³	Heat treatment
Martensitic transformation	Carbon steel	0.9–1C	0.90	4.0	1.6	Water quench from 800 °C
	Tungsten steel	6W, 0.3Cr, 0.7C	1.05	5.6	2.4	Water quench from 850 °C
	Chromium steel	3.5Cr, 0.9C	0.98	4.8	2.2	Water quench from 830 °C
	15% Cobalt steel	15Co, 7Cr, 0.5Mo, 1C	0.82	14.3	4.8	Air cool from 1150 °C
	35% Cobalt steel (KS)	35Co, 3–6Cr, 5–6W, 0.9C	0.90	20.0	8.0	Air cool from 1150 °C
	MT steel	8Al, 2C	0.60	16.0	3.6	
Dispersion (precipitation)	Vicalloy	52Co, 14V	1.0	36.0	24	Annealed at 600 °C
	Cunife	60Cu, 20Ni	0.54	43.8	12	Quenched from 1000 °C
	Cunico	50Cu, 21Ni, 29Co	0.34	52.5	7.2	Tempered at 600 °C
	Silmanal	87Ag, 9Mn, 4Al	0.055	477	0.6	Annealed at 250 °C
Diffusion (iron-rich) domains	Alni (a)	24Ni, 13Al, 3.5Cu	0.62	38	10	Air cool from 1150 °C
	Alni (b)	32Ni, 12Al, 0.5Ti	0.47	56	10	Temper at 600 °C
	Alnico (a)	17Ni, 10Al, 12Co, 6Cu	0.80	40	13.5	
	Alnico (b)	20Ni, 10Al, 13.5Co, 6Cu, 0.25Ti	0.65	50	13.5	Air cool from 1250 °C
	Alcomax II	11Ni, 8Al, 21Co, 4.5Cu	1.24	46	34	Temper at 600 °C
	Alcomax IV	13.5Ni, 8Al, 24.5Co, 2Nb	12.0	64	36	
	Alcomax IV (unidirectional)	13.5Ni, 8Al, 24.5Co, 2Nb	13.0	60	56	As above, but in strong magnetic field
Superlattice	FePt	78Pt	0.58	125	24	Quench from 1150 °C
	CoPt	23Co, 77Pt	0.45	207	30	
Fine particle	Iron powder		0.40	40	6.0	
	Elongated iron powder		0.57	61.3	12.8	
	Iron–cobalt powder	45Co	1.02	63.0	36	
	Bismanol (MnBi)	20Mn, 80Bi	0.42	263	33.4	
	Ferroxdure	$BaO.6Fe_2O_3$	0.20	120	8.0	
	Ferroxdure (oriented)	$BaO.6Fe_2O_3$	0.39	240	29.0	
	Co-ferrite	$(3CoO + FeO)Fe_2O_3$	0.25	51.8	9.6	Sintering at 1200–1300 °C

rather complex, but have been the subject of electron-microscopical investigations, see Fig. 15.26.

The Alni series of permanent magnet alloys are Fe_2NiAl with minor alloying additions. The remanence, through an increase in saturation magnetisation, may be increased by the addition of cobalt, leading to the Alnico and Alcomax series. The properties of these alloys are improved by magnetising in a strong magnetic field during cooling. Domain patterns established in the iron-rich regions are locked in place and demagnetisation occurs only by domain reversal. The anisotropy which opposes domain reversal, is increased if all the crystals in the alloy have a cube axis parallel to the applied field direction; this is achieved in the 'unidirectional' Alcomax alloys. A further development of Alcomax is Ticonal G, with 35 per cent cobalt and 8 per cent titanium. The largest ever energy product of 10^5 AW/m^3 has been reported for this material.

All the above diffusion-hardening alloys are mechanically very hard, and can only be fabricated by casting and grinding prior to heat treatment. This limitation is obviated in the precipitation-hardening alloys Cunife, Cunico and Vicalloy. Cunife and Cunico, discovered in 1935, are based on the compositions 60 per cent copper, 20 per cent nickel and 20 per cent iron, and 50 per cent copper, 20 per cent nickel and 30 per cent cobalt, respectively. Vicalloy, developed in 1940, is based on the composition 52 per cent cobalt, 10 per cent vanadium and 38 per cent iron. These alloys also develop a precipitation structure but unlike the diffusion hardening alloys the matrix has a higher magnetisation intensity than the precipitates, and hardening occurs by restriction of boundary movement in the matrix. They consist of a single high temperature fcc phase which on cooling decomposes into two other fcc phases. Because the crystal structure of

Fig. 15.26. Electronmicrograph of Alcomax III after ageing, showing the formation of iron-rich (whitish) and iron-depleted regions. (× 200 000. By courtesy of R. B. Nicholson.)

the three phases is similar, the precipitation is easy to nucleate and occurs on a very fine scale (Fig. 15.27). The alloys are quenched from 10 000 °C and are aged at 550 °C to 750 °C. Cold working prior to ageing refines the subsequent precipitate and can also introduce favourable anisotropy. As in the diffusion-hardening alloys, the optimum ageing condition is achieved with the formation of clusters or zones, the hardening effect being mainly due to coherency strains. Unlike most other permanent magnet materials these alloys are soft and ductile, and may easily be worked into finished shapes, for example wires and tapes for information recording and storage.

Two alloys which have the highest coercivities known (up to 3.2×10^5 A/m) are iron–platinum and cobalt–platinum alloys, with composition FePt and CoPt. Because of their price they exist as purely laboratory curiosities. These alloys undergo an order–disorder reaction on quenching from 1100 °C. The high temperature fcc structure on ordering becomes tetragonal; in different regions of the lattice any one of the three cube axes may become the tetragonal axis. The resulting internal strains as different regions order with different orientations gives rise to the very high values of H_c obtained; the partially ordered state giving optimum properties.

The small-particle (single domain) effect has also been utilised directly in magnets by the compacting of fine powders. The powders are produced either by the hydrogen reduction of organic compounds, usually oxalates, or by electro-deposition in mercury. Magnets have been made in this way from compacted iron, cobalt and iron–cobalt alloy particles. The particles produced by electro-deposition are dispersed in a lead matrix to form a commercial magnet material called Lodex. Another permanent magnet material is the compound MnBi

Fig. 15.27. Electronmicrograph of Vicalloy II. Homogenised at 1200 °C for two hours, water quenched and aged for $\frac{1}{4}$ hour at 650 °C. ($\times$ 128 000. By courtesy of R. B. Nicholson.)

known as Bismanol, which is dispersed as a powder in a lead or plastics matrix. The spacing of the manganese atoms in this compound is such that the exchange interaction is positive, and it becomes a ferromagnetic.

The magnetically hard ferrites are only so-called because they exhibit ferrimagnetism; crystallographically they bear no relation to the ferrite structure. Their structure is hexagonal magnetoplumbite and they have the chemical formula $MO.6Fe_2O_3$, where M is Pb, Ba or Sr. The favoured material is Magnadur (or Ferroxdur), $BaO.6Fe_2O_3$. The magnetically hard ferrites are essentially a postwar development of prewar French work.

Magnadur has a very high coercivity, and can be used with large air gaps. Again this material utilises the small-particle effect, but the critical particle size is of the order of one micron, compared with a critical size of one tenth of a micron or less for iron. The 'easy' direction of magnetisation is the hexagonal axis, [0001], and the anisotropy constant K is very large. Randomly oriented isotropic material has an energy product of about $8 \times 10^3 \, AW/m^3$, aligned elongated anisotropic material can be produced with an energy product of $2.8 \times 10^4 \, AW/m^3$. Remanence is low compared to metal magnets, but the raw materials are very cheap. Fabrication is by powder techniques.

Vectolite is a cobalt ferrite made by mixing roughly equal quantities of CoO, Fe_2O_3 and Fe_3O_4. Magnets are sintered at about $1000\,°C$ and cooled in a magnetic field to give a preferred direction remanence about twice that for Magnadur, but the coercivity is very much less. The properties of hard ferrites are included in Table 15.6.

Because of their high resistance to mechanical shock and to alternating fields, and because of the cheapness of their raw materials, ferrites are found in such applications as cycle dynamos, children's toys, magnetic oil filters and TV focussing.

QUESTIONS

1. Describe the various types of magnetic behaviour, and discuss their origin in terms of atomic and molecular structure.
2. The magnetic susceptibility of silver is -2.5×10^{-5}. What is the flux density within, and the magnetisation of, a silver wire in a magnetic field, parallel to the wire's axis, of $10^6 \, A/m$?
3. An alloy of copper and cobalt consists of spherical precipitates, averaging 10 nm diameter, of pure cobalt in a matrix of pure copper. The precipitates form 2 per cent by volume of the alloy. Cobalt is ferromagnetic, with a saturation magnetisation of $1.4 \, MA/m$. Each cobalt precipitate is a single domain, and acts as a strong dipole, which responds to any external field as a paramagnetic dipole. The effect is called 'superparamagnetism'. Calculate the susceptibility of the alloy at temperature of $300\,°K$.
4. Bcc iron is ferromagnetic, with a lattice parameter of 0.286 nm. The average magnetic moment of each atom is 2.2 Bohr magnetons. What is the saturation magnetisation of iron?
5. Calculate the saturation magnetisation of a sample of magnetite, given that each molecule is $Fe^{2+}Fe_2^{3+}O_4^{2-}$, and that the lattice parameter of the cubic unit cell, each cell consisting of eight molecules, is 0.837 nm. The magnetic moment of the Fe^{2+} ions is 4 Bohr magnetons; that of the Fe^{3+} ions cancel and their net moment is zero.
6. What strength of magnetic field is required to magnetise to saturation a sphere of

nickel? (The field is that required to overcome the demagnetisation field. The intensity of magnetisation of nickel at saturation is 0.55 MA/m.)

7. Define (a) demagnetising field, (b) magnetocrystalline anisotropy. Explain how these give rise to the formation of domain structures in ferromagnetics. How does the presence of domains affect the magnetisation of a ferromagnetic material?

8. A material of saturation intensity M_S, and magnetocrystalline anisotropy constant K, forms slab like domains of length L and width d, parallel to the easy direction of magnetisation. Deduce an expression for d in terms of the other quantities. If the domain structure is modified by the formation of closure domains, with magnetisation at 90° to the easy directions, what is the new value of the domain width?

9. Define magnetostriction. Discuss the relative effects of magnetostriction and magneto-crystalline anisotropy on the domain structure and on the B–H curve of a ferromagnetic material.

10. What is a soft magnetic material? Discuss the principles underlying the production of such a material.

11. What is a single domain particle? Define coercivity and remanence. Discuss the factors which determine the coercivity of hard magnetic materials.

12. A piece of iron contains non-magnetic inclusions in the form of spheres, of diameter 1 μm, arranged in a regular cubic array, a distance 10 μm apart. A domain structure forms in which 180° boundaries are parallel to one set of cube faces. Calculate (a) the initial susceptibility, and (b) the coercivity, of the material. ($M_S = 1.7$ MA/m, $K = 4 \times 10^4$ J/m^3, $a = 0.4$ nm, and $\theta_f = 10^3$ °K).

FURTHER READING

R. M. Bozorth: *Ferromagnetism*. Van Nostrand (1951).

F. Brailsford: *Magnetic materials*, 2nd edition. Methuen (1951).

K. H. Stewart: *Ferromagnetic Domains*. Cambridge University Press (1954).

D. Hadfield (ed.): *Permanent Magnets and Magnetism*. Iliffe (1962).

S. Chikazumi: *Physics of Magnetism*. Wiley (1964).

D. E. G. Williams: *The Magnetic Properties of Matter*. Longmans (1966).

J. E. Thompson: *The Magnetic Properties of Materials*. Newnes (1968).

R. S. Tebble and D. J. Craik: *Magnetic Materials*. Wiley-Interscience (1969).

A. E. Berkowitz and E. Kneller (eds.): *Magnetism and Metallurgy* (2 vols.). Academic Press (1969).

Index